A Visual Introduction to AutoCAD® and 3D Designing with Release 14

James D. Bethune
Boston University

Prentice Hall

Upper Saddle River, New Jersey   *Columbus, Ohio*

Library of Congress Cataloging-in-Publication Data

Bethune, James D., 1941 -
    A visual introduction to AutoCAD and 3D designing with Release 14
    / James D. Bethune.
        p. cm.
    Includes index.
    ISBN 0-13-020369-6
    1. Mechanical drawing. 2. AutoCAD (Computer file) 3. Three-
dimensional display systems.   I. Title.
T353.B459 2000
604.2 ′ 0285 ′ 5369--dc21   99-14592
                                                             CIP

Cover Art: Janoski Advertising Design
Editor: Stephen Helba
Production Editor: Louise N. Sette
Design Coordinator: Karrie Converse-Jones
Cover Designer: Janoski Advertising Design
Production Manager: Deidra Swartz
Illustrations: James D. Bethune
Marketing Manager: Chris Bracken

This book was set in Times New Roman and Arial by James D. Bethune and was printed and bound by Courier/Kendallville, Inc. The cover was printed by Phoenix Color Corp.

Printed in the United States of America

10 9 8 7 6 5 4 3 2 1

ISBN: 0-13-020369-6

Prentice-Hall International (UK) Limited, *London*
Prentice-Hall of Australia Pty. Limited, *Sydney*
Prentice-Hall of Canada, Inc., *Toronto*
Prentice-Hall Hispanoamericana, S. A., *Mexico*
Prentice-Hall of India Private Limited, *New Delhi*
Prentice-Hall of Japan, Inc., *Tokyo*
Simon & Schuster Asia Pte. Ltd., *Singapore*
Editora Prentice-Hall do Brasil, Ltda., *Rio de Janeiro*

Preface

The second edition of this book has been written for Release 14. The text and format of the book are essentially the same, but all the screen captures and step-by-step instructions have been updated to Release 14.

The book is divided into three parts. Chapters 1 through 9 introduce the fundamentals of AutoCAD Release 13 for Windows; Chapters 10 through 15 explain many of AutoCAD's three-dimensional capabilities; and Chapters 16 through 18 show how to apply these three-dimensional capabilities to design problems.

The fundamentals of AutoCAD are introduced using a tutorial approach, accompanied by many screen captures and other illustrations. This will help guide the reader through each example by using both text and visual instructions. It is hoped that the text will serve not only as an intsructional tool, but also as a reference for future work.

Each command, in a toolbar, is explained using a short tutorial that is independent of other command tutorials. This was done to avoid long, complicated tutorials that are sometimes difficult to complete. It also allows the text to be used to reference and review individual commands.

Chapters 1, 2, and 3 present an introduction to the drawing screen and the procedures used to set up, define, and create an AutoCAD drawing. Chapters 4, 5, and 6 demonstrate the commands of the Draw, Object, and Modify toolbars.

Chapter 7 includes three fairly long tutorials that show how commands can be used together to create more complicated shapes. The student is encouraged to work through one or all of these tutorials so they can learn how to create an AutoCAD drawing from defining the new drawing to saving it.

Chapter 8 discusses the Dimensioning toolbar and the general conventions associated with dimensioning a drawing. All the conventions presented are in compliance with ANSI Y14.5M. Both English and metric units are covered.

Chapter 9 demonstrates how to create blocks, layers, and attributes. The chapter ends with the example of a drawing's title block fitted with attributes.

Chapter 10 introduces the fundamentals of 3D drawing using AutoCAD, followed by Chapters 11 and 12 on surface and solid modeling. Again, each command in the appropriate toolbars is explained, and short tutorials at the end of the chapters show how to combine commands to form objects. There are design problems at the end of both chapters 11 and 12.

Rendering and viewpoints are covered in Chapters 13 and 14 to complete the understanding of how to create and present 3D drawings.

The last four chapters concentrate on 3D design concepts. Chapter 15 demonstrates how to apply dimensions to objects drawn in htree dimensions. Chapter 16 serves to develop 3D perception by showing how to design clips and brackets between specified 3D locations. This requires the reader to work with different planes while creating a single 3D object. Chapter 17 defines how to represent and use standard fasteners, and Chapter 18 culminates the design section by showing how to set up a family of working drawings, including assembly drawings and parts lists. This chapter includes an extensive tutorial that walks through the complete creation of a set of working drawings.

Thanks go to Steve Helba for his encouragement; Kimberly Gundling, associate editor; Bret Workman, copy editor; and Louise Sette, production editor. Thanks to David, Maria, Randy, Lisa, Hannah, and now Wil for their continued support. Also thanks to Sean, Claudette, and Sandy, and a special thanks to Cheryl, Hart and Heather.

James D. Bethune
Boston University
Boston, Massachusetts

To Mom, Martha Bethune, who started it all.

Contents

C H A P T E R

Looking Around the Drawing Screen

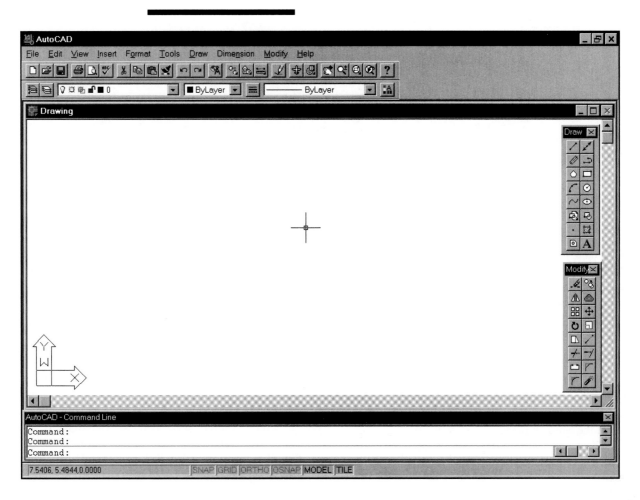

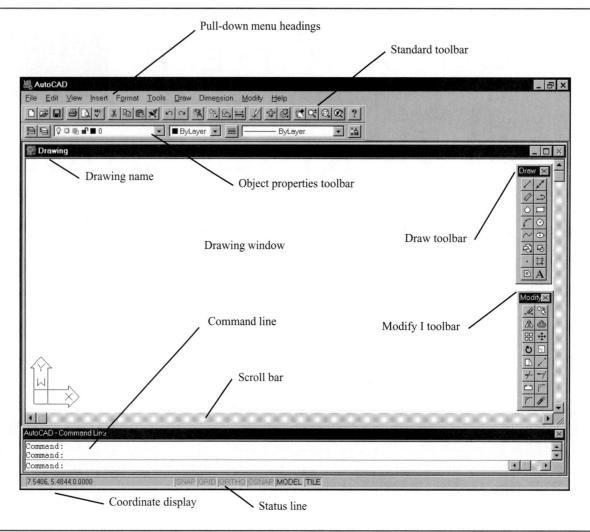

Figure 1-1

1-1 INTRODUCTION

This chapter explains the various aspects of AutoCAD's Windows drawing screen and shows how they can be manipulated. Figure 1-1 shows the initial AutoCAD Windows screen. Your screen may look slightly different because of your selected screen resolution values.

The dark line at the top of the screen displays the word

Drawing

This is the name of the current drawing file. As the drawing has not been named, the line reads "Drawing". Once a drawing name has been defined, it will appear in place of the word Drawing.

The second line from the top of the screen displays the pull-down menu headings. Clicking on any of the head-

ings will cause the pull-down menu to cascade below the heading.

The third line is the Standard toolbar and contains a group of the most commonly used commands.displays the Windows icons for exiting a program and changing a program along with several other often-used AutoCAD commands. It is assumed that the reader is familiar with basic Windows operations.

The fourth line contains some command icons and an area that shows the current or docked object properties that are active.

The bottom left corner of the screen shows the coordinate display position of the horizontal and vertical crosshairs in terms of an X,Y coordinate value, whose origin is the lower left corner of the drawing screen.

The commands listed on the bottom line are displayed in light gray when they are off and black when they are on.

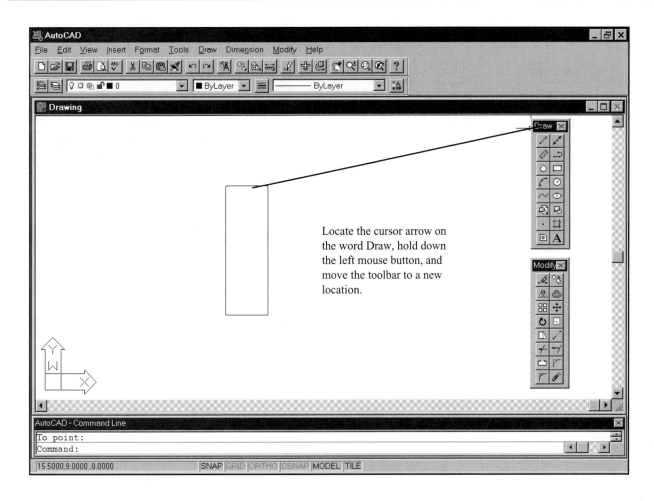

Locate the cursor arrow on the word Draw, hold down the left mouse button, and move the toolbar to a new location.

Figure 1-2

The horizontal and vertical scroll bars can be used to move around the drawing screen; they function as they do in other Windows applications.

The large open area in the center of the screen is called the drawing window or drawing editor. The two rectangular boxes of command icons, located along the right edge of the drawing window, are the Draw and Modify toolbars. They can be moved around the window as shown in Figure 1-2.

1-2 TOOLBARS

An AutoCAD toolbar is a group of command icons located under a common heading. The initial AutoCAD screen contains four toolbars: Standard, Objects Properties, Draw, and Modify. There are 13 additional predefined tool-

bars and you can create your own user-specific toolbars as needed.

To move a toolbar

See Figure 1-2.

1. Locate the cursor arrow on the heading Draw.
2. Press and hold down the left mouse button.

A light gray broken-line box will appear around the edge of the toolbox.

3. Still holding the left mouse button down, move the gray outline box to a new location on the screen.
4. Release the left button.

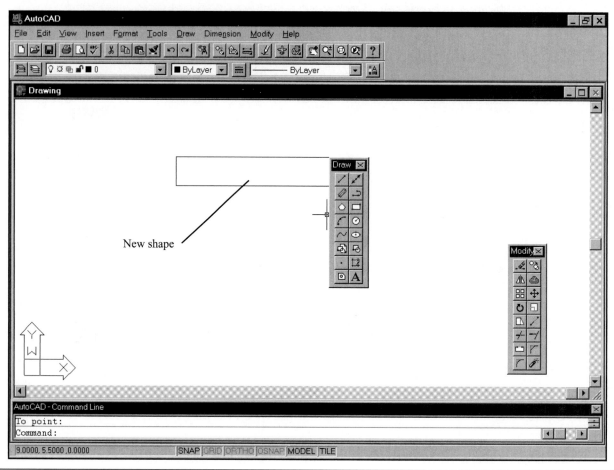

Figure 1-3

To change the shape of a toolbar

See Figure 1-3.

1. Locate the cursor arrow along the right edge of the Draw toolbox.

   A double opposing arrowhead will appear.

2. Press and hold the left mouse button.

   A light gray broken-line box will appear around the outside of the toolbox.

3. Still holding the left mouse button down, move the mouse around and watch how the gray box changes shape.
4. When the gray toolbox shape is a long vertical rectangle, release the left mouse button.

   A reshaped toolbox will appear.

To return the toolbar to its original location and shape

1. Locate the cursor arrow along the bottom or edge lines of the toolbox and return the toolbox to its original shape using the procedure outlined in Figure 1-3.
2. Move the reshaped toolbox to its original location along the right side of the drawing screen using the procedure outlined in Figure 1-2.

To add a new toolbar to the screen

See Figures 1-4 and 1-5.

1. Locate the cursor arrow on the View pull-down menu heading and press the left mouse button.
2. Select (locate the cursor arrow on the word Toolbar and press the left mouse button) the item Toolbars.

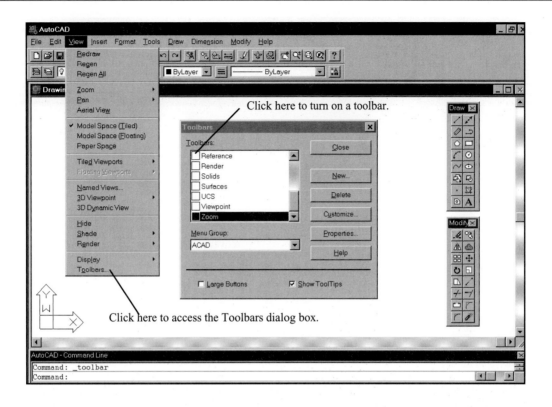

Figure 1-4

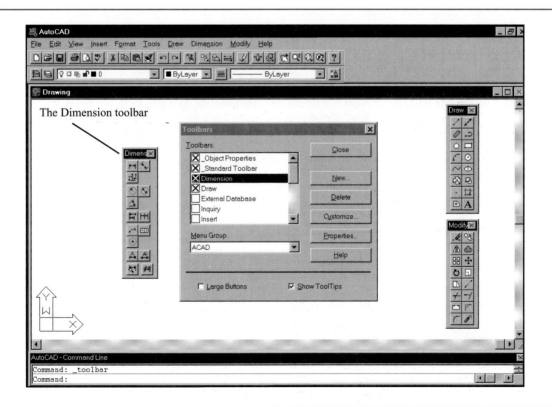

Figure 1-5

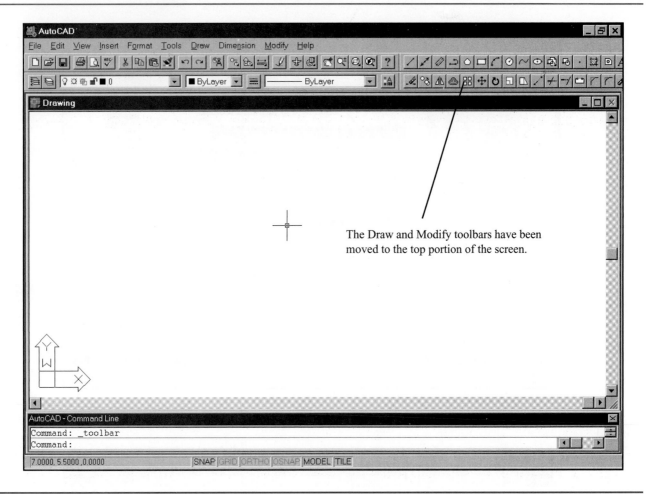

The Draw and Modify toolbars have been moved to the top portion of the screen.

Figure 1-6

The Toolbars dialog box will appear. This is a listing of available toolbars. See Figure 1-4. An X in the box to the left of a toolbar name indicates that the toolbar is on.

3. Select Dimension toolbar by clicking the box to the left of the word Dimension.

The Dimensioning toolbar will appear. See Figure 1-5. Any toolbar can be moved or have its shape changed as described in Figures 1-2 and 1-3.

To remove a toolbar from the screen

1. Locate the cursor arrow on the X located in the upper right corner of the toolbar and press the left mouse button.

Figure 1-6 shows the Draw and Modify toolbars docked horizontally at the top of the drawing screen. The Dimensioning toolbar has been located between them. To relocate the toolbars from the positions shown, locate the cursor arrow above the icons but still below the horizontal line that defines the toolbar area and press and hold down the left mouse button. A gray box will appear around the toolbars that will move with the cursor arrow.

1-3 THE COMMAND LINE BOX

The size of the command window, located at the bottom of the screen, may be changed to display more or fewer command lines. It is recommended that at least two command lines be visible at all times.

To move and resize the Command Line box

See Figure 1-7.

1. Locate the cursor arrow along the top edge of

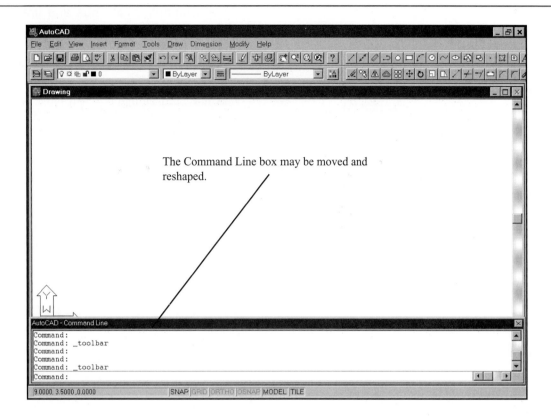

Figure 1-7

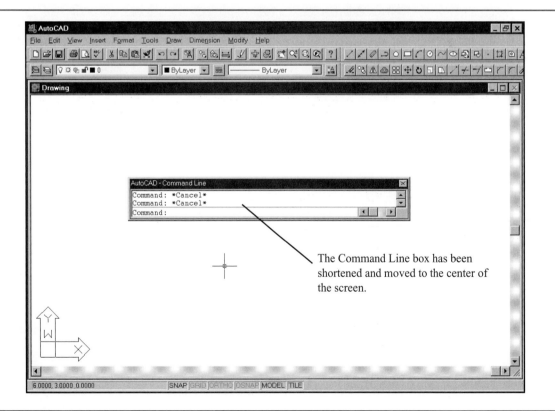

Figure 1-8

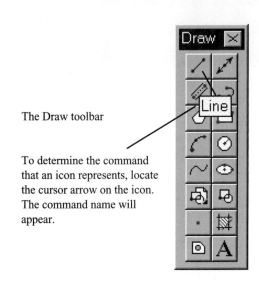

The Draw toolbar

To determine the command that an icon represents, locate the cursor arrow on the icon. The command name will appear.

Figure 1-9

the Command Line box and press and hold down the left mouse button.

2. Still holding the left mouse button down, move the cursor arrow to a new location on the drawing screen.

A gray, broken-line box will move with the cursor arrow and serves to display the new Command Line box location and shape.

3. Release the left mouse button to relocate the Command Line box.

The Command Line box may now be moved and reshaped just as a toolbar is moved. Figure 1-8 shows the Command Line box shortened and moved to the center of the drawing screen. The Command Line box may be returned to its original position using the same procedure.

1-4 COMMAND ICONS (TOOLS)

An icon is a picture that represents an AutoCAD command. Most commands have equivalent icons. Command icons are called "tools." A toolbox contains several tools.

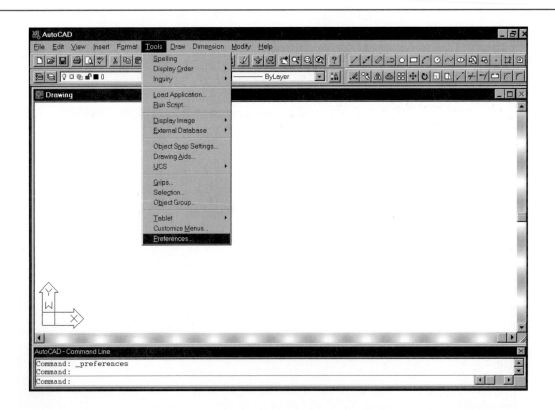

Figure 1-10

To determine the command an icon (tool) represents

See Figure 1-9.

1. Locate the cursor arrow on the selected icon (tool).

In the example shown, the Circle, Center, Radius command tool within the Draw toolbar was selected.

2. Hold the arrow still without pressing any mouse button.

The command name will appear below the tool. This is referred to as a "tooltip."

1-5 SCROLL BARS

The horizontal and vertical scroll bars can be removed from the screen. See Figures 1-10, 1-11, and 1-12.

1. Select the Tools pull-down menu.
2. Select Preferences.

See Figure 1-10. The Preferences dialog box will appear. See Figure 1-11.

3. Select the Display tab.

The Display options will appear as shown in Figure 1-12. Click on the box to the left of the "Display scroll bars in drawing window." A check mark will appear in the box indicating the scroll bar option is on. The scroll bar option may be turned off by clicking the box again. The check mark will disappear.

4. Select the OK boxes and return to the original drawing screen.

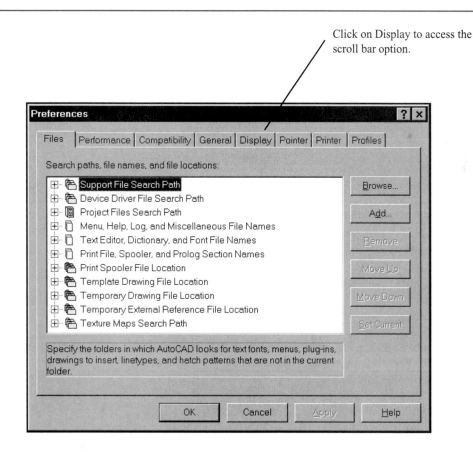

Click on Display to access the scroll bar option.

Figure 1-11

Click here to display the screen menus.

Click here to display scroll bars on the drawing screen.

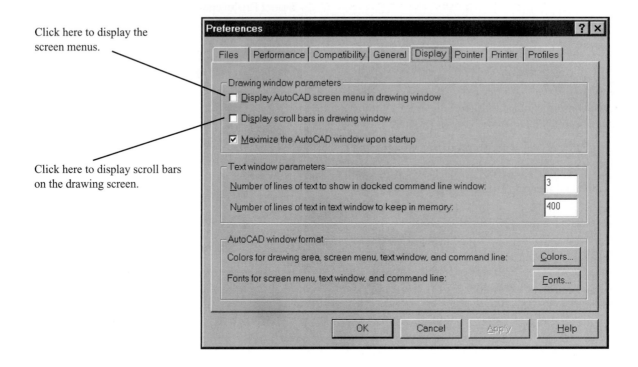

Figure 1-12

1-6 SCREEN MENUS

If you are used to working with AutoCAD's DOS versions or are just not comfortable with the command tools, you can activate screen menus. You may then select commands from the menus rather than using the tools.

To activate the screen menus

1. Select the Tools pull-down menu, then select Preferences.

   See Figure 1-11. The Preferences dialog box will appear.

2. Click the tab labeled Display.

   The Display options box will appear. See Figure 1-12.

3. Click the box to the left of "Display AutoCAD screen menu in drawing window."

The drawing window will appear with screen menus on the right side of the screen. See Figure 1-13.

1-7 EXERCISE PROBLEMS

EX1-1

How many tools are in the Object Snap toolbar?

EX1-2

How many tools are in the Surface toolbar?

EX1-3

Create a screen with the Draw and Modify toolbars located at the top of the drawing screen and the scroll bars removed.

EX1-4

Create a screen as shown in Figure EX1-4.

EX1-5

Create a screen as shown in Figure EX1-5.

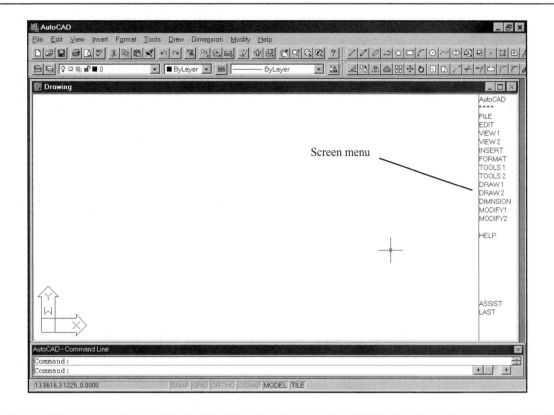

Figure 1-13

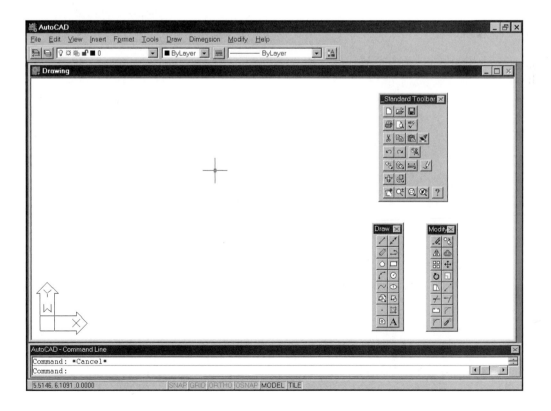

EX1-4

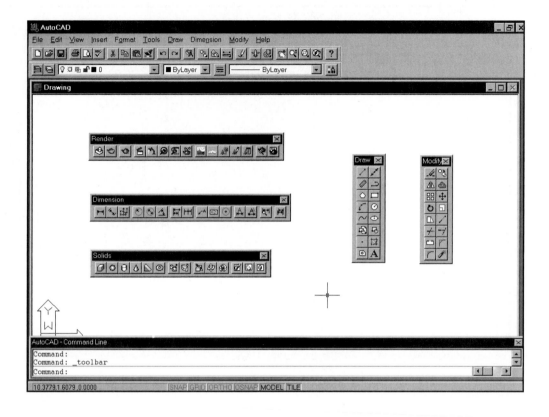

EX1-5

2

Setting Up a Drawing

2-1 INTRODUCTION

This chapter explains how to define a drawing's file name, to set the appropriate units, and to select the correct sized drawing sheet for a drawing. The chapter also shows how to use the Grid and Snap commands to apply a grid background to the drawing screen and how to limit the motion of the crosshairs. The Save and Save As commands are also demonstrated.

2-2 NAMING A DRAWING

Any combination of letters and numbers may be used as a file name. Either upper-or lower-case letters can be used as AutoCAD's file names are not character sensitive. The symbols $, -, and _ (underscore) may also be used. Other symbols, such as % and *, cannot be used. See Figure 2-1.

All AutoCAD drawing files will automatically have the extension ".dwg" added to the given file name. If you name a drawing FIRST, it will appear in the files as FIRST.dwg. Other extensions can be used but .dwg is the default setting. (A default setting is one that AutoCAD will use unless specifically told to use some other value.)

Unless otherwise specified, all drawings will be saved on the same drive on which the AutoCAD program files are located. For most applications this is the hard drive and is usually the C drive.

If you want to locate a file on another drive, specify the drive letter followed by a colon in front of the drawing name. For example, in Figure 2-1 the file specified A:FIRST will locate the drawing file FIRST on the A drive.

For large drawings, that is, drawings that have very large data bases likely to exceed the capacity of the specified drive, it is better to work on a drive that you are sure has sufficient space to accept the drawing file and then transfer the drawing to another drive as part of the saving process.

Correct drawing names:

    FIRST   EK-130-1   P1-1a

Incorrect drawing names:
    100%   *.*

To save a file on a disk in the A drive:

    A:FIRST

Figure 2-1

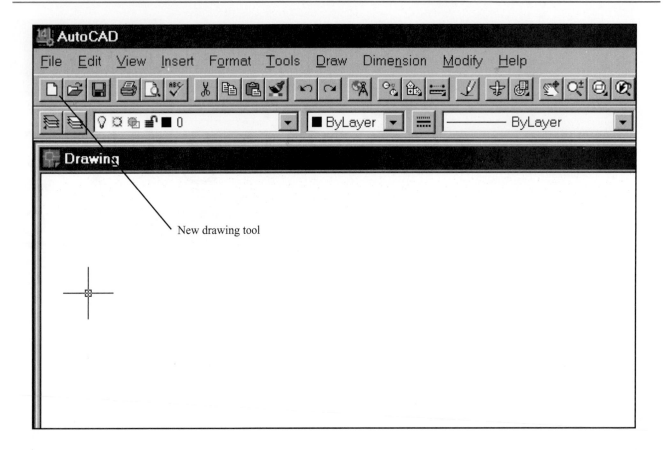

New drawing tool

Figure 2-2

To start a new drawing

There are three ways to access the Create New Drawing dialog box that is used to name a new drawing:

1. Select the New drawing tool in the Standard toolbar (see Figure 2-2).
2. Select the File pull-down menu, then select New (see Figure 2-3).
3. Type the word "new" in response to a Command prompt.

Any of these methods will cause the Create New Drawing dialog box to appear on the screen.

Figure 2-4 shows the Create New Drawing dialog box. Select the default units then click OK. The original drawing screen will reappear. Either inches or millimeters may be selected. Other units may be used as explained in the next section.

The drawing name may be assigned by clicking the Save As command listed under the File pull-down menu. The Save Drawing As dialog box will appear. See Figure 2-5. Type in the file name and click the Save box.

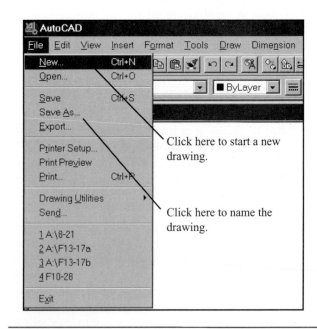

Click here to start a new drawing.

Click here to name the drawing.

Figure 2-3

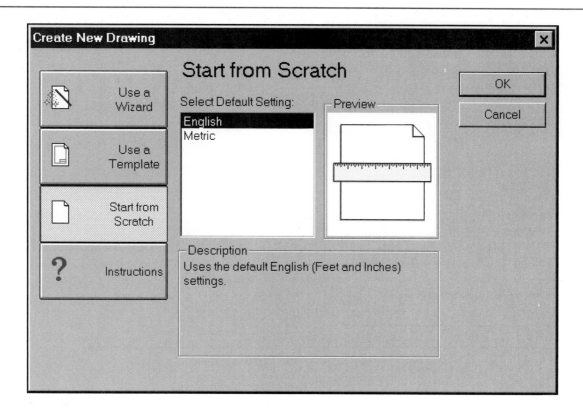

Figure 2-4

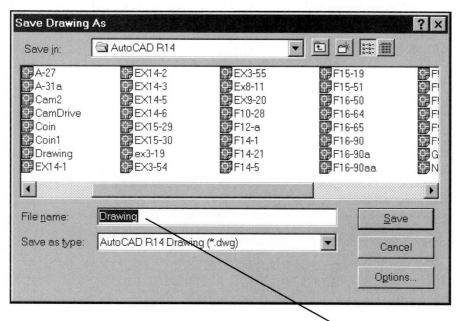

Backspace out the word Drawing and
type the new file name here.

Figure 2-5

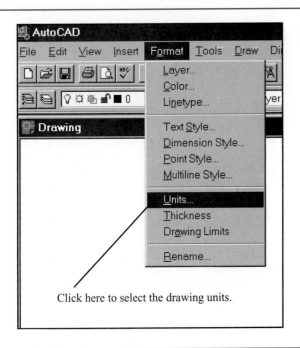

Click here to select the drawing units.

Figure 2-6

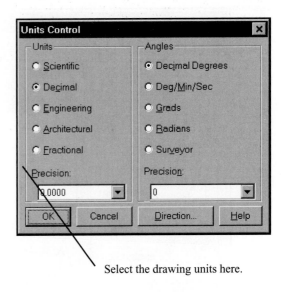

Select the drawing units here.

Figure 2-7

2-3 DRAWING UNITS

AutoCAD can work in any of five different unit systems: Scientific, Decimal, Engineering, Architectural, and Fractional. The default system is the Decimal system and it can be applied to either English (inches) or metric values (millimeters).

To specify or change the drawing units

1. Select the Format pull-down menu.
2. Select Units.

See Figure 2-6. The Units Control dialog box will appear. See Figure 2-7. Note that the circle to the left of the

word Decimal has a solid circle within it. This means that decimal units are on, and all the others are off. Only one unit system may be active at a time.

3. Select architectural units by clicking the circle next to the word Architectural.

A filled circle will appear within the circle next to Architectural, and will disappear from the word Decimal.

4. Select OK.

The original drawing screen will appear. Note that the coordinate display box now displays the cursor location in terms of feet and fractional inches. See Figure 2-8.

5. Repeat the above procedure and set the units back to decimal.

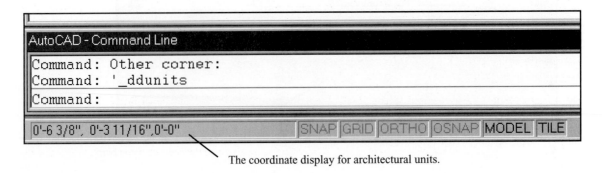

The coordinate display for architectural units.

Figure 2-8

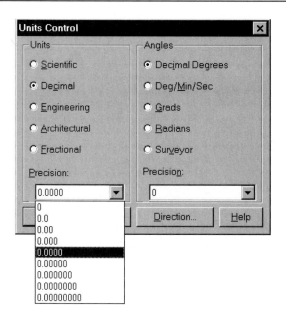

Figure 2-9

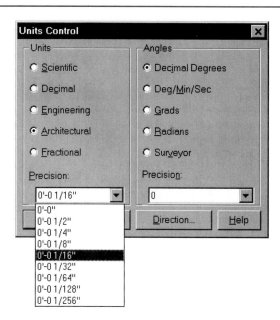

Figure 2-10

To specify or change the precision of the units system

Unit values can be expressed from zero to eight decimal places or from 0 to 1/256 of an inch.

1. Access the Units Control dialog box as explained above.
2. Select the arrow to the left of the current precision value display box below the word Precision.

A listing of the possible decimal precision values will cascade from the box. See Figure 2-9.

3. Select 0.00.

The values 0.00 will appear in the precision box.

4. Select OK.

The original drawing screen will appear. Note that the coordinate display box displays the cursor position in terms of two place values and not the default four places.

5. Change the units system to Architectural and set the precision to 0′-0 1/4″.

See Figure 2-10. Again note the difference in the coordinate display box.

6. Return the units to decimal places, that is, to the original default values.

To specify or change the Angles units values

Angle units may be specified in one of five different units: Decimal Degrees, Degrees/Minutes/Seconds, Gradians, Radians, or Surveyor units. Decimal Degrees is the default value.

Change the angle units by clicking the radio button to the left of the desired units. The precision of the angle units is changed as specified for linear units.

2-4 DRAWING LIMITS

Drawing limits are used to set the boundaries of the drawing. The drawing boundaries are usually set to match

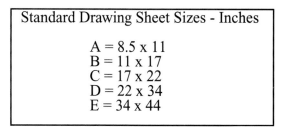

Standard Drawing Sheet Sizes - Inches
A = 8.5 x 11
B = 11 x 17
C = 17 x 22
D = 22 x 34
E = 34 x 44

Figure 2-11

Standard Drawing Sheet Sizes -
Millimeters

A4 = 210 x 297
A3 = 297 x 420
A2 = 420 x 594
A1 = 594 x 841
A0 = 841 x 1189

Figure 2-12

Standard Drawing Sheet Sizes -
Architectural U.S.

A = 9 x 12
B = 12 x 18
C = 18 x 24
D = 24 x 36
E = 36 x 48

Figure 2-13

the size of a sheet of drawing paper. This means that when the drawing is plotted and a hard copy is made, it will fit on the drawing paper.

Figure 2-11 shows a listing of standard flat sized drawing papers for engineering applications, Figure 2-12 shows standard metric sizes, and Figure 2-13 shows standard architectural sizes.

A standard 8.5 x 11-inch letter-sized sheet of paper as used by most impact and laser printers is referred to as an A-sized sheet of drawing paper.

To align the drawing limits with a standard A4 (metric) paper size

1. Select the Format pull-down menu.
2. Select Drawing Limits.

See Figure 2-14. The following prompts will appear in the Command Line box.

Reset Model space limits:
ON/OFF/<Lower left corner> <0.0000,0.0000>:

3. Press the Enter key or the right button on the mouse.

This means that you have accepted the default value of 0.0000,0.0000 or that the lower left of the drawing screen is the origin of an X,Y axis.

Upper right corner <12.0000,9.0000>:

4. Type 297,210; press Enter.

This changes the upper right corner of the drawing screen to an X,Y value of 297,210. There will be no visible change to the screen. The new limit extends beyond the

current screen calibration. The screen must be recalibrated to align with the new drawing limits.

5. Select the Zoom All tool from the standard tool-bar, or select the View pull-down menu; then select Zoom, then All.

See Figure 2-15. The Zoom All commands will align whatever drawing limits have been defined to the drawing screen. You can verify that the new drawing limits are in place by moving the cursor around the screen and noting the coordinate display box in the lower left corner of the screen. The values should be much larger than they were with the original 12,9 default screen settings. See Figure 2-15.

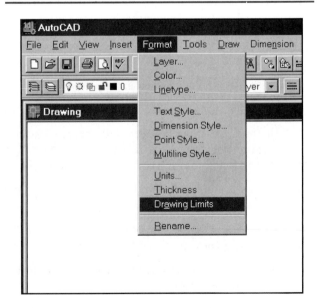

Figure 2-14

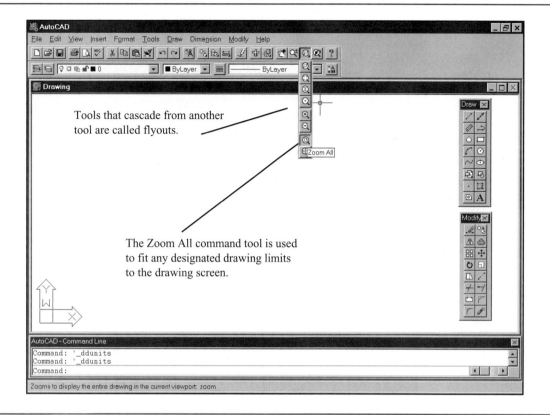

Tools that cascade from another tool are called flyouts.

The Zoom All command tool is used to fit any designated drawing limits to the drawing screen.

Figure 2-15

When larger-sized drawings are created, the scroll bars can be used to move about the drawing, or the Zoom All commands may be used to reduce the drawing to fit the screen and individual sections may be zoomed up as needed.

2-5 GRID AND SNAP

The GRID command is used to place a dotted grid background on the drawing screen. This background grid is helpful for establishing visual reference points for sizing and locating points and lines.

The SNAP command limits the movement of the cursor to predefined points on the screen. For example, if the SNAP command values were set to match the GRID values, the cursor would snap from grid point to grid point. It could not be located at a point between the grid points.

To set the Grid and Snap values

1. Select the Tools pull-down menu.
2. Select Drawing Aids.

See Figure 2-16.

Click here to set the Grid and Snap values.

Figure 2-16

A check mark indicates the command is on.

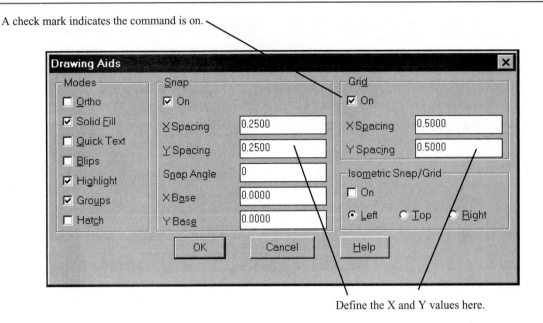

Define the X and Y values here.

Figure 2-17

The Drawing Aids dialog box will appear. See Figure 2-17.

3. Click On the Grid and Snap boxes.
4. Select the cursor on the X spacing box to the right of the given value under the Snap heading.

A vertical flashing cursor will appear.

5. Backspace out the existing value and type in 0.2500.
6. Click the Y spacing box.

The Y spacing will automatically be made equal to the X spacing value. Nonrectangular Snap spacing can be created by specifying different X and Y spacing values.

7. Select the X spacing box under the Grid heading.
8. Backspace out the existing value and type in 0.5000.
9. Click the Y spacing box to make the X and Y values equal.
10. Select OK.

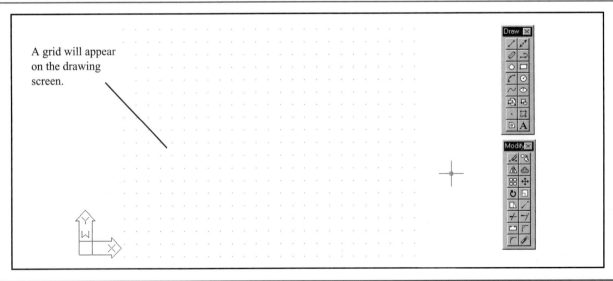

A grid will appear on the drawing screen.

Figure 2-18

The GRID and SNAP commands set for metric values.

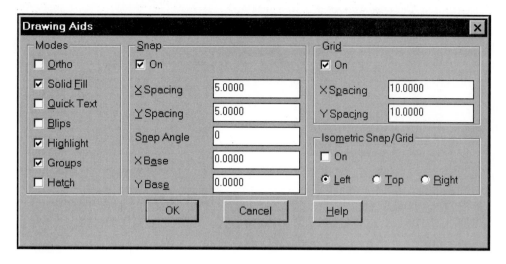

Figure 2-19

The original drawing screen will appear with a dotted grid. See Figure 2-18. The Snap values have been set to exactly half the Grid values so the cursor can be located directly on grid points or halfway between them only.

To set Grid and Snap spacing values for metric units

First set the drawing limits for metric values as presented in Section 2-4, then use Grid spacing values of 10 and Snap spacing values of 5. See Figure 2-19. The grid will not cover the entire screen, as the 297,210 limit values are not the same proportions as the drawing screen.

To set the Grid and Snap values for architectural units

Figure 2-20 shows a Drawing Aids dialog box with Grid and Snap spacing values set for 6″ and 3″ respectively. The procedure to access and change the dialog box is as explained above for decimal inch values.

The GRID and SNAP commands set for architectural values.

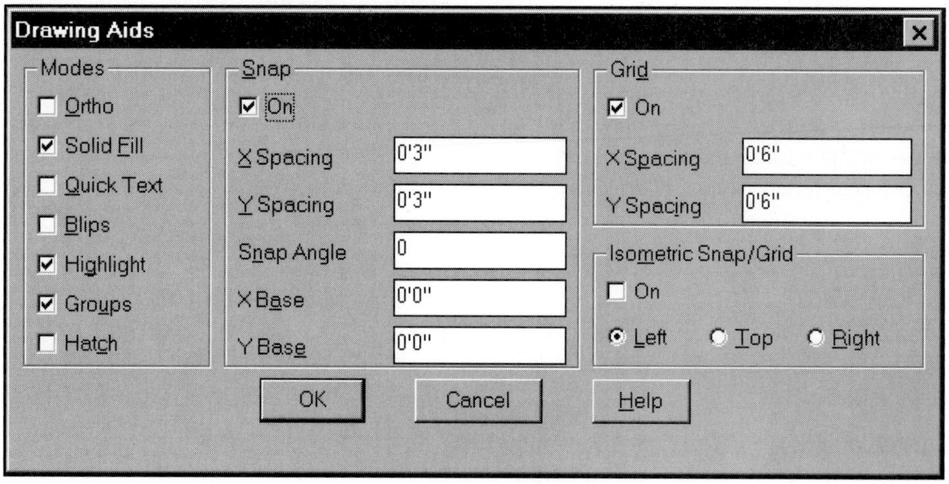

Figure 2-20

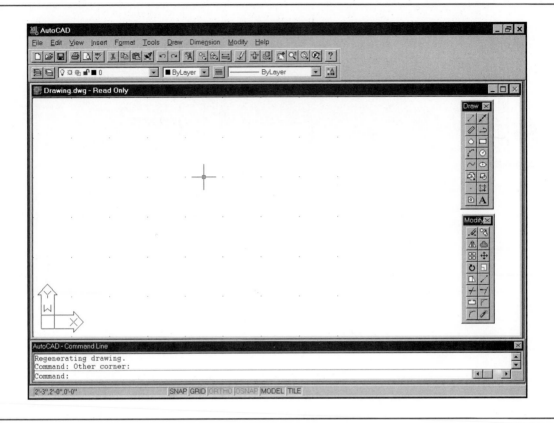

Figure 2-21

2-6 SAMPLE PROBLEM SP2-1

Set up a drawing for an architectural application as follows:

Sheet size = 48 x 36
Grid = 6″ spacing
Snap = 3″ spacing

To calibrate the sheet size

1. Select the Format pull-down menu, then Units, and turn on the Architectural units. Select OK.
2. Select the Tools pull-down menu, then Drawing Aids.
3. Turn on the Grid and Snap functions, then set them as shown in Figure 2-20.

Note the use of feet (′) and inches (″) symbols in Figure 2-20. The feet and inches symbols are created using the apostrophe and quotation marks on the keyboard.

4. Select the Format pull-down menu, then Drawing Limits. Set the lower limit for 0.0000,0.0000 (accept the default values), and the upper right limit for 48,36.

The limits are given in terms of their X,Y coordinate values.

5. Select the Zoom All tool.

The screen should appear as shown in Figure 2-21. Verify that the new drawing limits are in place by moving the crosshairs to the upper right portion of the drawing screen and noting the coordinate display values. The display should read approximately 48 (4′), 36(3′) in architectural units.

2-7 SAVE AND SAVE AS

The Save and Save As commands are used to save a drawing. The Save command is also called the "quick save" command. It is used primarily while working on a drawing to save your work as you go. Save can also be used to save a finished drawing, but only under the original file name of the drawing.

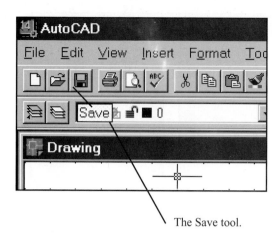

The Save tool.

Figure 2-22

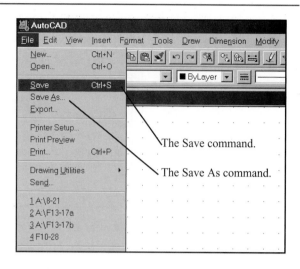

The Save command.

The Save As command.

Figure 2-23

To access the Save command

The SAVE command can be accessed in one of two ways:

1. Use the Save icon on the Standard toolbar (see Figure 2-22).
2. Select the File pull-down menu and select Save (see Figure 2-23).

There is no visible change to the drawing when Save is activated, but the line

Command: _qsave

will appear in the Command Line box.

The SAVE AS command is used to save a drawing just as the SAVE command does, and to save a drawing under a name that is different from the original drawing name. For example, you can name a drawing FIRST and then decide to save it on a floppy disk on the A drive under the name BOX.

To access the Save As command

1. Select the File pull-down menu.
2. Select Save As (See Figure 2-23).

The Save Drawing As dialog box will appear. See Figure 2-24. If you wish to save a drawing under its current name, that is, the name that appears in the File Name box, click the OK box.

The Save As dialog box also provides a preview of the drawing to be saved. The drawing, screen2, listed in the File name box, is the artwork used for Figure 1-2 in this book.

Enter the new drawing name here.

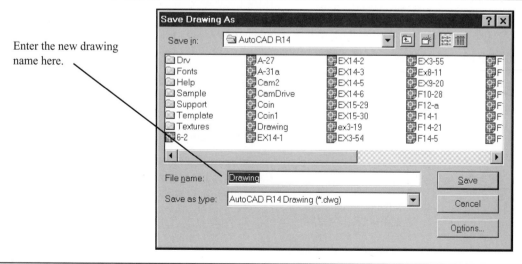

Figure 2-24

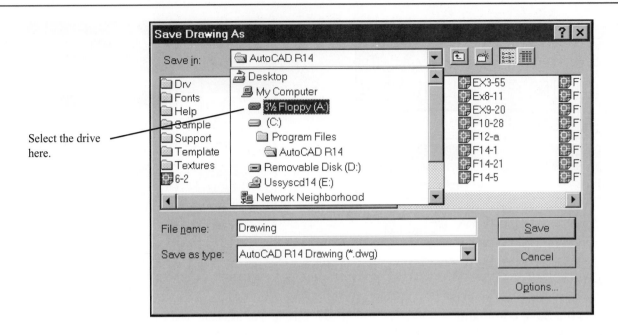

Figure 2-25

To save a drawing under a different name on a different drive

1. Access the Save Drawing As dialog box.
2. Select the arrow on the right side of the box under the Drives heading.

A selection of available drives will cascade from the box. See Figure 2-25.

3. Select A.

The files listed in the Directories box will change to a listing of the files on drive A.

4. Select the box under the File name box.

A flashing vertical cursor should appear in the box.

5. Type in the new file name.

In this example, the name BOX was used.

6. Select OK.

The original drawing screen will appear with the name BOX at the top of the screen.

2-8 OPEN

The OPEN command is used to call up an existing drawing so that you may continue working on it, or to revise it.

To access the Open command

The OPEN command may be accessed in one of two ways:

1. Use the Open tool on the Standard toolbar (see Figure 2-21).
2. Select the File pull-down menu and select Open (see Figure 2-22).

The Select File dialog box will appear. See Figure 2-26. Drawings will be listed under the heading File name. You may also select files from other directories or files with extensions other than .dwg. If you know the specific file name you are looking for, remove the * symbol in the fFile name box and type in the desired file name, then select the Open box.

If you forget which file name has been assigned to which drawing, AutoCAD provides a Browse/Search option that allows you to visually search the files by directories, or you can initiate a specific file search. The Browse/Search option can be accessed using the Find File button located on the Select File dialog box.

The Browse/Search dialog box

Figure 2-27 shows the Browse/Search dialog box. Each file in the selected directory will be listed in the box. In addition there will be a thumbnail preview picture of the saved drawing. Any listed drawing file can be accessed by double-clicking its name.

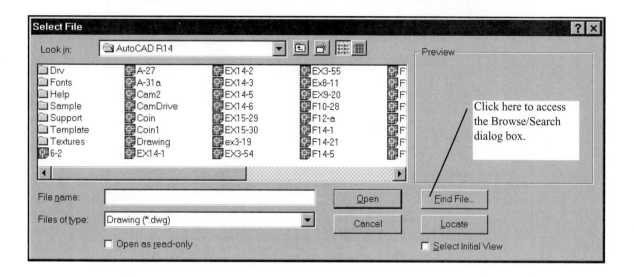

Figure 2-26

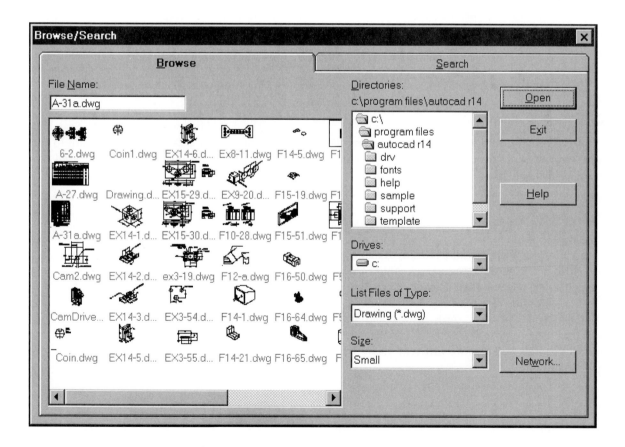

Figure 2-27

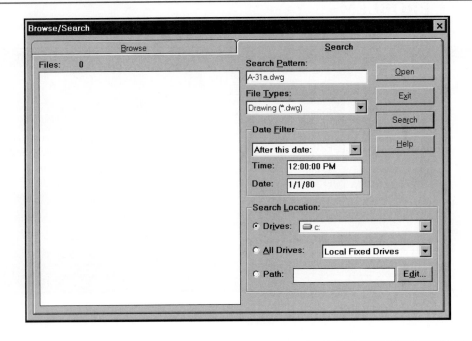

Figure 2-28

To select a file from the Browse picture listing:

1. Select a drawing by clicking its picture.

The drawing file name will appear in the File name box.

2. Click the Open button.

You can browse different directories by using the Drives box, or you can browse different drawing file extensions by using the List Files of Type box.

To search for a file

The Search option allows you to be more specific when you are looking for an existing file. For example, you can specify the date a drawing was created; AutoCAD will locate and present a preview of all drawings created on that date.

Note in Figure 2-27, the words Browse and Search are located on what appear to be the tabs of file folders. This means there are actually two different dialog boxes within the shown box.

To access the Search dialog box

1. Locate the cursor on the heading Search on what looks like a file folder tab near the top of the box.
2. Click the word Search.

The dialog box will change to the Search dialog box as shown in Figure 2-28. You can return to the Browse portion of the dialog box by clicking the word Browse near the top of the box.

2-9 EXERCISE PROBLEMS

EX2-1

Create the following drawing screen setup.

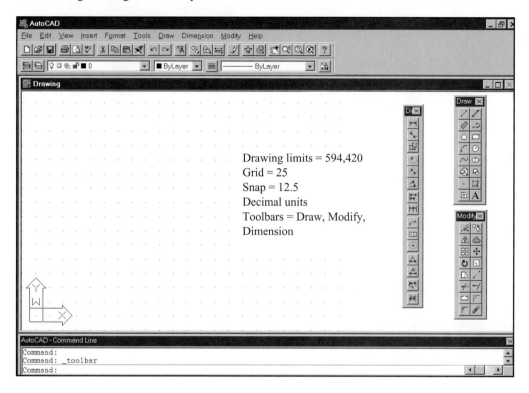

EX2-2

Create the following drawing screen setup.

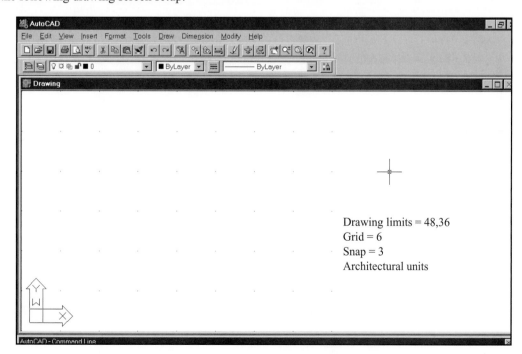

EX2-3

Create the following drawing screen setup.

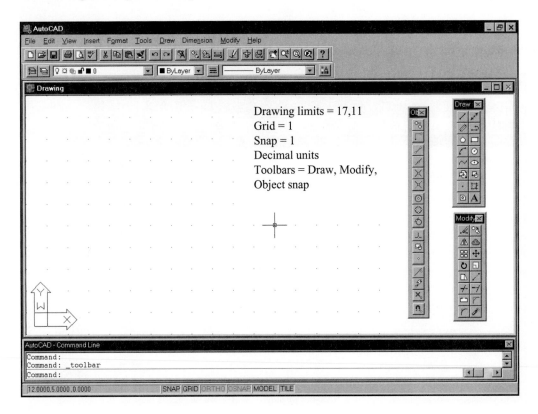

EX2-4

Create the following drawing screen setup.

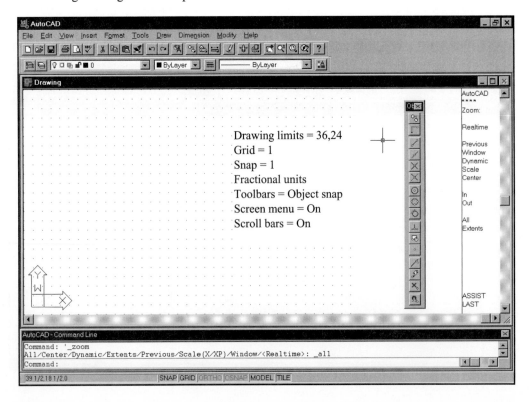

EX2-5

Create the following drawing screen setup.

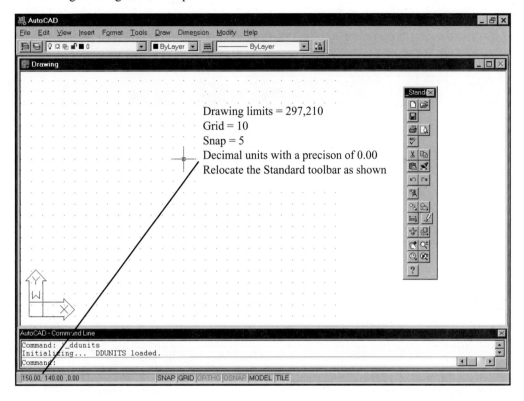

Drawing limits = 297,210
Grid = 10
Snap = 5
Decimal units with a precison of 0.00
Relocate the Standard toolbar as shown

EX2-6

Create the following drawing screen setup.

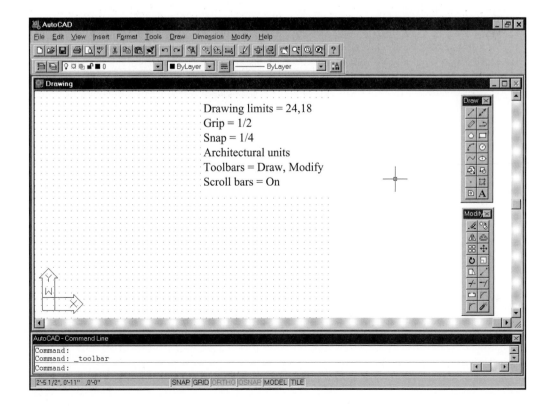

Drawing limits = 24,18
Grip = 1/2
Snap = 1/4
Architectural units
Toolbars = Draw, Modify
Scroll bars = On

EX2-7

Create the following drawing screen setup.

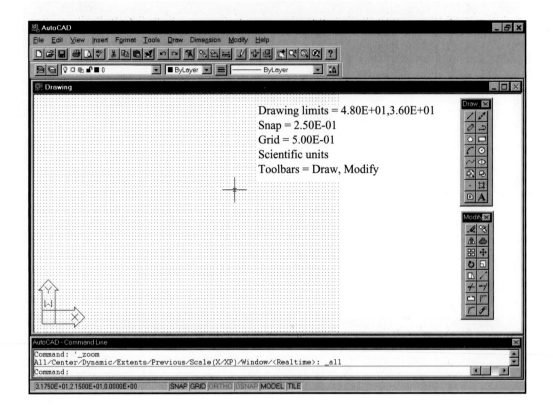

CHAPTER 3

Working with AutoCAD Commands

3-1 INTRODUCTION

This chapter demonstrates how to work with AutoCAD commands. AutoCAD commands operate using command tools, menus, and a series of prompts. The prompts appear in the Command Line box and ask for a selection or numerical input so a command sequence can be completed. The LINE command was chosen to demonstrate the various input and prompt sequences that are typical with AutoCAD commands.

The ERASE command is also included in the chapter, as well as instructions for scaling a drawing to fit a standard-sized sheet of drawing paper.

3-2 LINE — RANDOM POINTS

The LINE command is used to draw straight lines between two defined points. Figure 3-1 shows the Line tool on the Draw toolbar.

There are four ways to define the length and location of a line: randomly select points; set a specific value for the Snap function and select points using the Snap spacing values; enter the coordinate values for the start and end points; or use relative inputs and specify the starting point and the length and direction of the line.

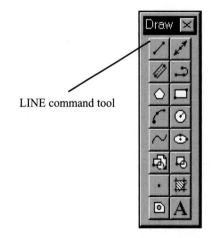

LINE command tool

Figure 3-1

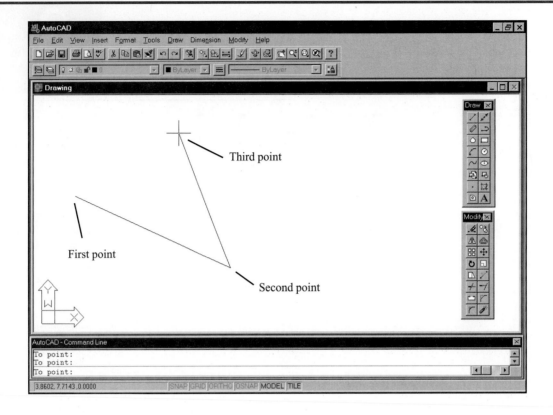

Figure 3-2

To randomly select points

See Figure 3-2.

1. Select the Line tool in the Draw toolbar.

You can also access the LINE command using the Draw 1 listing on the screen menus.

The following command sequence will appear in the Command Line box.

Command: _line From point:

2. Locate the cross hairs randomly on the drawing screen and press the left mouse button.

Note:

A colon at the end of a Command prompt indicates that a response line is required.

To point:

As you move the cross hairs, a rubber band-type line will extend from the designated point to the crosshairs.

This type of dynamic motion is called the drag mode and it occurs in many AutoCAD commands.

3. Randomly pick another point on the screen.

To point:

AutoCAD will keep asking for another point until you either press the Enter key or the right mouse button, that is, give the prompt a null entry.

4. Press the right mouse button to exit the LINE command.

You should have two lines on the screen similar to those shown in Figure 3-2.

5. Press the right mouse button again.

This second Enter response will start the LINE command again.

Note:

Any AutoCAD command can be reactivated after the last step in the sequence is completed by pressing the right mouse button or the Enter key.

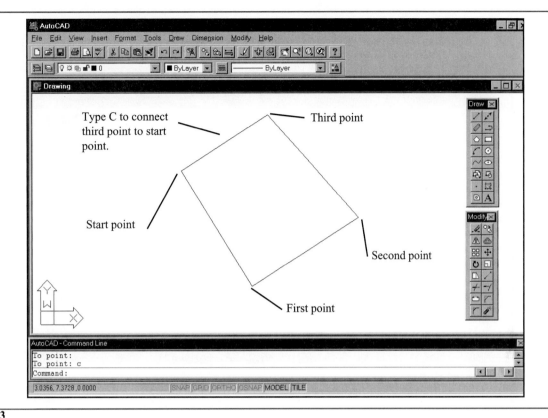

Figure 3-3

To create a closed area

See Figure 3-3.

1. Select the Line tool from the Draw toolbar.

   Command: _line From point:

2. Select a random point.

   To point:

3. Select a second random point.

   To point:

4. Select a third random point.

To point:

5. Type C, then press Enter to activate the Close option.

A line will be drawn from the third point to the starting point, creating a closed area.

3-3 ERASE

You can erase any line by using the ERASE command. There are two ways to erase lines: selecting individual lines or windowing a group of lines.

To erase individual lines

See Figure 3-4.

1. Select the Erase tool on the Modify toolbar.

The following prompt will appear in the Command Line box.

   Command: _erase
   Select objects:

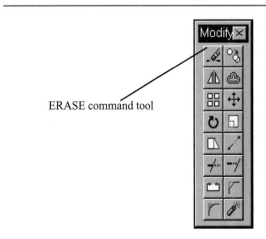

ERASE command tool

Figure 3-4

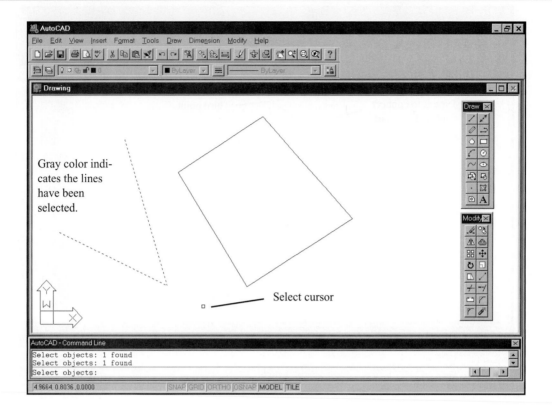

Figure 3-5

The cross hairs will be replaced by a rectangular cursor. This is the select cursor (pickbox). Any time you see this cursor, AutoCAD expects you to select an entity, in this example, a line. See Figure 3-5.

If you change your mind and do not want to erase anything, select the next command you want to use and the Erase command will be terminated.

2. Select the two open lines by locating the rectangular cursor on each line, one at a time, and pressing the left mouse button.

The lines will change colors from black to a broken gray pattern. This color change indicates that the line has been selected. The selection can be confirmed by a change in the prompt.

Select objects: 1 found
Select objects:

3. Press the right mouse button or the Enter key to complete the erase sequence.

The two lines should disappear from the screen.

To return an erased line to the screen

If, after you erase a line, you realize that you did not want to erase the line, you can return the line to the screen using the UNDO command. Figure 3-6 shows the UNDO command tool.

1. Select the Undo tool.

The lines should reappear on the screen with the following command prompt line.

Command:_u ERASE GROUP

The Undo command can be used to reverse any individual command or group of commands.

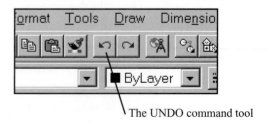

The UNDO command tool

Figure 3-6

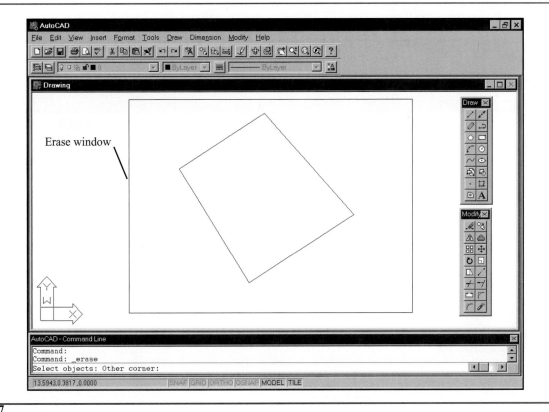

Figure 3-7

To erase a group of lines simultaneously

See Figure 3-7.

1. Select the Erase tool.

    Command: _erase
    Select objects:

2. Locate the rectangular select cursor above and to the left of the lines to be erased and press the left mouse button.
3. Move the mouse and a window will drag from the selected first point.
4. When all the lines to be erased are completely within the window, press the left mouse button.

All the lines completely within the window will be selected and will change to broken gray lines. The number of lines selected will be referenced in the Command Line box.

5. Press the right mouse button or the Enter key to complete the command sequence.

The lines will disappear from the screen. The Undo tool will return all the lines, if needed. If the erase window had been created from right to left, any line even partially within the erase window would be erased.

Note:

A line must be completely within the erase window for it to be erased, if the erase window is created from left to right. An erase window created from right to left will erase all lines within the window, even if only part of the line appears in the window. See Figure 3-8.

Erase window created
from left to right

The line is not completely within the window, so it will not be erased.

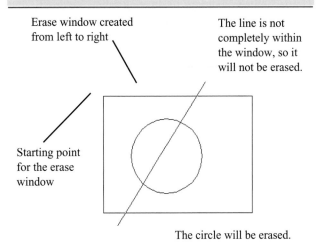

Starting point for the erase window

The circle will be erased.

Figure 3-8

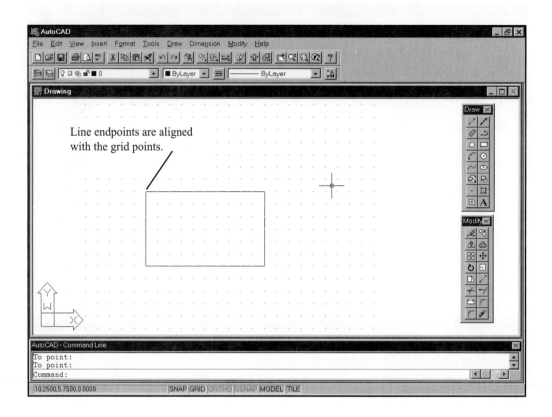

Figure 3-9

3-4 LINE — SNAP POINTS

Lines can be drawn using calibrated snap spacing. The technique is similar to drawing points randomly, but because the cursor is limited to specific points, the length of lines can be determined accurately. See Figure 3-9.

Problem: Draw a 5″ x 3″ rectangle.

1. Set the Grid spacing for .5 and the Snap spacing for .25 and turn both commands on.
2. Select the Line tool from the Draw toolbar.

   Command: _line From point:

3. Select a grid point.

   To line:

4. Move the cursor horizontally to the right 10 grid spaces and press the left mouse button.

The Grid spacing has been set at .5, so 10 spaces equals 5 units. Watch the coordinate display as you move the cursor. The X value should increase by 5 and the Y value should stay the same.

   To line:

5. Move the cursor vertically 6 spaces and press the left mouse button.

   To line:

6. Move the cursor horizontally to the left 10 spaces and press the left mouse button.

   To point:

7. Type C; press Enter.

   Save or Erase the drawing as desired.

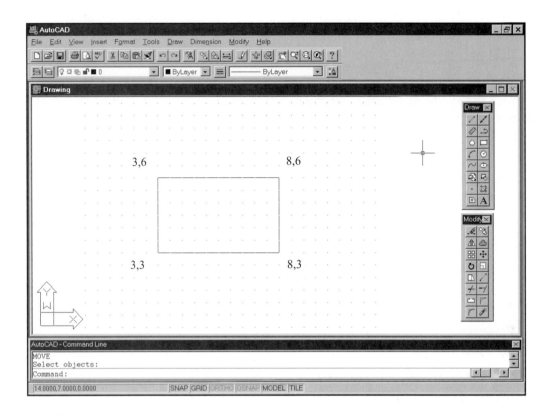

Figure 3-10

3-5 LINE — COORDINATE VALUES

AutoCAD will also accept locational inputs, the start and end points of a line, in terms of X,Y coordinate values. The 0,0 origin is located in the lower left of the drawing screen.

Problem: Draw a 3″ x 5″ rectangle starting at the 3,3 coordinate point. See Figure 3-10.

1. Select the Line tool from the Draw toolbar.

   Command: _line From point:

2. Type 3,3; press Enter.

   To point:

3. Type 8,3; press Enter.

   The X value has been increased 5 inches.

   To point:

4. Type 8,6; press Enter.

   The Y value has been increased 3 inches.

   To point:

5. Type 3,6; press Enter.

   To point:

6. Type c; press Enter.

   Save or Erase the drawing as desired.

3-6 LINE — POLAR VALUES

Polar coordinate values can be used to specify a length and direction for a line. The input is in the following format.

@distance<angle

No units need to be assigned for either distance or angle. The current specified units and precision will be assumed. Both positive and negative values can be used for either input.

Problem: Draw a 151.00 x 132.00 millimeter rectangle starting at point 50,50. See Figure 3-11.

Note in this example, the required values could not easily be aligned with a Snap spacing.

1. Set the drawing screen limits for 297,210.

   See Section 2-4.

2. Set the Grid spacing for 10 and the Snap spacing for 5.
3. Select the Line tool from the Draw toolbar.

   Command: _line From point:

4. Type 50,50; press Enter.

   To point:

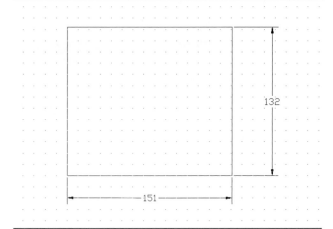

Figure 3-11

5. Type @151<0; press Enter.

   To point:

6. Type @132<90; press Enter.

   To point:

7. Type @151<180; press Enter.

   To point:

8. Type @132<-90 (or @132<270); press Enter.

The last line could have been drawn using the close, or C, input. Figure 3-12 shows the results.

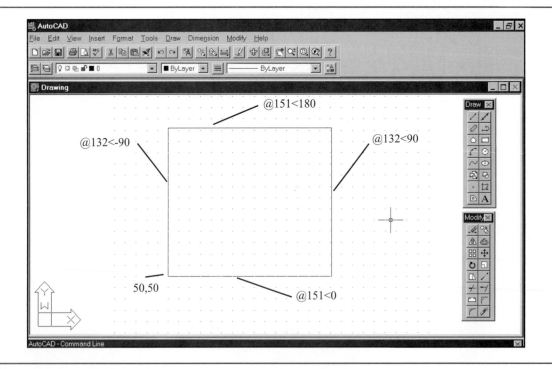

Figure 3-12

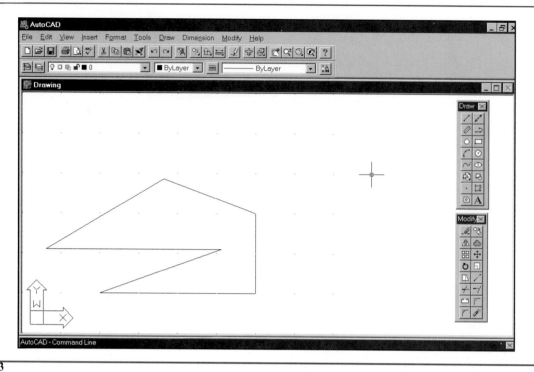

Figure 3-13

3-7 SAMPLE PROBLEM SP3-1

Problem: Draw the figure shown in Figure 3-13.

1. Select the Format pull-down menu, then Units.
2. Select Architectural from the units control dialog box.
3. Set the drawing limits for 48″ x 36″ and use Zoom All to calibrate the limits to match the area of the drawing screen. The Zoom Window command may also be used.
4. Set the Grid spacing for 6″ and the Snap spacing for 3″.
5. Move the Draw and Modify toolbars to the right side of the drawing screen, off the background grid.
6. Select the Line tool on the Draw toolbar.

Command:_line From point:

7. Type 1′0″,0′6″; press Enter

The 1′0″,0′6″ starting point could also have been located using the coordinate display in the lower left corner of the screen.

To point:

8. Move the cursor 4 grid marks to the right and press the left mouse button.

The line is 24″ long, so it will match up with the defined grid points. The input @2′<0 could also have been used.

To point:

9. Move the cursor vertically 2 grid spaces.

This line could also have been created using the input, @1′<90.

To point:

10. Type @ 1′3″<160; press Enter.

To point:

11. Type @ 1′9″<-150; press Enter.

To point:

12. Type @ 2′3″<0; press Enter.

To point:

13. Type c; press Enter.

To point:

14. Press Enter.

The drawing is now complete. Use the Save As command to save the drawing. Use the file name SP3-1.

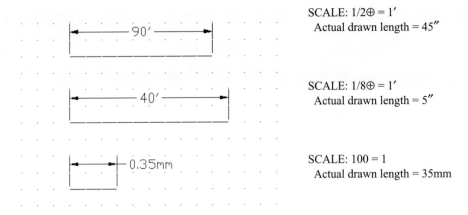

SCALE: 1/2⊕ = 1′
Actual drawn length = 45″

SCALE: 1/8⊕ = 1′
Actual drawn length = 5″

SCALE: 100 = 1
Actual drawn length = 35mm

Figure 3-14

3-8 DRAWING SCALES

The actual sizes of objects are often too big to fit on a standard-sized sheet of drawing paper. The overall size for a house may be 30′ x 40′ or an aircraft may be 60′ long. Conversely, some lines may be too short to be visible. The conductor paths within an integrated circuit may be measured in 0.001 millimeters.

Drawing scales are used to fit large drawings to standard drawing sheet sizes or to make very small drawings large enough to be seen. Drawing scales are designated on a drawing using the following format.

SCALE: 1/2 = 1
SCALE: 1/4″ = 1′
SCALE: 100 = 1

A scale of $1/2 = 1$ means that every 1/2 unit on a drawing represents 1 unit on an object. In other words, the drawing of the object is exactly half the size of the object.

A scale of $100 = 1$ means that the drawing is 100 times larger than the object.

To determine the drawing length of a scaled line using decimal units

Figure 3-14 shows a line with a dimension of 90′. If the line were to be drawn at a scale of $1/2″ = 1′$, multiply the actual length of the line by the scale factor to determine the drawn length.

(scale factor)(actual length) = drawing length
$(1/2)(90′) = 45″$

If the scale were $1/4″ = 1′$, then the calculations would be

$(1/4)(90′) = 22.5″$

If the scale were $.1″ = 1′$, then the calculation would be

$(1/10)(90′) = 9″$

To determine the drawing length of a scaled line using architectural units

Figure 3-14 also shows a line with a dimension of 40′. The drawing length is determined as follows.
If the scale were $1/2″ = 1′$

$(1/2)(40′) = 20″$

If the scale were $1/4″ = 1′$

$(1/4)(40′) = 10″$

If the scale were $1/8″ = 1′$

$(1/8)(40′) = 5″$

To determine the drawing length of a very short line using decimal units

Figure 3-14 also shows a line with a dimension of 0.35mm. The drawing length is determined as follows.
If the scale were $10 = 1$

$(10)(.35) = 0.35$

If the scale were $1000 = 1$

$(100)(.35) = 35.0$

Choose a scale factor that is easy to apply and understand, such as 1/2, 1/4, or multiples of 10. A scale of $1.32 = 1$ or $1/7.2 = 1$ would be difficult to calculate and difficult to visualize.

To determine the scale needed to match an object to a standard sheet size

Figure 3-15 shows a dimensioned base plate. The overall dimensions of 60 x 55 mean that the object is larger than any standard-sized flat sheet. The object must be scaled down.

If a scale of 1/2 = 1 were used, the overall drawing size of the object will be 30 x 27.5, which would fit on an E-sized (34 x 44) drawing sheet. See Figure 2-11 through 2-13 for listings of standard drawing sheet sizes.

If a scale of 1/4 = 1 were used, the overall drawing size of the object would be 15 x 13.75. This size would fit comfortably on a D-sized (22 x 34) sheet of drawing paper, and would just fit on a C-sized (17 x 22) sheet. It is best to scale the drawing so it fits comfortably on a drawing sheet, that is, there is open area around the object for dimensions, notes, and other supplementary information.

A scale of 0.1 = 1 could be used to fit the object on an A-sized (8.5 x 11) sheet of drawing paper. The drawing would be drawn at the reduced size, but the dimensions must be drawn at a size that can be read easily. This means that the line portion of the drawing would be drawn at a different scale than the dimension portion.

Figure 3-16 shows a partial floor plan for a house. A scale of 1/4″ = 1′ was used to draw the object on a standard B-sized sheet of drawing paper.

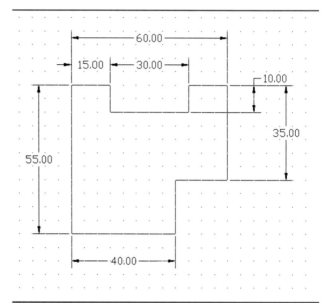

Figure 3-15

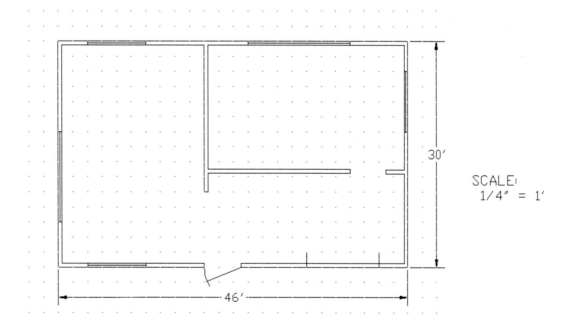

Figure 3-16

3-9 EXERCISE PROBLEMS

Redraw the following figures. Do not include dimensions on the drawing.

EX3-1 INCHES

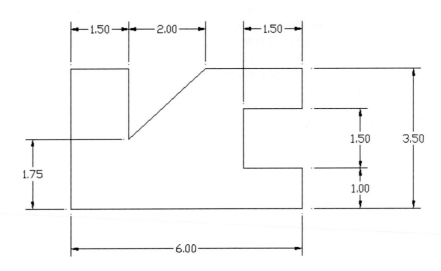

GUIDE PLATE

EX3-2 MILLIMETERS

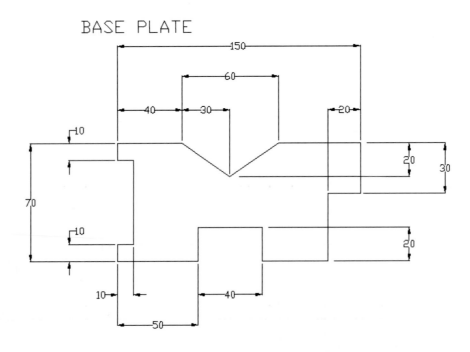

BASE PLATE

EX3-3 INCHES

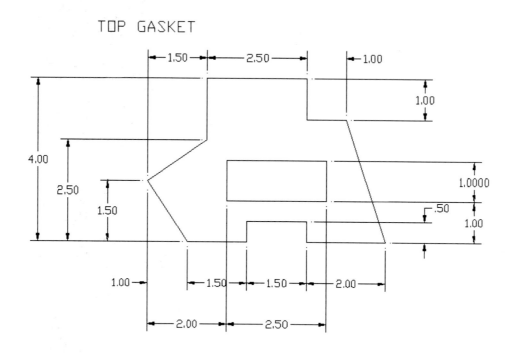

TOP GASKET

EX3-4 MILLIMETERS

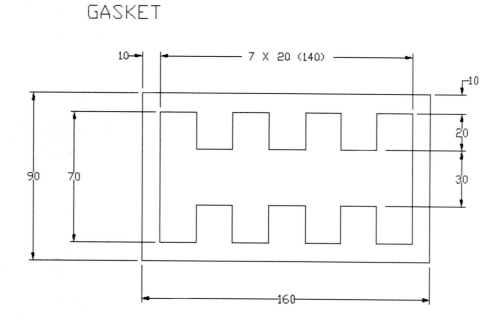

GASKET

EX3-5 INCHES

A-B = 5.0000
B-C = 1.9526
C-D = 2.3049
D-E = 1.8039
E-F = 3.2500
F-G = 4.1003

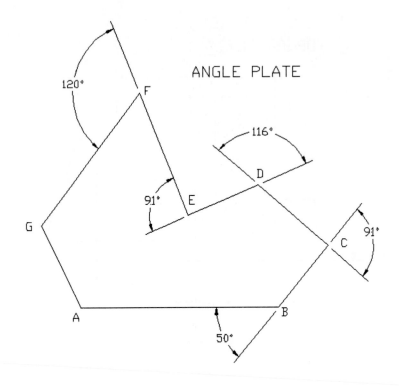

ANGLE PLATE

EX3-6 MILLIMETERS

A-B = 67.0820
B-C = 50.0000
C-D = 64.0312
D-E = 50.0000
E-F = 42.4264
F-G = 64.0312
G-H = 76.1577

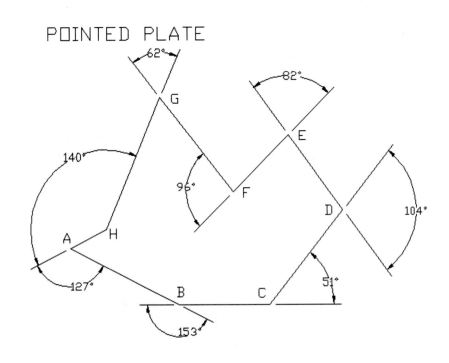

POINTED PLATE

EX3-7 INCHES

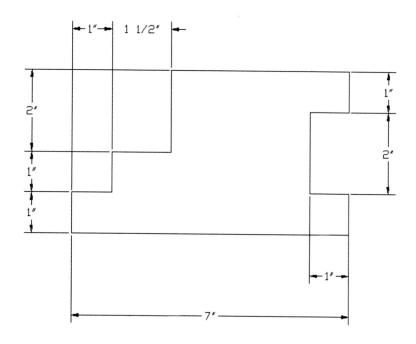

EX3-8 INCHES

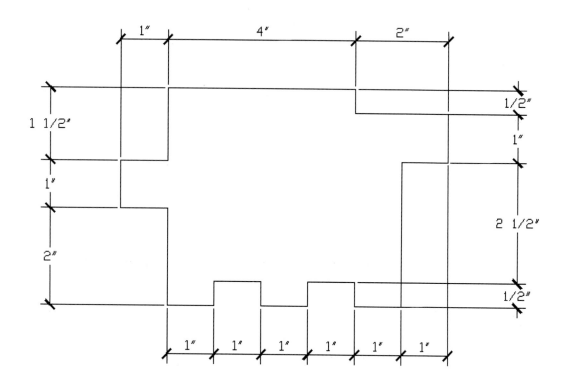

EX3-9 INCHES

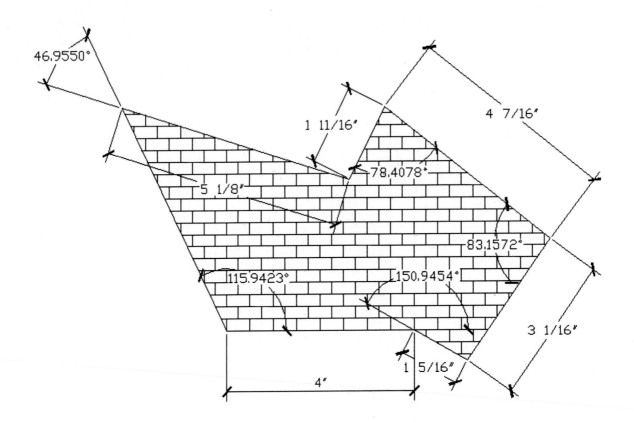

CHAPTER 4

The Draw Toolbar

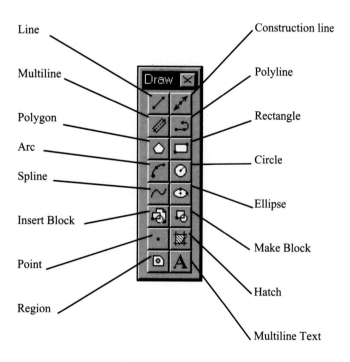

Line

Multiline

Polygon

Arc

Spline

Insert Block

Point

Region

Construction line

Polyline

Rectangle

Circle

Ellipse

Make Block

Hatch

Multiline Text

Figure 4-1

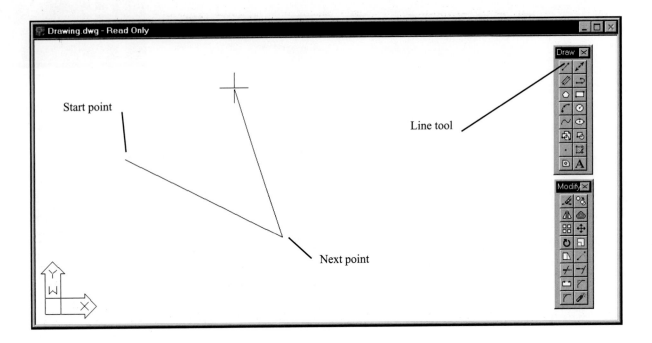

Figure 4-2

4-1 INTRODUCTION

This chapter explains the commands found in the Draw toolbar. See Figure 4-1. There are ten command tools in the Draw toolbar and each tool has flyout tools. All commands will be demonstrated.

4-2 LINE

There are four line tools: point-to-point, which was demonstrated in chapter 3; construction, multiline, and polyline. Each linetype has its own tool in the Draw toolbar.

To use the LINE Point-to-point command

See Figure 4-2. See Chapter 3.

1. Select the Point-to-point tool on the Draw toolbar.

   Command: Line From point:

2. Select a start point.

To point:

3. Select the endpoint of the line.

To point:

4. Press Enter.

The Enter command ends the drawing sequence. Pressing the Enter command a second time will reactivate the point-to-point command sequence.

To use the CONSTRUCTION LINE command

The CONSTRUCTION LINE command is used to draw lines of infinite length. Construction lines are very helpful during the initial layout of a drawing. They can be trimmed during the creation of the drawing as needed.

1. Select the Construction line tool from the Draw toolbar.

   Command:_ xline/Hor/Ver/Angl/Bisect/Offset/ <From point >:

   The From point command is the default setting.

2. Select a starting point.

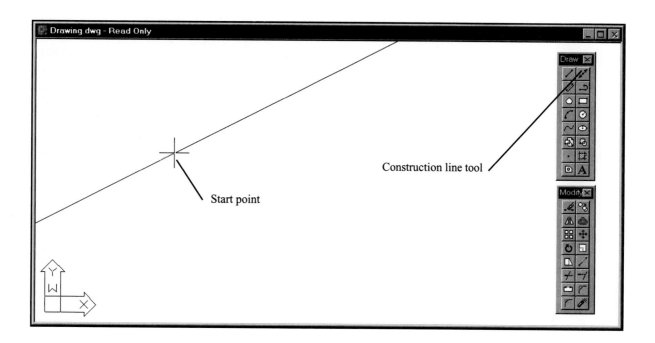

Figure 4-3

Through point:

A line will pivot about the designated starting point and extend at infinite length in a direction through the cross hairs. You can position the direction of the line by moving the cross hairs. See Figure 4-3.

3. Select a through point.

An infinite length line will be drawn through the two designated points.

Through point:

Another infinite line will appear through the starting point through the cross hairs.

4. Press Enter.

This will end the Construction line command sequence. The sequence may be reactivated and another construction line drawn by pressing the Enter key a second time.

Note:

AutoCAD often presents a listing of options in association with a command, for example, the command line generated when the Construction line command is active.

Command:_xline/Hor/Ver/Ang/Bisect?Offset/<F rom point>:

These options may be activated by typing the capital letters in their headings. Typing a response of letter h will activate the Hor (horizontal) command option.

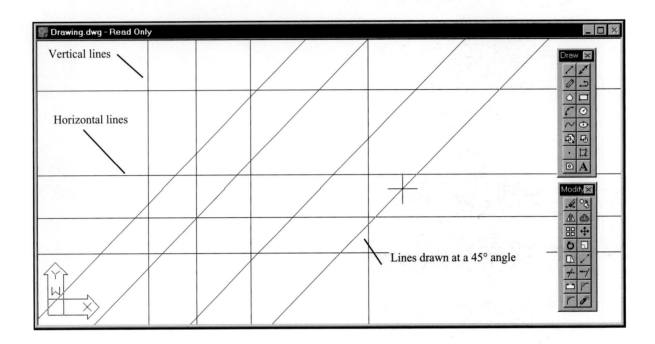

Figure 4-4

Other Construction line commands:
HOR/VER/ANG

The lines shown in Figure 4-4 were created using the Hor (horizontal), Ver (vertical), and Ang (angular) options. A grid was created and the snap option was turned on so the lines could be drawn through known points. Figure 4-5 was created as follows:

1. Set up the drawing screen with Grid and Snap spacing of .5.
2. Select the Construction line tool in the Draw toolbar.

   *Command:_ xline/Hor/Ver/Ang/Bisect/Offset/
   <From point>:*

3. Type H; press Enter.

   A horizontal line will appear through the crosshairs.

   Through point:

4. Select a point on the drawing screen.

   A horizontal line will appear through the point. As you move the mouse, another horizontal line will appear through the cross hairs.

Through point:

5. Select a second point.

   Through point:

6. Select a third point.

   Through point:

7. Double-click the Enter key.

   *Command:_ xline/Hor/Ver/Ang/Bisect/Offset/
   <From point>:*

8. Type V; press Enter.

   Through point:

9. Draw three vertical lines, then double-click the Enter key.

   *Command:_ xline/Hor/Ver/Ang/Bisect/Offset/
   <From point>:*

10. Type A; press Enter.

    Reference/<enter angle(0.0000)>:

11. Type 45; press Enter.

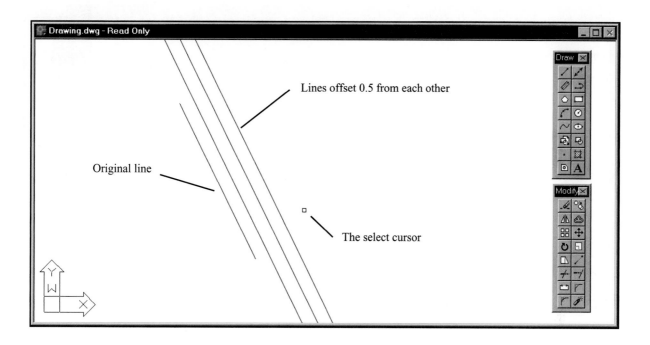

Figure 4-5

Through point:

An infinite line at 45 degrees will appear through the crosshairs.

12. Draw three 45-degree lines.

Through point:

13. Press Enter.

Your drawing should look approximately like Figure 4-4.

Other Construction line commands: OFFSET

The Offset option allows you to draw a line parallel to an existing line, regardless of the line's orientation, at a predefined distance. See Figure 4-5.

1. Select the Construction line tool in the Draw toolbar.

Command:_ xline/Hor/Ver/Ang/Bisect/Offset/ <From point>:

2. Draw a line approximately 15° to the vertical (See Figure 4-6).

Through point:

3. Double-click the Enter key to restart the Construction line command sequence.

Command:_ xline/Hor/Ver/Ang/Bisect/Offset/ <From point>:

4. Type O; press Enter.

Offset distance or through <Through>:

5. Type .5; press Enter.

Select a line object:

6. Select the line.

Side to offset:

7. Select the right side of the line.

Select a line object:

8. Select the line just created by the offset option.

Side to offset:

9. Again select the right side of the selected line.

Select a line object:

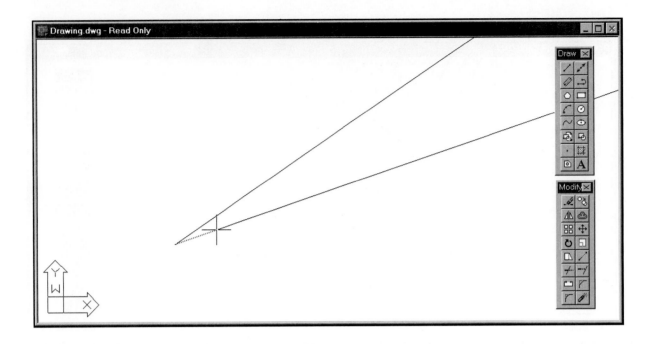

Figure 4-6

10. Draw several more lines.

Select a line object:

11. Press the right mouse button.

This will end the OFFSET command sequence and return a command prompt to the command prompt line.

To use the RAY command

The RAY command creates a line from a defined point of infinite length, but unlike the Construction line command, the line will extend only in the direction of the crosshairs. See Figure 4-6.

Not all AutoCAD commands have tools. To access these commands, type the command name in response to a command prompt.

1. Type the word "ray" in response to a command prompt; press Enter.

Command:_ray From point:

2. Select the starting point for the Ray line.

Through point:

3. Select a point.

Another line will appear from the starting point through the crosshairs.

Through point:

4. Select another point.

Through point:

5. Press Enter, ending the command sequence.

This will end the RAY command sequence and return you to a command prompt.

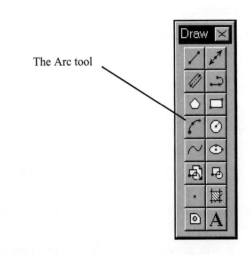

The Arc tool

Figure 4-7

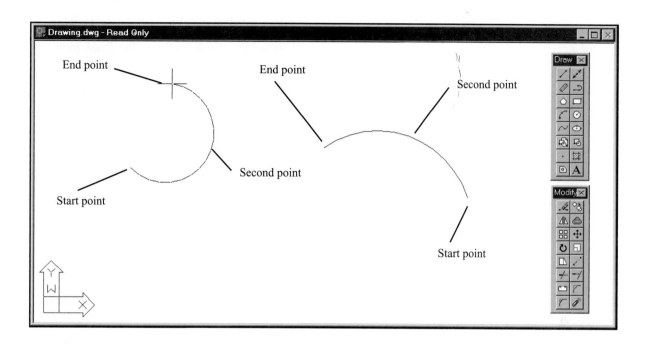

Figure 4-8

4-3 ARC

The Arc tool is located on the Draw toolbar. See Figure 4-7. Arcs can be defined using combinations of start, end, and center points, and by defining chord lengths or angles.

To draw an arc: 3 points

See Figure 4-8.

1. Select the Arc tool from the Draw toolbar.

   Command:_ arc Center/<Start point>:

2. Select a start point.

   Center/End/<Second point>:

3. Select a second point.

   End point:

Move the cursor around and note how the arc changes.

4. Select a third point.

   Command:

Figure 4-8 also shows a second arc created using the 3-point option.

To draw an arc: Start Center End

See Figure 4-9.

1. Select the Arc tool from the Draw toolbar.

   Command:_ arc Center/<Start point>:

2. Type c; press Enter.

   Center:

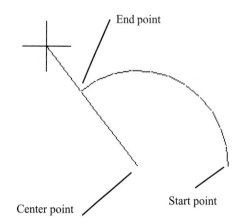

Figure 4-9

Arc has an included angle of 136.5°.

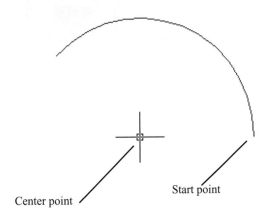

Center point

Start point

Figure 4-10

3. Select a center point.

   Start point:

4. Select a start point.

   Angle/Length of Cord/<Endpoint>:

4. Select an end point.

   Command:

In the example shown the endpoint was selected by moveing the cursor to a point. The actual end point of the arc need not be the same as the selected screen point. A ray will appear from the center point to the cursor as the endpoint is selected.

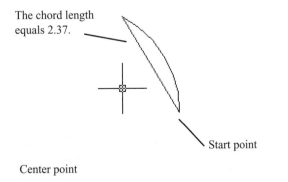

The chord length equals 2.37.

Center point

Start point

Figure 4-11

To draw an arc: Start Center Angle

See Figure 4-10.

1. Select the Angle tool from the Draw toolbar.

   Command:_arc Center/<Start point>:

2. Type c; press Enter.

   Center:

3. Select a center point.

   Start point:

4. Select a start point.

   Angle/Length of chord/<End point>:

5. Type a; press Enter.

   Included angle:

6. Type 136.5; press Enter.

To draw an arc: Start Center Chord Length

See Figure 4-11.

1. Select the Arc tool from the Draw toolbar.

   Command:_ arc Center/<Start point>:

2. Type c; press Enter.

   Center:

3. Select a center point.

   Angle/Length of chord/<End point>:

4. Type l; press Enter.

   Length of chord:

5. Type 2.37; press Enter.

4-4 CIRCLE

The CIRCLE tool is on the Draw toolbar. See Figure 4-12. Circles may be defined by a center point and either a radius or diameter, by two or three points on the diameter, or by defining the tangents to two existing lines or arcs and then defining a radius value.

To draw a circle: Center Point Radius

See Figure 4-13.

1. Select the Circle tool from the Draw toolbar.

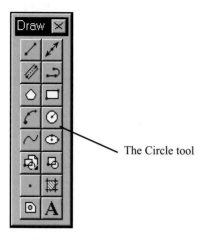

The Circle tool

Figure 4-12

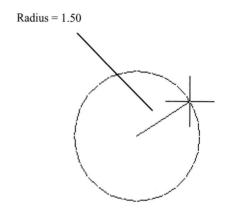

Radius = 1.50

Figure 4-13

Command: _circle 3P/2P/TTR/<Center point>:

2. Select a center point.

Command: _circle 3P/2P/TTR/<Center point>: Diameter/<radius>:

3. Type 1.50; press Enter.

To draw a circle: Center Point Diameter

See Figure 4-14.

1. Select the Circle tool from the Draw toolbar.

Command: _circle 3P/2P/TTR/<Center point>:

2. Select a center point.

Command: _circle 3P/2P/TTR/<Center point>: Diameter/<Radius>:

3. Type d; press Enter.

Diameter:

4. Type 3.00; press Enter.

To draw a circle: 2 Points

See Figure 4-15.

1. Select the Circle from the Draw toolbar.

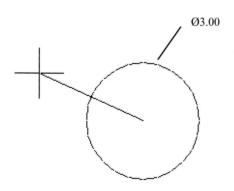

Ø3.00

Figure 4-14

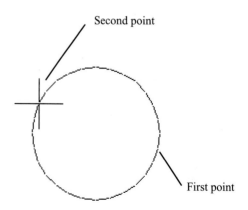

Second point

First point

Figure 4-15

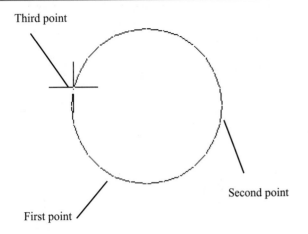

Third point

First point

Second point

Figure 4-16

Command: _circle 3P/2P/TTR/<Center point>:

2. Type 2p; press Enter.

   First point on the diameter:

3. Select a first point.

   Second point on the diameter:

4. Select a second point.

The circle will automatically be drawn through the first and second points.

To draw a circle: 3 Points

See Figure 4-16.

1. Select the Circle tool from the Draw toolbar.

   Command: _circle 3P/2P/TTR/<Center point>:

2. Type 3p; press Enter.

   First point:

3. Select a first point.

   Second point:

4. Select a second point.

   Third point:

5. Select a third point.

To draw a circle: Tangent Tangent Radius

The Tangent Tangent Radius option allows you to draw a circle tangent to two other entities.

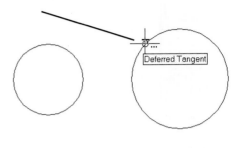

The first tangent spec

Deferred Tangent

Given two existing circles, draw a third circle tangent to the existing two.

Figure 4-17

1. Select the Circle tool from the Draw toolbar.

   Command: _circle 3P/2P/TTR/<Center point>:

2. Type ttr; press Enter.

   Enter tangent spec:

This prompt is asking you to identify one of the entities you want the circle drawn tangent to. See Figure 4-17.

3. Select an entity.

   Enter second tangent spec:

4. Select the other entity.

   Enter second tangent spec: Radius <2.0000>:

5. Type 1.50; press Enter.

   See Figure 4-18.

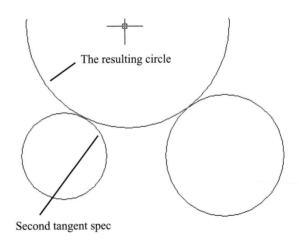

The resulting circle

Second tangent spec

Figure 4-18

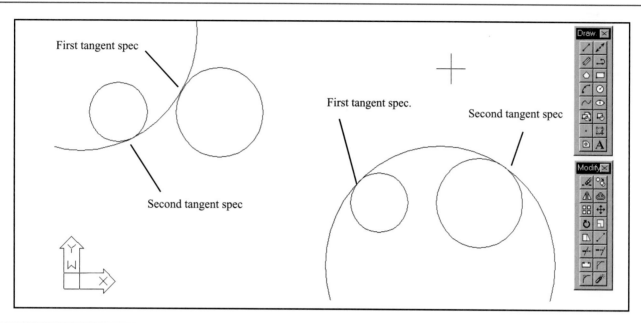

First tangent spec

Second tangent spec

First tangent spec.

Second tangent spec

Figure 4-19

The Tangent, Tangent, Radius command is quadrant sensitive, that is, the resulting circle position will depend on which quadrants of the existing circles are selected. Figure 4-19 shows two examples of circles drawn using the TTR command through different quadrants.

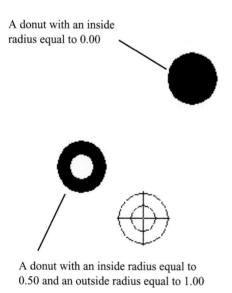

A donut with an inside radius equal to 0.00

A donut with an inside radius equal to 0.50 and an outside radius equal to 1.00

Figure 4-20

To draw a donut

See Figure 4-20. The Donut command does not have a tool.

1. Type donut; press Enter.

   Inside diameter <.5000>:

2. Press Enter, accepting the default value.

   Outside diameter <1.0000>:

3. Press Enter, accepting the default value.

A donut will appear on the screen in drag mode, that is, the donut will move with the crosshairs. To locate a donut, press the left mouse button.

Figure 4-20 also shows a donut with an inside radius value of 0.0000. It appears as a filled circle.

4-5 POLYLINES

A polyline is a line made from a series of individual, connected line segments that act as a single entity. Polylines are used to generate curves and splines and can be used in three-dimensional applications to produce solid objects.

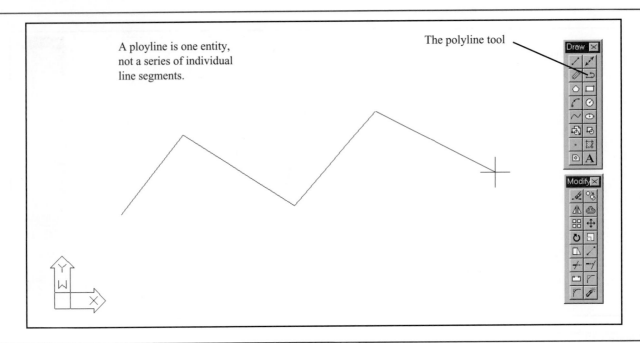

A ployline is one entity, not a series of individual line segments.

The polyline tool

Figure 4-21

To draw a polyline

See Figure 4-21.

1. Select the Polyline tool from the Draw toolbar.

   Command: _pline
   From point:

2. Select a start point.

   Arc/Close/Halfwidth/Length/Undo/Width/
   <Endpoint of line>:

3. Select a second point.

   Arc/Close/Halfwidth/Length/Undo/Width/
   <Endpoint of line>:

4. Select several more points.

   Arc/Close/Halfwidth/Length/Undo/Width/
   <Endpoint of line>:

5. Press Enter.

To verify that a polyline is a single entity

Figure 4-22 shows a polyline.

1. Select the Erase tool from the Modify toolbar or type the word Erase.

   Select objects:

2. Select any one of the line segments in the polyline.

   Select objects:

   The entire polyline, not just the individual line segment, will be selected. This is because, to AutoCAD, the polyline is a single entity.

3. Press Enter.

   The entire polyline will disappear.

4. Select the Undo tool from the Standard toolbar.

   The object will reappear.

To draw a polyline: Arc

See Figure 4-23.

1. Select the Polyline tool from the Draw toolbar.

   From point:

2. Select a start point.

   Arc/Close/Halfwidth/Length/Undo/Width/
   <Endpoint of line>:

3. Type A; press Enter.

   Angle/CEnter/CLose/Direction/Halfwidth/Line/
   Radius/Second pt/Undo/Width/<Endpoint of arc>:

4. Select a second point.

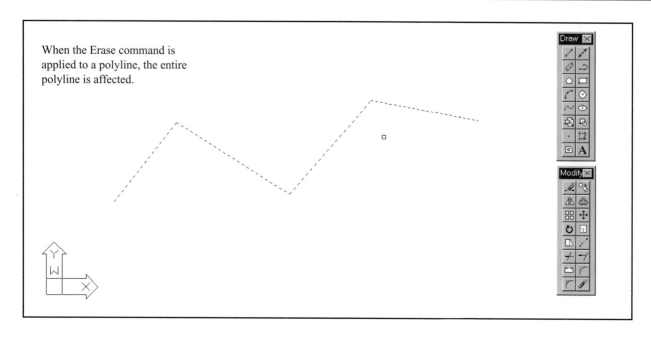

When the Erase command is applied to a polyline, the entire polyline is affected.

Figure 4-22

Angle/CEnter/CLose/Direction/Halfwidth/Line/Radius/Second pt/Undo/Width/<Endpoint of arc>:

5. Select two more points.

Angle/CEnter/CLose/Direction/Halfwidth/Line/Radius/Second pt/Undo/Width/<Endpoint of arc>:

6. Press Enter to end the command sequence.

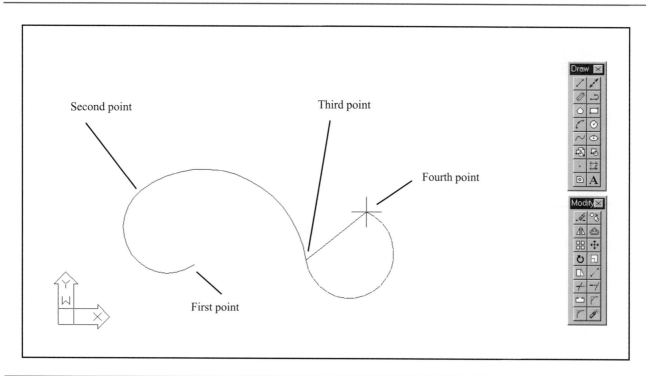

Second point

Third point

Fourth point

First point

Figure 4-23

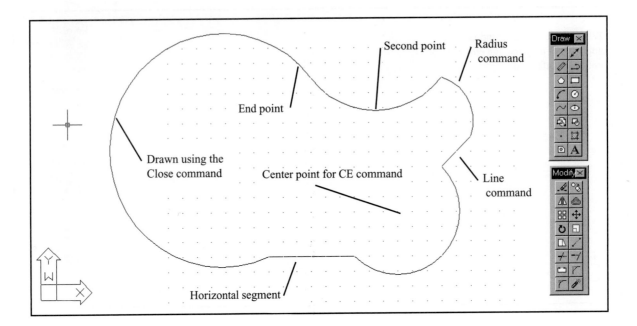

Figure 4-24

Other options with a polyline arc

See Figure 4-24. The example shown includes a background grid with .5 spacing. All points selected in this example are aligned with the grid points.

1. Select the Polyline tool from the Draw toolbar.

Command: _pline
From point:

2. Select a start point.

Arc/Close/Halfwidth/Length/Undo/Width/ <Endpoint of line>:

3. Draw a short horizontal line segment.

Arc/Close/Halfwidth/Length/Undo/Width/< Endpoint of line>:

4. Type A; press Enter.

Angle/CEnter/CLose/Direction/Halfwidth/Line/ Radius/Second pt/Undo/Width/<Endpoint of arc>:

5. Type CE; press Enter.

CE activates the center command. You can now define an arc that will be part of the polyline by defining the arc's center point and its angle, chord length, or end point.

Center point:

6. Select a center point.

Angle/Length/<End point>:

7. Select an end point.

Angle/CEnter/CLose/Direction/Halfwidth/Line/ Radius/Second pt/Undo/Width/<Endpoint of arc>:

8. Type L; press Enter.

The Line option is used to draw straight line segments.

Arc/Close/Halfway/Length/Undo/Width/<End-point>:

9. Select an end point.

Arc/Close/Halfway/Length/Undo/Width/<End-point>:

10. Type A; press Enter.

Angle/CEnter/CLose/Direction/Halfwidth/Line/ Radius/Second pt/Undo/Width/<Endpoint of arc>:

11. Type R; press Enter.

Radius

12. Type 1.5; press Enter.

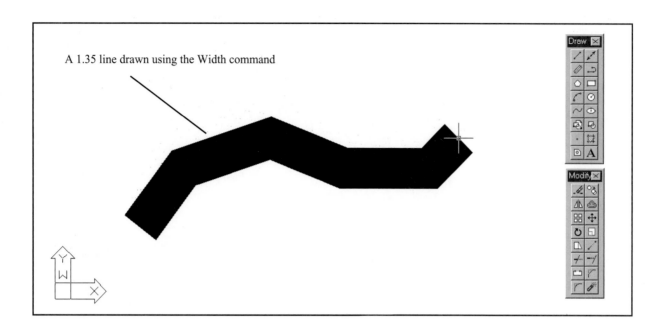

A 1.35 line drawn using the Width command

Figure 4-25

Angle/<Endpoint>:

13. Select an endpoint.

Angle/CEnter/CLose/Direction/Halfwidth/Line/ Radius/Second pt/Undo/Width/<Endpoint of arc>:

14. Type S; press Enter.

Second point:

15. Select a point.

Endpoint:

16. Select an end point.

Angle/CEnter/CLose/Direction/Halfwidth/Line/Ra dius/Second pt/Undo/Width/<Endpoint of arc>:

17. Type CL; press Enter.

The CL (close) option will join the last point drawn to the first point of the polyline using an arc.

To draw different line thicknesses

See Figure 4-25.

1. Select the Polyline tool from the Draw toolbar.

From point:

2. Select a start point.

Current line width is 0.000
Arc/Close/Halfwidth/Undo/Width/<Endpoint of arc>:

3. Type W; press Enter.

The Width option is used to define the width of a line. The Halfwidth option is used to define half the width of a line.

Starting width <0.0000>:

4. Type 1.35; press Enter.

Ending width <1.3500>:

5. Press Enter.

Arc/Close/Halfwidth/Length/Undo/Width/<End-point of line>:

6. Draw several line segments.

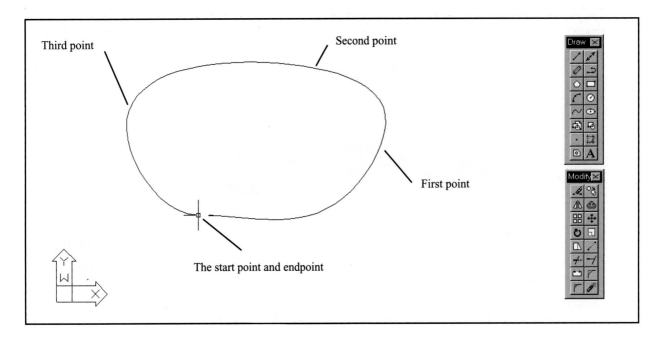

Third point

Second point

First point

The start point and endpoint

Figure 4-26

4-6 Spline

A spline is a curved line created through a series of predefined points. If the curve forms an enclosed area it is called a closed spline. Curved lines that do not enclose an area are called open splines. See Figure 4-26.

To draw a spline

1. Select the Spline tool from the Draw toolbar.

   Command:_spline
   Object/<Enter first point>:

2. Select a start point.

   Enter point:

3. Select a second point.

   Close/Fit tolerance/<Enter point>:

4. Select three more points.

   Close/Fit tolerance/<Enter point>:

5. Type C; press Enter.

   Enter tangent

6. Select a point.

Note how the shape of the spline changes as you move the cursor to locate the tangent point. These changes are based on your selection of a tangent point, which, in turn, affects the mathematical calculations used to create the curve.

4-6 MULTILINE

A multiline is a group of parallel lines that can be drawn together as if they were a single line. The default setting is two parallel lines located at a unit scale of 1.00 apart. See Figure 4-27.

To draw a multiline

1. Select the Multiline tool from the Draw toolbar

   Justification = Top, Scale = 1.00, Style = STAN-DARD
   Justification/Scale/STyle/<Front point>:

2. Select a start point.

   To point:

3. Select a second point.

   Undo<to point>:

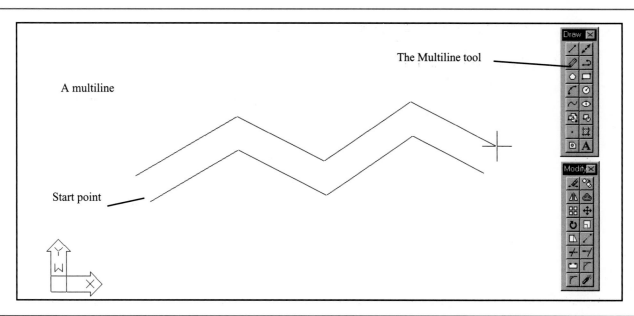

Figure 4-27

4. Select other points.

   Close/Undo<to point>:

5. Press Enter.

   The Close option will draw a multiline from the last point entered to the first point entered for the current drawing sequence.

To change a line's justification

   The justification line allows you to use the top, bottom, or an imaginary line (zero) between the two parallel lines to define the multiline's location. See Figure 4-28.

1. Select the Multiline tool from the Draw toolbar.

   Justification = Top, Scale = 1.00, Style = STANDARD
   Justification/Scale/STyle/<Front point>:

2. Type j; press Enter.

   Top/Zero/Botttom <bottom>:

3. Select the desired justification by typing the first letter of the desired justification style.

   Justification/Scale/STyle/<Front point>:

4. Draw the multiline by selecting points.

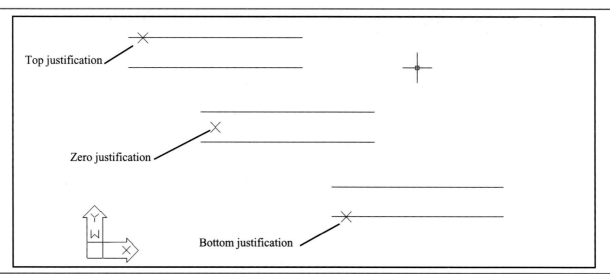

Figure 4-28

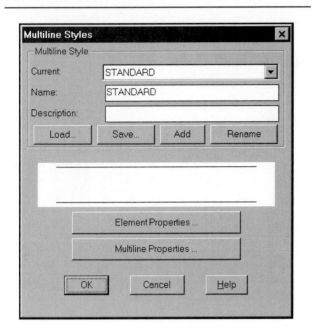

Figure 4-29

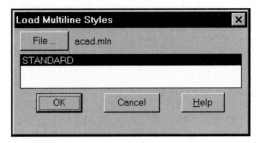

No new multiline styles have been created so only the default Standard style is listed.

Figure 4-30

Type in the new style name here.

Type a description of the style here.

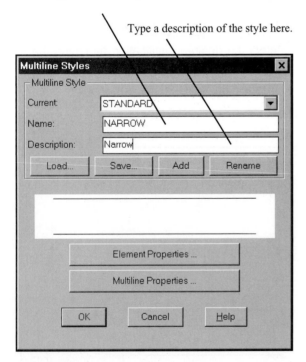

Figure 4-31

To create a new style of a multiple line

Different parallel line patterns may be created and saved for future use.

Command:

1. Type mlstyle; press Enter.

The Multiline Styles dialog box will appear. See Figure 4-29. As different multiline styles are created they may be saved for future use. The Load option is used to load all existing multilines saved. Multilines have a file extension of .mln.

2. Select the Load box.

The Load Multiline Styles dialog box will appear. See Figure 4-30. No other multilines exist at this time, so only the Standard option will be listed. Return to the Multiline Styles dialog box.

3. Select the Cancel box.

The Multiline Styles dialog box will appear and will now be used to create a new multiline style, Narrow.

4. Type in the new multiline style name and a description of the line in the appropriate boxes.

See Figure 4-31.

5. Select the Add box.

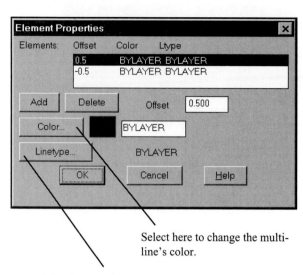

Select here to change the multi-line's color.

Select here to change the multiline's linetype.

Figure 4-32

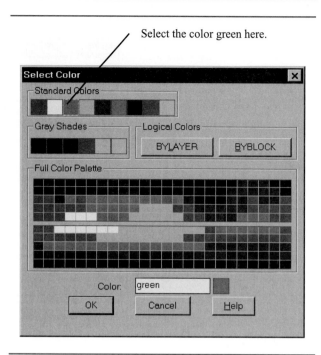

Select the color green here.

Figure 4-33

The new style should become the current style. If it does not, make it the current style by using the scroll bar at the right side of the Current box.

To select a multiline color

1. Select the Element Properties box.

The Element Properties dialog box will appear. See Figure 4-32. This dialog box is used to define the new multiline called Narrow where the two parallel lines of the multiline are 0.25 apart.

2. Select the Color box to set the color of the new line.

The Select Color dialog box will appear. See Figure 4-33.

3. Select the color green; select the OK box.

The Element Properties dialog box will reappear.

To select a multiline linetype

1. Select the Linetype box on the Element Properties dialog box.

The Select Linetype dialog box will appear. See Figure 4-34. In the example presented only three options are available. Other linetypes may be loaded and then selected.

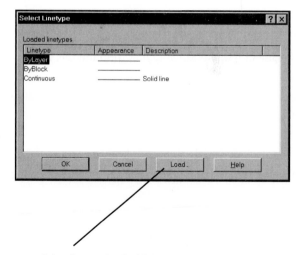

Select here to load additional linetypes.

Figure 4-34

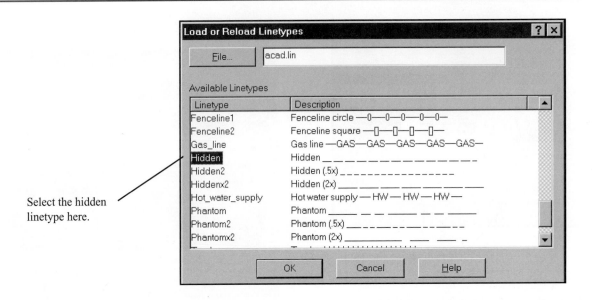

Select the hidden linetype here.

Figure 4-35

2. Select the Load box.

The Load or Reload Linetypes dialog box will appear. See Figure 4-35.

3. Scroll down the linetype options and select the Hidden linetype.

The Select Linetype dialog box will reappear with the Hidden linetype listed.

4. Select the Hidden linetype, then the OK box.

The Element Properties dialog box will reappear. See Figure 4-36. Note that only the upper line of the multiline is changed.

5. Repeat the procedure to define the bottom line of the multiline as a green hidden line.

To change the width of a multiline

1. Select the Offset box and change the current - 0.500 to - 0.125.

See Figure 4-37.

2. Select the upper line and change its offset value from 0.500 to 0.125, then select OK.

The Multiline Styles dialog box will reappear.

The color and linetype have been changed.

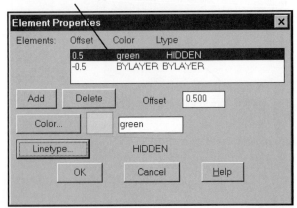

Figure 4-36

The lower portion of the line has been changed

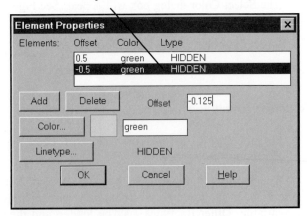

Figure 4-37

Click here to display the line's joints.

Click here and type 45 here.

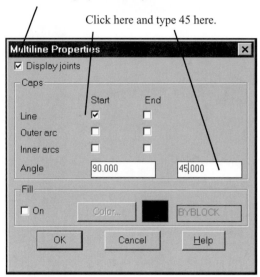

Figure 4-38

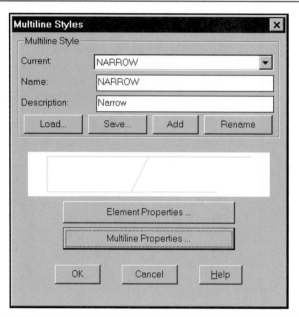

Figure 4-39

To change multiline properties

The Multiline Styles box will reappear.

1. Select the Multiline Properties box.

The Multiline Properties dialog box will appear. See Figure 4-38. This box is used to set the joints, end lines and angles, and fill pattern for the line.

2. Select the Display joints box (a check mark will appear in the box when it is on), the Start-Line box, and change the End angle from 90 to 45.
3. Select the OK box.

The Multiline Styles dialog box will reappear. See Figure 4-39. Note the appearance of the Narrow multiline in the preview box.

To save a new multiline

1. Select the Save box.

The Save Multiline Style dialog box will appear. See Figure 4-40. Multiline formats are saved using the .mln extension.

2. Type the file name for the multiline in the File Name box, then select the OK box.

The Multiline Styles dialog box will reappear with Narrow displayed in the Current box.

3. Select the OK box and draw a multiline using the Multiline tool in the Draw toolbar.

Type the file here.

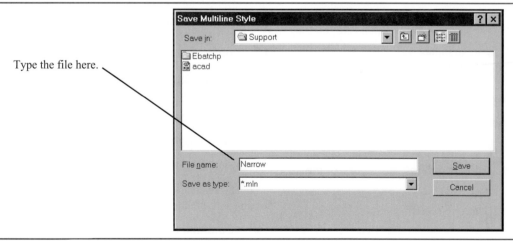

Figure 4-40

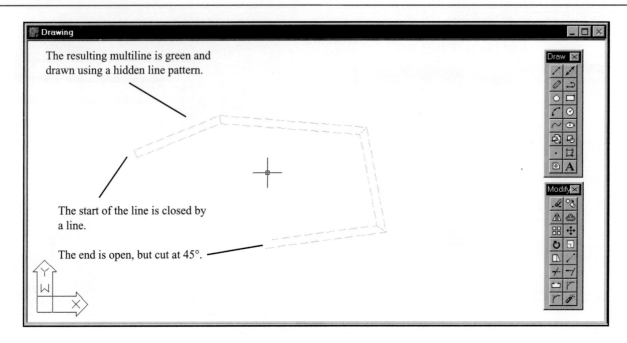

Figure 4-41

Figure 4-41 shows a multiline created using the Narrow style. It is a green line with joints displayed and a line across the start point. The end of the line is open and cut at 45°.

4-7 ELLIPSE

The ELLIPSE command can be used to draw an ellipse by defining the lengths of its major and minor axes, by defining an angle of rotation about the major axis, or by defining an included angle.

To draw an ellipse using three points

See the top left figure in Figure 4-42.

1. Select the Ellipse tool from the Draw toolbar.

   Arc/Center/<Axis endpoint 1>:

2. Select a start point for one of the axes.

   Axis endpoint 2:

3. Select an endpoint that defines the length of the axis.

   <Other axis distance>/Rotation:

4. Select a point that defines half the length of the other axis.

This distance is the radius of the axis. In the example shown, points 1 and 2 were used to define the major axes, and point 3 defines the minor axis. The bottom figure in Figure 4-41 shows an object where points 1 and 2 were used to define the minor axes, and point 3 the major axis.

To draw an ellipse Center

See the top figure in Figure 4-42.

1. Select the Ellipse tool from the Draw toolbar.

   Arc/Center/<Axis endpoint 1>:

2. Type c; press Enter.

   Arc/Center/<Axis endpoint>: c
   Center of ellipse:

3. Select the center point of the ellipse.

   Axis endpoint:

4. Select one of the end points of one of the axes.

The distance between the center point and the end-point is equal to the radius of the axis.

   <Other axis distance>Rotation:

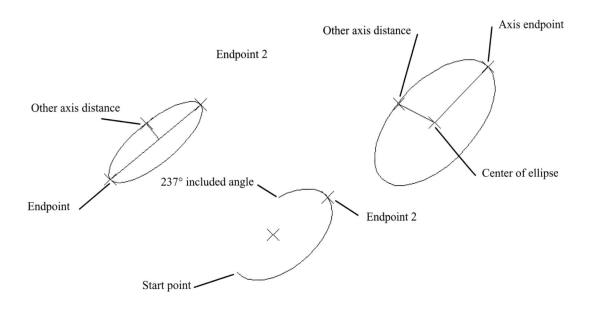

Endpoint 2

Other axis distance

Axis endpoint

Other axis distance

Endpoint

237° included angle

Endpoint 2

Center of ellipse

Start point

Figure 4-42

5. Select a point that defines half the length of the other axis.

To draw an ellipse Arc

See the figure in the bottom center of Figure 4-42.

1. Select the Ellipse arc tool from the Draw toolbar.

Arc/Center/<Axis endpoint 1>:

2. Type a; press Enter.

Arc/Center/<Axis endpoint 1>: a
<Axis endpoint 1>/Center:

3. Select a point.

Axis endpoint 2:

4. Select a second point.

The distance between points 1 and 2 defines the length of the major axis.

<Other axis distance>/Rotation

5. Select a point that defines the minor axis.

Parameter/<start angle>:

6. Type 0; press Enter.

Parameter/Included/<end angle>:

7. Type 237; press Enter.

To draw an ellipse by defining its angle of rotation about the major axis

An ellipse may also be defined in terms of its angle of rotation about the major axis. See Figure 4-43. An ellipse with 0 degrees rotation is a circle, an ellipse of constant radius. An ellipse with 90 degrees rotation is a straight line, an end view of an ellipse.

1. Select the ellipse axis endpoint from the Draw toolbar.

Arc/Center/<Axis endpoint 1>:

2. Select an endpoint of the major axis.

Axis endpoint 2:

3. Select the other endpoint of the major axis.

<Other axis distance>:Rotation:

4. Type R; press Enter.

Rotation around major axis:

5. Type 60; press Enter.

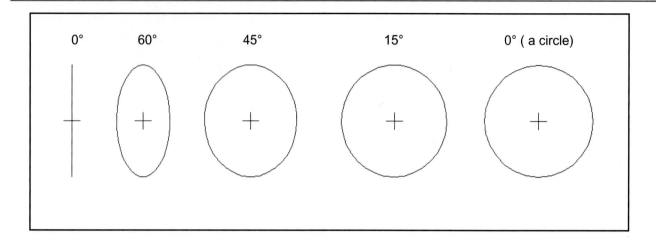

Figure 4-43

4-8 RECTANGLE

There are five flyout tools associated with the rectangle tool. See Figure 4-43. The rectangle and polygon tools will be discussed in this section. The 2D solid, Region, and Boundary tools will be discussed later in the book.

To draw a rectangle

See Figure 4-44.

1. Select the Rectangle tool from the Draw toolbar.

Command: rectang
Chamfer/Elevation/Fillet/Thickness/Width/
<First corner>:

2. Select a point.

Other corner:

3. Select a point.

The distance between the two points is the diagonal distance across the rectangle's corners.

A rectangle drawn using the Rectangle command is considered to be one entity. If the same sized rectangle were drawn using the Line command, the resulting rectangle would be considered four individual unrelated lines. This is similar to a line drawn from line segments and one drawn using Polyline.

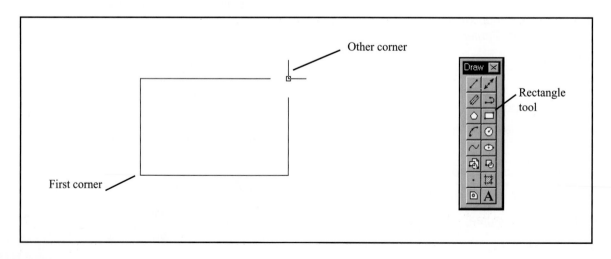

Figure 4-44

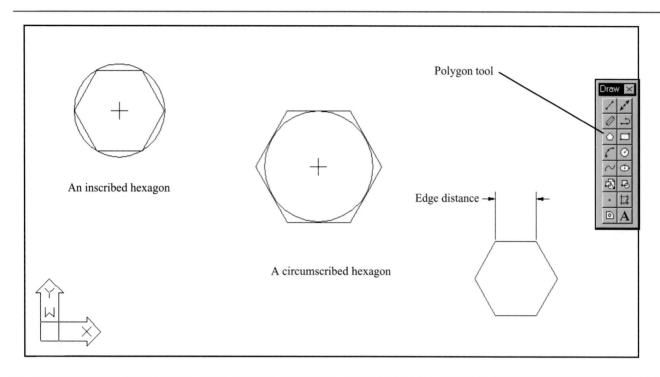

Polygon tool

An inscribed hexagon

A circumscribed hexagon

Edge distance

Draw

Figure 4-45

4-9 POLYGON

A polygon is a closed figure bound by straight lines. The Polygon command will draw only regular polygons, where all sides are equal. A regular polygon with four equal sides is a square.

To draw a polygon centerpoint

See Figure 4-45. A six-sided polygon or hexagon will be drawn in this example.

1. Select the Polygon tool from the Draw toolbar.

   Command: _polygon Number of sides <4>:

2. Type 6; press Enter.

   Edge/<Center of polygon>:

3. Select a center point.

   Inscribed in circle/Circumscribed about a circle (I/C)<I>:

4. Press Enter.

   Radius of circle:

5. Type 1.25; press Enter.

The Circumscribe option is used in a similar manner.

To draw a polygon edge distance

See Figure 4-45.

1. Select the Polygon tool from the Draw toolbar.

   Command: -polygon Number of sides <6>:

2. Press Enter.

   Edge/<Center of polygon>:

3. Type E; press Enter.

   First endpoint of edge:

4. Select a point.

   First endpoint of edge: Second endpoint of edge:

5. Select a point.

4-10 POINT

The POINT command is used to draw points on a drawing. The default setting for a point shape is a small dot, the kind used to display background grids. Other point

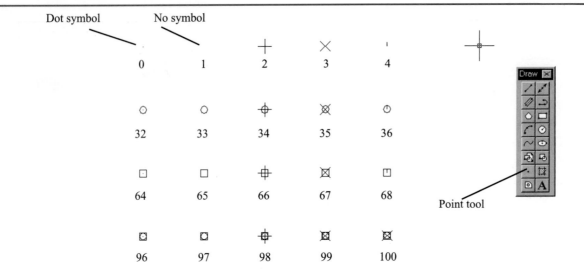

Figure 4-46

shapes are available. See Figure 4-46. The default small dot point shape is listed as 0. Note that shape 1 is a void. The five available shapes can be enhanced by adding the numbers 32, 64, and 96 to the point's assigned value. For example, shape 2 is crossed horizontal and vertical lines. Shape 34 is a shape 2 with a circle around it, shape 66 is shape 2 with a square overlay, and shape 98 is shape 2 with both a circle and a square around it.

To change the shape of a point

See Figure 4-47. Start with a command prompt.

Command:

1. Type pdmode and press Enter in response to a command prompt.

   Command: pdmode

New value for PDMODE<0>:

2. Type 2; press Enter.

   Command: _point POINT

3. Select several points to verify that the shape has been changed.

An Enter command should end the command sequence. If it does not, draw a line and use Enter to return to a command prompt. Erase the line and continue working.

To change the size of a point

See Figure 4-47.

1. Type pdsize in response to a command prompt.

   Command: _pdsize

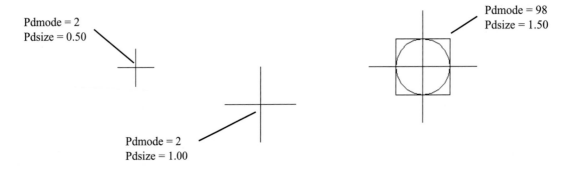

Figure 4-47

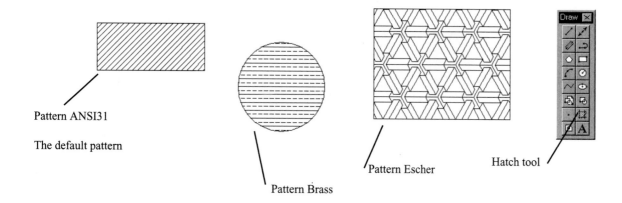

Pattern ANSI31

The default pattern

Pattern Brass

Pattern Escher

Hatch tool

Figure 4-48

New value for PDSIZE<0.0000>:

2. Type 0.5; press Enter.

The default value is 0. Figure 4-38 shows points drawn at pdsize values of 0.5 and 1.

4-11 HATCH

The HATCH command is used to draw a pattern within an enclosed area. There are many different patterns available within the Hatch command. Figure 4-48 shows an application of three different patterns. The most commonly used pattern is ANSI31, evenly spaced parallel lines at 45 degrees to the horizontal.

To select a hatch pattern

1. Select the Hatch tool on the Draw toolbar.

The Boundary Hatch dialog box will appear. See Figure 4-49.

2. Click the arrow button to the right of the word Pattern.

Preview of selected pattern

Select a new pattern here.

Used to change the spacing of a pattern

Used to define area to be hatched

Used to access the hatch Style Codes

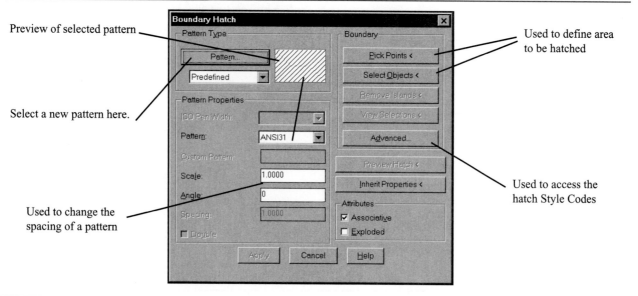

Figure 4-49

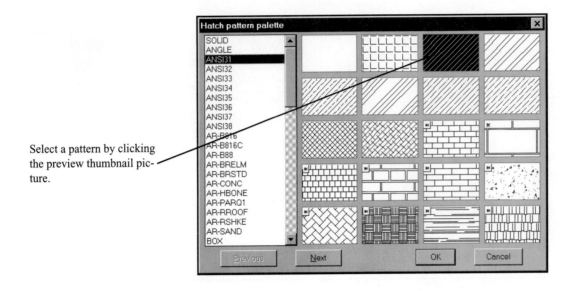

Select a pattern by clicking the preview thumbnail picture.

Figure 4-50

A listing of patterns will cascade from the box with a scroll bar. See Figure 4-50.

3. Scroll down the list until you find the pattern you want, then click the pattern name.

A sample of the pattern selected will appear in the Pattern Type box. ANSI31 is the first pattern listed, so it can be accepted without scrolling.

To hatch a given enclosed shape

See Figure 4-50.

1. Select the Hatch tool from the Draw toolbar.

Command:

The ANSI31 pattern will be displayed. If it is not, select it as described above.

2. Pick the Pick Points box.

Command: _bhatch
Select internal point:

3. Locate the crosshairs within the boundary to be hatched and click the left mouse button.

The boundary lines of the selected area will change to a broken-line gray pattern.

Select internal point:

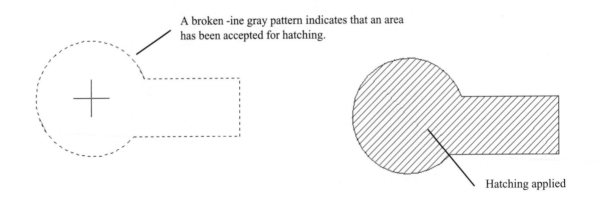

A broken -ine gray pattern indicates that an area has been accepted for hatching.

Hatching applied

Figure 4-51

This is a modified ANSI31 pattern.
Scale = 1.00, Angle = 15

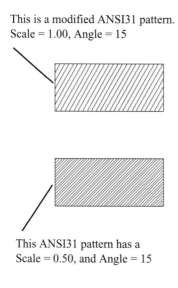

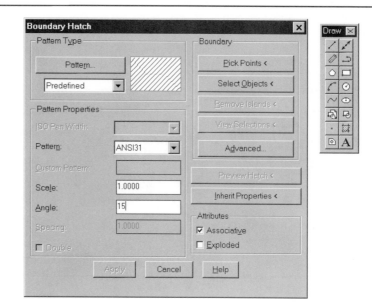

This ANSI31 pattern has a
Scale = 0.50, and Angle = 15

Figure 4-52

4. Press Enter.

The Boundary Hatch dialog box will reappear.

5. Select the Apply box.

The hatch pattern will appear in the selected area.

To change the spacing and angle of a hatch pattern

See Figure 4-52.

1. Select the Hatch tool from the Draw toolbar.

Command:

The Boundary Hatch dialog box will appear.

2. Select the Scale box.

A flashing cursor will appear.

3. Backspace to remove the current value and type in a new value.

In this example, a value of 1 was used.

4. Select the Angle box, backspace to remove the current value, and type in a new value.

In this example, 15 was used.

5. Select the Pick Points box.

Select internal points:

6. Select the area to be hatched.
7. Press Enter.

The Boundary Hatch dialog box will reappear.

8. Select the Apply box.

The area will be hatched. Compare the two hatched areas shown in Figure 4-52.

To change the Hatch Style Codes

See Figure 4-53.

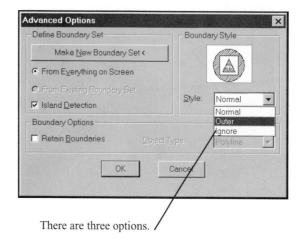

There are three options.

Figure 4-53

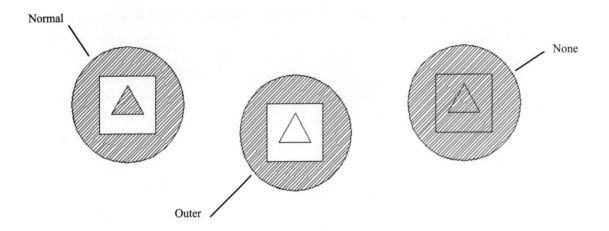

Figure 4-54

1. Select the Hatch tool from the Draw toolbar.

   Command:

   The Boundary Hatch dialog box will appear.

2. Select the Advanced box.

   The Advanced Options dialog box will appear.

3. Select the Style box.

   There are three options: Normal, Outer, and Ignore. The box just to the right of the Style box will preview the selected option.

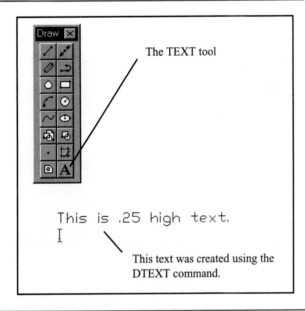

The TEXT tool

This is .25 high text.

This text was created using the DTEXT command.

Figure 4-55

4. Select the arrow button to the right of the Style box.

   A listing of the options will cascade from the box.

5. Scroll down to the option you want.
6. Select the OK box.

   The Boundary Hatch dialog box will reappear. Apply the hatching as described above. Figure 4-54 shows examples of the different Hatch Style Codes applications.

4-12 TEXT

The Text tool is located on the Draw toolbar.

To use the DTEXT command

   See Figure 4-55. The is no tool for the DTEXT command.

1. Type DTEXT; press Enter.

   Command: _dtext Justify/Style/<Start point>:

2. Select a start point.

   The text will start at the selected point and run left to right.

   Height <.1250>:

3. Type .25; press Enter.

   This will change the text height from .1250 to .2500.

   Rotation angle <0.00>:

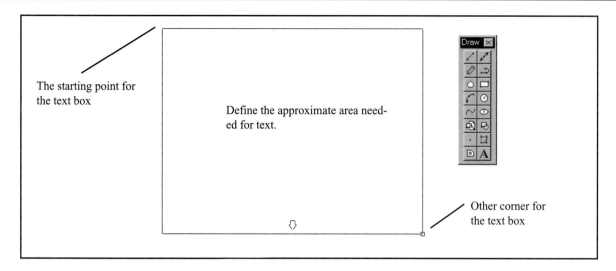

The starting point for the text box

Define the approximate area needed for text.

Other corner for the text box

Figure 4-56

4. Press Enter.

A box will appear with its lower left corner at the start point.

Text:

5. Type: This is a sample text; press Enter.

Text:

6. Press Enter.

The DTEXT command always assumes you will be typing another line of text regardless of how many lines you type. An Enter response signals that you are done entering text.

To use TEXT

The Text command is used if you are going to enter a large amount of text. See Figures 4-56 and Figure 4-57. First define the area in which the text is to be entered; the Multiline Text Editor dialog box will appear on the screen. Text is typed into the dialog box, just as with word processing programs, then transferred back to the drawing screen.

1. Select the Text tool from the Draw toolbar.

Command:_mtext Current text style: STANDARD. Text height: 0.2500
   Specify first corner:

2. Select a point.

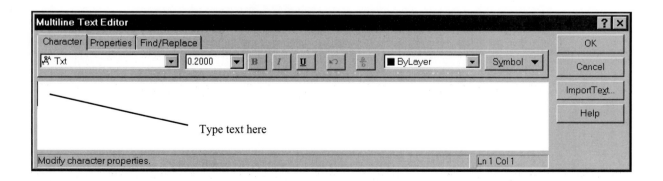

Type text here

Figure 4-57

Figure 4-58

The area you are about to define is for the entire text entry.

Specify opposite corner or [Height/ Justify/ Rotation/Style/Width]:

3. Define the other corner of the area.

The Multiline Text Editor dialog box will appear. See Figure 4-57.

4. Type in your text, then select the OK box.

See Figure 4-58. The text will be located on the drawing screen within the specified area. See Figure 4-59.

Justify text

Text may be justified in one of nine different formats using one of two different techniques. The standard text justification is left, that is, text will be entered from left to right. Other justifications are sometimes needed on engineering drawings. For example to add text to the end of a leader line that extends to the left can most easily be done using the middle right format. Justification formats may be entered from the keyboard or by using the Multiline Text Editor dialog box.

To change the text justification - Method I

This method shows how to change a text's justification using keyboard inputs.

1. Select the Text tool from the Draw tool box.

Command: _mtext
Specify first corner:

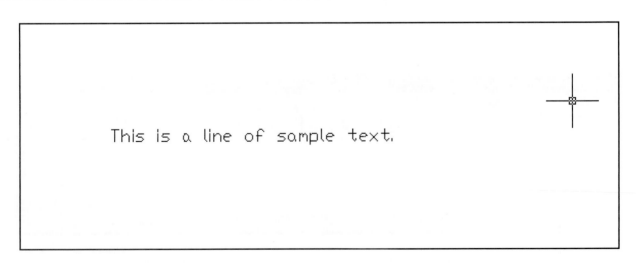

Figure 4-59

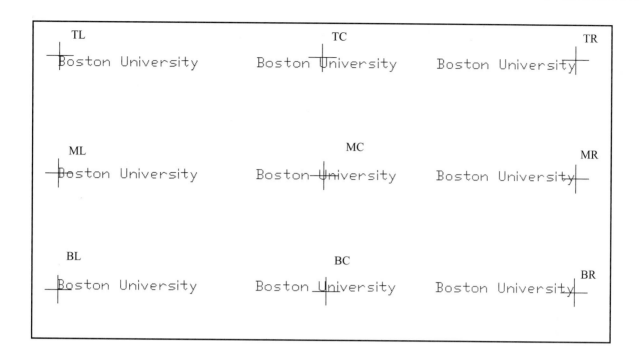

Figure 4-60

2. Select the upper left corner for the text box.

Specify opposite corner or [Height/ Justify/ Rotation/Style/Width]:

3. Type j; press Enter.

Enter justification [TL/ TC/ TR/ ML/ MC/ MR/ BL/ BC/ BR]:

4. Type the letters for the justification format; press Enter.

5. Type in the text.

Figure 4-60 shows the meaning of the above notations.

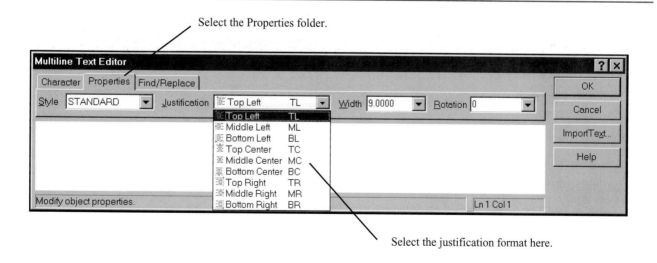

Figure 4-61

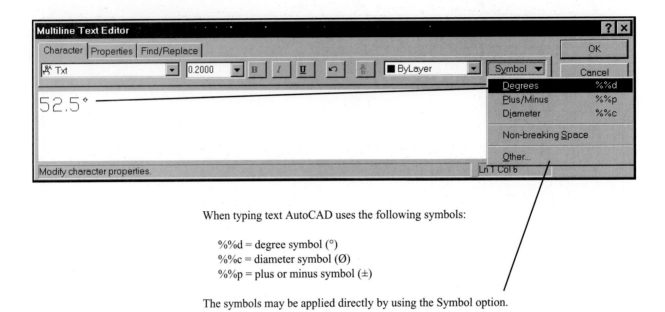

When typing text AutoCAD uses the following symbols:

%%d = degree symbol (°)
%%c = diameter symbol (Ø)
%%p = plus or minus symbol (±)

The symbols may be applied directly by using the Symbol option.

Figure 4-62

To change the text justification - Method II

The Multiline Text Editor dialog box can also be used to change text height, color, font style (if available), and justification. The default justification is top left. The text will start in the upper left corner of the text area and proceed to the right and down, line by line, from the top to the bottom. Other justifications are available.

1. Select the Text tool from the Draw toolbox.

Command: _mtext
Attach/Rotation/Style/Height/Direction/<Insertion point>:

2. Select a start point.

Attach/Rotate/Style/Height/Direction/Width/2 points/<Other corner>:

3. Select a second point.

The Edit MText dialog box will appear.

4. Select the Properties box.

The Multiline Text Editor Properties dialog box will appear. See Figure 4-61.

5. Select the arrow box next to the Attachment box.
6. Select the Top Right option.
7. Select the OK box.

Note that the text height, width, rotation, and font may also be changed using the MText Properties dialog box. The symbols may also be applied directly using the Symbol option. Figure 4-62 shows how to use the symbol option.

To edit existing text

Text that has been created and added to a drawing may be edited using the Properties tool found on the Object Properties toolbar. Figure 4-63 shows some sample text that contains a spelling error. The error may be corrected using the following procedure.

Figure 4-63

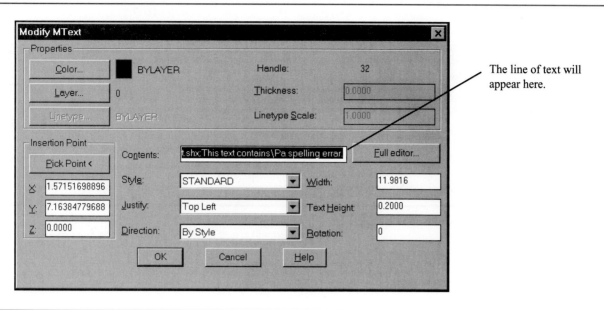

The line of text will appear here.

Figure 4-64

1. Select the Properties tool from the Object Properties toolbar.

   Select object

2. Select the line of text; press Enter.

   The Modify MText dialog box will appear. See Figure 4-64. The line of text will appear in the Contents box.

3. Locate the cursor behind the spelling error and backspace out the error.
4. Type in the correct spelling.

   See Figure 4-65.

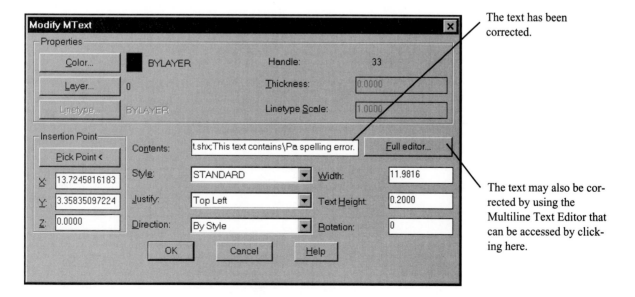

The text has been corrected.

The text may also be corrected by using the Multiline Text Editor that can be accessed by clicking here.

Figure 4-65

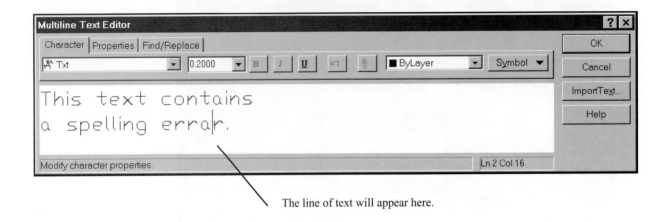

The line of text will appear here.

Figure 4-66

The text may also be corrected by using the Full Editor option on the Edit MText dialog box. If the Full Editor box is selected, the Multiline Text Editor will appear as shown in Figure 4-66. The text may then be edited as previously explained.

4-13 EXERCISE PROBLEMS

EX4-1 INCHES

Redraw the following text as shown below. All dimensions are in inches.

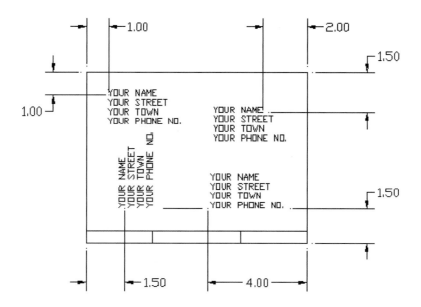

EX4-2 MILLIMETERS

Redraw the following text as shown below. All dimensions are in millimeters.

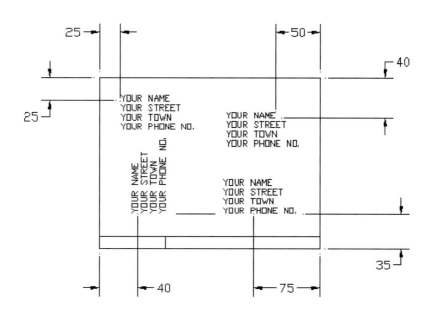

EX4-3 INCHES

Redraw the following text as shown below.

1. This text is 0.25 high.

2. This text is 0.50 high.

3. This text is 0.375 high and underlined.

4. This text is 0.2000 high and is red.

EX4-4 FRACTIONAL INCHES

Redraw the following text as shown below. The text height is 1/2″. The X and Y spacing on the background grid equals 1/2″.

This is STANDARD text, justified to the left.

This is STANDARD text, justified to the right.

This is Times New Roman text.

This is text rotated 90°.

EX4-5

Use the CIRCLE command to draw the shape shown below. The background grid spacing may be .5″, 10mm, or 6″.

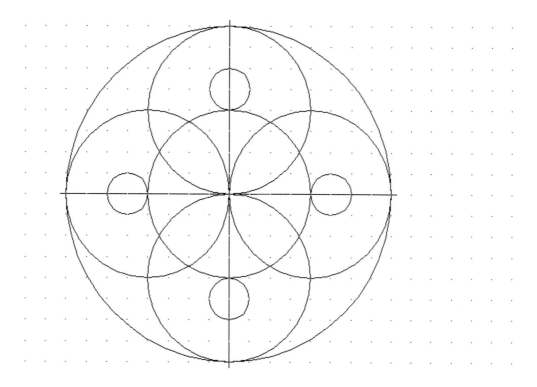

EX4-6

Use the POLYGON command to draw the shapes shown below. Each shape is circumscribed about a circle. The radius values for the circumscribed circles are as follows.

Square = .5″, 10mm, 6″
Pentagon = 1.0″, 50mm, 1′-0″
Hexagon = 1.5″, 75mm, 1′-6″
Heptagon = 2.0″, 100mm, 2′-0″
Octagon = 2.5″, 125mm, 2′-6″
Nonagon = 3.0″, 150mm, 3′-0″
Decagon = 3.5″, 175mm, 3′-6″

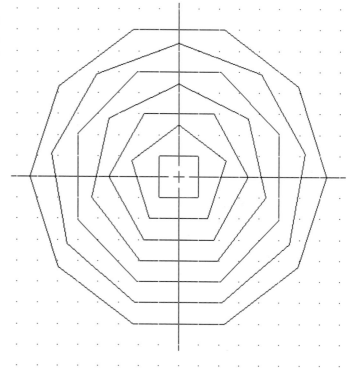

EX4-7

Use the ARC command to redraw the following shape. The background grid spacing equals .50″, 10mm, or 1′-0″.

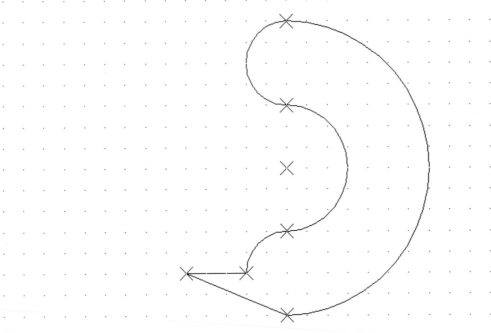

EX4-8

Use the ARC command to redraw the following shape. The background grid spacing equals .50″, 10mm, or 1′-0″.

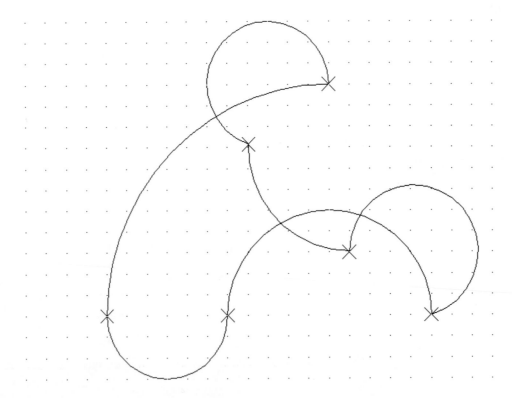

EX4-9 INCHES

Redraw the following figure. All dimensions are in inches.

GUIDE PLATE

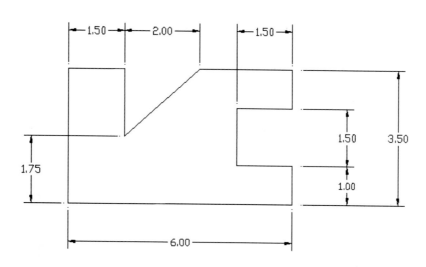

EX4-10 INCHES

Redraw the following figure. All dimensions are in inches.

CENTER DUCT

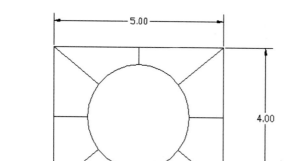

EX4-11 MILLIMETERS

Redraw the following figure. All dimensions are in millimeters.

CENTER COVER

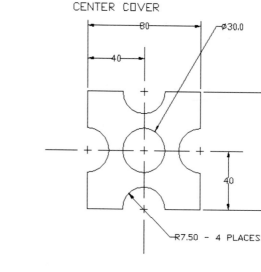

EX4-12 INCHES

Redraw the following figure. All dimensions are in inches. Do not include dimensions on the final drawing.

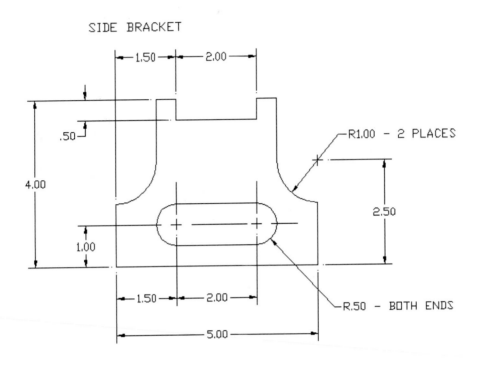

SIDE BRACKET

EX4-13 INCHES

Redraw the following figure. All dimensions are in inches.

EX4-14 MILLIMETERS

Redraw the following figure. All dimensions are in inches.

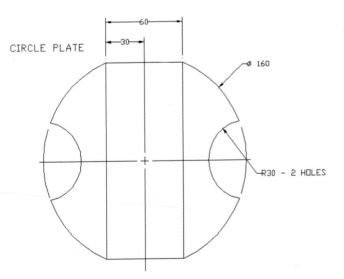

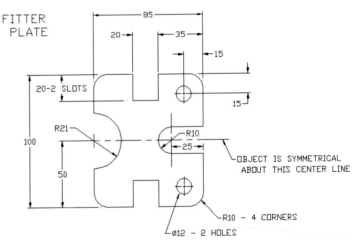

EX4-15 MILLIMETERS

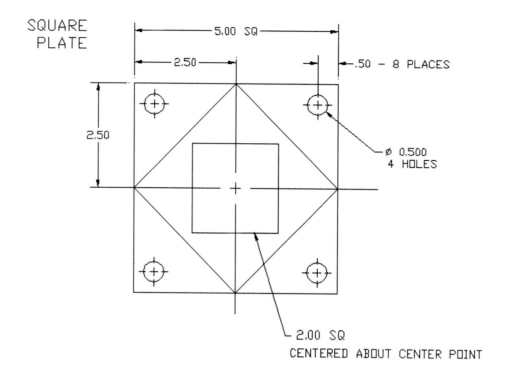

SQUARE
PLATE

5.00 SQ

2.50

.50 – 8 PLACES

2.50

ø 0.500
4 HOLES

2.00 SQ
CENTERED ABOUT CENTER POINT

EX4-16 INCHES

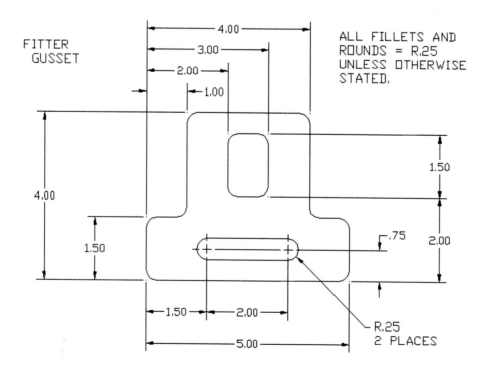

FITTER
GUSSET

4.00

3.00

2.00

1.00

ALL FILLETS AND
ROUNDS = R.25
UNLESS OTHERWISE
STATED.

1.50

4.00

1.50

.75

2.00

1.50

2.00

5.00

R.25
2 PLACES

EX4-17 INCHES

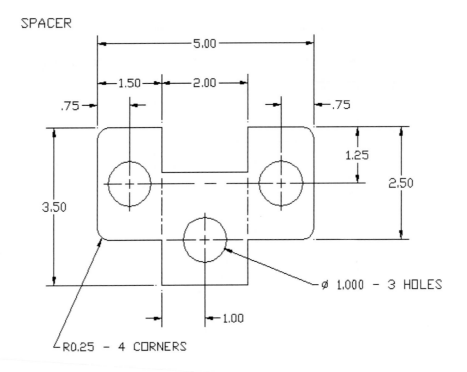

SPACER

EX4-18 MILLIMETERS

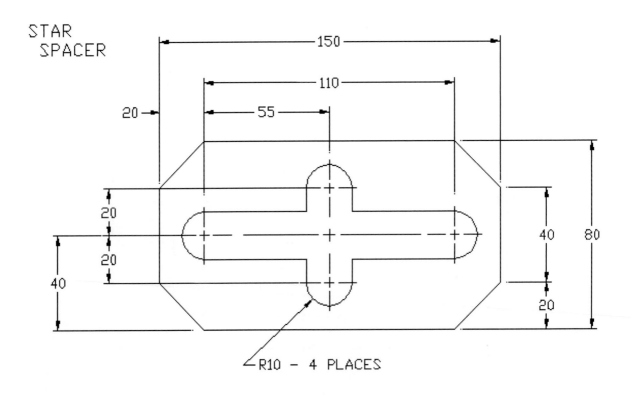

STAR SPACER

EX4-19 MILLIMETERS

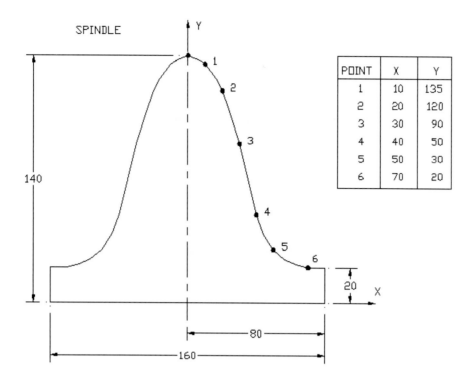

POINT	X	Y
1	10	135
2	20	120
3	30	90
4	40	50
5	50	30
6	70	20

EX4-20

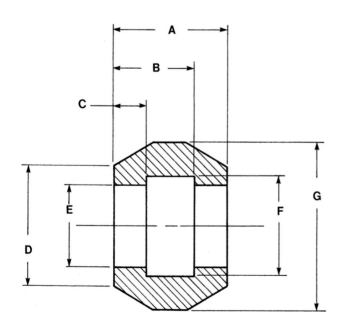

DIMENSIONS	INCHES	mm
A	.50	12
B	1.25	32
C	1.75	44
D	Ø1.75	44
E	Ø 1.25	32
F	Ø1.50	38
G	Ø2.50	64

EX4-21

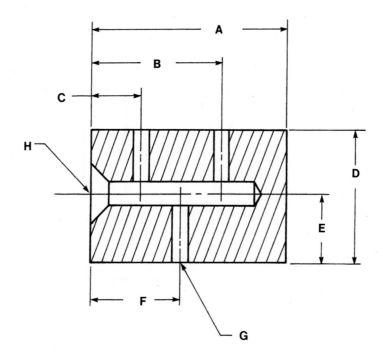

DIMENSIONS	INCHES	mm
A	3.00	72
B	2.00	48
C	.75	18
D	2.00	48
E	1.00	24
F	1.38	33
G	Ø .25	6
H	Ø .375 X 2.50 DEEP Ø .875 X 82° CSINK	Ø 10 X 60 DEEP Ø 24 X 82° CSINK

CHAPTER 5

The Object Snap Toolbar

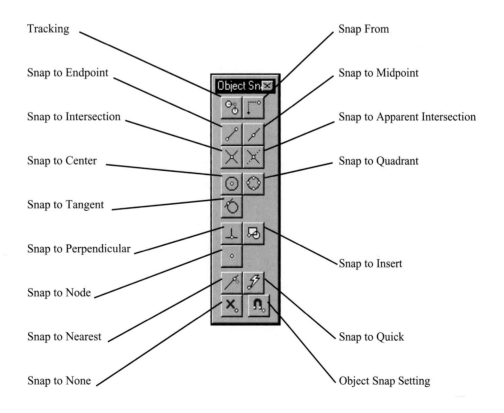

Tracking

Snap to Endpoint

Snap to Intersection

Snap to Center

Snap to Tangent

Snap to Perpendicular

Snap to Node

Snap to Nearest

Snap to None

Snap From

Snap to Midpoint

Snap to Apparent Intersection

Snap to Quadrant

Snap to Insert

Snap to Quick

Object Snap Setting

Figure 5-1

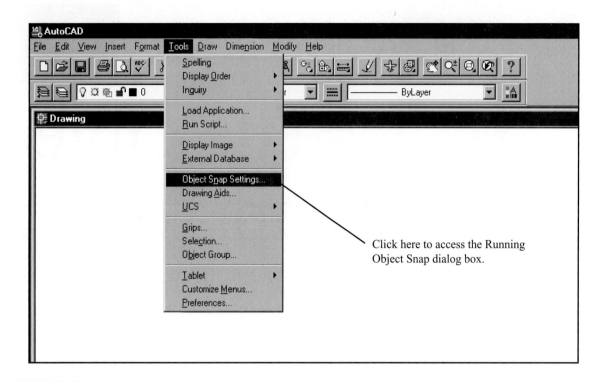

Click here to access the Running
Object Snap dialog box.

Figure 5-2

5-1 INTRODUCTION

This chapter explains how to use the Object Snap toolbar. Object Snap, called OSNAP, is an important AutoCAD command as it allows you to grab and position drawing elements very accurately. You can, for example, use the OSNAP command to position the center point of a circle exactly on the intersection of two existing lines. You can also draw a line to the midpoint of an existing line without calculating the line's midpoint dimension.

Figure 5-1 shows the Object Snap toolbar. There are no flyout icons associated with the Object Snap toolbar. The Quick, Insertion, and Calculator commands will not be included in this chapter.

To access the Object Snap toolbar using menus

See Figure 5-2.

1. Select the Tools heading.
2. Select Object Snap Settings.

The Osnap Settings dialog box will appear on the screen. See Figure 5-3.

5-2 RUNNING OBJECT SNAP

The Object Snap Setting tool located on the Object Snap toolbar allows you to access the Running Object Snap dialog box. The Running Object Snap dialog box can be used to turn an object snap option on permanently; that is, the command will remain on until you turn it off. If the Endpoint option is on, the cursor will snap to the nearest endpoint every time a point selection is made. Turning an Object Snap option on is very helpful when you know you are going to select several points of the same type. However, in most cases, it is more practical to activate an object snap option as needed using the Object Snap toolbar or the Object Snap screen menu (shift/right button).

To turn on an Object Snap command option

1. Select the Object Snap Setting tool on the Object Snap toolbar.

   The Osnap Settings dialog box will appear.

2. Select the Object Snap command option by clicking the box to the left of the command name.

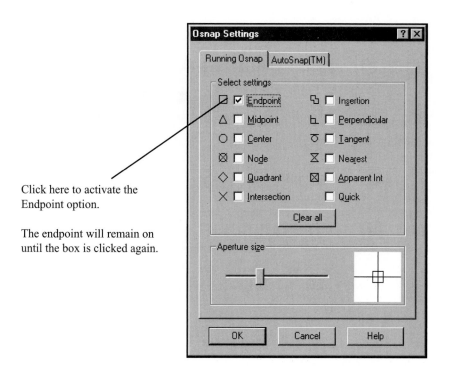

Click here to activate the
Endpoint option.

The endpoint will remain on
until the box is clicked again.

Figure 5-3

If the Endpoint box were clicked, a check mark
would appear in the box indicating the command is on.
When you return to the drawing screen the cursor will
include a rectangle and will automatically snap to the endpoint of any entity selected.

If an OSNAP command is on, you do not have to use
the OSNAP icons or the Shift/right button option to activate the commands. However, if a command is on and you
wish to use another command (e.g., the Endpoint command is on and you want to use Intersection), you will
have to turn the Endpoint command off, then activate the
other command.

To turn off an Object Snap command

There are two methods that can be used to turn off
an Object Snap command.

1. Select the Running Object Snap icon and click
   the appropriate command option.

The check mark will disappear from the box, indicating the command is off.

2. Select the Snap to None tool on the Object Snap
   toolbar.

All Object Snap commands will be turned off.

To change the size of the Object Snap aperture box

The Osnap Settings dialog box also contains an
Aperture size option. This option allows you to change the
size of the rectangular box on the cursor. A larger box
makes it easier to grab objects, but too large a box may
grab more than one object, or the wrong object.

To access the Object Snap commands using the keyboard and mouse

The Object Snap command options box may be
accessed in conjunction with other commands by pressing
the Shift key and the right mouse button simultaneously.
The OSNAP commands will appear as a column of commands.

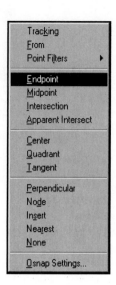

This Object Snap command box may be accessed by pressing the Shift key and the right mouse button simultaneously.

Figure 5-4

1. Press the Shift key of the keyboard and the right mouse button simultaneously.

The Object Snap command box will appear. See Figure 5-4.

5-3 ENDPOINT

The Endpoint option is used to snap to the endpoint of an existing entity. Figure 5-5 shows an existing line.

To snap to the endpoint of a line

1. Select the Line icon from the Draw toolbar and draw a line.
2. Double-click the left mouse button to end the Line command sequence and to start a new sequence.

   Command:_line From point:

3. Select the Endpoint icon from the Object Snap toolbar.

   Command: _line From point _endp of

A box around the intersection of the cursor indicates that an Object Snap option is on.

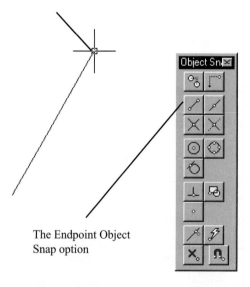

The Endpoint Object Snap option

Figure 5-5

When the line is selected, the midpoint is identified.

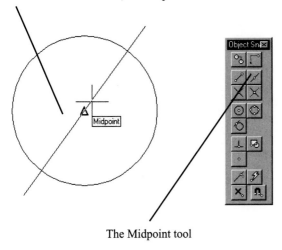

The Midpoint tool

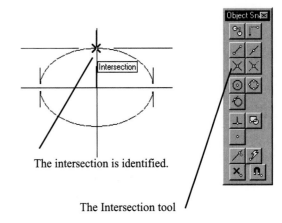

The intersection is identified.

The Intersection tool

Figure 5-6

Figure 5-7

The OSNAP commands do not work independently, but in conjunction with other commands. In this example, the LINE command had to first be activated, then the OSNAP command. Note that the cursor changes shape from cross hairs to cross hairs with a rectangle. Locate the cursor rectangle over the end of the line and press the left mouse button. The starting point of the line will be at the endpoint of the existing line. The ENDPOINT command also works with arcs, mlines, and 3D applications.

5-4 MIDPOINT

The Midpoint option is used to snap to the midpoint of an existing entity. In the example presented in Figure 5-6, a circle is to be drawn with its center point at the midpoint of the line.

To draw a center about the midpoint of a line

1. Select the Circle tool from the Draw toolbar.

   Command: _circle 3P/2P/TTR/<Center point>:

2. Select the Midpoint tool from the Object Snap toolbar.

   Command: _circle 3P/2P/TTR/<Center point>: _mid of

3. Select the line.

   Command: _circle 3P/2P/TTR/<Center point>: _mid of: <snap off> Diameter/<Radius>:

4. Select a radius value.

5-5 INTERSECTION

The Intersection option is used to snap to the intersection of two or more entities. Figure 5-7 shows a set of projection lines that are to be used to define an ellipse.

To use the OSNAP intersection command to define an ellipse

1. Select the Ellipse Center option from the Draw toolbar.

   Center of ellipse:

2. Select the Intersection tool from the Object Snap toolbar.

   Center of ellipse: _appint of

3. Select the center point for the ellipse.

   Axis endpoint

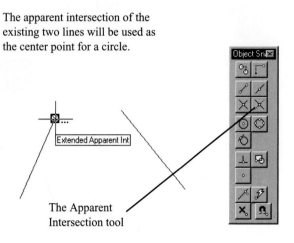

The apparent intersection of the existing two lines will be used as the center point for a circle.

Extended Apparent Int

The Apparent Intersection tool

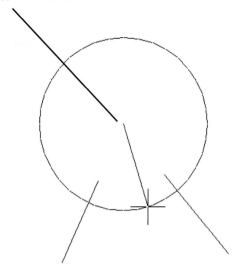

The apparent intersection is the center for a new circle.

Figure 5-8

4. Select the Intersection tool from the Object Snap toolbar and select the intersection that defines the length of one of the axes from the ellipse center point.

   <Other axis distance>/Rotation: int of

5. Select the Intersection tool from the Object Snap toolbar and select the intersection that defines the other axis length.

5-6 APPARENT INTERSECTION

The Apparent Intersection option is used to snap to an intersection that would be created if the two entities were extended to create an intersection. Figure 5-8 shows two lines that do not intersect but, if continued, would intersect.

To draw a circle centered about an Apparent Intersection

1. Select the Circle tool from the Draw toolbar.

   Command: _circle 3P/2P/ TTR/ <Center point>:

2. Select the Apparent Intersection tool from the Object Snap toolbar.

   Command: _circle 3P/2P/ TTR/ <Center point>: _appint of

3. Select one of the lines.

   Command: _circle 3P/2P/ TTR/ <Center point>: _appint of and

4. Select the other line.

   <Snap off> Diameter/:<Radius>:

5. Select a radius value for the circle.

5-7 CENTER

The Center option is used to draw a line from a given point directly to the center point of a circle. See Figure 5-9.

To draw a line to the center point of a circle

1. Select the Line tool from the Draw toolbar.

   Command: _line From point:

2. Select the start point for the line.

   To point:

3. Select the Center tool from the Object Snap toolbar.

   To point: _cen of

4. Select any point on the circle.

   Do not try to select the center point directly. Select

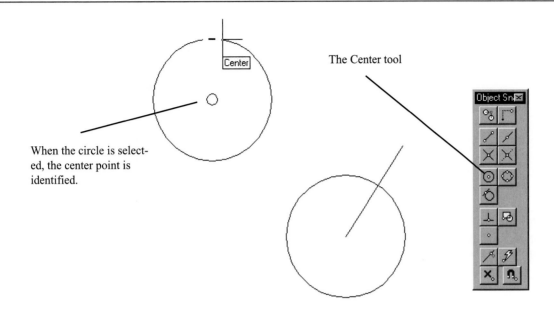

When the circle is selected, the center point is identified.

The Center tool

Figure 5-9

any point on the edge of the circle or arc; the center point will be calculated automatically.

5-8 QUADRANT

The Quadrant option is used to snap directly to one of the quadrant points of an arc or circle. Figure 5-10 shows the quadrant points for an arc and a circle.

To draw a line to one of a circle's quadrant points

See Figure 5-10.

1. Select the Line tool from the Draw toolbar.

   Command: _line From point:

2. Select a start point for the line.

   To point:

3. Select the Quadrant tool from the Object Snap toolbar.

   To point: _qua of

4. Select a point on the circle near the desired quadrant point.

The quadrant marks for an arc and circle are indicated here using an X.

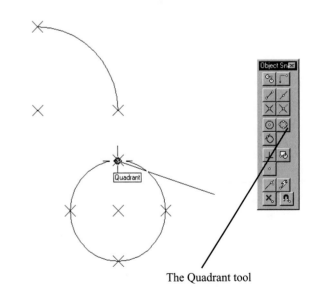

The Quadrant tool

Figure 5-10

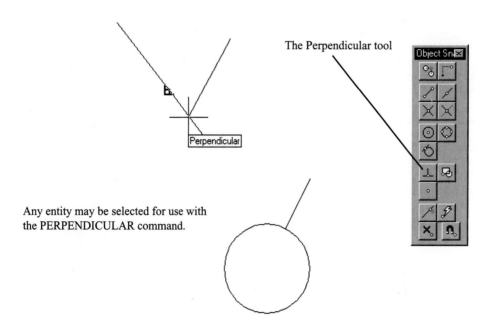

The Perpendicular tool

Any entity may be selected for use with the PERPENDICULAR command.

Figure 5-11

5-9 PERPENDICULAR

The Perpendicular option is used to draw a line perpendicular to an existing entity. See Figure 5-11.

To draw a line perpendicular to a line

1. Select the Line tool on the Draw toolbar.

   Command: _line From point:

2. Select a start point for the line.

   To point:

3. Select the Perpendicular tool on the Object Snap toolbar.

   To point: _per to

4. Select the line or entity that will be perpendicular to the drawn line.

   Figure 5-11 also shows a line drawn perpendicular to a circle.

5-10 TANGENT

The Tangent option is used to draw lines tangent to existing circles and arcs. See Figure 5-12.

To draw a line tangent to a circle

1. Select the Line tool from the Draw toolbar.

   Command: _line From point:

2. Select the start point for the line.

   To point:

3. Select the Tangent tool from the Object Snap toolbar.

   To point: _tan to

4. Select the circle.

5-11 NODE

The Node option is used to draw a line to an existing point. The point must have been defined previously as a

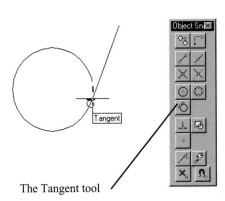

The Tangent tool

Figure 5-12

point. The Node option will not snap to a random point within an existing entity, for example, a point on a line. See Figure 5-13.

To draw a line to an existing point

1. Select the Line tool from the Draw toolbar.

   Command: _line From point:

2. Select a start point for the line.

   To point:

3. Select the Node tool from the Object Snap toolbar.

   To point: _ nod of

4. Select the existing point.

5-12 NEAREST

The Nearest option is used to snap to the nearest available point on an existing entity. See Figure 5-14.

To draw a line from a point to the nearest selected point on an existing line

1. Select the Line tool from the Draw toolbar.

   Command: _line From point

2. Select a start point for the line.

   To point:

The Node tool

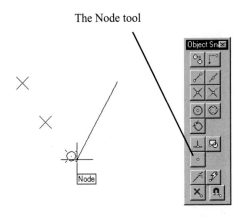

Figure 5-13

3. Select the Nearest tool from the Object Snap toolbar.

   To point: _nea to

4. Select the existing line.

The existing line need be only within the rectangular box on the cursor. The line endpoint will be snapped to the nearest available point on the line.

The Nearest tool

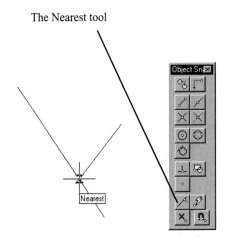

Figure 5-14

5-13 EXERCISE PROBLEMS

Redraw the following shapes. Do not include the dimensions.

EX5-1 INCHES

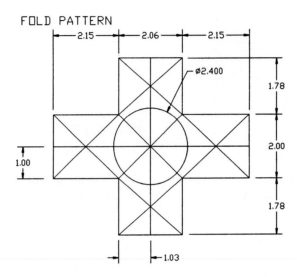

EX5-2 MILLIMETERS

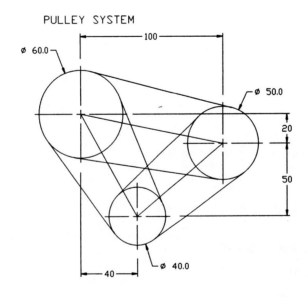

EX5-3 MILLIMETERS

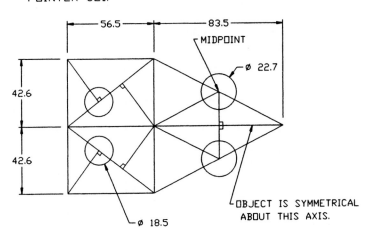

POINTER CLIP

EX5-4 MILLIMETERS

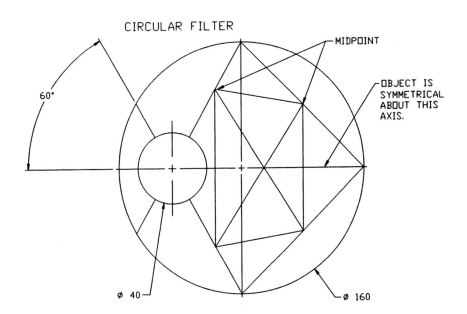

CIRCULAR FILTER

EX5-5 MILLIMETERS

CROSS FILTER

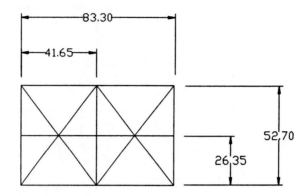

EX5-6 INCHES

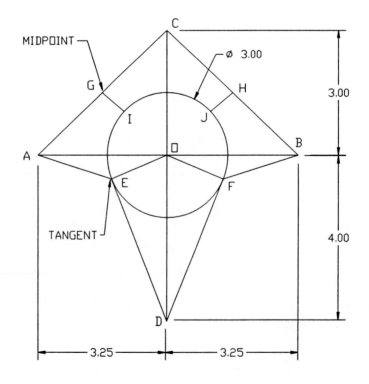

EX5-7 FEET AND INCHES

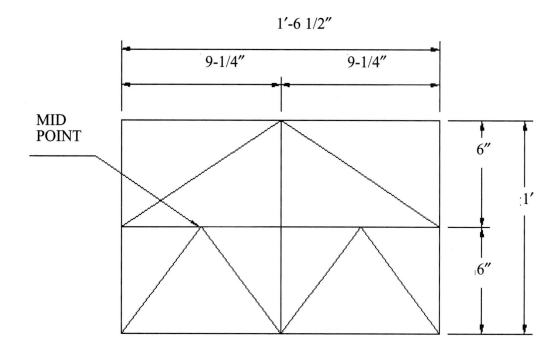

EX5-8 FEET AND INCHES

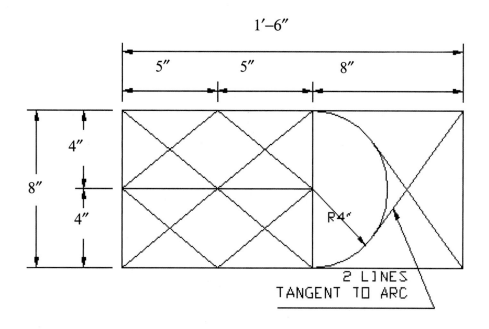

EX5-9 FEET

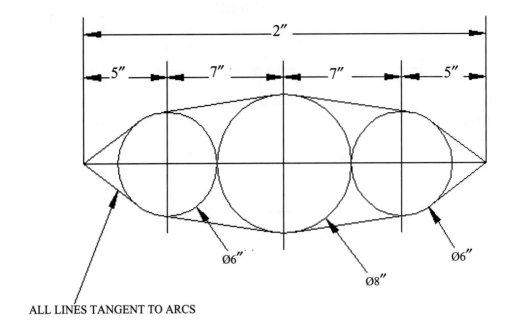

ALL LINES TANGENT TO ARCS

C H A P T E R 6

The Modify Toolbars

The Modify I toolbar

The Modify II toolbar

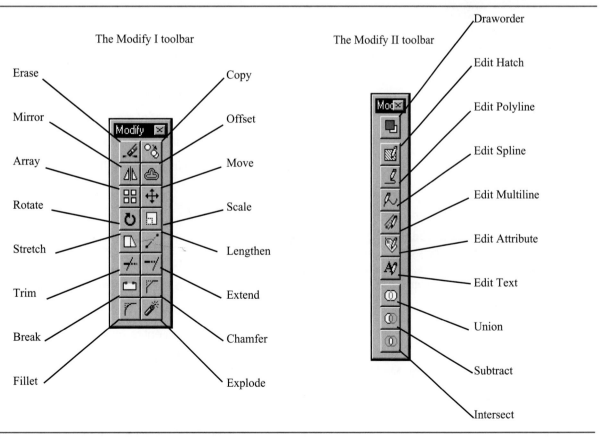

Erase
Mirror
Array
Rotate
Stretch
Trim
Break
Fillet

Copy
Offset
Move
Scale
Lengthen
Extend
Chamfer
Explode

Draworder
Edit Hatch
Edit Polyline
Edit Spline
Edit Multiline
Edit Attribute
Edit Text
Union
Subtract
Intersect

Figure 6-1

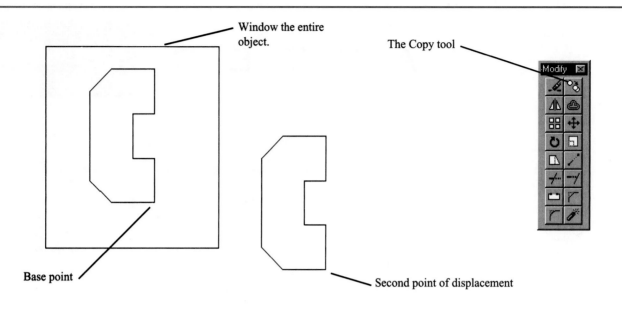

Figure 6-2

6-1 INTRODUCTION

This chapter explains the two Modify toolbars; Modify I and Modify II. Figure 6-1 shows the Modify I and Modify II toolbars. The Modify II toolbar commands are presented later in the chapter.

6-2 ERASE

The ERASE command was explained in Section 3-3.

6-3 COPY

The COPY command is used to make an exact copy of an existing line or object. The COPY command can also be used to create multiple copies, that is, more than one copy without reactivating the command.

To copy an object

See Figure 6-2.

1. Select the Copy tool from the Modify I toolbar.

   Select objects:

2. Window the entire object.

Select objects:

3. Press Enter.

   <Base point or displacement>/Multiple:

4. Select a base point.

In the example shown in Figure 6-3 the lower right corner of the object was selected as the base point.

Second point of displacement:

5. Select a second displacement point.

The original object will remain in its original location, and a new object will appear at the second displacement point.

To draw multiple copies

See Figure 6-3.

1. Select the Copy tool from the Modify I toolbar.

   Select objects:

2. Window the entire object.

   Select objects:

3. Press Enter.

   <Base point or displacement>/Multiple:

4. Type m; press Enter.

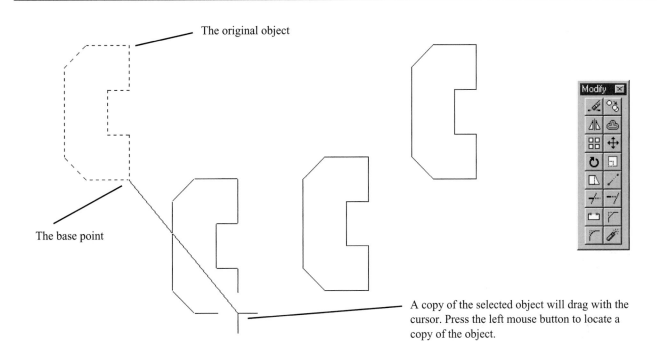

The original object

The base point

A copy of the selected object will drag with the cursor. Press the left mouse button to locate a copy of the object.

Figure 6-3

Base point:

5. Select a base point.

In the example shown, the lower right corner of the object was selected as the base point.

Second point or displacement:

AutoCAD will now shift to drag mode and a copy of the object will follow the cursor as you move them about the screen. A copy of the object will be deposited at each cursor location where you press the left mouse button.

6. Select additional locations for the copies.

Second point or displacement:

7. Press Enter.

6-4 MIRROR

See Figure 6-4.

1. Select the Mirror tool from the Modify I toolbar.

Select objects:

2. Window the object.

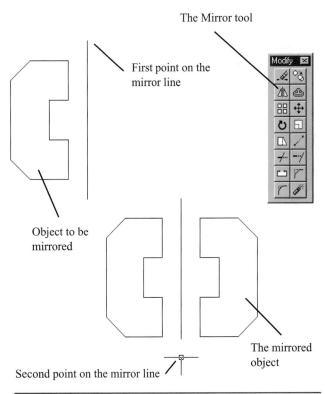

The Mirror tool

First point on the mirror line

Object to be mirrored

The mirrored object

Second point on the mirror line

Figure 6-4

An object mirrored about one of its edge lines

First point on the mirror line

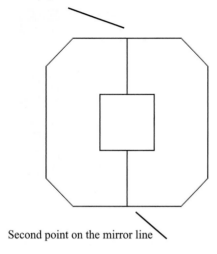

Second point on the mirror line

Figure 6-5

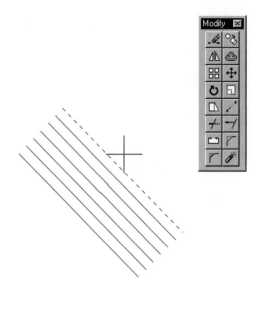

Figure 6-6

First point of mirror line:

3. Select a point on the mirror line.

Second point:

4. Select a second point on the mirror line.

Delete old object? <N>:

5. Press Enter.

Any line may be used as a mirror line including lines within the object. Figure 6-5 shows an objected mirrored about one of its edge lines. The MIRROR command is very useful when drawing symmetrical objects. Only half the object need be drawn. The second half can be created using the MIRROR command.

6-5 OFFSET

See Figure 6-6.

1. Select the Offset tool from the Modify I toolbar.

Offset distance or Through<0.0000>:

2. Specify the offset distance.

In this example a distance of .25 was selected.

Offset distance or through<0.2500>

Select object to offset:

3. Select a line.

Side to offset?:

4. Select a point to the right of the line.

The process may be repeated by selecting the same object or another object by pressing the left mouse button. The offset object will be added by again pressing the left mouse button.

Figure 6-7 shows the OFFSET command applied to a circle.

6-6 ARRAY

See Figure 6-8.

1. Select the Array tool from the Modify I toolbar.

Select objects:

2. Select all the lines to be arrayed.

Rectangular or polar array (<R>/P):

3. Type r; press Enter.

Number of rows (—)<1>:

4. Type 2; press Enter.

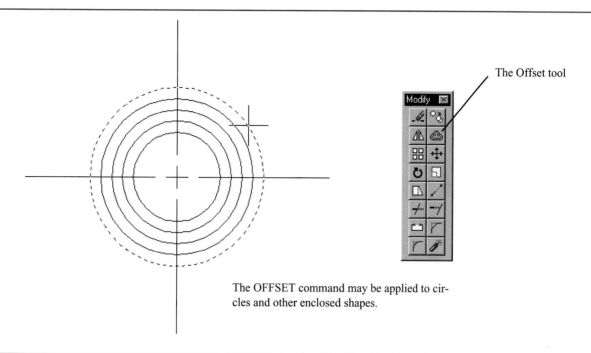

The Offset tool

The OFFSET command may be applied to circles and other enclosed shapes.

Figure 6-7

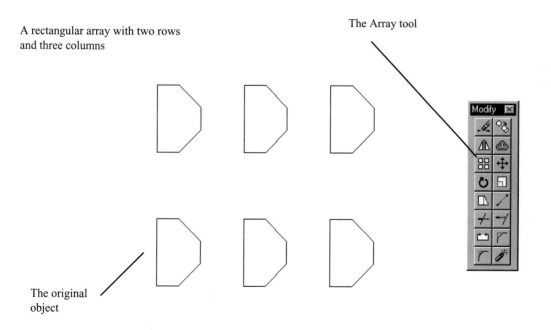

A rectangular array with two rows and three columns

The Array tool

The original object

Figure 6-8

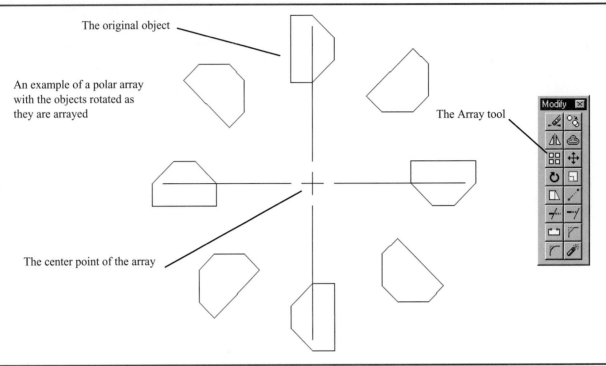

The original object

An example of a polar array
with the objects rotated as
they are arrayed

The Array tool

The center point of the array

Modify

Figure 6-9

Number of columns (|||)<1>:

5. Type 3; press Enter.

Unit cell or distance between rows:

6. Type 3; press Enter.

Distance between colomns:

7. Type 2; press Enter.

To use the Polar Array command

See Figure 6-9.

1. Select the POLAR Array tool from the Modify I toolbar.

Select objects:

2. Select all lines to be arrayed.

Rectangular or polar array (<R>/P)":

3. Type p; press Enter.

Base/ Specify center point of array:

4. Select a center point.

Number of items:

5. Type 8; press Enter.

The number entered is the total number of items to appear on the screen, including the original object. In this example, seven objects were added to the existing object for a total of eight objects in the array.

Angle to fill (+=ccw, -=cw)<360>:

6. Press Enter.

The default value is a full 360 degrees. Any angular value can be entered.

Rotate objects as they are copied?<Y>:

7. Press Enter.

The objects will be rotated as they are arrayed. Figure 6-10 shows what would happen if the objects were not rotated as they were arrayed (type n; press Enter).

6-7 MOVE

The MOVE command is used to move a line or object to a new location on the drawing.

To move an object

See Figure 6-11.

1. Select the Move tool from the Modify I toolbar.

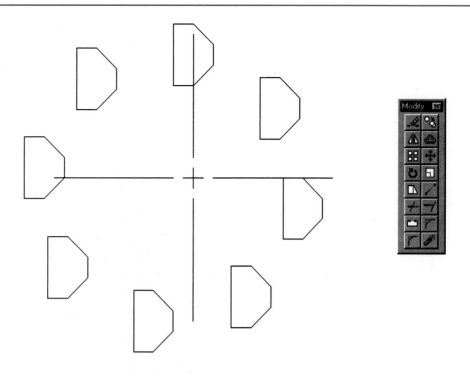

Figure 6-10

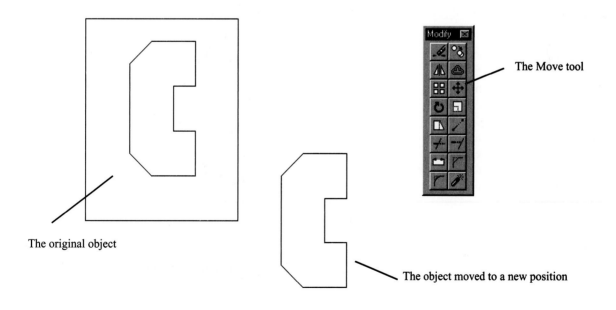

The Move tool

The original object

The object moved to a new position

Figure 6-11

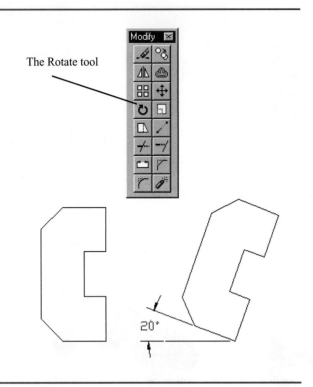

The Rotate tool

Figure 6-12

Select objects:

2. Window the entire object.

Select objects:

3. Press Enter.

Base point or displacement:

4. Select a base point.

Any point may be selected. Snap points are usually used as base points, because they can be used to define precise displacement distances and accurate new locations. Object Snap point can also be used. Object Snap points are discussed in Chapter 6.

Second point of displacement:

5. Select a second displacement point.

The object will now be located relative to the second displacement point.

6-8 ROTATE

The ROTATE command is used to rotate a given object about a specified base point. The base point need not be on the object. The 3D ROTATE command will be covered later in the book.

To rotate an object

See Figure 6-12.

1. Select the Rotate tool from the Modify I toolbar.

Select objects:

2. Window the object.

Select objects:

3. Press Enter.

Base point:

4. Select a base point.

The base point may be anywhere on the screen. In the example shown the lower right corner of the object was selected as the base point.

<Rotation angle>/Reference:

5. Type -20; press Enter.

The object will rotate about the base point 20 degrees in the clockwise direction. AutoCAD's default settings use the counterclockwise direction as positive.

6-9 SCALE

See Figure 6-13.

1. Select the Scale tool from the Modify I toolbar.

Select objects:

2. Window the object.

Select objects:

3. Press Enter.

Base point:

4. Select a base point.

In the example shown, the lower right corner of the object was selected as the base point.

<Scale factor>/Reference:

5. Type 2.5; press Enter.

A scale factor of 2.5 will create a drawing two and a half times as large as the original. A scale factor of 2 will generate a drawing twice as large as the original.

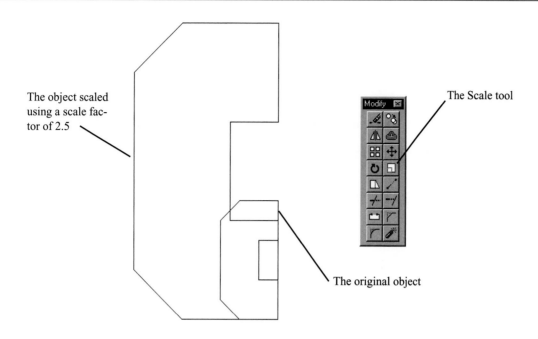

The object scaled using a scale factor of 2.5

The Scale tool

The original object

Figure 6-13

6-10 STRETCH

Figure 6-14 shows the Stretch tool. The Stretch tool is used to change the shape of an existing object.

To stretch an object

1. Select the Stretch tool from the Modify I toolbar.

   Select objects:

2. Window the portion of the object that is to stretched.

   Be sure that the window includes the entire object to be stretched.

   Select objects:

3. Press Enter.

   Base point or displacement:

4. Select a base point for the stretch.

   In the example shown, the right upper corner of the hexagon was selected.

   Second point of displacement:

5. Move the cross hairs to a selected point.

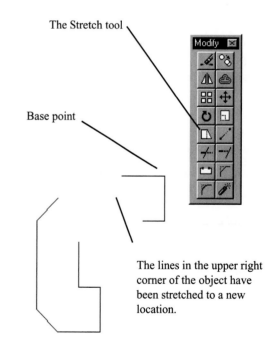

The Stretch tool

Base point

The lines in the upper right corner of the object have been stretched to a new location.

Figure 6-14

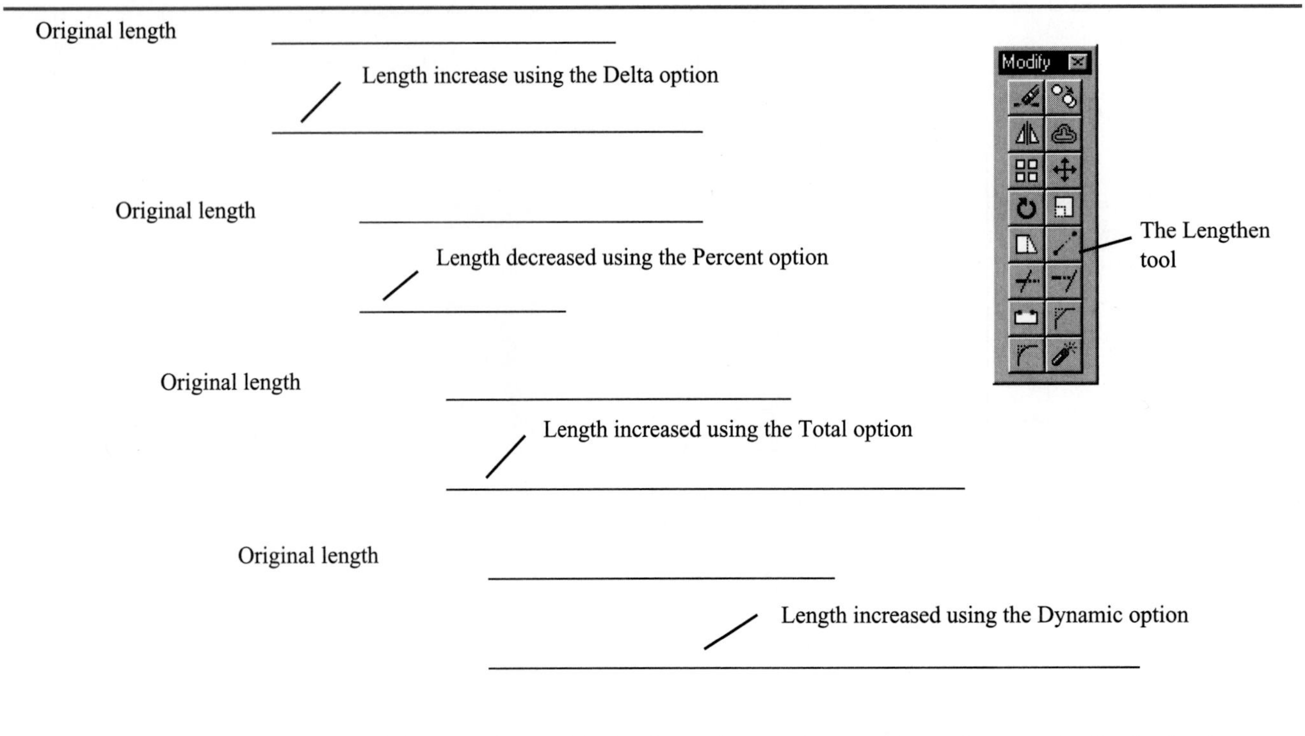

Original length _____

　　　　　　　　　　　Length increase using the Delta option

　　　　　　　／ _____

Original length _____

　　　　　　　Length decreased using the Percent option

　　　　　／ _____

Original length _____

　　　　　　Length increased using the Total option

　　　／ _____

Original length _____

　　　　　　Length increased using the Dynamic option

　　／ _____

Figure 6-15

6-11 LENGTHEN

The LENGTHEN command has several options. See Figure 6-15. Each of the four lines shown was created from the same length original line.

To use the Delta option

1. Select the Lengthen tool from the Modify toolbar.

   DElta/Percent/Total/DYnamic/<Select object>:

2. Type DE; press Enter.

   Angle/<Enter delta length (0.0000):

3. Type 1.0;press Enter.

This defines the change of length as 1.0. This means that a new line will be created, 1.0000 longer than the original. Note that the location of the select point will determine which end of the line is lengthened. In the above example a select point near the right end of the line was selected, so the line lengthened to the right.

   <Select object to change>Undo

4. Select the line.

The line will change lengths.

5. Press Enter again to restart the Lengthen command.

To use the Percent option

   DElta/Percent/Total/DYnamic/<Select object>:

1. Type P; press Enter

   Enter percent length <100.00>:

2. Type 60; press Enter

   <Select object to change>Undo

3. Select the line.

The line's length will be reduced by 60% or slightly more than half of its original length.

4. Press Enter.

To use the Total option

   DElta/Percent/Total/DYnamic/<Select object>:

1. Select the line.

Original objects

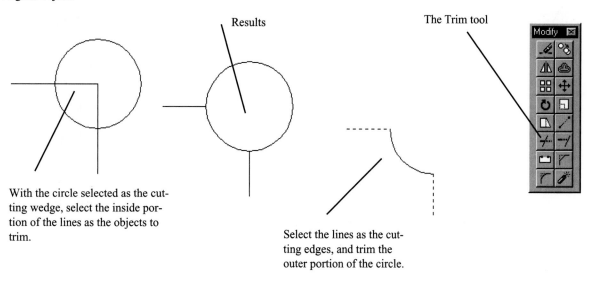

Results

The Trim tool

With the circle selected as the cutting wedge, select the inside portion of the lines as the objects to trim.

Select the lines as the cutting edges, and trim the outer portion of the circle.

Figure 6-16

Angle/<Enter total length (4.0000)>:

The default value is the current length of the line.

2. Type 6; press Enter.

   <Select object to change>Undo

3. Select the line.

   The new line will be 6.0000 long.

4. Press Enter.

To use the Dynamic option

   DElta/Percent/Total/DYnamic/<Select object>:

1. Type DY; press Enter.

   <Select object to change>/Undo:

2. Select the line.

   Specify new end point:

3. Select a new end point; press Enter.

   The length of the line will vary as the cross hairs are moved about the screen. The Angle option is used with arcs to increase their length in terms of their included angle.

6-12 TRIM

   Figure 6-16 shows the Trim tool. The Trim tool is used to trim lines to a new length by first establishing a cutting plane line, then trimming lines.

To use the TRIM command

1. Select the TRIM tool from the Modify toolbar.

   Select cutting edges:

2. Select the circle.

   <Select object to trim>/Project/Edge/Undo:

3. Select the ends of the horizontal and vertical lines within the circle.

4. Press Enter.

   Select cutting edges:

5. Select the horizontal and vertical lines.

   <Select object to trim>/Project/Edge/Undo:

6. Select the portion of the circle above the lines.

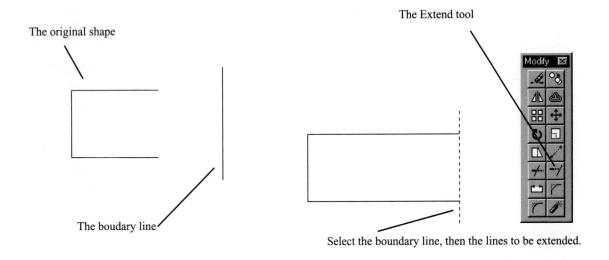

The original shape

The Extend tool

The boudary line

Select the boundary line, then the lines to be extended.

Figure 6-17

6-13 EXTEND

See Figure 6-17.

1. Select the Extend tool from the Modify I toolbar.

   Select boundary edges:

2. Select a line that can be used as a boundary edge.

   A boundary line must be positioned so that all of the lines to be extended will intersect it. If you have an object that you want to make longer, draw a boundary line or use the MOVE command and move one of the object's edge lines to the new position and use it as a boundary line.

   Select objects:

3. Select the lines to be extended.

6-14 BREAK

The BREAK command is used to divide a line or other entity into two parts. There are two options; one that defines the two points of the break directly by two select points, and one that first selects the line or object and then selects the two break points. Figure 6-18 shows examples of each option.

To use BREAK - 2 point option

1. Select the Break tool from the Modify toolbar.

   Select object:

   The point used to select the line will also be the first point of the break.

2. Select a point on the line.

   The second point selected will be the second point of the break. The line will be divided into two parts at the points selected.

To use BREAK - 3 point option

1. Select the Break selected tool from the Modify I toolbar.

   Command:_break Select object:

2. Select the line.

   Enter second point (or F for first point): _f

3. Type F; press Enter.

   Enter first point:

4. Select the first point of the break.

   Enter second point:

5. Select the second point.

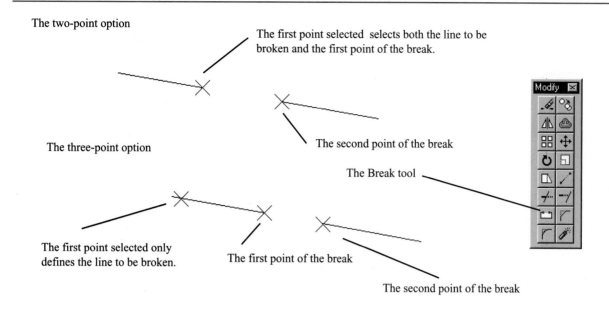

The two-point option

The first point selected selects both the line to be broken and the first point of the break.

The second point of the break

The three-point option

The Break tool

The first point selected only defines the line to be broken.

The first point of the break

The second point of the break

Figure 6-18

6-15 CHAMFER

Figure 6-19 shows examples of how the Chamfer tool is used. A chamfer is a straight-lined corner cut, where-as a fillet is a rounded corner.

To create a chamfer using the distance option

Chamfers are usually cut at 45 degrees, but other angles can be drawn by defining different lengths for Dist1 and Dist 2.

1. Select the Chamfer tool from the Modify toolbar.

   (TRIM mode) Current chamfer Dist1 = 0.6000, Dist2 = 0.6000
   Polyline/Distance/Angle/Trim/Method/<Select first line>:

2. Type d; press Enter.

   Enter first chamfer distance <0.5000>:

3. Type 1.00; press Enter.

   Enter second chamfer distance <1.0000>:

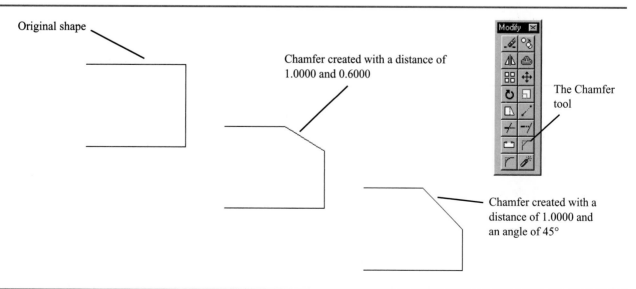

Original shape

Chamfer created with a distance of 1.0000 and 0.6000

The Chamfer tool

Chamfer created with a distance of 1.0000 and an angle of 45°

Figure 6-19

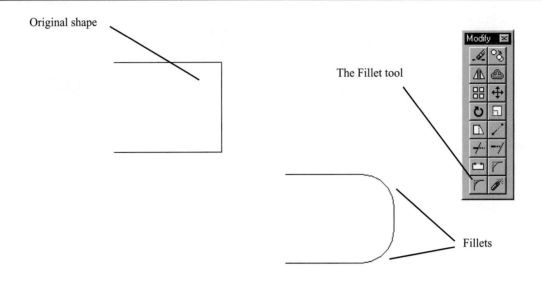

Original shape

The Fillet tool

Fillets

Figure 6-20

AutoCAD assumes that the chamfer will be at 45 degrees so it automatically sets the second distance equal to the first.

4. Type 0.60; press Enter.

Command:

5. Press Enter.

(TRIM mode) Current chamfer Dist1 = 0.6000, Dist2 = 0.6000
Polyline/Distance/Angle/Trim/Method/<Select first line>:

6. Select a line.

Select a second line:

7. Select another line.

To create a chamfer using the angle option

1. Select the Chamfer tool from the Modify I toolbar.

(TRIM mode) Current chamfer Dist1 = 0.6000, Dist2 = 0.6000
Polyline/Distance/Angle/Trim/Method/<Select first line>:

2. Type a; press Enter.

Enter first chamfer distance <1.0000>:

3. Press Enter or enter a new value.

Enter chamfer angle from the first line <0>:

4. Type 45; press Enter.

6-16 FILLET

See Figure 6-20. A fillet is a rounded corner. The FILLET command can be applied to either an internal or external corner.

1. Select the Fillet tool from the Modify I toolbar.

(TRIM mode) Current fillet radius = 0.6000
Polyline/Radius/Trim/<Select first object>:

2. Type R; press Enter.

Enter fillet radius <0.5000>:

3. Type .75; press Enter.

Command:

4. Again select the Fillet tool from the Modify toolbar or press the Enter key.

(TRIM mode) Current fillet radius = 1.2600
Polyline/Radius/Trim/<Select first object>:

5. Select a line.

Select second object:

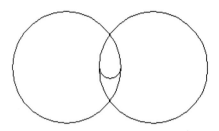

The FILLET command applied to circles

Figure 6-21

Selected points   Resulting shapes

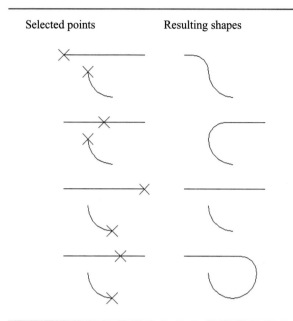

Figure 6-22

6. Select another line.

Fillets can also be drawn between circles. See Figure 6-21. The FILLET command is location sensitive, meaning that the location of your select points will affect the shape of the resulting fillet. Figure 6-22 shows how selection points can be used to create different fillet shapes.

6-17 EXPLODE

The EXPLODE command is used to change polylines and blocks into individual line segments. For example, to remove two lines of a hexagon drawn using the Polygon command, as shown in Figure 6-23, you have to first explode the hexagon.

1. Select the Polygon tool from the Draw toolbar.
2. Draw a hexagon.
3. Select the Explode tool from the Modify I toolbar.

    Command: _explode
    Select objects:

4. Select the hexagon.

    Select objects:

There will be no visual difference in the hexagon.

5. Press Enter.
6. Use the Erase command to remove the two lines.

The hexagon is a polyline. It is one entity, not six line segments. The EXPLODE command will change the polyline to individual line segments.

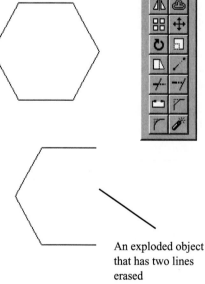

An exploded object that has two lines erased

Figure 6-23

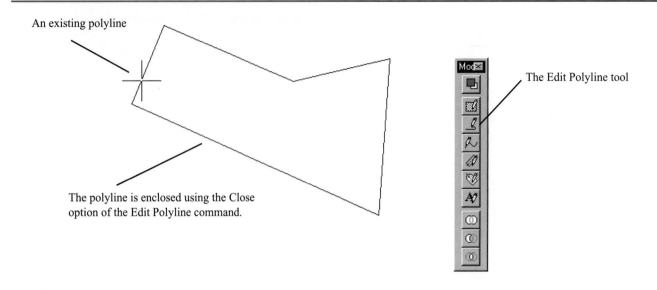

An existing polyline

The Edit Polyline tool

The polyline is enclosed using the Close
option of the Edit Polyline command.

Figure 6-24

6-18 EDIT POLYLINE

Figure 6-24 shows the Edit Polyline tool located on the Modify II toolbar.

To close an existing polyline

The Close option can be used with the LINE command, Section 3-2, and with the POLYLINE command, Section 4-5. In each of these commands, the close option is applied as part of the command sequence. Once the sequence is ended, the Close option can not be applied. The Edit Polyline Close option can be applied to an existing polyline. See Figure 6-24.

1. Select the Edit Polyline tool from the Modify II toolbar.

   Command: _pedit Select polyline:

2. Select an existing polyline.

   Close/ Join/ Width/ Edit vertex/ Fit/ Spline/ Decurve/ Ltype gen/ Undo/ eXit/<X>:

3. Type C; press Enter.

To create a polyline from existing line segments

See Figure 6-25.

1. Select the Edit Polyline tool from the Modify II toolbar.

Command: _pedit Select polyline:

2. Select one line of the group you want to change to a polyline.

   Object selected is not a polyline
   Do you want to turn it into one? <Y>

3. Press Enter accepting the default Y response.

   Close/ Join/ Width/ Edit vertex/ Fit/ Spline/ Decurve/ Ltype gen/ Undo /eXit/ <X>:

4. Type J; press Enter.

   Select objects:

5. Select all the lines you want incorporated into a polyline.

   4 Segments added to polyline
   Close/ Join /Width/ Edit vertex /Fit/ Spline/ Decurve /Ltype gen/ Undo/ eXit /<X>:

   The line segments are now a polyline.

To change the width of an existing polyline

See Figure 6-26.

1. Select the Edit Polyline tool from the Modify II toolbar.

   Command: _pedit Select polyline:

2. Select the polyline.

Select the Join option, then window all the line segments to create a polyline.

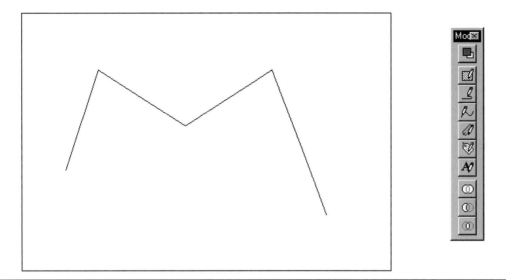

Figure 6-25

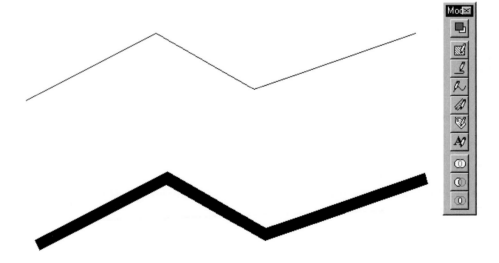

Use the Width option to change the thickness of a polyline.

Figure 6-26

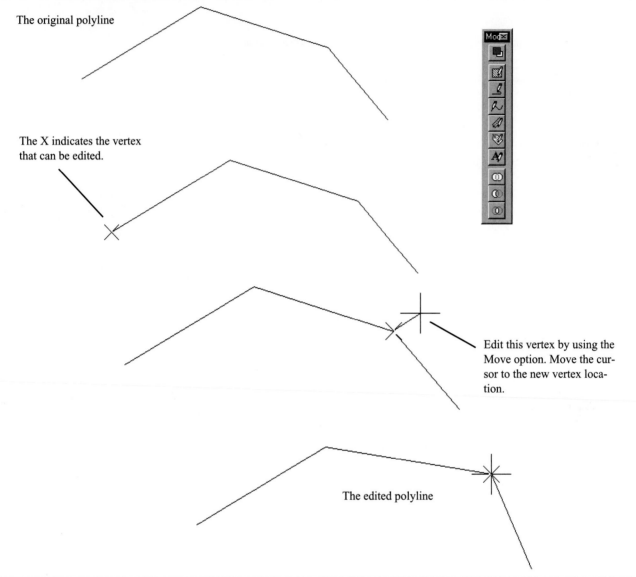

The original polyline

The X indicates the vertex
that can be edited.

Edit this vertex by using the
Move option. Move the cursor to the new vertex location.

The edited polyline

Figure 6-27

Close/ Join/ Width/ Edit vertex /Fit/ Spline/ Decurve/ Ltype gen/ Undo /eXit/ <X>:

3. Type W; press Enter.

Enter new width for all segments:

4. Type .25; press Enter.

To use the Edit Vertex command

See Figure 6-27.

1. Select the Edit Polyline tool from the Modify II toolbar.

Command: _pedit Select polyline:

2. Select the polyline.

Close/ Join/ Width/ Edit vertex/ Fit/ Spline/ Decurve/ Ltype gen/ Undo/ eXit/ <X>:

3. Type E; press Enter.

Next/Previous/Break/Insert/Move/Regen/Straigh ten/Tangent/Width/eXit<N>:

An X will appear at the first point of the polyline. See Figure 6-27. The X indicates which vertex will be edit-ed. The X can be advanced to the next vertex using the default setting N or next option.

Original polyline

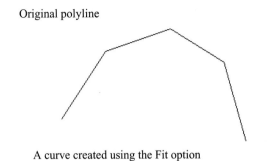

A curve created using the Fit option

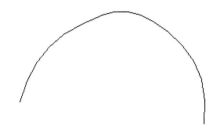

Figure 6-28

Original polyline

A curve created using the Spline option

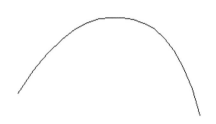

Figure 6-29

4. Press Enter until you reach the vertex you want to edit.

Next/Previous/Break/Insert/Move/Regen/Straigh ten/Tangent/Width/eXit<N>:

5. Type M; press Enter.

In the example shown, the vertex was moved to change the shape of the polyline.

Enter new location

6. Move the cross hairs to the new location.

A line will drag between the original vertex location and the new location.

7. Press the left mouse button to set the new vertex location.
8. Type X; press Enter.

To fit a curve to a polyline

See Figure 6-28.

1. Select the Edit Polyline tool from the Modify II toolbar.

Command: _pedit Select polyline:

2. Select the polyline.

Close/ Join/ Width/ Edit vertex/ Fit/ Spline/ Decurve/ Ltype gen/ Undo/ eXit/ <X>:

3. Type F; press Enter.

The polyline will change to a curve. The shape of the curve can be changed using the Edit Vertex option as explained above.

To change a polyline to a spline

See Figure 6-29.

1. Select the Edit Polyline tool from the Modify toolbar.

Command: _pedit Select polyline:

2. Select the polyline.

Close/ Join/ Width/ Edit vertex/ Fit/ Spline/ Decurve/ Ltype gen/ Undo/ eXit/< X>:

3. Type S; press Enter.

The polyline will change to a spline. The Spline option produces a different shaped curve than does the Fit option. Compare Figures 6-28 and 6-29. Each curve is shaped with the same polyline. The Fit option generates a curve that passes through each vertex point on the polyline. The Spline option uses the vertex points as frames and generates a smoother curve than the Fit option. Splines may be edited using the Edit Vertex option.

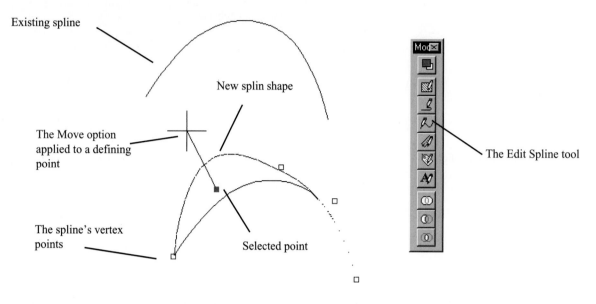

Existing spline

New splin shape

The Move option applied to a defining point

The spline's vertex points

Selected point

The Edit Spline tool

Figure 6-30

To change an existing fitted curve or spline back to a polyline

1. Select the Edit Polyline tool from the Modify II toolbar.

   Command: _pedit Select polyline:

2. Select the curve or spline.

   Close/ Join/ Width/ Edit vertex/ Fit/ Spline/ Decurve/ Ltype gen/ Undo/ eXit/ <X>:

3. Type D; press Enter.

   The curve or spline will return to its original straight line segment polyline.

6-19 EDIT SPLINE

The Edit Spline option is used to change the shape of an exisiting spline.

To use the Move option

   See Figure 6-30.

1. Select the Edit Spline tool from the Modify II toolbar.

   Command: _splinedit
   Select spline:

2. Select the spline.

   Close/Move Vertex/Refine/rVerse/Undo/eXit <X>:

   A series of blue boxes will appear indicating the original location of the spline's vertex points.

3. Type M: press Enter.

   The blue box will change to a solid red box. The cursor will now have a line extending from the red box to the cursor. As the cursor is moved the spline's shape will change.

4. Press Enter.

   The Edit option will move to the next vertex point of the spline. The selected box will turn from open blue to solid red.

5. Select a new location for the point.
6. Type X; press Enter.

   The shape of the spline will be redefined.

To use the Close option

   See Figure 6-31.

1. Select the Edit Spline tool from the Modify II toolbar.

   Command: _splinedit
   Select spline:

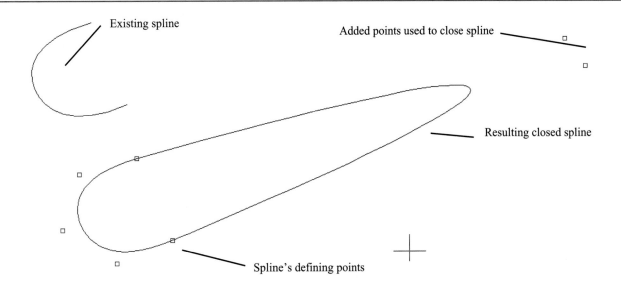

Existing spline

Added points used to close spline

Resulting closed spline

Spline's defining points

Figure 6-31

2. Select the spline.

Close/Move Vertex/Refine/rVerse/Undo/eXit <X>:

3. Type C; press Enter.

AutoCAD will created a closed spline based on the existing spline's vertex points.

To use the Refine option

The Refine option allows you to adjust the shape of spline by adding new points or by changing the weighted scale factor of a point's value in the spline calculation.

1. Select the Edit Spline tool from the Modify II toolbar.

*Command: _splinedit
Select spline:*

2. Select the spline.

Close/Move Vertex/Refine/rVerse/Undo/eXit <X>:

3. Type R; press Enter.

Add control point/Evaluate Order/Weight/eXit <X>:

4. Type W; press Enter.

Next/Previous/Select point/eXit/ <Enter new weight> <1.0000> <N>:

5. Press the Enter key until the desired vertex point is selected.
6. Type 3; press Enter.

Notice the change in the spline's shape in Figure 6-32.

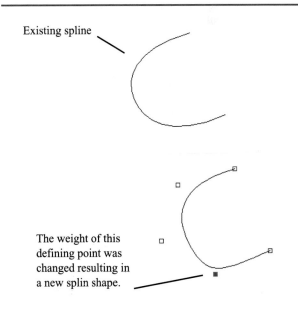

Existing spline

The weight of this defining point was changed resulting in a new splin shape.

Figure 6-32

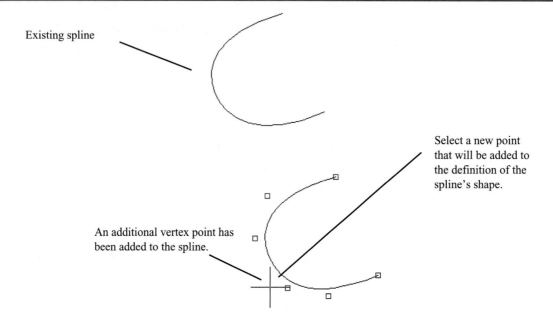

Existing spline

Select a new point that will be added to the definition of the spline's shape.

An additional vertex point has been added to the spline.

Figure 6-33

To use the Add control point option

See Figure 6-33.

1. Select the Edit Spline tool from the Modify II toolbar.

   Command: _splinedit
   Select spline:

2. Select the spline.

   Close/Move Vertex/Refine/rVerse/Undo/eXit <X>:

3. Type R; press Enter.

   Add control point/Evaluate Order/Weight/eXit <X>:

4. Type A; press Enter.

   Select a point on the spline:

5. Select a new point on the spline.

The new point will be added to the spline calculation and a new shape will result. Note in Figure 6-33 that the spline now has six vertex points. The existing spline had five vertex points.

6-20 EDIT MULTILINE

The EDIT MULTILINE command is used to change intersection patterns between existing multilines. Figure 6-34 shows an intersection between two multilines and three different patterns created using the EDIT MULTILINE command. The pattern changes are selected from the Multiline Edit Tools dialog box.

1. Select the Edit Multiline tool from the Modify II toolbar.

The Multiline Edit Tools dialog box will appear on the screen. See Figure 6-35.

2. Select a pattern and then click the OK box.

   Select first mline:

3. Select one of the multilines.

   Select the second mline:

4. Select the other multiline.

The first column of options in the Multiline Edit Tools dialog box is used for multilines that cross, the second for multilines that form a tee, and the third column is for corners.

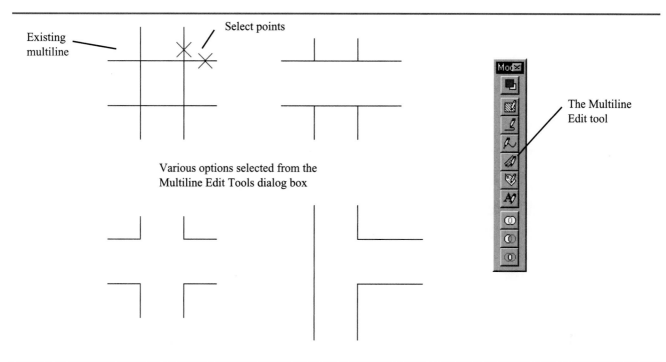

Existing multiline

Select points

The Multiline Edit tool

Various options selected from the Multiline Edit Tools dialog box

Figure 6-34

This column is for multilines that intersect.

This column is for multilines that form a tee.

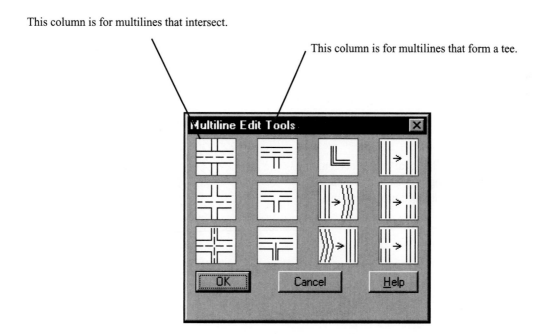

Figure 6-35

Existing text

This is a line of existing text.

The Edit Text tool

The selected text will appear in the Multiline Text Editor dialog box.

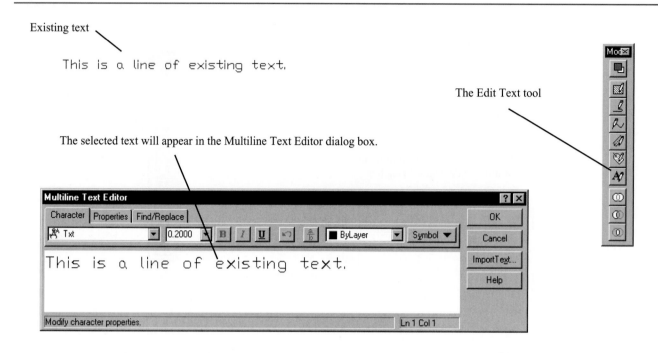

Figure 6-36

6-21 EDIT TEXT

Text already entered on the screen can be edited using the Edit Text tool. See Figure 6-36.

1. Select the Edit Text tool from the Modify II toolbar.

   <Select an annotation object>/Undo:

2. Select the text to be edited.

The Edit Mtext dialog box will appear on the screen. The text will appear within the dialog box and can be edited by moving the cursor as with most word processing programs.

6-22 EDIT HATCH

See Figure 6-37. An existing hatch pattern can be changed using the Edit Hatch command.

1. Select the Edit Hatch tool from the Modify II toolbar.

   Select hatch object:

2. Select the hatch pattern you want to change.

The Hatchedit dialog box there will appear. Click the arrow on the right side of the Pattern box to cascade a listing of available hatch patterns. In the example shown, the original ANSI31 pattern was changed to ANSI36.

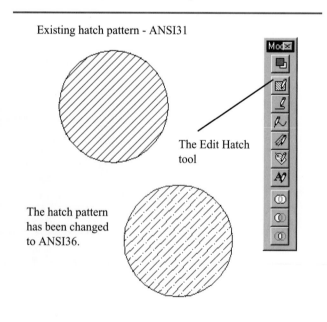

Existing hatch pattern - ANSI31

The Edit Hatch tool

The hatch pattern has been changed to ANSI36.

Figure 6-37

6-23 EXERCISE PROBLEMS

Redraw the following figures. Do not include dimensions. Dimensional values are as indicated.

EX6-1 INCHES

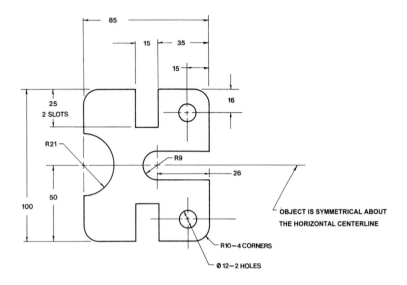

EX6-2 MILLIMETERS

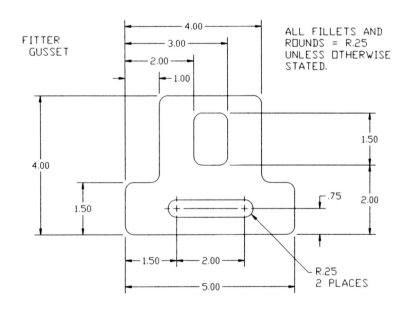

EX6-3 MILLIMETERS

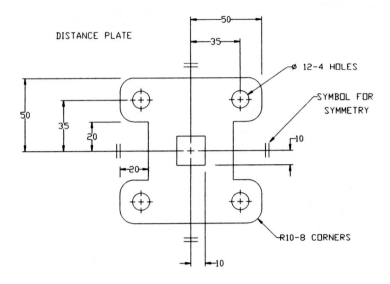

EX6-4 MILLIMETERS

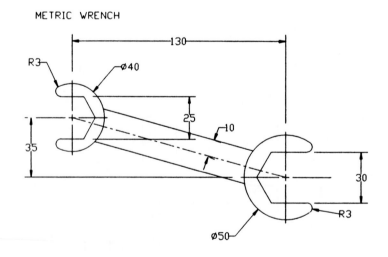

EX6-5 INCHES

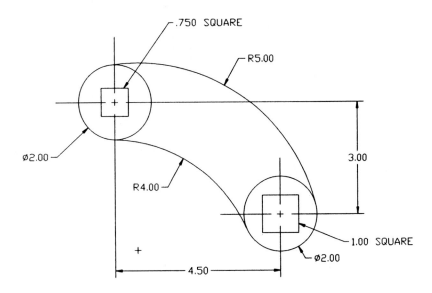

EX6-6 MILLIMETERS

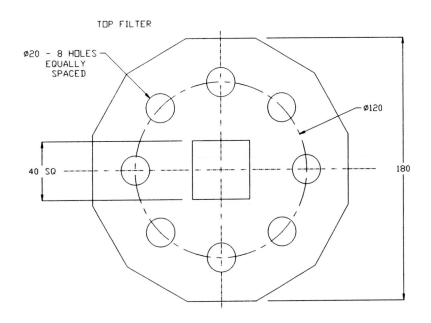

EX6-7 FEET OR INCHES

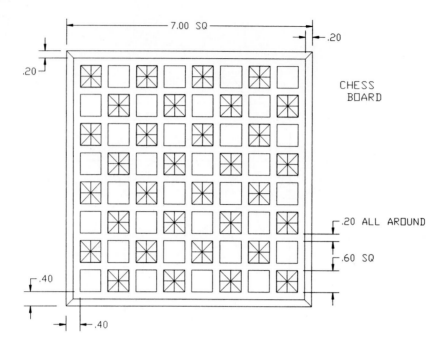

EX6-8 FEET OR INCHES

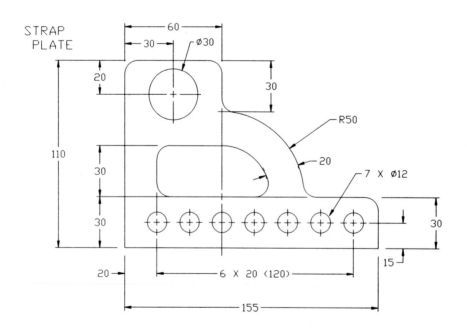

EX6-9 MILLIMETERS

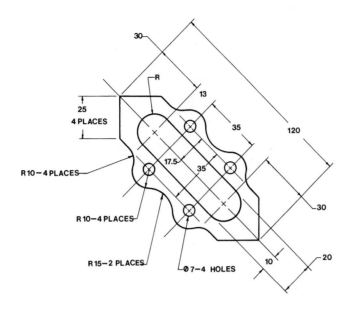

EX6-10 MILLIMETERS

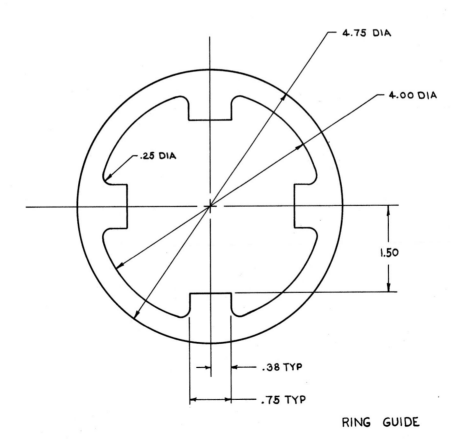

RING GUIDE

EX6-11 MILLIMETERS

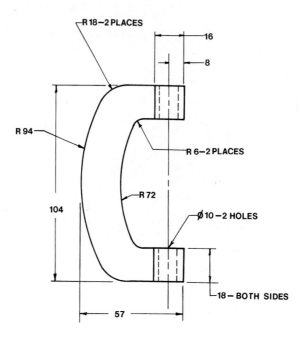

EX6-12 MILLIMETERS

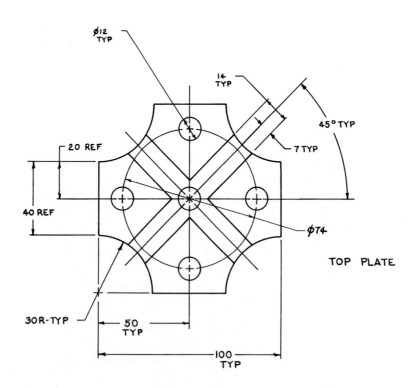

TOP PLATE

EX6-13 MILLIMETERS

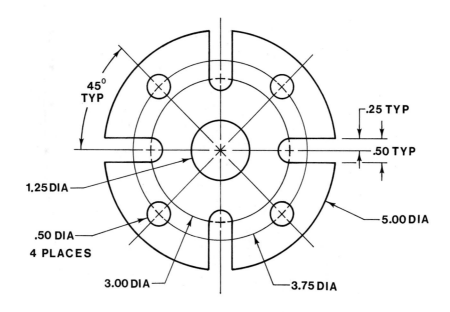

45°
TYP

.25 TYP

.50 TYP

1.25 DIA

.50 DIA
4 PLACES

5.00 DIA

3.00 DIA

3.75 DIA

MATL .25 STEEL

EX6-14 FEET OR INCHES

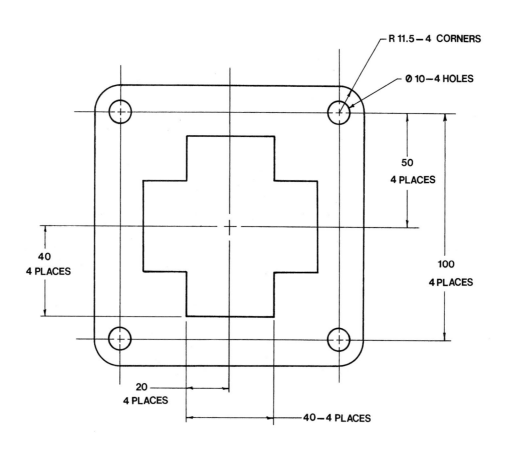

R 11.5 — 4 CORNERS

Ø 10 — 4 HOLES

50
4 PLACES

40
4 PLACES

100
4 PLACES

20
4 PLACES

40 — 4 PLACES

EX6-15 MILLIMETERS

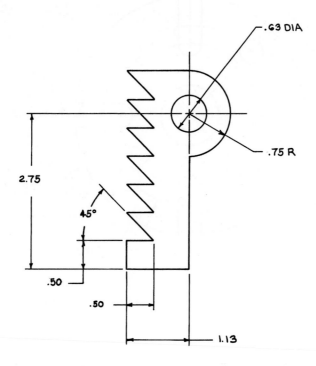

7

Two-Dimensional Tutorials

7-1 INTRODUCTION

This chapter presents three tutorials designed to show how to apply the fundamental commands introduced in the last six chapters. The intent is for you to first review the previous chapters, so that you are familiar with AutoCAD's basic operations and commands. Then work through the tutorials. If a command sequence is not clear, you can refer to the command explanations in the previous chapters.

7-2 TUTORIAL T7-1

Draw the shape shown in Figure 7-1.

Toolbar/Commands used:

Draw/Line
Draw/Circle Center Diameter
Draw/Circle/Tangent Tangent Radius
Modify/Trim

To set the GRID and SNAP commands

1. Select Tools heading, then Drawing Aids.

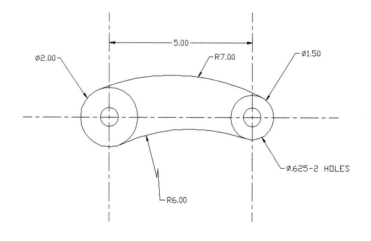

Figure 7-1

139

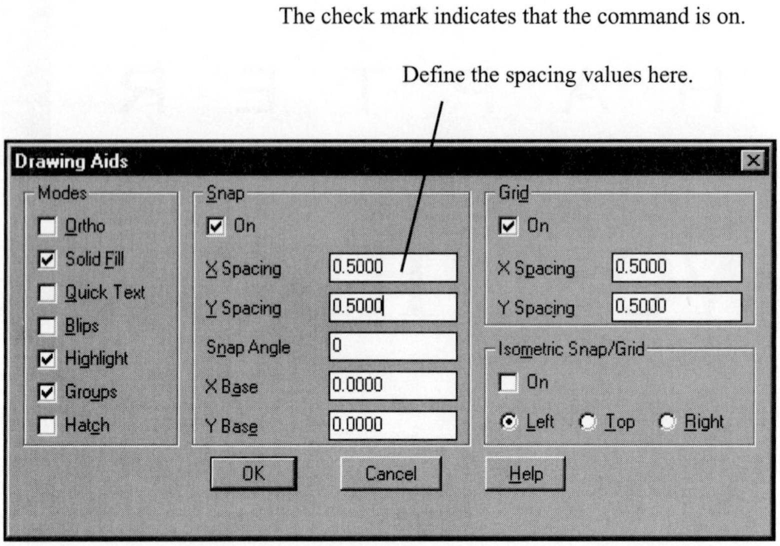

The check mark indicates that the command is on.

Define the spacing values here.

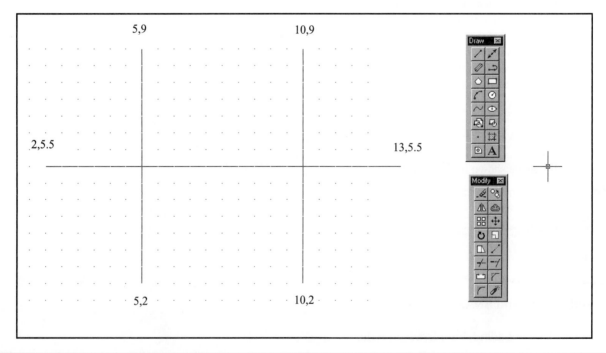

Figure 7-2

The Drawing Aids dialog box will appear. See Figure 7-2.

2. Turn both Grid and Snap on and set their X,Y spacing to 0.5000.

To define the centerlines

Draw a horizontal line and two vertical lines 5.00 apart as shown in Figure 7-3.

1. Select Line tool from the Draw toolbar.

Command: _line From point:

Figure 7-3

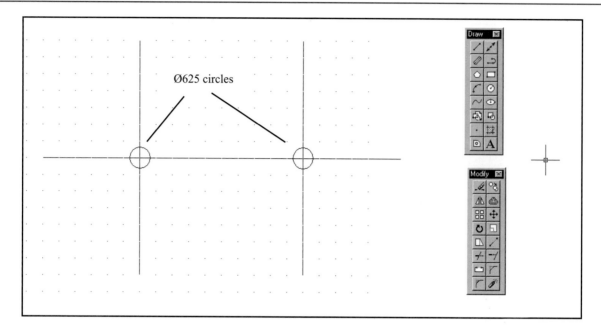

Ø625 circles

Figure 7-4

2. Type 2,5.5; press Enter.

To point:

3. Type 13,5.5; press Enter.

As the endpoints of all three lines are on grid/snap points, the endpoints could be located by using the coordinate display in the lower left corner of the screen.

To point:

4. Press Enter twice or double-click the Enter key.

Command: _line From point:

5. Type 5,9; press Enter.

To point:

6. Type 5,2; press Enter.

To point:

7. Press Enter twice.

Command: -line From point:

8. Type 10,9; press Enter.

To point:

9. Type 10,2; press Enter.

To point:

10. Press Enter.

Command:

Use the PAN command if necessary to center the drawing on the drawing screen.

To draw the Ø.625 circles

See Figure 7-4.

1. Select the Circle tool from the Draw toolbar.

Command: _circle 3P/2P/TTR/<Center point>:

2. Select the left intersection (5,5.5) as the circle center point.

Command: _circle 3P/2P/TTR/<Center point>: Diameter/<Radius>:

3. Type d; press Enter.

Diameter:

4. Type .625; press Enter.

Command:

5. Press Enter.

CIRCLE 3P/2P/TTR/<Center point>:

6. Select the right intersection (10,5.5) as the center point.

CIRCLE/ 3P/ 2P/ TTR/ <Center point> :Diameter/<Radius> <.3125>:

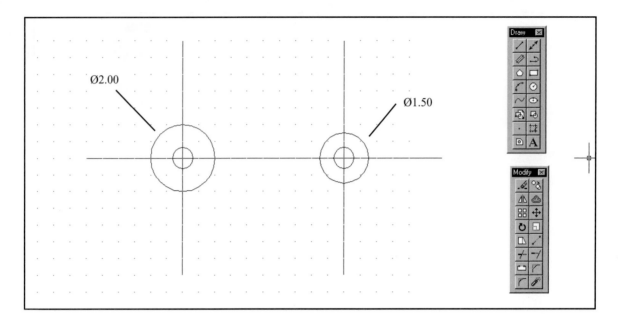

Figure 7-5

The .625 diameter value has become the default value expressed in terms of a radius value.

7. Press Enter.

Command:

To draw the Ø2.00 and Ø1.50 circles

See Figure 7-5.

1. Select the Circle tool from the Draw toolbar.

Command: _circle 3P/2P/TTR/<Center point>:

2. Select the left intersection as the center point.

Command: _circle 3P/2P/TTR/<Center point>: Diameter/<Radius><0.3125>:

3. Type D; press Enter.

Diameter:

4. Type 2.00; press Enter.

Command:

5. Press Enter.

Command: _circle 3P/2P/TTR/<Center point>:

6. Select the right intersection as the center point.

CIRCLE 3P/2P/TTR/ < Center point >: Diameter/Radius> <1.0000>:

7. Type D; press Enter:

Diameter:

8. Type 1.50; press Enter.

Command:

To draw arcs tangent to the circles

See Figure 7-6.

1. Select the Circle tool from the Draw toolbar.

Command: _circle 3P/2P/TTR/<Center point>:

2. Type TTR: press Enter.

Enter Tangent spec:

3. Select the upper left quadrant of the 2.00 diameter circle.

Turn the SNAP command off to make it easier to select the circle edge line; press the F9 key.

Enter second Tangent spec:

4. Select the upper right quadrant of the 1.50 diameter circle.

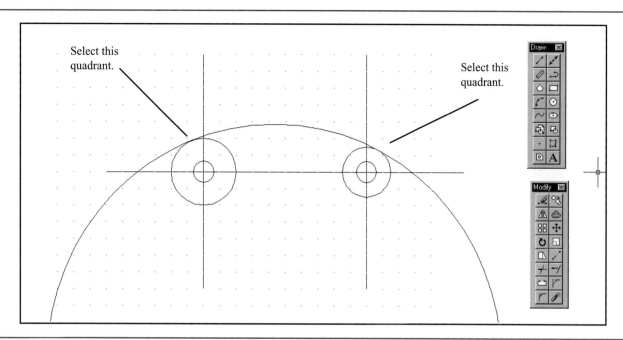

Figure 7-6

Enter second tangent spec: Radius <0.7500>:

5. Type 7; press Enter.

Command:

6. Use the Tangent Tangent Radius option of the Circle tool to draw the 6.00 radius arc.

Remember that the TTR option is quadrant sensitive when selecting the tangent specs. See Figure 7-7.

Command:

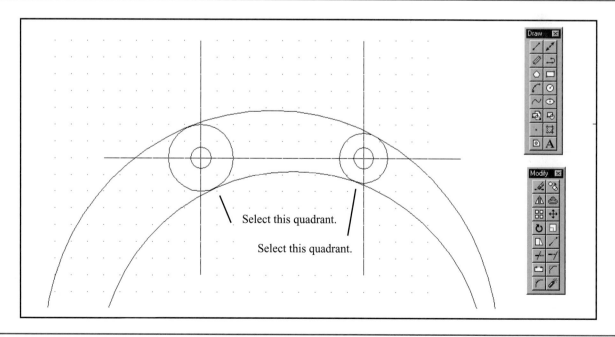

Figure 7-7

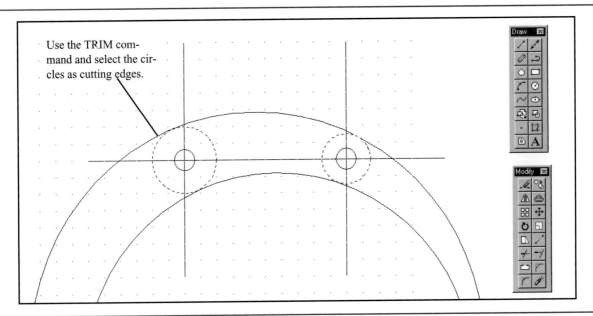

Figure 7-8

To trim the large arcs

See Figures 7-8 and 7-9.

1. Select the Trim tool from the Modify toolbar.

   Select cutting edges:
   Select objects:

2. Select the 2.00 and 1.50 diameter circles as cutting edges; press Enter.

<Select object to trim>/Project/Edge/Undo:

3. Select the excess portions of the 7.00 and 6.00 radius circles; press Enter.

   Command:

4. Select the View heading, then Redraw All to remove excess marks from the screen.

   Your screen should look like Figure 7-9.

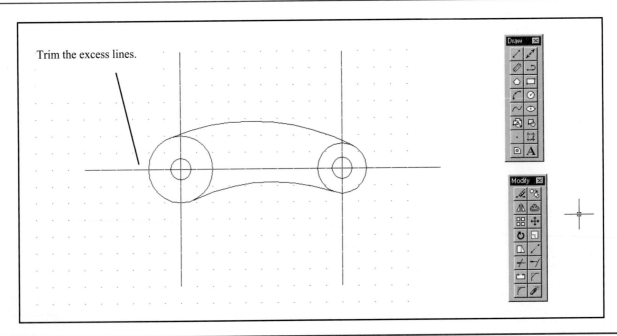

Figure 7-9

Select the linetype box.

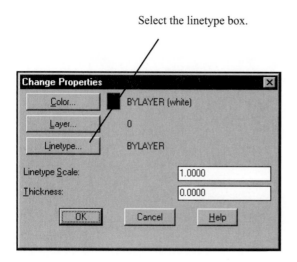

Figure 7-10

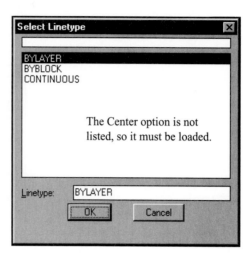

Figure 7-11

To create centerlines

Change the continuous lines used to define the hole's center points to centerlines.

1. Select the Modify heading, then Properties.

   Command: (ai_propchk)
   Select Objects:

2. Select the horizontal and two vertical lines and press the right mouse button (the same as pressing the Enter key).

The Change Properties dialog box will appear. See Figure 7-10.

3. Select Linetype.

The Select Linetype dialog box will appear. See Figure 7-11. The listing of line types does not include a centerline, so it must be loaded into the Select Linetype dialog box.

4. Select Format, then Linetype.

The Layer & Linetype Properties dialog box will appear. See Figure 7-12.

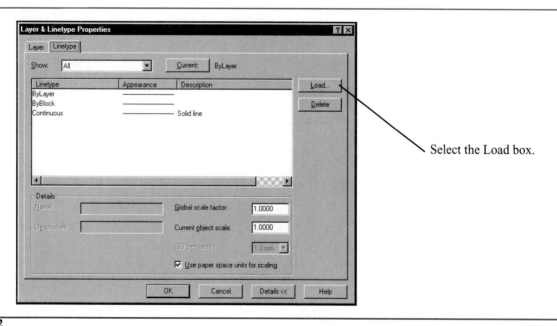

Select the Load box.

Figure 7-12

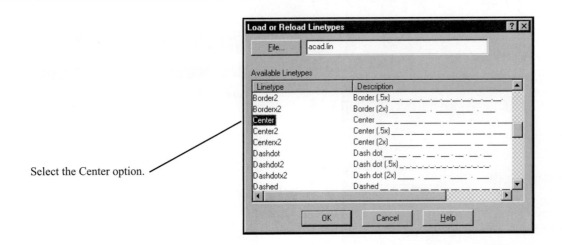

Select the Center option.

Figure 7-13

5. Select the load box.

The Load or Reload Linetypes dialog box will appear. See Figure 7-13.

6. Scroll down the listing and select the Center option.
7. Select OK

The Layer & Linetype Properties dialog box will reappear with the Center linetype listed. See Figure 7-14.

8. Select OK.

The drawing will reappear.

9. Again select the Modify heading and then Properties.

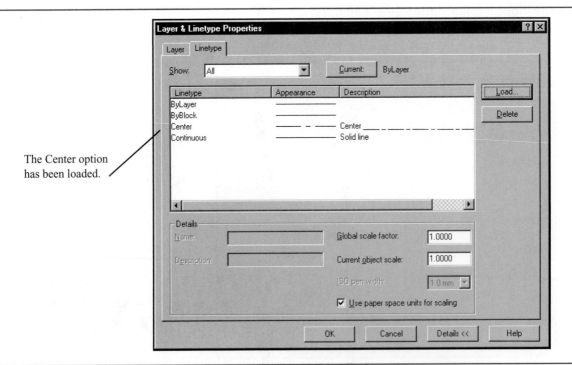

The Center option has been loaded.

Figure 7-14

Command: (ai_propchk)
Select Objects:

10. Select the horizontal and two vertical lines and press the right mouse button (the same as pressing the Enter key).

The Change Properties dialog box will appear. See Figure 7-11.

11. Select Linetype.

The Select Linetype dialog box will appear. It will now include a Center option. See Figure 7-15.

12. Select Center; select OK.

The Change Properties dialog box will reappear.

13. Change the Linetype Scale from 1.0000 to 0.5000.

Locate the arrow cursor within the Linetype Scale value and the cursor will change from an arrow to a flashing I-bar. Click the left mouse button and type or backspace as needed.

14. Click OK.

Your screen should look like Figure 7-16. Save the shape if desired.

A centerline can now be created.

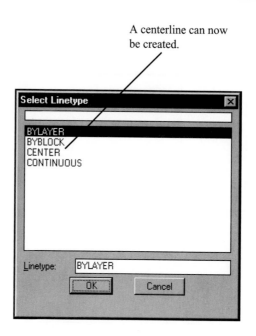

Figure 7-15

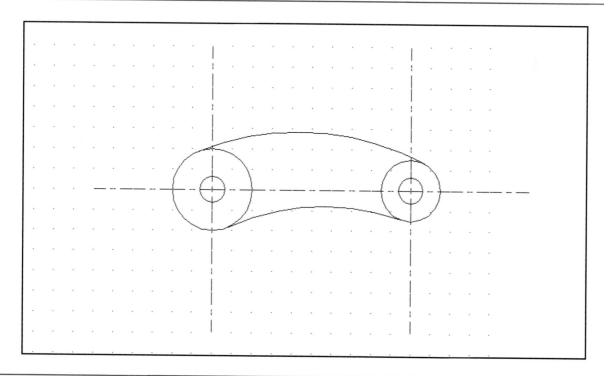

Figure 7-16

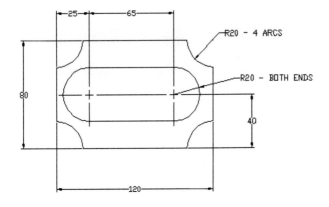

Figure 7-17

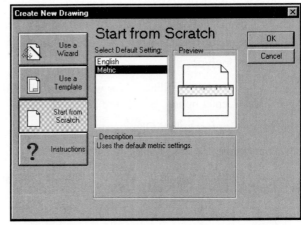

Figure 7-18

7-3 TUTORIAL T7-2

Draw the shape shown in Figure 7-17.

Toolbar/Commands used:

Draw/Line
Draw/Circle Center Radius
Modify/Copy
Modify/Trim

To set the drawing screen for a drawing using millimeter dimensions

The dimensional values are in millimeters, so the drawing screen must be calibrated for metric values.

1. Select the New tool from the Standard toolbar.

The Create New Drawing dialog box will appear. see Figure 7-18.

2. Select the Metric option, then OK.

This option will set the drawing limits to approximately 420 x 297 or an A3 sheet of drawing paper.

To name the drawing

1. Select the File heading, then Save As.

The Save Drawing As dialog box will appear.

2. Type the name of the drawing in the File name box.

3. Select the Save box.

The drawing name will appear just above the upper left corner of the drawing screen. See Figure 7-19.

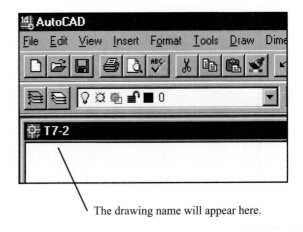

The drawing name will appear here.

Figure 7-19

Turn the Grid command on here.

Enter spacing values here.

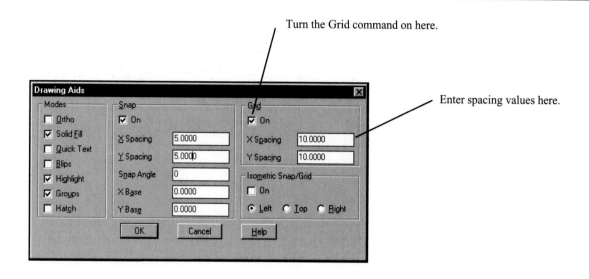

Figure 7-20

To set the Grid and Snap spacing

Create a grid with 10 spacing and set the snap spacing for 5.

1. Select the Tools heading, then Drawing Aids.

The Drawing Aids dialog box will appear. See Figure 7-20.

2. Turn the Snap and Grid commands on and change the grid X and Y spacing to 10; change the snap X and Y spacing to 5.

To draw the base rectangle

Draw a rectangular box 80 x 120. See Figure 7-21.

1. Select the Line tool from the Draw toolbar.

Command: _line From point:

2. Select point 40,50.

To point:

3. Type @120<0; press Enter.

To point:

4. Type @80<90; press Enter.

To point:

5. Type @120<180; press Enter.

To point:

6. Type C; press Enter.

Command:

Use the PAN command to position the object near the center of the drawing screen.

Draw an 80 x 120 millimeter rectangle.

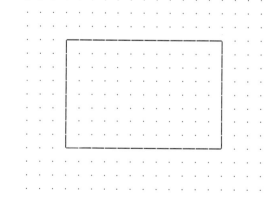

Figure 7-21

Draw six Ø40 circles.

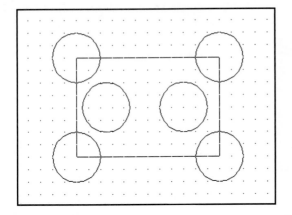

Figure 7-22

First trim the circles then use the resulting arcs as cutting edges to trim the lines.

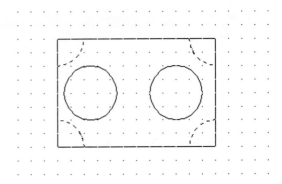

Figure 7-23

To create the crescent-shaped corners

Create the crescent-shaped corners by first drawing circles, then trimming the excess. See Figures 7-22 and 7-23.

1. Select the Circle tool from the Draw toolbar.

   Command: _circle 3P/2P/TTR/<Center point>:

2. Select the upper left corner of the rectangle.

   Command: _circle 3P/2P/TTR/<Center point>: Diameter/<Radius>:

3. Type 20; press Enter.

   Command:

4. Select the Copy tool from the Modify toolbar.

   Select objects:

5. Select the circle.

   Select objects:

6. Press Enter.

   <Base point or displacement>/Multiple:

7. Type M; press Enter.

   Base point:

8. Select the circle's center point as the base point.

   Base point: Second point of displacement:

The resulting shape.

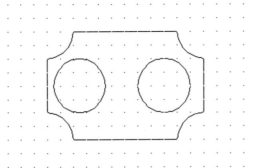

Figure 7-24

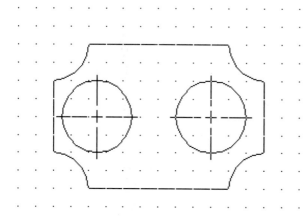

These centerlines were created using the System Variable "dimcen." Set dimcen for a value of -3, then use the Center option within the DIM command to create the centerlines.

Figure 7-25

9. Locate a circle on each of the rectangle's corners and locate two circles within the rectangle as shown.

   See Figure 7-22.

   Command:

10. Select the Trim tool from the Modify toolbar.

    Select cutting edges:
    Select objects:

11. Select the four lines of the rectangle; press Enter.

    <Select object to trim>/Project/Edge/Undo:

12. Select the outside edge of each of the four circles.

    Command:

13. Press Enter.

    Select objects:

14. Select the four remaining arcs; press Enter.

    See Figure 7-23.

15. Select the corner lines on each of the four corners.

    See Figure 7-24.

    Command:

To add centerlines to the circles — System variable method

1. Type Setvar; press Enter.

   Variable name or?:

2. Type Dimcen; press Enter.

   New value for DIMCEN <2.5000>:

3. Type -3; press Enter.

   Command:

4. Type Dim; press Enter.

   Dim:

5. Type Center; press Enter.

   Select arc or circle:

6. Select the left circle.

   Dim:

7. Press Enter.

   Select the second circle:

8. Press Enter.

   See Figure 7-25.

9. Press the Escape key.

   Command:

Draw these lines, then use them as cutting edges to trim the circles.

Results

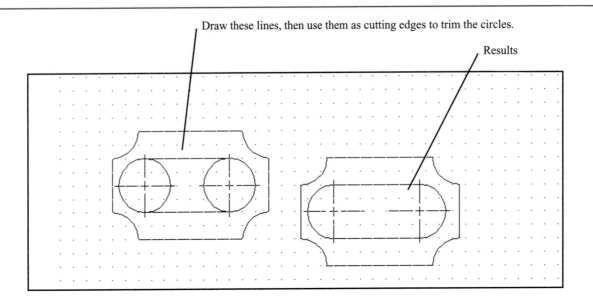

Figure 7-26

To create the internal slot

Create the slot by drawing lines between the vertical centerlines and trimming away excess lines. See Figures7-26.

1. Select the Line tool from the Draw toolbar.

    Command: _line From point:

2. Select the intersection between the vertical centerline and the top edge of the left circle.

    The intersection happens to be on a snap point in this example. If it were not, you could have used OSNAP Intersection to ensure accuracy.

    To point:

3. Select the intersection between the vertical centerline and the top edge of the right circle.

    To point:

4. Press Enter twice.

    Line From point:

5. Select the intersection between the vertical centerline and the bottom edge of the left circle.

    To point:

6. Select the intersection between the vertical centerline and the bottom edge of the other circle.

    See Figure 7-26.

    To point:

7. Press Enter.

    Command:

8. Select the Trim tool from the Modify toolbar.

    Select cutting edges:
    Select objects:

9. Select the two lines between the circles.

    <Select objects to trim>/Project/Edge/Undo:

10. Select the inside edges of the circles.

    <Select objects to trim>/Project/Edge/Undo:

11. Press Enter.

    Command:

To join the horizontal centerlines of the two circles to make one centerline

See Figure 7-27.

1. Select the Line tool from the Draw toolbar.

    Command: Line From point:

2. Activate the OSNAP menu by pressing the shift key and right mouse button simultaneously.
3. Select the Endpoint option.
4. Select the right endpoint of the left horizontal centerline.

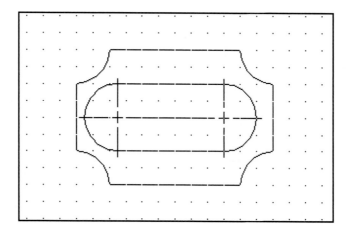

Final drawing

Figure 7-27

To point:

5. Again activate the OSNAP Endpoint option and select the left endpoint of the right horizontal centerline.

7-4 TUTORIAL T7-3

Draw the shape shown in Figure 7-28. Start a new drawing.

Toolbar/Commands used:

Units
Draw/Line
Modify/Trim
Modify/Rotate
Draw/Polygon
Modify/Fillet

To set up the drawing for fractional units.

1. Select the Format heading, then Units.

The Units Control dialog box will appear. See Figure 7-29.

2. Select Fractional units with a 1/8 precision, then OK.

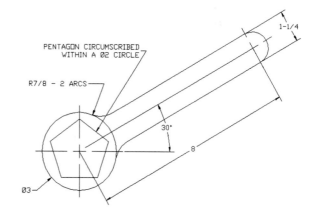

Figure 7-28

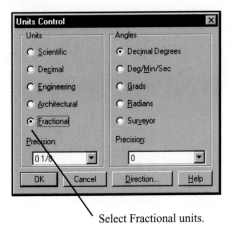

Select Fractional units.

Figure 7-29

3. Select the Format heading, then Drawing limits.

ON/OFF/<Lower left corner><0,0,>:

4. Press Enter.

Upper right corner <420,297>:

5. Type 12,9; press Enter.
6. Select the Tools heading, then Drawing Aids.

The Drawing Aids dialog box will appear. See Figure 7-30.

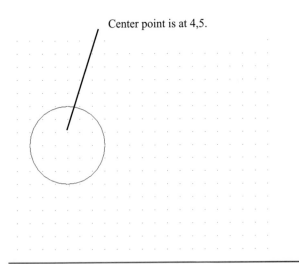

Center point is at 4,5.

Figure 7-31

Turn on the Grid and Snap.

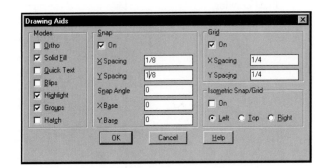

Figure 7-30

7. Turn Snap and Grid on and set the Snap spacing to 1/4 and the Grid spacing to 1/2.

To draw the large circle

See Figure 7-31.

1. Select the Circle tool from the Draw toolbar.

Command: _circle 3P/2P/TTR/<Center point>;

2. Select a center point (4,5).

Command: _circle 3P/2P/TTR/<Center point>;

Add the centerlines.

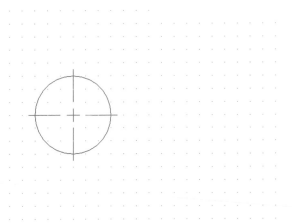

Figure 7-32

Draw the pentagon.

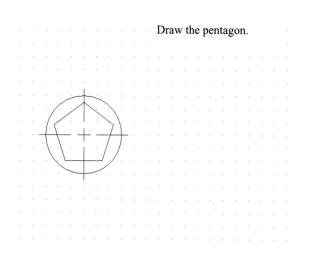

Figure 7-33

Offset the vertical centerline, then add the circle.

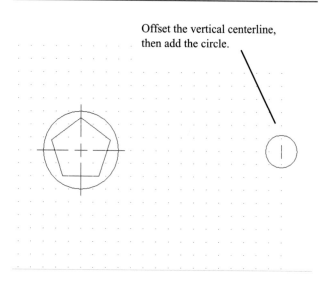

Figure 7-34

Diameter/<Radius>:

3. Type 1-1/2; press Enter.

Command:

To draw the circle's centerline

Draw centerlines for the circle using the DIMCEN command. See Figure 7-32.

1. Type Setvar; press Enter.

Variable name or ?:

2. Type Dimcen; press Enter.

New value for DIMCEN <1/8>:

3. Type -1/4; press Enter.

Command:

4. Type Dim; press Enter.

Dim:

5. Type Center; press Enter.

Select arc or circle:

6. Select the circle.

To draw the pentagon

Draw a pentagon using the POLYLINE command. See Figure 7-33.

1. Select the Polyline tool from the Draw toolbar.

Command: _polygon Number of sides <4>:

2. Type 5; press Enter.

Edge/<Center of polygon>:

3. Select the center point of the circle (4,5).

Inscribed in circle/Circumscribed about circle (I/C) <I>:

4. Type C; press Enter.

Radius of circle:

5. Type 1; press Enter.

Command:

To draw the handle

Locate the end of the figure using the OFFSET command, then draw the end shape using the CIRCLE command. See Figure 7-34.

1. Select the Offset tool from the Modify toolbar.

Command: _offset
Offset distance or Through <Through>:

2. Type 8; press Enter.

Select object to offset:

3. Select the vertical centerpoint line of the circle.

Side to offset:

4. Select the right side.

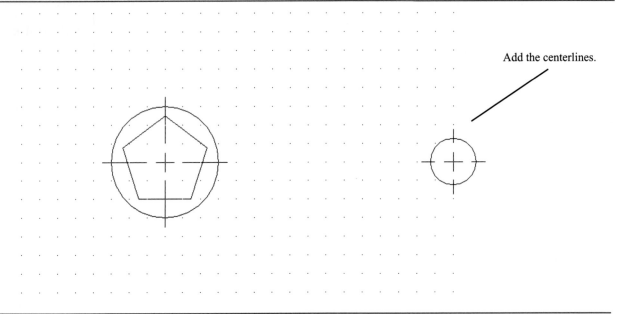

Add the centerlines.

Figure 7-35

Command:

5. Select the Circle Center Radius tool from the Draw toolbar.

   Command: _circle 3P/2P/TTR/<Center point>:

6. Select the center of the offset line segment.

   The line's center point is also a grid point so it can easily be selected. If it were not, you could use OSNAP Midpoint to select the point.

   Command: _circle 3P/2P/TTR/<Center point>: Diameter/<Radius><1-1/2>:

7. Type 5/8; press Enter.

   The 5/8 value is half the given 1 1/4 diameter value.

Draw the centerlines

   Add the centerlines to the small circle. See Figure 7-35.

1. Type Dim; press Enter.

   Dim:

2. Type Center; press Enter.

   Select arc or circle

3. Select the small circle.

To draw the rounded end of the handle

   Complete the extended portion of the shape using the LINE command, then use the TRIM command to remove excess lines. See Figure 7-36.

1. Turn the ORTHO command on and the SNAP command off.

   Ortho is turned on by pressing the F8 key or by double-clicking the word Ortho at the bottom of the screen.

2. Select the Line tool from the Draw toolbar.

   Command: _line From point:

3. Use Osnap and select the intersection point between the small circle's top edge and the vertical centerline.

   To point:

4. Drag a line from the start point to a point within the large circle.

   To point:

5. Double-click the Enter key and draw another line from the other intersection of the bottom edge of the circle and the vertical centerline.

6. Select the Trim tool from the Modify toolbar.

   Select cutting edges:
   Select objects:

7. Select the two horizontal lines just drawn.

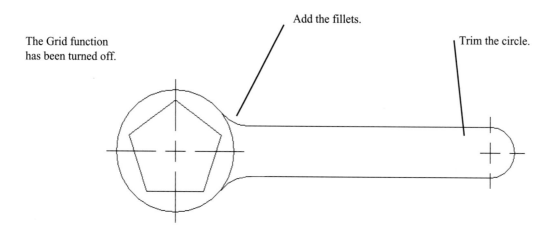

The Grid function has been turned off.

Add the fillets.

Trim the circle.

Figure 7-36

<Select objects to trim>/Project/Edge/Undo:

8. Select the left edge of the small circle.

   Command:

9. Select the Erase tool from the Modify toolbar and erase the far left horizontal line segment of the small circle's centerline.

To draw the fillets

Draw fillets between the horizontal lines and the large circle. Turn the ORTHO command off.

1. Select the Fillet tool from the Modify toolbar.

   (TRIM mode) Current fillet radius = 1/2 Polyline/Radius/Trim/<Select first object>:

2. Type R; press Enter.

   Enter fillet radius <1/2>:

3. Type 7/8; press Enter.

   Command:

4. Press Enter.

   -TRIM mode) Current fillet radius = 7/8 Polyline/Radius/Trim/<Select first object>:

5. Select the top horizontal line.

   Select second object:

6. Select the large circle.

   Command:

7. Press Enter.

   (TRIM mode) Current fillet radius = 7/8 Polyline/Radius/Trim/<Select first object>:

8. Select the bottom horizontal line.

   Select second object:

9. Select the large circle.

To Rotate the handle into position

Rotate the extended portion of the shape 30 degrees.

1. Select the Rotate tool from the Modify toolbar.

   Select objects:

2. Select the two fillets, the two horizontal lines, the semicircle, and all the centerline segments.

   (Turn Snap off, if needed, to make it easier to select the entities.)

   Select objects:

3. Press Enter.

   Base point:

4. Select the large circle's center point.

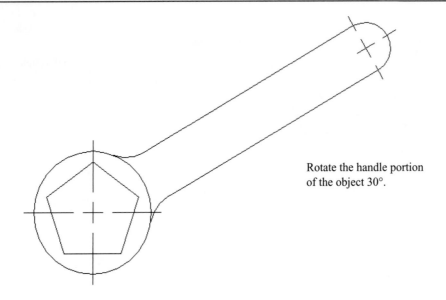

Rotate the handle portion
of the object 30°.

Figure 7-37

The finished drawing

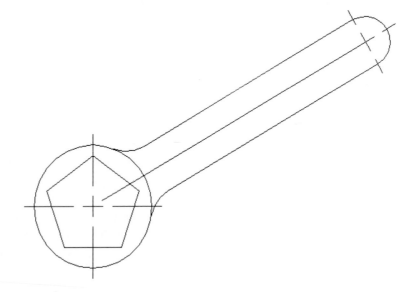

Figure 7-38

<Rotation angle>/Reference:

Note that the system is now in Drag mode and the extended portion of the shape can be rotated using the cursor. See Figure 7-37.

5. Type 30; press Enter.

Command:

To draw a slanted centerline

Draw a slanted centerline through the extended portion of the shape. See Figure 7-38.

1. Select the Line tool from the Draw toolbar.

   Command: _Line From point:

2. Press the shift and right mouse button simultaneously and select the OSNAP Endpoint option.
3. Select the end of the formerly horizontal line segment of the small circle's center point.

   To point:

4. Select the large circle's center point.

   To point:

5. Press Enter.

   Command:

6. Use the BREAK command to create the gaps on the slanted centerline.
7. Save the shape if desired.

7-5 EXERCISE PROBLEMS

Redraw the following shapes. Do not include the dimensions.

EX7-1 MILLIMETERS

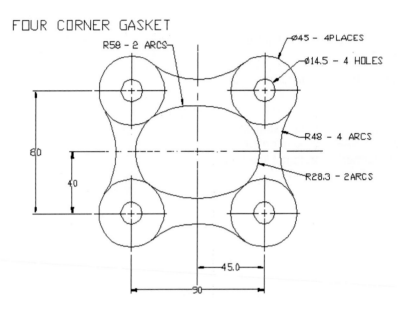

EX7-2 MILLIMETERS

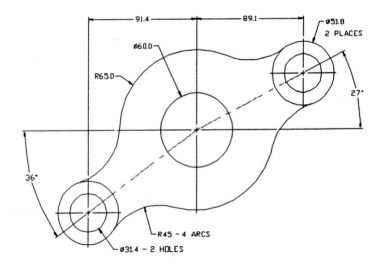

EX7-3 INCHES

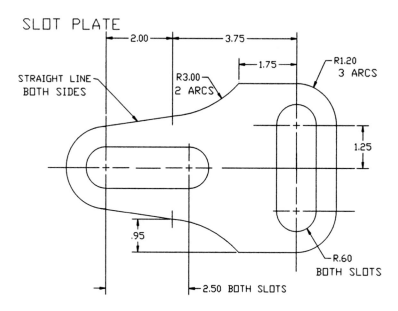

SLOT PLATE

EX7-4 INCHES

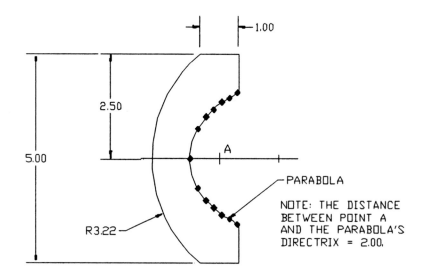

PARABOLA

NOTE: THE DISTANCE
BETWEEN POINT A
AND THE PARABOLA'S
DIRECTRIX = 2.00,

EX7-5 MILLIMETERS

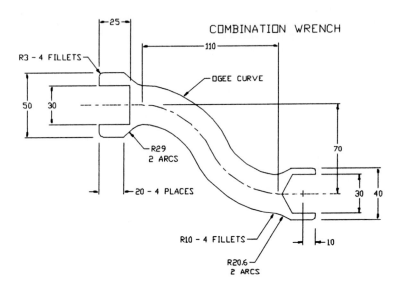

EX7-6 MILLIMETERS

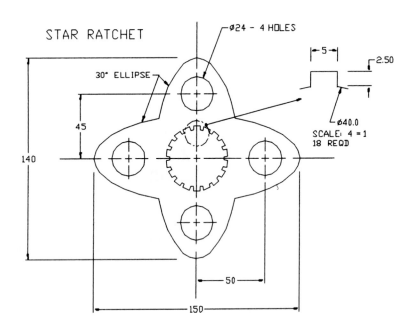

EX7-7 MILLIMETERS

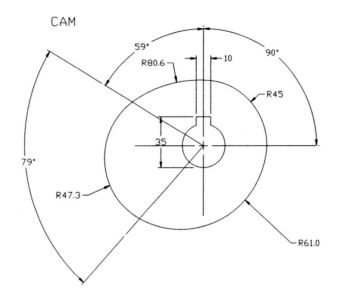

EX7-8 MILLIMETERS

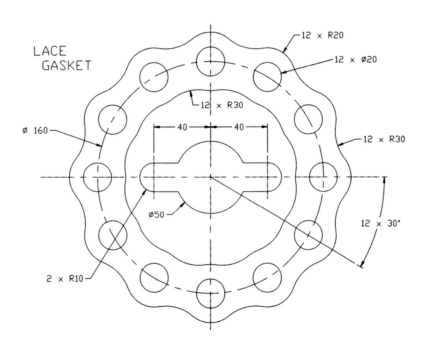

EX7-9 INCHES

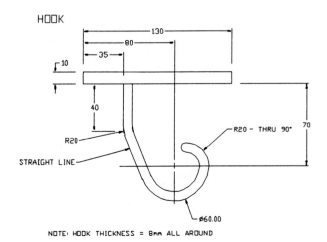

HOOK

NOTE: HOOK THICKNESS = 8mm ALL AROUND

EX7-10 MILLIMETERS

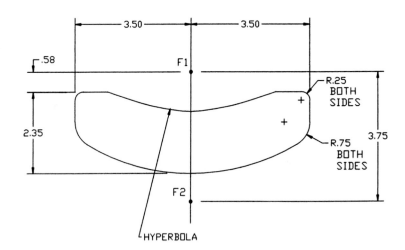

EX7-11 MILLIMETERS

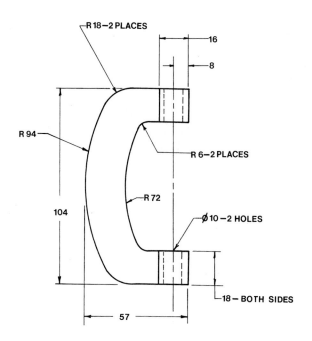

EX7-13 MILLIMETERS

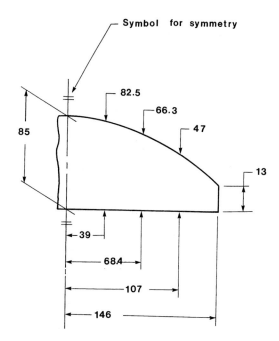

EX7-12 INCHES

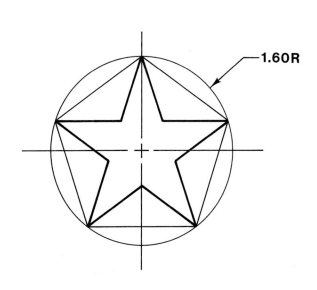

EX7-14 MILLIMETERS

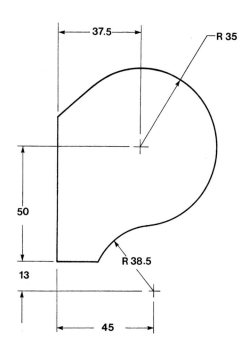

EX7-15 MILLIMETERS

R 48

R 102

R 13

R 4

38   51

EX7-16 INCHES

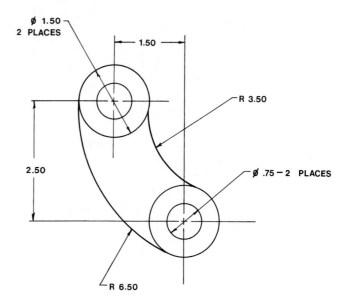

Ø 1.50
2 PLACES

1.50

R 3.50

2.50

Ø .75 – 2 PLACES

R 6.50

EX7-17 MILLIMETERS

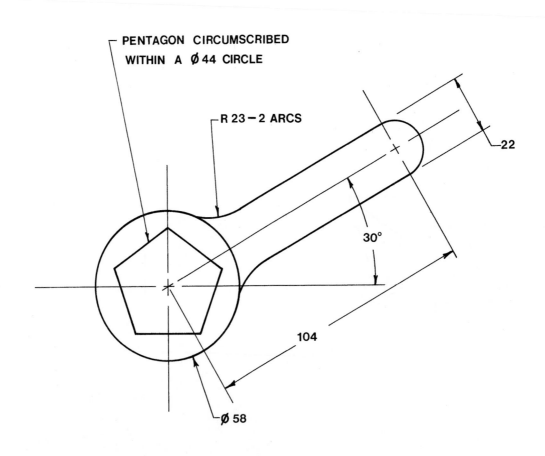

PENTAGON CIRCUMSCRIBED
WITHIN A Ø 44 CIRCLE

R 23 – 2 ARCS

22

30°

104

Ø 58

EX7-18 INCHES

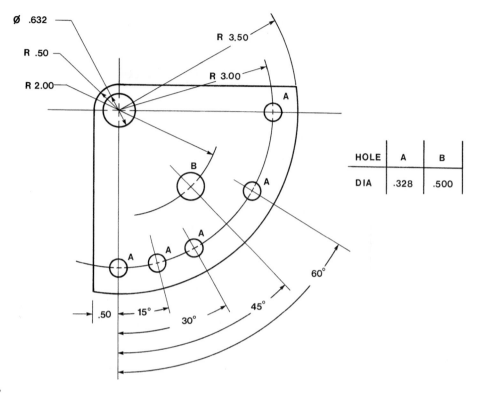

HOLE	A	B
DIA	.328	.500

EX7-19 INCHES

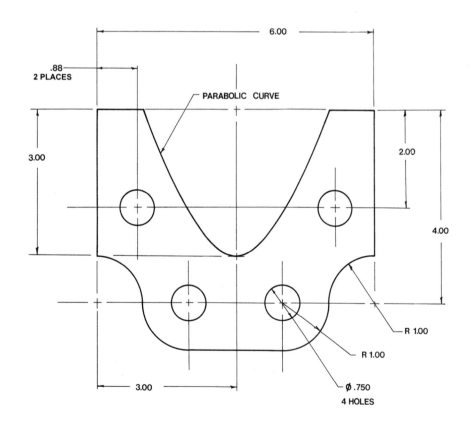

EX7-20 MILLIMETERS

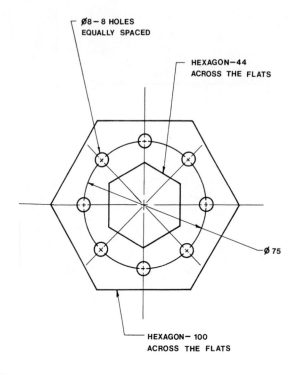

EX7-21 MILLIMETERS

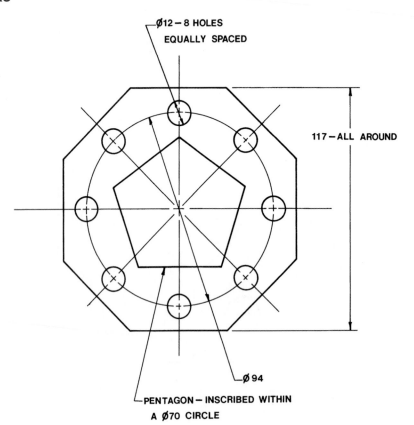

EX7-22 MILLIMETERS

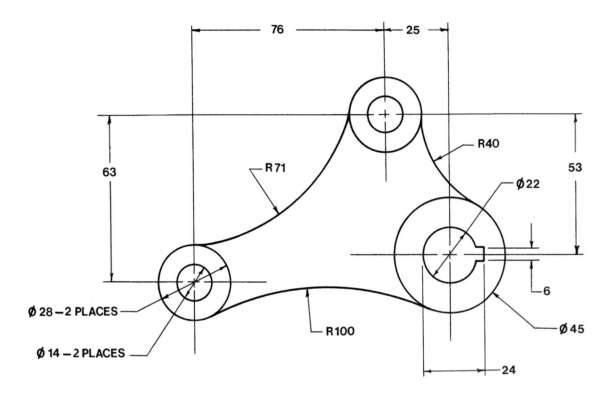

EX7-23 MILLIMETERS

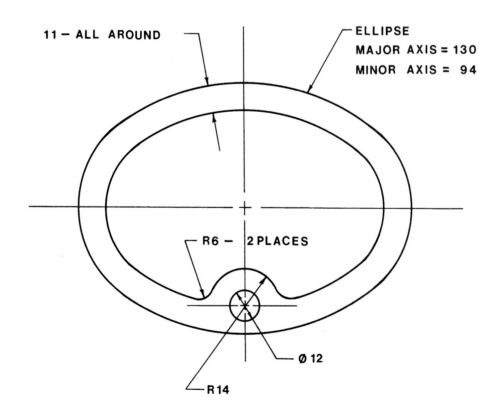

EX7-24 MILLIMETERS

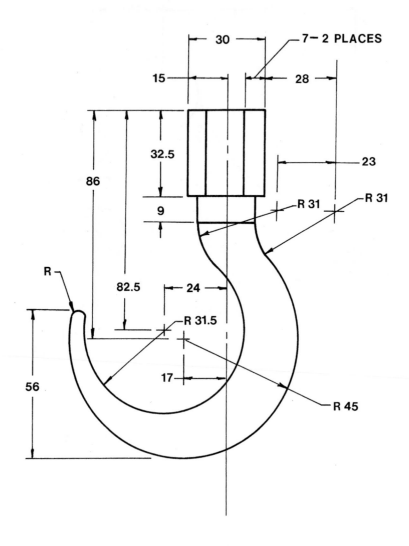

CHAPTER 8

The Dimensioning Toolbar

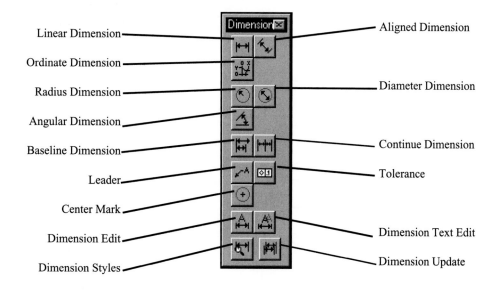

Linear Dimension

Ordinate Dimension

Radius Dimension

Angular Dimension

Baseline Dimension

Leader

Center Mark

Dimension Edit

Dimension Styles

Aligned Dimension

Diameter Dimension

Continue Dimension

Tolerance

Dimension Text Edit

Dimension Update

Figure 8-1

8-1 INTRODUCTION

This chapter explains the Dimensioning toolbar. See Figure 8-1. The chapter first explains dimensioning terminology and conventions, then presents an explanation of each tool within the Dimensioning toolbar. The chapter also demonstrates how dimensions are applied to drawings and gives examples of standard drawing conventions and practices.

8-2 TERMINOLOGY AND CONVENTIONS

Some common terms

See Figure 8-2.

Dimension lines: Mechanical drawings; lines between extension lines that end with arrowheads and include a numerical dimensional value located within the line.

Architectural drawings; Lines between extension lines that end with tick marks and include a numerical dimensional value above the line.

Extension lines: Lines that extend away from an object and allow dimensions to be located off the surface of an object.

Leader lines: Lines drawn at an angle, not horizontal or vertical, that are used to dimension specific shapes such as holes. The start point of a leader line includes an arrowhead. Numerical values are drawn at the end opposite the arrowhead.

Linear dimensions: Dimensions that define the straight line distance between two points.

Angular dimensions: Dimensions that define the angular value, measured in degrees, between two straight lines.

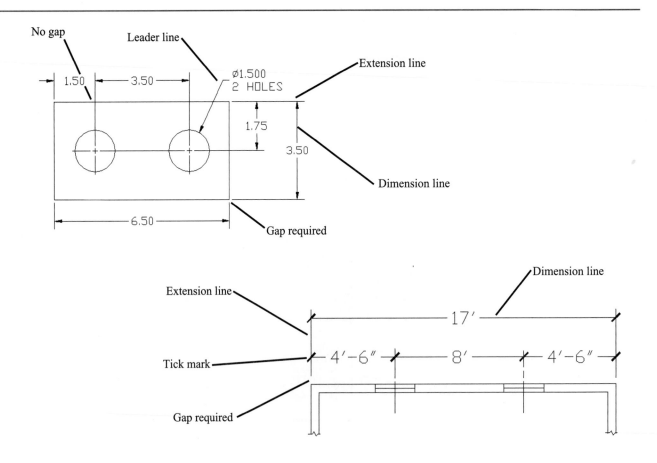

Figure 8-2

Some dimensioning conventions

See Figure 8-3.

1. Dimension lines should be drawn evenly spaced, that is, the distance between dimension lines is uniform. A general rule of thumb is to locate dimension lines about 1/2 inch or 15 millimeters apart.
2. There should a noticeable gap between the edge of a part and the beginning of an extension line. This serves as a visual break between the object and the extension line. The visual difference between the line types can be furthered by using different colors for the two types of lines.
3. Leader lines are used to define the size of holes and should be positioned so that the arrowhead points at the center of the hole.
4. Centerlines may be used as extension lines. No gap is used when a centerline is extended beyond the edge lines of an object.
5. Align dimension lines whenever possible to give the drawing a neat, organized appearance.

Some common errors

See Figure 8-4.

1. Avoid crossing extension lines. Place longer dimensions farther away from the object than shorter dimensions.
2. Do not locate dimensions within cutouts, always use extension lines.
3. Do not locate any dimension close to the object. Dimension lines should be at least 1/2 inch or 15 millimeters from the edge of the object.
4. Avoid long extension lines. Locate dimensions in the same general area as the feature being defined.

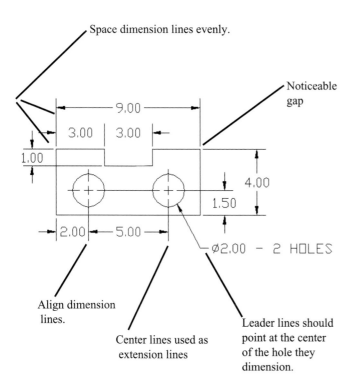

Figure 8-3

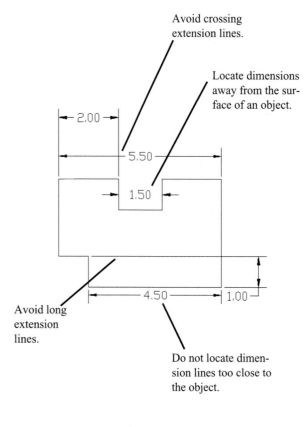

Figure 8-4

8-3 LINEAR DIMENSION

The LINEAR DIMENSION command is used to create either horizontal or vertical dimensions.

To create a horizontal dimension by selecting extension lines

See Figure 8-5.

1. Select the Linear Dimension tool from the Dimensioning toolbar.

   Command: _dimlinear
   First extension line origin or RETURN to select:

2. Select the starting point for the first extension line.

   Second extension line origin:

3. Select the starting point for the second extension line.

   Dimension line location (MText/ Text/ Angle /Horizontal/ Vertical/ Rotated):

4. Locate the dimension line location by moving the cross hairs.

5. Press the left mouse button when the desired dimension line location has been selected.

The dimensional value locations shown in Figure 8-5 are the default settings locations. The location and style may be changed using the Dimension Styles command discussed in Section 8-4.

To create vertical dimensions

The vertical dimension shown in Figure 8-5 was created using the same procedure demonstrated for the horizontal dimension, except different extension line origin points were selected. AutoCAD will automatically switch from horizontal to vertical dimension lines as you move the cursor around the object.

If there is confusion between horizontal and vertical lines when adding dimensions, that is, you don't seem to be able to generate a vertical line, type V in response to the following prompt.

Dimension line location (MText/ Text/ Angle /Horizontal/ Vertical/ Rotated):

Respond by typing V and then pressing Enter. The system will now draw vertical dimension lines.

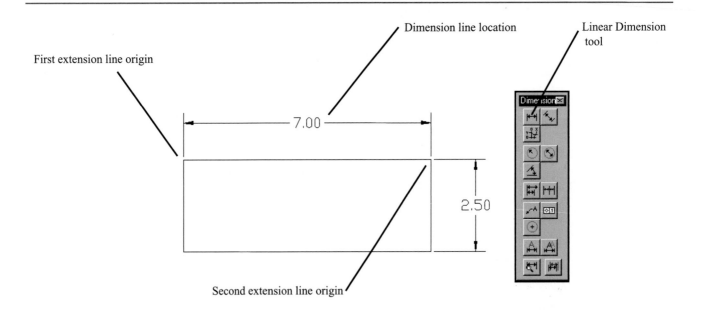

Dimension line location

Linear Dimension tool

First extension line origin

7.00

2.50

Second extension line origin

Figure 8-5

To create a horizontal dimension by select the distance to be dimensioned

See Figure 8-6.

1. Select the Linear Dimension tool from the Dimensioning tool.

   Command: _dimlinear
   First extension line origin or RETURN to select:

2. Press the right mouse button.

   Select object to Dimension:

This option allows you to select the distance to be dimensioned directly. The option applies only to horizontal and vertical lengths. Aligned dimensions, although linear, are created using the Aligned Dimension tool.

To change the default dimension text

AutoCAD will automatically create a text value for a given linear distance. A different value or additional information may be added as follows. See Figure 8-7.

1. Select the Linear Dimension tool from the Dimensioning toolbar.

   Command: _dimlinear
   First extension line origin or RETURN to select:

2. Select the starting point for the first extension line.

   Second extension line origin:

3. Select the starting point for the second extension line.

   Dimension line location (MText/ Text/ Angle/ Horizontal/ Vertical/ Rotated):

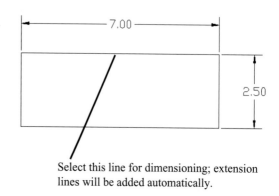

Select this line for dimensioning; extension lines will be added automatically.

Figure 8-6

4. Type M; press Enter.

The Multiline Text Editor dialog box will appear. See Figure 8-8. The greater than, less than symbols (<>) that appears in the open area at the top of the dialog box represents the default text. To remove the text, locate the cursor to the right of the symbols and backspace to remove the symbols, then type in the new text. In the example shown, the default value of 7.00 was replaced with a value of 7.000.

Information can be added before or after the default text by locating the cursor in the appropriate place and typing the additional information. For example, locating the cursor to the right of the <> symbols and typing -2 PLACES would yield a final value of 7.00-2 PLACES on the drawing.

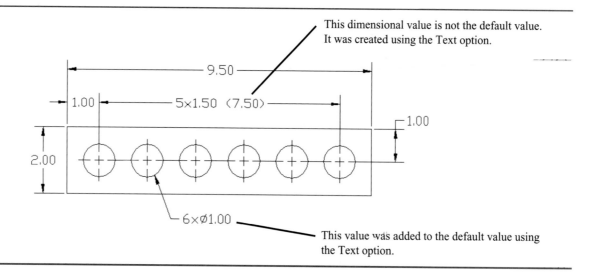

This dimensional value is not the default value. It was created using the Text option.

This value was added to the default value using the Text option.

Figure 8-7

The Multiline Text Editor dialog box can be used to edit the dimension text as it was used to edit drawing screen text - that was explained in Chapter 4-. Figure 8-9 shows the font pull-down menu and Figure 8-10 shows the symbol pull-down menu. The symbols can be selected from the listing and added directly to the dimension text.

The Properties option is used change the text justification or style. See Figure 8-11. The Find/Replace can be used to change dimension text as with screen text. It is very useful when making extensive dimension text changes.

Represents the default text value

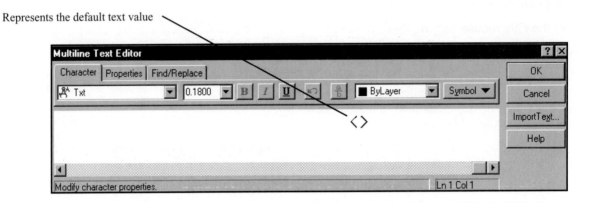

Figure 8-8

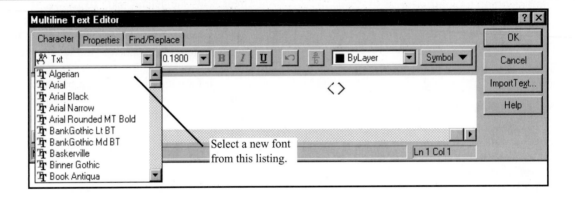

Figure 8-9

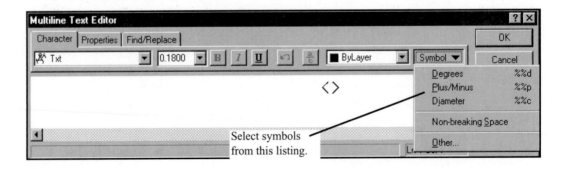

Figure 8-10

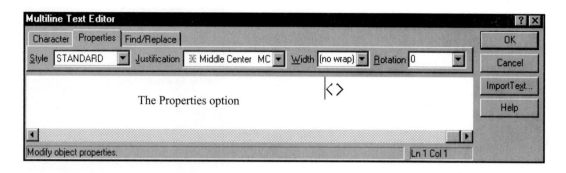

Figure 8-11

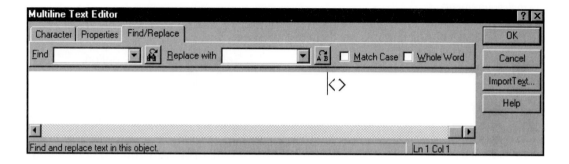

Figure 8-12

To use the Text option

The Text option is different from the Mtext option in that it is used to change a single line of text. When using the Linear dimension command, AutoCAD will automatically calculate the length of the specified distance and include that value as the text value for the dimension. If this value is not the value desired, the Text option may be used to enter a different value. For example, if the calculated value were 2.0000 and you only wanted two decimal places, 2.00, the text option could be used to type in the new value. See Figure 8-13.

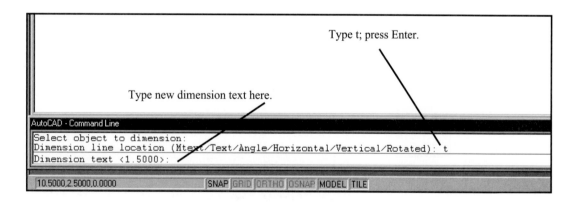

Figure 8-13

8-4 DIMENSION STYLES

The DIMENSION STYLES command opens a group of dialog boxes that are used to control the appearance of dimensions. Figure 8-14 shows the Dimension Styles tool on the Dimensioning toolbar

There is a great variety of styles used to create technical drawings. The style difference may be the result of different drawing conventions. For example, architects locate dimensions above the dimensions lines, and mechanical engineers locate the dimensions within the dimension lines. AutoCAD works in decimal units for either millimeters or inches, so parameters set for inches would not be usable for millimeter drawings. The DIMENSION STYLES command allows you to conveniently choose and set dimension parameters that suit your particular drawing requirements.

Figure 8-15 shows the Dimension Styles dialog box. The Geometry, Format, and Annotation options are used to change the default style settings. The changes can be saved as a new style, then recalled for future use.

To dimension using the mechanical format and decimal inches

Dimension the shape shown in Figure 8-16. The units for the object shown in Figure 8-16 are decimal inches. The Grid is set at 0.50 spacing.

1. Select the Dimension Styles tool on the Dimensioning toolbar.

The Dimension Styles dialog box will appear.

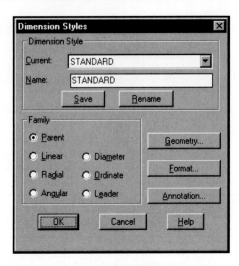

Figure 8-15

2. Select the Geometry option.

The Geometry dialog box will appear. See Figure 8-17.

3. Change the Overall Scale from 1.0000 to 2.0000 by locating the cursor arrow in the value box and pressing the left mouse button, then backspacing to remove the number 1 and typing in the number 2.

All the dimensioning parameters will now be increased by a factor of two.

4. Select the OK box.

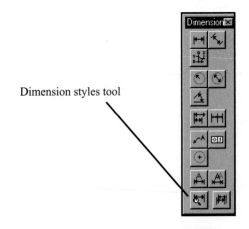

Dimension styles tool

Figure 8-14

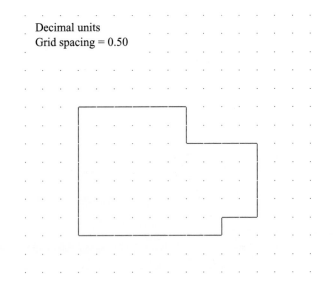

Decimal units
Grid spacing = 0.50

Figure 8-16

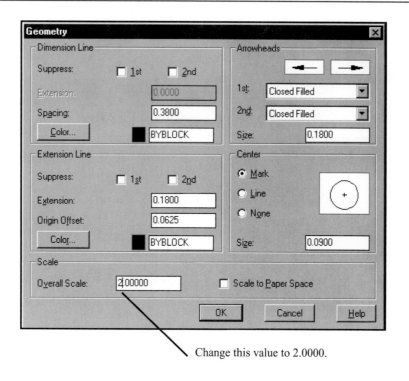

Change this value to 2.0000.

Figure 8-17

The Dimension Styles dialog box will reappear.

5. Select the Format option.

The Format dialog box will appear. See Figure 8-18.

6. Select the arrow box in the Vertical Justification box, then select the Centered option.
7. Select the OK box.

The Dimension Styles options will reappear.

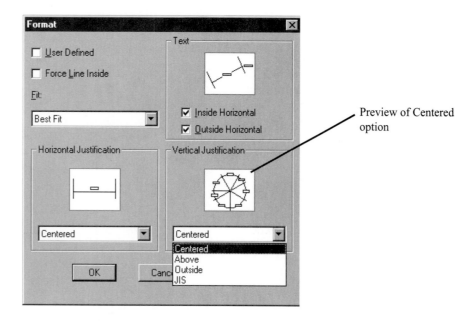

Preview of Centered option

Figure 8-18

Create a new style named MECHINCH.

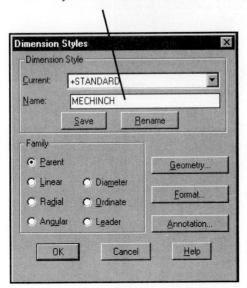

Figure 8-19

8. Locate the cursor arrow in the box to the right of the word ANSI in the Name box and press the left mouse button.

A flashing cursor will appear in the box.

9. Backspace out ANSI and type in MECHINCH.

See Figure 8-19. The name stands for a mechanical drawing done in decimal inches.

10. Select the Save box.

11. Select OK.

The dimensional parameters are now set. The Linear Dimension command was used to add the dimensions as shown in Figure 8-20.

To dimension using the mechanical format and millimeters

Dimension the shape shown in Figure 8-21. The units for the shape are millimeters, and the Grid spacing equals 10 millimeters.

1. Create a new drawing and select the metric option in the Select Default Setting box.
2. Select the Dimension Styles tool.

The Dimension Styles dialog box will appear.

3. Select the Geometry option.

The Geometry dialog box will appear.

4. Change the Overall Scale factor from 1.000 to 2.000.

See Figure 8-22. Note the difference between this metric Geometry dialog box and the inch dialog box shown in Figure 8-17. The size values are now calibrated for metric units.

5. Select the OK box.

The Dimension Styles dialog box will appear.

6. Change Dimension Style Name to MECHMM.

See Figure 8-23.

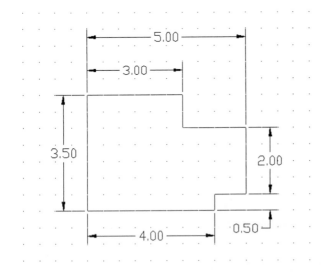

Figure 8-20

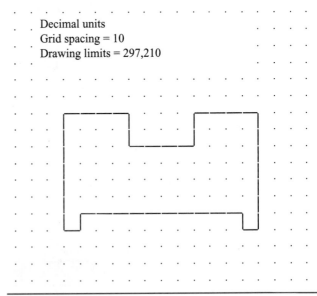

Figure 8-21

Change the Overall Scale to 2.0000.

The size values are now calibrated for metric units.

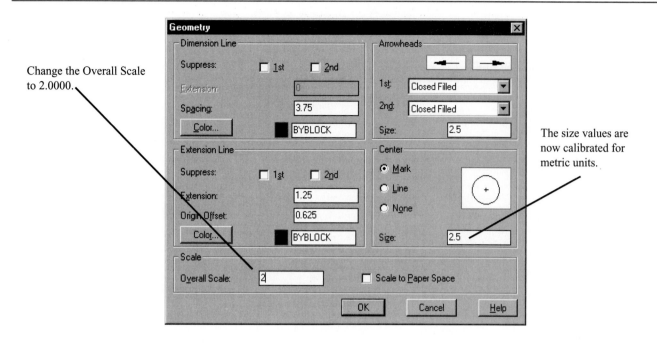

Figure 8-22

7. Select Save.
8. Select OK.

Dual dimension option

Figure 8-24 shows an example of dual dimensioning. Dual dimensions are used to include both inch and mil-limeter values for a dimension. When industry first started to shift from exclusively inch values to metric values, dual dimensions were used to convert many existing drawings to millimeter drawings. It is now considered better to dimension a distance using either inches or millimeters, not both. The American National Standard publication ANSI Y14.5M - 1982 states "Dual dimensioning is no longer featured in this Standard."

Enter new style name here.

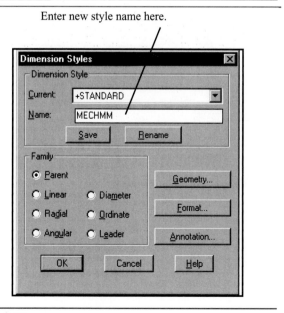

Figure 8-23

Dual dimensions

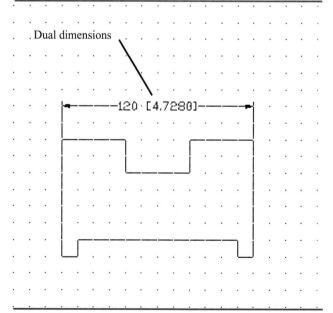

Figure 8-24

To turn on Alternate Units

Dual dimension option;
click here for other options.

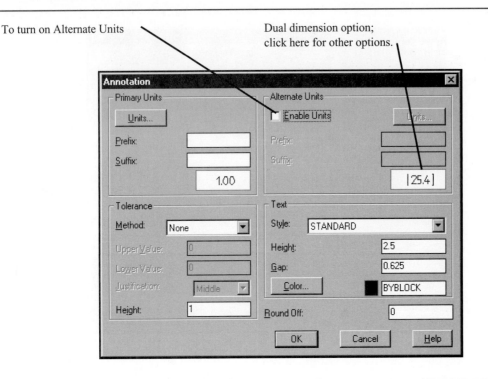

Figure 8-25

Dual dimensions can be created automatically by using the Alternative Units box located on the Annotation dialog box. See Figure 8-26. The Annotation dialog box is access edfrom the Dimension Styles dialog box.

Figure 8-26 shows the shape presented in Figure 8-21 dimensioned using millimeters.

To dimension using architectural units

Figure 8-27 shows a partial floor plan dimensioned using architectural units. The drawing sheet size has been set to 24″ x 18″ and the Grid spacing equals 1/2″. The drawing was created at a scale of 1/4″ = 1′.

1. Select the Format heading, then the Units command.

The Units Control dialog box will appear. See Figure 8-28.

2. Select the Architectural units option, and set the precision for 1/8; press OK.
3. Select the Dimension Styles tool.

The Dimension Styles dialog box will appear. See Figure 8-31. Note that the values are calibrated in architectural units.

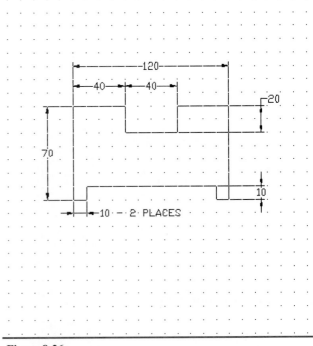

Figure 8-26

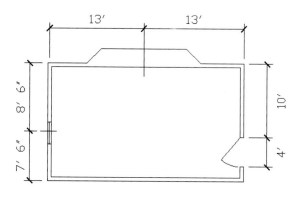

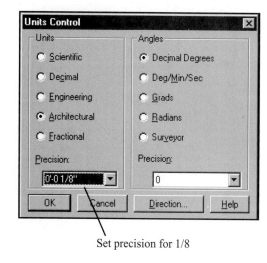

Set precision for 1/8

Figure 8-27

Figure 8-28

4. Change the Overall Scale factor to 3.0000.

The Overall Scale factor needed will vary according to the drawing limits selected. If your dimensions appear too small or too large, change the scale factor until a correct appearance is achieved.

To change the arrowheads to tick marks

1. Select the arrow box to the right of the heading Closed Filled within the Arrowheads box.

A listing of arrowhead options will cascade.

2. Select Oblique.

The Geometry dialog box should look like Figure 8-29. The Dimension Styles dialog box will reappear.

To save the new Dimension Style

1. Change the Dimension Style Name to ARCH.
2. Select Save.

New Overall Scale

The Oblique option will produce tick marks rather than arrowheads.

Architectural units

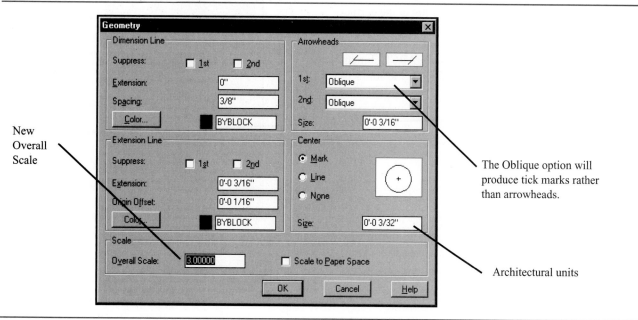

Figure 8-29

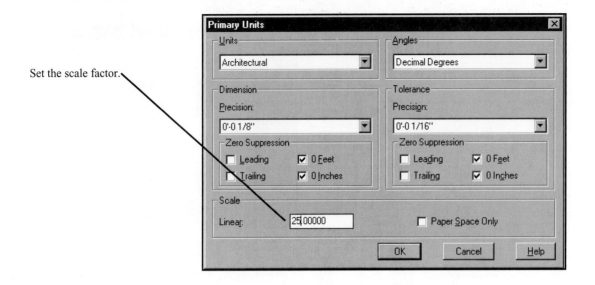

Set the scale factor.

Figure 8-30

To set the scale factor for the Linear units

1. Select ANNOTATION.

   The Annotation dialog box will appear.

2. Select Units.

   The Primary Units dialog box will appear. See Figure 8-30.

3. Change the Linear value in the Scale box from 1.0000 to 25.0000.

   The scale factor will vary according to the scale and size of the drawing. If the dimensions are not at the correct scale, change the Linear scale factor until the correct scale is achieved.

   If you have not previously created an ARCH dimension style, use the Units option to change the units to Architectural.

4. Select OK.

   The Annotation dialog box will reappear.

5. Select OK.

   The Dimension Styles dialog box will appear.

6. Select Save.

   Figure 8-31 shows the Dimension Styles dialog box. The shape can now be dimensioned using the Linear Dimension command.

Save the settings for architectural drawings.

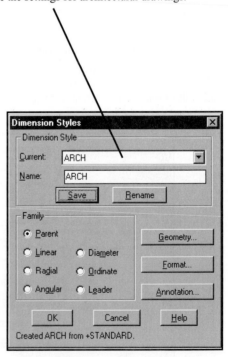

Figure 8-31

STANDARD TOLERANCES

$$X = \pm 1$$
$$X.X = \pm 0.1$$
$$X.XX = \pm 0.01$$
$$X.XXX = \pm 0.001$$
$$X.XXXX = \pm 0.0005$$

$$X^{\circ} = \pm 0.1^{\circ}$$

Figure 8-32

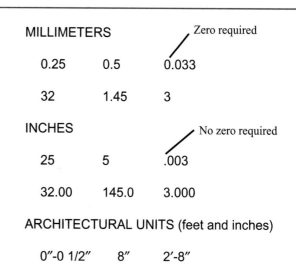

MILLIMETERS		Zero required
0.25	0.5	0.033
32	1.45	3
INCHES		No zero required
25	5	.003
32.00	145.0	3.000
ARCHITECTURAL UNITS (feet and inches)		
0″-0 1/2″	8″	2′-8″

Figure 8-34

8-5 UNITS

It is important to understand that dimensional values are not the same as mathematical units. Dimensional values are manufacturing instructions and always include a tolerance, even if the tolerance value is not stated. Manufacturers use a predefined set of standard dimensions that are applied to any dimensional value that does not include a written tolerance. Standard tolerance values differ from organization to organization. Figure 8-32 shows a chart of standard tolerances.

In Figure 8-33 a distance is dimensioned twice: once as 5.50 and a second time as 5.5000. Mathematically these two values are equal but they are not the same manufacturing instruction. The 5.50 value could, for example, have a standard tolerance of +/- .01, whereas the 5.5000 value could have a standard tolerance of +/- .0005. A tolerance of +/- .0005 is more difficult and, therefore, more expensive to manufacture than a tolerance of +/- .01.

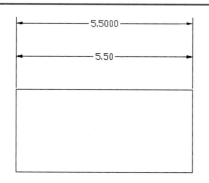

Figure 8-33

Figure 8-34 shows examples of units expressed in millimeters, decimal inches, and architectural units. A zero is not required to the left of the decimal point for decimal inch values less than one. Millimeter values do not require zeros to the right of the decimal point. Architectural units should always include the feet (′) and inch (″) symbols. Millimeter and decimal inch values never include symbols; the units will be defined in the title block of the drawing.

To prevent a 0 from appearing to the left of the decimal point

1. Select the Dimension Styles tool.

   The Dimensions Styles dialog box will appear.

2. Select Annotation.

   The Annotation dialog box will appear.

3. Select Units.

   The Primary Units dialog box will appear. See Figure 8-35.

4. Click the box to the left of the word Leading within the Zero Suppression box.

   An X will appear in the box indicating that the function is on.

5. Select the OK boxes to return to the drawing.

   Save the change if desired. You can now dimension using any of the dimension commands; no zeros will appear to the left of the decimal point. Figure 8-35 shows the results.

Turn the Leading option on to prevent zeros to the left of the decimal point.

Turn the Trailing option on to prevent zeros to the right of the decimal point.

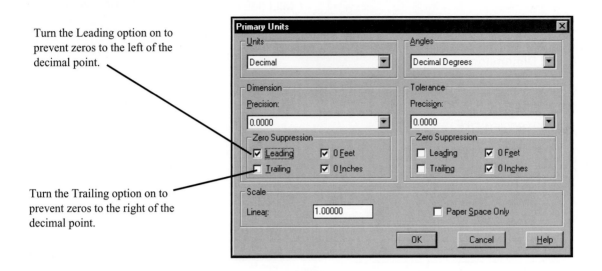

Figure 8-35

To prevent a 0 from appearing to the right of the decimal point

1. Select the Dimension Styles tool.

    The Dimensions Styles dialog box will appear.

2. Select Annotation.

    The Annotation dialog box will appear.

3. Select Units.

    The Primary Units dialog box will appear.

4. Click the box to the left of the word Trailing within the Zero Suppression box.

    An X will appear in the box indicating that the function is on.

5. Select the OK boxes to return to the drawing.

    Save the changes if desired. You can now dimension using any of the dimension commands; no zeros will appear to the right of the decimal point. Figure 8-36 shows the results.

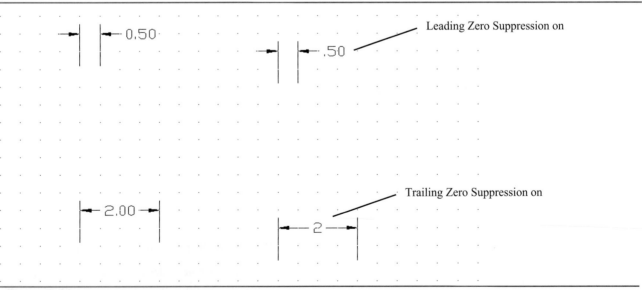

Figure 8-36

8-6 ALIGNED DIMENSIONS

See Figure 8-37.

To create an Aligned Dimensions

1. Select the Aligned Dimension tool.

   Command: _dimaligned
   First extension line origin or RETURN to select:

2. Select the first extension line origin point.

   Second extension line origin:

3. Select the second extension line origin point.

   Dimension line location (MText/Text/Angle):

4. Select the location for the dimension line.

The RETURN option

1. Select the Aligned Dimension tool.

   Command: _dimaligned
   First extension line origin or Return to select:

2. Press the Return key.

   Select object to dimension:

3. Select the line.

   Dimension line location (MTExt/Text/Angle):

4. Select the dimension line location.

A response of M to the last prompt line will activate the Multiline text option. The Multiline Text Editor dialog box will appear. The Text option can be used to replace or supplement the default text generated by AutoCAD. The Multiline Text option is discussed in Section 8-3.

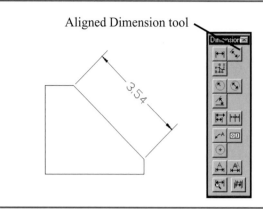

Aligned Dimension tool

Figure 8-38

A response of A to the prompt will activate the angle option. The angle option allows you to change the angle of the text within the dimension line. See Figure 8-38. The default angle value is 0 degrees or horizontal. The example shown in Figure 8-38 used an angle of -45 degrees. The prompt responses are as follows.

Dimension line location (Text/Angle): A (ENTER)
Enter text angle: -45 (ENTER)

8-7 RADIUS AND DIAMETER DIMENSIONS

Figure 8-39 shows an object that includes both arcs and circles. The general rule is to dimension arcs using a radius dimension and circles using diameter dimensions. This convention is consistent with the tooling required to produce the feature shape. Any arc greater than 180 degrees is considered a circle and is dimensioned using a diameter.

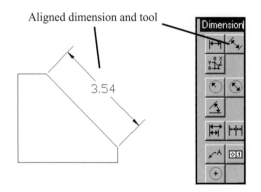

Aligned dimension and tool

Figure 8-37

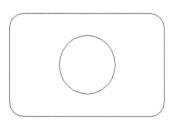

Figure 8-39

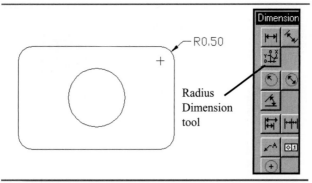

Figure 8-40

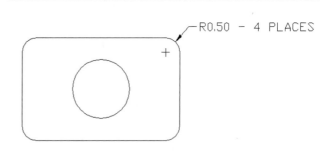

Figure 8-42

To create a Radius dimension

1. Select the Radius Dimension tool.

   Command: _dimradius
   Select arc or circle:

2. Select the arc to be dimensioned.

   Dimension line location (Mtext/Text/Angle)

3. Position the radius dimension so that its leader line is not horizontal or vertical.

Figure 8-40 shows the resulting dimension. The dimension text and the angle of the text can be altered using the (MText/Text/Angle) options in the last prompt. In the example shown, it would be better to add the words 4 PLACES to the radius dimension than to include four radius dimensions.

To alter the default dimensions

1. Select the Radius Dimension tool.

   Command: _dimradius
   Select arc or circle:

2. Select the arc to be dimensioned.

   Dimension line location (MText/Text/Angle)

3. Type M; press Enter.

   The Edit MText dialog box will appear.

4. Locate the flashing cursor just to the right of the <> symbols and type -4 PLACES.

   See Figure 8-41.

5. Select OK.

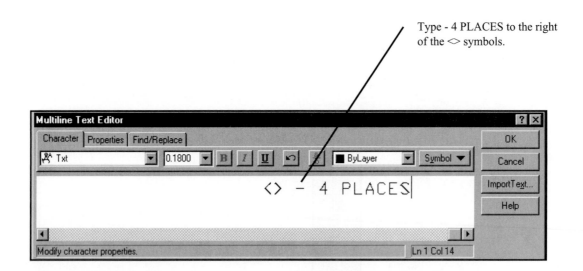

Figure 8-41

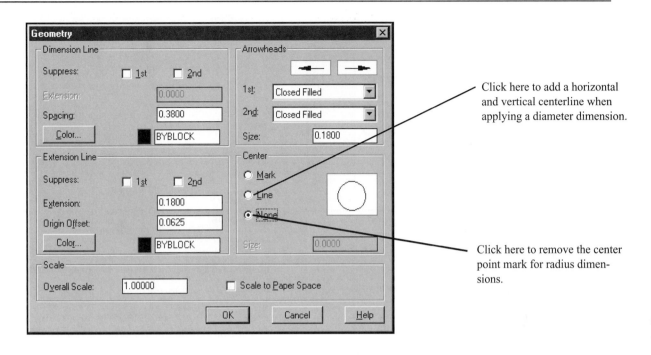

Figure 8-43

Figure 8-42 shows the resulting dimension. The RADIUS DIMENSION command will automatically include a center point with the dimension. The center point can be excluded from the dimension as follows.

To remove the center mark from a radius dimension

1. Select the Dimension Styles tool.

   The Dimension Styles dialog box will appear.

2. Select Geometry.

   The Geometry dialog box will appear.

3. Click the radio button to the left of the word None within the Center box.

   A solid circle will appear in the button, indicating that it is on. The center mark is the circle shown in the preview box to the right of the word None. See Figure 8-43.

4. Select the OK box to return to the drawing.

   You will have to redimension the arc, including the text alteration. Figure 8-44 shows the results.

To create a diameter dimension

Circles require three dimensions: a diameter value plus two linear dimensions used to locate the circle's center point. AutoCAD can be configured to automatically add horizontal and vertical centerlines as follows.

1. Select the Dimension Styles tool.

   The Dimension Styles dialog box will appear.

2. Select Geometry.

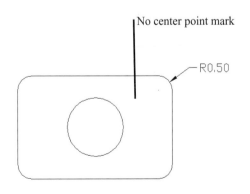

Figure 8-44

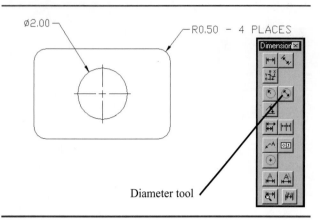

Figure 8-45

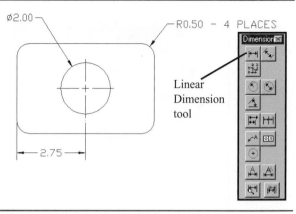

Figure 8-46

The Geometry dialog box will appear.

3. Click the radio button to the left of the word Line within the Center box.

   See Figure 8-43.

4. Select the OK box to return to the drawing.
5. Select the Diameter Dimension tool.

   Command: _dimdiameter
   Select arc or circle:

6. Select the circle.

   Dimension line location (MText/Text/Angle):

7. Locate the dimension away from the object so that the leader line is neither horizontal nor vertical.

   Figure 8-45 shows the results.

To add linear dimensions to given centerlines

1. Select the Linear Dimension tool.

   Command: _dimlinear
   First extension line origin or RETURN to select:

2. Press the Shift key and the right mouse button simultaneously (or the middle button on a three-button mouse) to access the Osnap menu.
3. Select Endpoint from the OSNAP menu.

   First extension line origin or RETURN to select:
   _endp of

4. Select the lower endpoint of the circle's vertical centerline.

   Second extension line origin:

5. Use the Shift key/right button option to access the OSNAP menu, then select the Endpoint option.

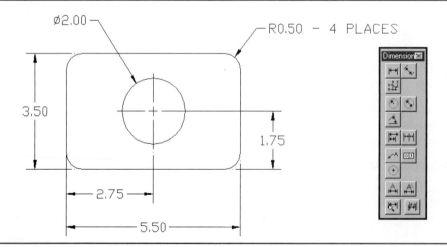

Figure 8-47

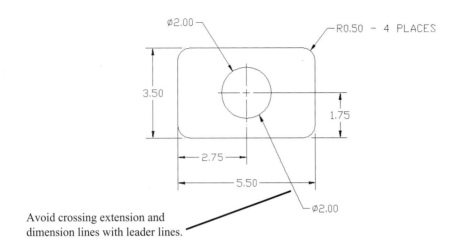

Avoid crossing extension and
dimension lines with leader lines.

Figure 8-48

6. Select the endpoint of the vertical edge line (the end point that joins with the corner arc).

Figure 8-46 shows the results.

7. Repeat the above procedure to add the vertical dimension needed to locate the circle's center point.
8. Add the overall dimensions using the Linear Dimension command.

Figure 8-47 shows the results. Radius and diameter dimensions are usually added to a drawing after the linear dimensions because they are less restricted in their locations. Linear dimensions are located close to the distance they are defining, whereas radius and diameter dimensions can be located farther away and use leader lines to identify the appropriate arc or circle.

Avoid crossing extension and dimension lines with leader lines. See Figure 8-48.

NOTE

The diameter symbol can be added when using the Multiline Text Editor dialog box by typing %%c or by using the Symbols option. See Figure 8-10. The characters %%c will appear on the Multiline Text Editor screen, but will be converted to the diameter symbol Ø when the text is applied to the drawing.

Angular dimensions

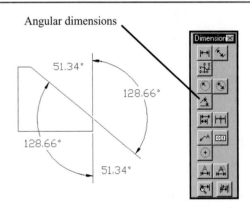

Figure 8-49

Dimension this slanted surface using an angular dimension.

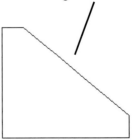

Figure 8-50

8-8 ANGULAR DIMENSIONS

Figure 8-49 shows four possible angular dimensions that could be created using the ANGULAR DIMENSIONS command. The extension lines and degree symbol will be added automatically.

To create an angular dimension

See Figure 8-50.

1. Select the Angular Dimension tool.

Command: _dimangular
Select arc, circle, line or press RETURN:

> **NOTE**
>
> The degree symbol can be added when using the Multiline Text Editor dialog box by using the Symbols option or by typing %%d. The characters %%d will appear on the Edit MText screen, but will be converted to ° when the text is applied to the drawing.

2. Select the short vertical line on the lower right side of the object.

Second line:

3. Select the slanted line.

Dimension arc line location (MText/Text/Angle):

4. Locate the text away from the object.

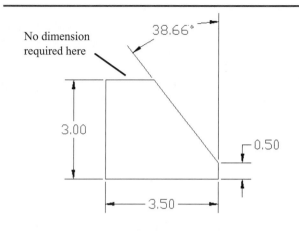

Figure 8-52

Figure 8-51 shows the results. It is considered better to use two extension lines for angular dimensions and not have one of the arrowheads touch the surface of the part.

Avoid over-dimensioning

Figure 8-52 shows a shape dimensioned using an angular dimension. The shape is completely defined. Any additional dimension would be an error. It is tempting, in an effort to make sure a shape is completely defined, to add more dimensions, such as a horizontal dimension for the short horizontal edge at the top of the shape. This dimension is not needed and is considered double dimensioning.

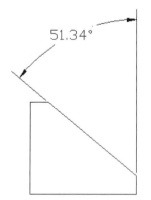

Figure 8-51

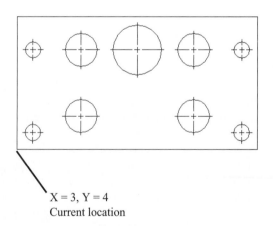

Figure 8-53

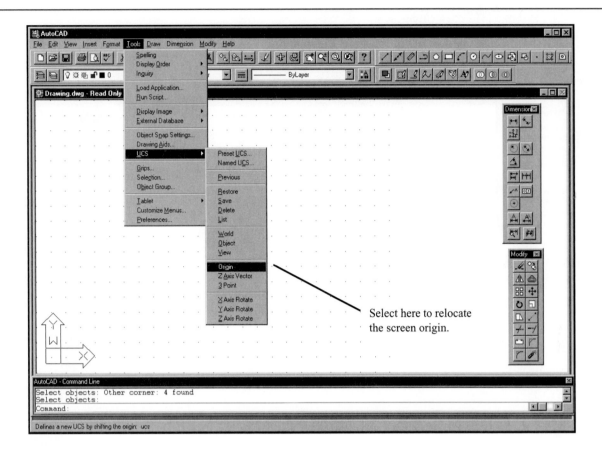

Select here to relocate the screen origin.

Figure 8-54

8-9 ORDINATE DIMENSIONS

Ordinate dimensions are dimensions based on an X,Y coordinate system. Ordinate dimensions do not include extension, dimension lines, or arrowheads, but simply horizontal and vertical leader lines drawn directly from the features of the object. Ordinate dimensions are particularly useful when dimensioning an object that includes many small holes.

Figure 8-53 shows an object that is to be dimensioned using ordinate dimensions. Ordinate dimensions are automatically calculated from the X,Y origin, or in this example, the lower left corner of the screen. If the object had been drawn with its lower left corner on the origin, you could proceed directly to the Ordinate Dimension command; however, the lower left corner of the object is currently located at X=3, Y=4. First move the origin to the corner of the object, then use the Ordinate Dimension command.

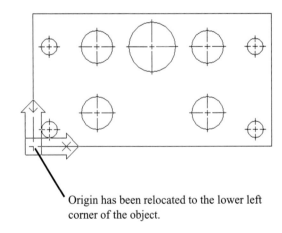

Origin has been relocated to the lower left corner of the object.

Figure 8-55

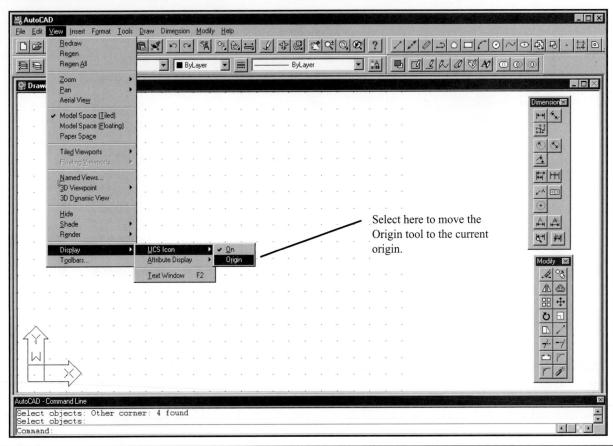

Figure 8-56

To move the origin

1. Select the Tools pull-down menu.

The Tools pull-down-menu will cascade.

2. Select UCS.

The UCS menu will appear next to the Tools pull-down menu. See Figure 8-54.

3. Select Origin.

Origin point <0,0,0>:

4. Select the lower left corner of the object.

The origin (0,0) is now located at the lower left corner of the object. This can be verified by looking at the coordinate display at the lower left corner of the screen.

The origin tool may move to the lower left corner as shown in Figure 8-55, depending on your computer's settings. The tool can be moved back to the original screen location as follows.

To move the ORIGIN tool

1. Select the View pull-down menu, then Display.

The Display menu will appear next to the Options pull-down menu. See Figure 8-56.

2. Select the UCS icon, then Origin.

The Origin tool will go to the current origin.

To add ordinate dimensions to an object

The following procedure assumes that you have already used the DIMENSION STYLES command (Section 8-4) to set the style of the dimensions to what you want and that you have moved the origin to the lower left corner of the object as shown.

1. Turn the Ortho command on (Press the F8 key).
2. Select the Ordinate Dimensions tool from the Dimensioning toolbar.

Command: _dimordinate
Select feature:

Ordinate dimensions

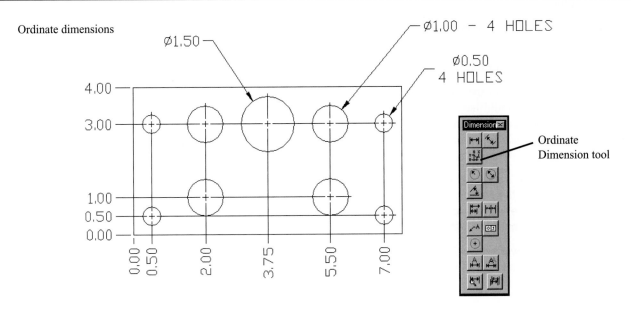

Figure 8-57

Baseline dimensions

The first extension of line of this dimension, created using the Linear Dimension command, will become the base-line for the other dimensions.

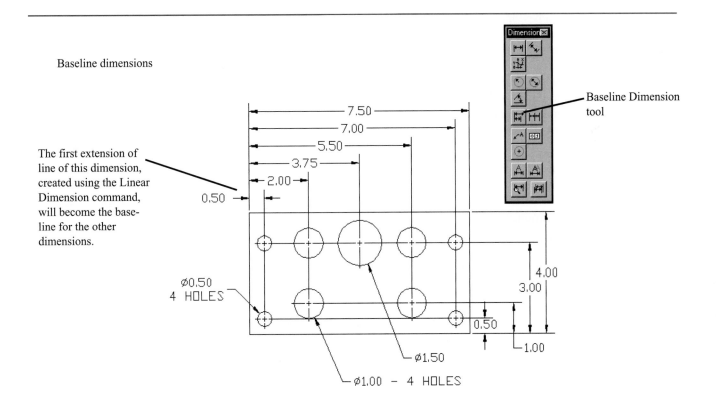

Figure 8-58

3. Select the first circle's vertical centerline.

Leader endpoint (Xdatum/Ydatum/MText/Text):

4. Select a point along the X axis directly below the vertical centerline of the circle.

The ordinate value of the point will be added to the drawing. This point should have a 0.50 value. The text value may be modified using either the MText or Text option, or by using the Dimension Styles Annotation box to define the precision of the text.

5. Press the right button to restart the command and dimension the object's other features.
6. Extend the centerlines across the object and add the diameter dimensions for the holes.

Figure 8-57 shows the completed drawing. The Text option of the prompt shown in step 3 can be used to modify or remove the default text value.

8-10 BASELINE DIMENSIONS

Baseline dimensions are a series of dimensions that originate from a common baseline or datum line. Baseline dimensions are very useful because they help eliminate tolerance buildup that is associated with chain-type dimensions.

The Baseline command can be used only after an initial dimension has been drawn. AutoCAD will define the first extension line origin of the initial dimension selected as the baseline for all baseline dimensions.

To use the Baseline Dimension command

See Figure 8-58.

1. Select the Linear Dimension tool.

Command: _dimlinear
First extension line origin or RETURN to select:

2. Select the upper left corner of the object.

This selection determines the baseline.

Second extension line origin:

3. Select the endpoint of the first circle's vertical centerline; use Osnap if needed to ensure accuracy.

Dimension line location (Text/ Angle/ Horizontal/ Vertical/ Rotated):

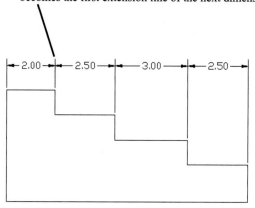

The second extension line of the previous dimension becomes the first extension line of the next dimension.

Figure 8-59

4. Select a location for the dimension line.

Command:

5. Select the Baseline Dimension tool.

Second extension line origin or Return to select:

6. Select the endpoint of the next circle's vertical centerline.

Second extension line origin or Return to select:

7. Continue to select the circle centerlines until all circles are located.

Second extension line origin or Return to select:

8. Select the upper right corner of the object.
9. Press Enter.

This will end the BASELINE DIMENSION command.

10. Repeat the above procedure for the vertical baseline dimensions.
11. Add the circles' diameter values.

The Baseline Dimension option can also be used with the ANGULAR DIMENSION and ALIGNED DIMENSION commands.

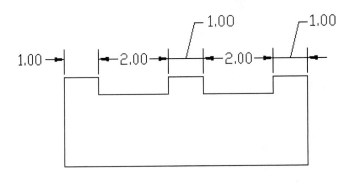

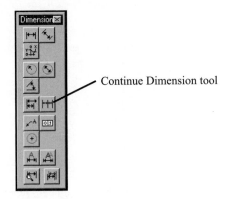

Continue Dimension tool

Figure 8-60

8-11 CONTINUE DIMENSION

The CONTINUE DIMENSION command is used to create chain dimensions based on an initial linear, angular, or ordinate dimension. The second extension line's origin becomes the first extension line origin for the continued dimension.

To use the CONTINUE DIMENSION command

See Figure 8-59.

1. Select the Linear Dimension tool.

    Command:
    First extension line origin or RETURN for select:

2. Select the upper left corner of the object.

Second extension line origin:

3. Select the right end point of the uppermost horizontal line.

    Dimension line location (Text/ Angle/ Horizontal/ Vertical/ Rotated):

4. Select a dimension line location.

    Command:

5. Select the Continue Dimension tool.

    Command: _dimcontinue
    Second extension line origin or Return to select:

6. Select the next linear distance to be dimensioned.

    Second extension line origin or Return to select:

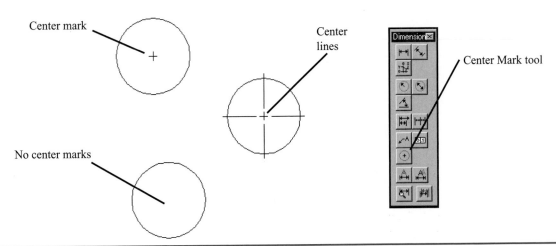

Center mark

Center lines

Center Mark tool

No center marks

Figure 8-61

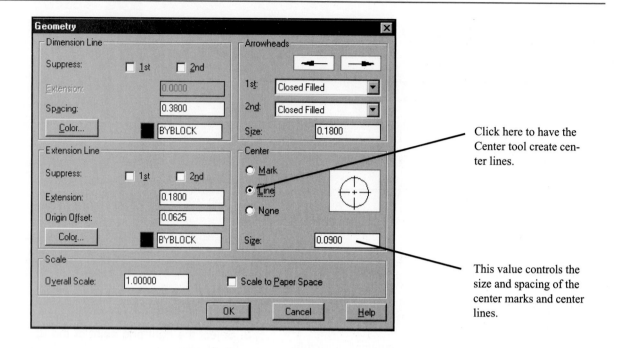

Figure 8-62

7. Continue until the object's horizontal edges are completely dimensioned.

AutoCAD will automatically align the dimensions. Figure 8-60 shows how the Continue Dimension command dimensions distances that are too small for both the arrowhead and dimension value to fit within the extension lines.

8-12 CENTER MARK

When AutoCAD first draws a circle or arc, a center mark appears on the drawing. However, these marks will disappear when the Redraw View or Redraw All commands are applied.

To add a permanent center mark to a given circle

See Figure 8-61.

1. Select the Center Mark tool.

   Command: _dimcenter
   Select arc or circle:

2. Select the circle.

   Command:

The Center Mark command can also be used to add horizontal and vertical centerlines to a circle.

To add centerlines to a given circle

1. Select the Dimension Styles tool.

   The Dimension Styles dialog box will appear.

2. Select Geometry.

   The Geometry dialog box will appear. See Figure 8-62.

3. Select the radio button to the left of the word Line in the Center box.

   The preview display will show a horizontal and vertical centerline.

4. Select OK to return to the drawing.
5. Select the Center Mark tool.

   Select arc or circle:

6. Select the circle.

Horizontal and vertical centerlines will appear. The size of the center mark can be controlled using the Size option in the Geometry dialog box. If the center ine's size appears unacceptable, try different sizes until an acceptable size is achieved.

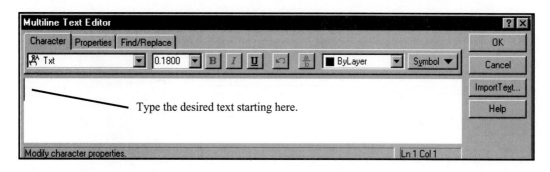

Figure 8-63

8-13 LEADER

Leader lines are slanted lines that extend from notes or dimensions to a specific feature or location on the surface of a drawing. They usually end with an arrowhead or dot. The RADIUS and DIAMETER DIMENSION commands automatically create a leader line. The LEADER command can be used to add leader lines not associated with radius and diameter dimensions.

To create a leader line with text

1. Select the Leader tool.

   Command: _leader
   From point:

2. Select the starting point for the leader line.

   This is the point where the arrowhead will appear.

   To point:

3. Select the location of the endpoint of the slanted line segment.

To point (Format/ Annotation/ Undo) <Annotation>:

4. Press Enter.

This response accepts the default Annotation response.

Annotation (or RETURN for options):

5. Press Enter.

Tolerance/Copy/Block/None/<MText>:

6. Press Enter.

The Edit MText dialog box will appear. See Figure 8-63.

7. Type the desired text.

The flashing cursor may be repositioned as needed. The initial location for the flashing cursor, at the upper left corner of the text box, represents the end of the leader line. Figure 8-64 shows the resulting leader line note.

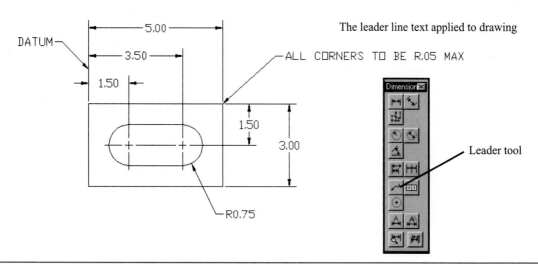

The leader line text applied to drawing

Leader tool

Figure 8-64

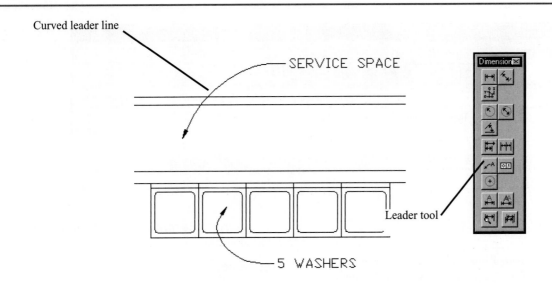

Curved leader line

SERVICE SPACE

Leader tool

5 WASHERS

Figure 8-65

The Leader command can be used to draw curved leader lines and leader lines that end with dots. See Figure 8-65.

To draw a curved leader line

1. Select the Leader tool.

   Command: _leader
   From point:

2. Select the starting point for the leader line.

   This is the point where the arrowhead will appear.

   To point:

3. Select the location of the endpoint of the slanted line segment.

   Select a point just past the arrowhead.

   To point (Format/ Annotation/ Undo) <Annotation>:

4. Type F; press Enter.

   Spline/STraight/Arrow/None/<Exit>:

5. Type S; press Enter.

   To point (Format /Annotation /Undo) <Annotation>:

   AutoCAD will shift to the Drag mode that allows you to move the cursor around and watch the changes in shapes of the leader line. More than one point may be selected to define the shape.

   To point (Format/ Annotation /Undo) <Annotation>:

6. Press Enter.

   Annotation (or Return for options):

7. Press Enter.

   Tolerance/Copy/Block/None/<Mtext>:

8. Press Enter.

   The Edit MText dialog box will appear.

9. Type in the appropriate text.
10. Select OK.

   The text will appear on the drawing.

To draw a leader line with a dot at its end

1. Select the Dimension Styles tool.

   The Dimension Styles dialog box will appear.

2. Select Geometry.

   The Geometry dialog box will appear.

3. Select the box with an arrow to the right of the word Closed in the Arrowheads box.

   A listing of the shape options will cascade. See Figure 8-66.

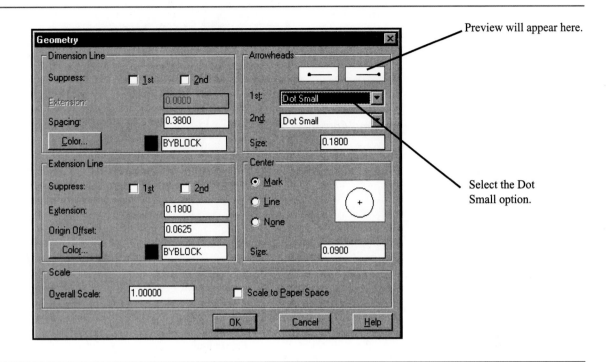

Figure 8-66

4. Select Dot Small.

A preview of the dot will appear in the Arrowheads preview box. The size of the dot can be controlled using the Size box.

5. Select OK to return to the drawing.
6. Use the Leader tool to create leader lines as described above.

8-14 Dimension Text Edit

The Dimension Text Edit command is used to edit existing dimensioning text. Existing text can be moved or rotated. There are four positions associated with the Dimension Text Edit; Left, Right, Home, and Angle.

To use the Dimension Text Edit option

1. Select the Dimension Text Edit tool.

   Command: _dimtedit
   Select dimension:

2. Select the dimension.

   Enter text location (Left/Right/Home/Angle):

Each of these options may be accessed directly by using the appropriate letter; L for left, H for home, R for right and A for angle.

3. Type L; press Enter.

The text will be repositioned at the left of the dimension line. See Figure 8-67. Text can also be positioned by eye by moving the text using the cursor. When the text is located in its new position, press the left mouse button.

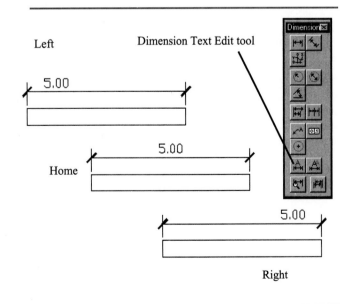

Figure 8-67

To change the angle of existing dimension text

1. Select the Dimension Text Edit tool.

   Command: _dimtedit
   Select dimension:

2. Select the dimension.

   Enter text location (Left/Right/Home/Angle):

3. Type A; press Enter.

   Enter text angle:

4. Type 90; press Enter.

   Figure 8-68 shows the results.

8-15 TOLERANCES

Tolerances are numerical values assigned with the dimensions that define the limits of manufacturing acceptability for a distance. AutoCAD can create four types of tolerances: symmetrical, deviation, limits, and basic. See Figure 8-69. Many companies also use a group of standard tolerances that are applied to any dimensional value that is not assigned a specific tolerance. See Figure 8-32. Tolerances for numerical values expressed in millimeters are applied using a different convention than that used for inches. Figure 8-70 shows the different conventions. The conventions presented are consistent with the conventions described in Section 8-5 for dimensional unit presentations.

Tolerances for millimeter dimensions may include a single 0 for a zero limit tolerance. Dimensional values may be expressed using whole numbers such as 32, and do not require decimal values to the right of the decimal point even if the tolerance includes values to the right of the decimal point.

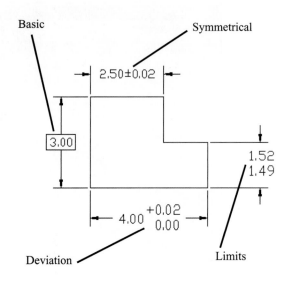

Figure 8-69

Tolerances for inch values must be consistent with the stated numerical dimensional value. A dimensional value expressed using three decimal places must have a tolerance value expressed in three decimal places, even if one of the tolerance values is zero. The zero value would be expressed as 0.000.

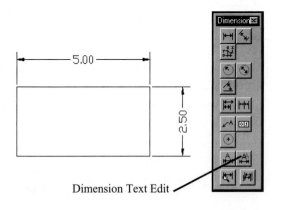

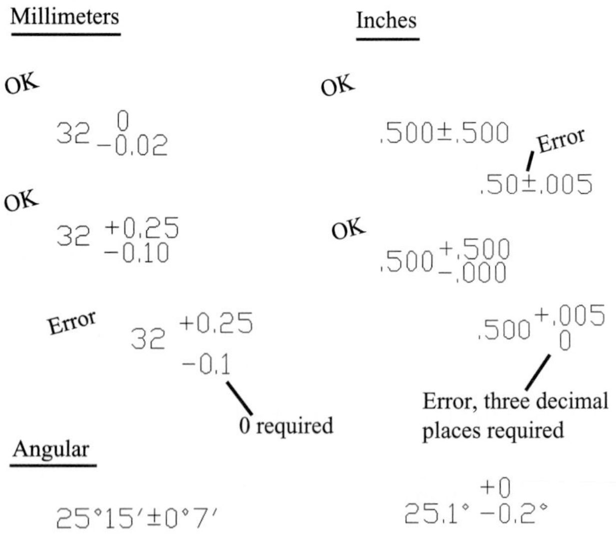

Figure 8-68

Figure 8-70

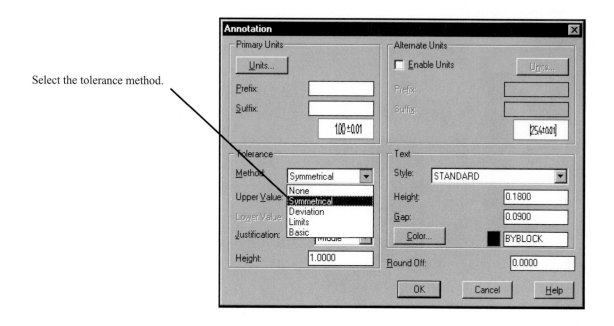

Select the tolerance method.

Figure 8-71

To create a symmetrical tolerance: inches

1. Select the Dimension Styles tool.

   The Dimension Styles dialog box will appear.

2. Select Annotation.

   The Annotation dialog box will appear.

3. Click the arrow box to the right of the word None in the Method box located in the Tolerance box.

   A listing of options will cascade. See Figure 8-71.

4. Select Symmetrical.
5. Change the Upper Value to 0.0030 by locating the cursor within the Upper Value box, backspacing to remove the existing value, and typing in the new value.
6. Check that the Justification box reads Middle.

   If it does not read Middle, use the arrow box to the right of the Justification box to change the reading. See Figure 8-72. The value displayed in the Units box will not change at this time, but the tolerance placed on the drawing will change.

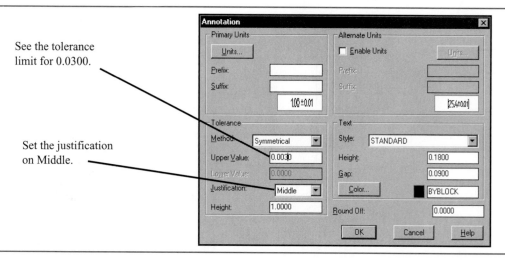

See the tolerance limit for 0.0300.

Set the justification on Middle.

Figure 8-72

7. Select OK to return to the drawing.
8. Select the Linear Dimension tool.

Command: _dimlinear
First extension line or Return to select

9. Press Enter.

Select object to dimension:

10. Select the line.

Dimension line location (Text /Angle/ Horizontal/ Vertical/ Rotated):

11. Locate the dimension line and press the left mouse button.

Figure 8-73 shows the resulting dimension.

To create a symmetrical tolerance: millimeters

The object shown in Figure 8-74 is 80 millimeters long and is to have a tolerance of +/- 0.02. This example assumes that the Overall Scale in the Geometry dialog box has been changed to 25.4 as explained in Section 8-4.

1. Select the Dimension Styles tool.

The Dimension Styles dialog box will appear.

2. Select Annotation.

The Annotation dialog box will appear.

3. Click the arrow box to the right of the word None in the Method box located in the Tolerance box.

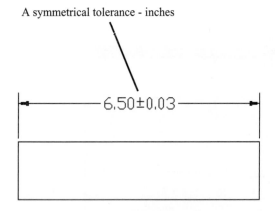

A symmetrical tolerance - inches

6.50±0.03

Figure 8-73

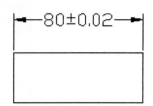

80±0.02

Figure 8-74

A listing of options will cascade. See Figure 8-71.

4. Select Symmetrical.
5. Change the Upper Value to 0.0200 by locating the cursor within the Upper Value box, backspacing to remove the existing value, and typing in the new value.

The value displayed in the Units box will not change at this time, but the tolerance placed on the drawing will change.

6. Select Units.

The Primary Units dialog box will appear. See Figure 8-75.

7. Click the box to the left of the word Trailing in the Zero Suppression box.

An X will appear in the box, indicating that the option is on. AutoCAD will now suppress all zeros to the right of the decimal point on the primary dimension value. It will not suppress the tolerance values.

8. Select OK, OK, OK to return to the drawing.
9. Use the Linear Dimension tool to create the dimension and tolerance values.

To create a deviation tolerance

The object shown in Figure 8-76 is to have a horizontal tolerance of +.02, -.01, and a vertical tolerance of +.03, -.00.

1. Select the Dimension Styles tool.

The Dimension Styles dialog box will appear.

2. Select Annotation.

The Annotation dialog box will appear.

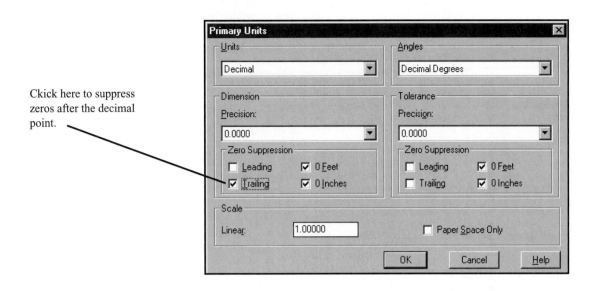

Ckick here to suppress zeros after the decimal point.

Figure 8-75

3. Click the arrow box to the right of the word None in the Method box located in the Tolerance box.

A listing of options will cascade. See Figure 8-71.

4. Select Deviation.
5. Change the Upper Value to 0.0200 by locating the cursor within the Upper Value box, backspacing to remove the existing value, and typing in the new value.

The value displayed in the Units box will not change at this time, but the tolerance placed on the drawing will change.

6. Change the Lower Value to 0.0100 by locating the cursor within the Lower Value box, backspacing to remove the existing value, and typing in the new value.
7. Check that the Justification box reads Middle.

If it does not read Middle, use the arrow box to the right of the Justification box to change the reading. Also use the Units option to check that the Zero Suppression for Trailing values has been turned off, that is, there is no X in the box to the left of the word Trailing.

8. Select OK, OK to return to the drawing.
9. Use the Linear Dimension tool to add the appropriate dimension and tolerance.
10. Select the Dimension Styles tool, then Annotation as explained above.

11. Change the Upper Value to 0.03 and the Lower Value to 0.00.
12. Select OK, OK to return to the drawing.
13. Use the Linear Dimensions tool to add the vertical dimension.

Figure 8-76 shows the resulting dimensions.

In the Deviation dialog box, change the Upper Value to 0.02 and the Lower Value to 0.01.

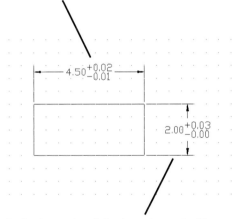

In the Annotation dialog box, change the Upper Value to 0.03 and the Lower Value to 0.00.

Figure 8-76

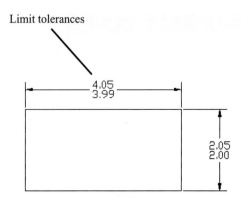

Limit tolerances

Figure 8-77

To create a limit tolerance

The object shown in Figure 8-77 is to be dimensioned using limit tolerances. The horizontal dimension limits are 4.05 to 3.99 and the vertical limits are 2.o5 to 2.00.

1. Select the Dimension Styles tool.

The Dimension Styles dialog box will appear.

2. Select Annotation.

The Annotation dialog box will appear.

3. Click the arrow box to the right of the word None in the Method box located in the Tolerance box.

A listing of options will cascade. See Figure 8-69.

4. Select Limits.

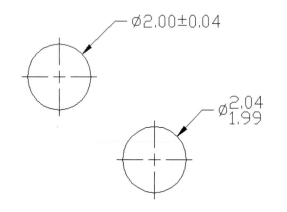

Figure 8-78

5. Change the Upper Value to 0.0500 by locating the cursor within the Upper Value box, backspacing to remove the existing value, and typing in the new value.

The value displayed in the Units box will not change, but the tolerance placed on the drawing will change.

6. Change the Lower Value to 0.0100 by locating the cursor within the Lower Value box, backspacing to remove the existing value, and typing in the new value.

7. Check that the Justification box reads Middle.

If it does not read Middle, use the arrow box to the right of the Justification box to change the reading. Also use the Units option to check that the Zero Suppression for Trailing values has been turned off, that is, there is no X in the box to the left of the word Trailing.

8. Select OK to return to the drawing.

9. Use the Linear Dimension tool to add the appropriate dimension and tolerance.

10. Select the Dimension Styles tool, then the Annotation option, and change the upper limit to 0.05 and the lower limit to 0.00.

11. Select OK, OK to return to the drawing.

12. Use the Linear Dimension tool to add the vertical dimension to the drawing.

Figure 8-78 shows examples of tolerances applied to diameter dimensions.

8-16 GEOMETRIC TOLERANCES

Geometric tolerances are tolerances that limit dimensional variations based on the geometric properties of an object. Figure 8-79 shows an object dimensioned using geometric tolerances. The geometric tolerances were created as follows.

To define a datum

1. Select the Tolerance tool.

The Symbol dialog box will appear. See Figure 8-80.

2. Select OK.

The Geometric Tolerance dialog box will appear. See Figure 8-81.

3. Click the Datum Identifier box and type -A-, then select OK.

Command: _tolerance
Enter tolerance location:

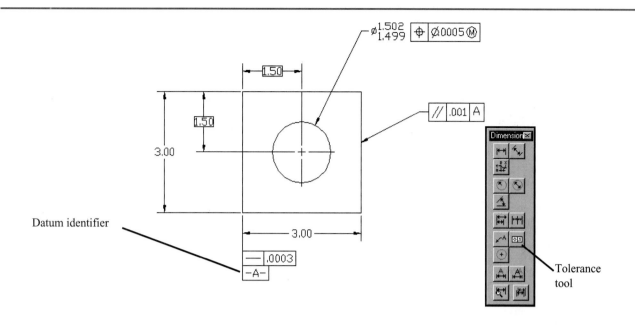

Figure 8-79

4. Position the Datum Identifier and press the left mouse button.

See Figure 8-79.

To define a straightness value

1. Select the Tolerance tool.

The Symbol dialog box will appear. See Figure 8-80.

2. Select OK.

The Geometric Tolerance dialog box will appear. See Figure 8-81.

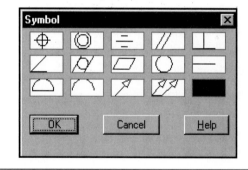

Figure 8-80

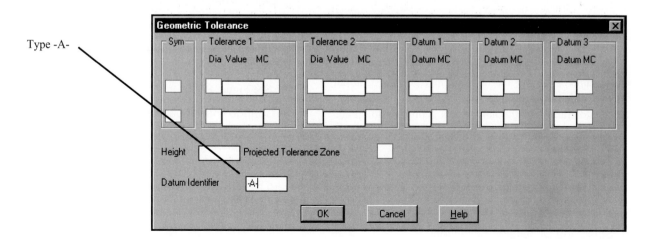

Figure 8-81

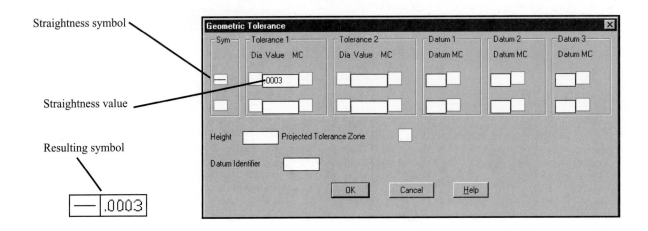

Straightness symbol

Straightness value

Resulting symbol

Figure 8-82

3. Select the top open box under the heading Sym.

The Symbol dialog box will reappear.

4. Select the straightness symbol, then OK.

The Geometric Tolerance dialog box will reappear with the straightness symbol in the first box under the heading Sym.

5. Click the open box under the word Value in the Tolerance 1 box, type .0003, and select OK.

See Figure 8-80.

Command: _tolerance
Enter tolerance location:

6. Position the straightness tolerance and press the left mouse button.

Use the MOVE and OSNAP commands if necessary to reposition the tolerance box. See Figure 8-79.

To create a positional tolerance

A positional tolerance is used to locate and tolerance a hole in an object. Positional tolerances require basic locating dimensions for the hole's center point. Positional tolerances also require a feature tolerance to define the diameter tolerances of the hole, and a geometric tolerance to define the position tolerance for the hole's center point.

To create a base dimension

See the two 1.50 dimensions in Figure 8-79 used to locate the center position of the hole.

1. Select the Dimension Styles tool.

The Dimension Styles dialog box will appear.

2. Select Annotation.

The Annotation dialog box will appear.

3. Select the Basic option next to the heading Method in the Tolerance box.

See Figure 8-72.

4. Select OK,OK to return to the drawing.
5. Use the Linear Dimension tool to add the appropriate dimensions.

See Figure 8-79.

To add a feature tolerance to a hole

1. Select the Dimension Styles tool.

The Dimension Styles dialog box will appear.

2. Select Annotation.

The Annotation dialog box will appear.

3. Select the Limits option in the Method box located in the Tolerance box.

See Figure 8-72.

4. Change the Upper Value to 0.0020 by locating the cursor within the Upper Value box, backspacing to remove the existing value, and typing in the new value.

The value displayed in the Units box will not change, but the tolerance placed on the drawing will change.

5. Change the Lower Value to 0.0010 by locating

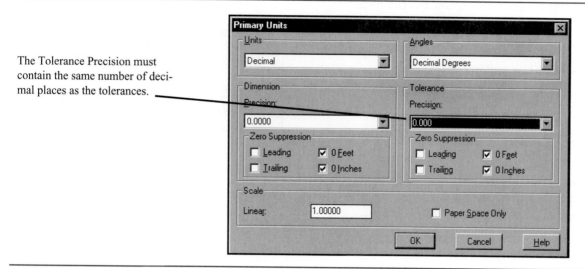

The Tolerance Precision must contain the same number of decimal places as the tolerances.

Figure 8-83

the cursor within the Lower Value box, backspacing to remove the existing value, and typing in the new value.

6. Check that the Justification box reads Middle.
7. Select the Units box.

The Primary units dialog box will appear.

8. Change the precision for the Dimension and Tolerance boxes to three decimal places (0.000).

See Figure 8-83. AutoCAD will truncate any input according to the number of decimal places allowed by the precision settings. If the precision settings had been two decimal places (0.00), the resulting limit dimensions would have been 1.50 for both. The values defined in the third decimal place would have been ignored.

9. Select OK, OK, OK to return to the drawing.

10. Select the Diameter Dimension tool.

Select arc or circle:

11. Select the hole.

Dimension line location (MText/Text/Angle):

12. Locate the diameter dimension.

To add a positional tolerance to the hole's feature tolerance

1. Select the Tolerance tool.

The Symbol dialog box will appear.

2. Select the positioning tolerance symbol, then OK.

The Geometric Tolerance dialog box will appear with the positioning symbol in the first box under the Sym heading. See Figure 8-84.

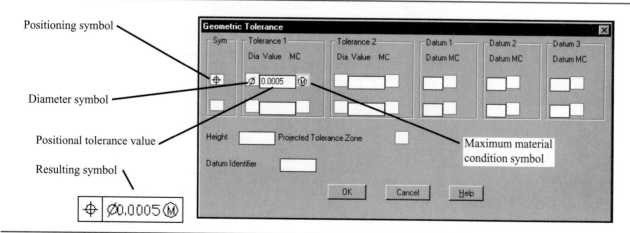

Positioning symbol

Diameter symbol

Positional tolerance value

Resulting symbol

Maximum material condition symbol

Figure 8-84

3. Select the top left open box under the Dia heading in the Tolerance 1 box.

A diameter symbol will appear.

4. Select the Value box and type 0.0005.

The numbers will appear in the box.

5. Select the open box under the MC heading.

The Material Condition dialog box will appear. See Figure 8-85.

6. Select the maximum material condition (MMC) symbol (the circle with an M in it).
7. Select OK.

The MMC symbol will appear in the MC box in the Geometric Tolerance dialog box.

8. Select OK.

Enter tolerance location:

9. Locate the tolerance box.

Use the Move and Osnap commands to position the box if necessary.

To add a geometric tolerance with a leader line

1. Select the Leader tool.

Command: _leader
From point:

2. Select the start point for the leader line (the arrowhead end).

To point:

3. Select the other end of the leader line.

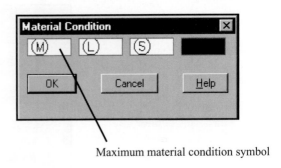

Maximum material condition symbol

Figure 8-85

To point: (Format/ Annotation/ Undo) <Annotation>:

4. Press Enter.

Annotation (or Return for options):

5. Press Enter.

Tolerance/Copy/Block/None/<MText>:

6. Type T; press Enter.

The Symbol dialog box will appear

7. Select the parallel symbol, then OK.

The Geometric Tolerance dialog box will appear. See Figure 8-86.

8. Select the open box under the heading Value in the Tolerance 1 box and type 0.0010.
9. Select the open box under the heading Datum in the Datum 1 box, type A, then select OK.

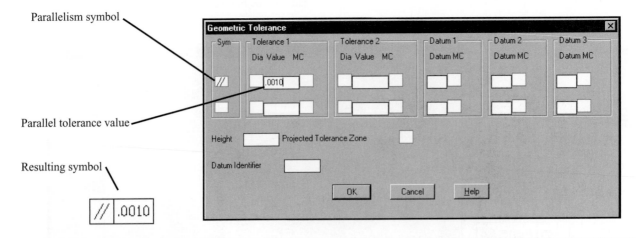

Figure 8-86

8-17 OBLIQUE DIMENSIONS

The OBLIQUE command is used to create oblique extension lines. See Figure 8-87. Oblique extension lines allow you to move dimensions away from areas that contain many extension lines positioned close together. The dimension lines associated with oblique extension lines are drawn in the direction in which they apply. In the example shown, the distances defined are vertical distances, so the dimension lines drawn with the oblique extension lines are vertical.

To draw oblique extension lines

The OBLIQUE command requires an existing dimension.

1. Select the Dimension Edit tool.

Command: _dimedit Dimension Edit (Home/ New/ Rotate/ Oblique) <Home>: _o
Select objects:

2. Select the existing dimension.

Select objects:

3. Press Enter.

Enter obliquing angle (press Enter for none):

4. Type the required angular value; press Enter.

In the example shown in Figure 8-87, an angle of 45 degrees was used.

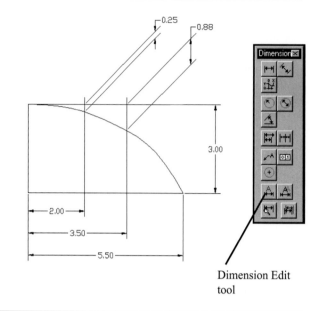

Dimension Edit tool

Figure 8-87

8-18 EXERCISE PROBLEMS

Redraw the following shapes. Locate the dimensions and tolerances as shown.

EX8-1 INCHES

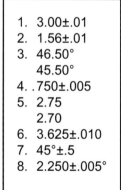

1. 3.00±.01
2. 1.56±.01
3. 46.50°
   45.50°
4. .750±.005
5. 2.75
   2.70
6. 3.625±.010
7. 45°±.5
8. 2.250±.005°

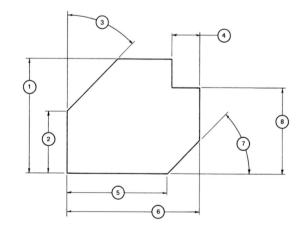

EX8-2 MILLIMETERS

1. 38±0.05
2. 10±0.1
3. 5±0.05
4. 45.50°
    44.50°
5. 40±0.1
6. 22±0.1
7. 12 $^{+0}_{-.1}$
8. 25 $^{+.05}_{-0}$
9. 51.50
    50.75
10. 76±0.1

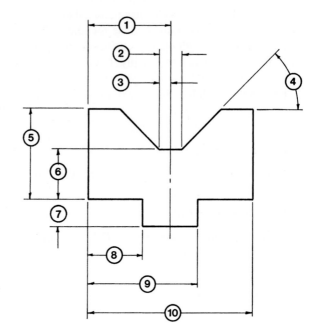

EX8-3 MILLIMETERS

1. 34±0.25
2. 17±0.25
3. 25±0.05
4. 15.00
    14.80
5. 50±0.05
6. 80±0.1
7. R5±0.1 - 8 PLACES
8. 45±0.25
9. 60±0.1
10. x Ø14 - 3 HOLES
11. 15.00
    14.80
12. 30.00
    29.80

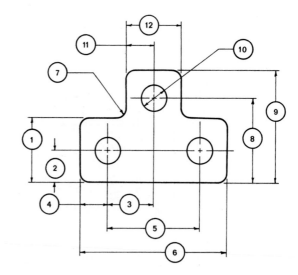

EX8-4 INCHES

1. 1.50±.02

2. 1.50±.03

3. .625±.001

4. .754 / .749

5. .625±.001

6. 2.253 / 2.249

7. ⌀ .502 / .500

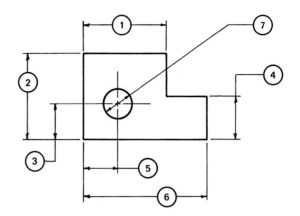

EX8-5 MILLIMETERS

1. ⌀30±0.5

2. ⌀ 15.04 / 15.00

3. 10±1

4. 20±1

5. 66.50 / 65.03

6. 15±0.3

7. 35±.02

8. 70±0.2

NOTE: ALL FILLETS AND ROUNDS = R5 UNLESS OTHERWISE STATED

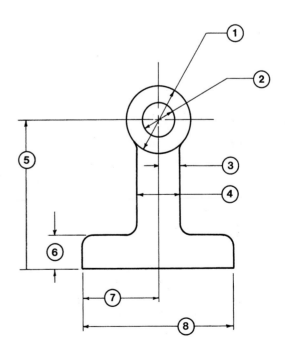

EX8-6 MILLIMETERS

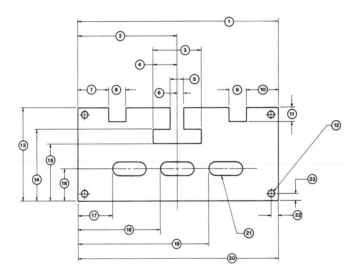

1. 184±0.5	7. 28±0.1	13. 83±0.4	18. 76±0.2
2. 91.5±0.1	8. 16.00 / 15.96	14. 63.50 / 62.48	19. 120±0.2
3. 44.4 / 44.0	9. 16.00 / 15.06	15. 50.03 / 49.97	20. (184)
4. 22±.03	10. 28±0.1	16. 28.5±0.2	21. 12 × 31 / R - 3 PLACES
5. 13.00 / 12.94	11. 12.5±0.1	17. 32±0.2	22. 6±0.01
6. 6.5±0.2	12. ⌀6.00 / 5.96 4 HOLES		23. 6±0.01

EX8-7

Measure and redraw the shapes in EX8-7 through EX8-23. Add the appropriate dimensions. Specify the units and the scale of the drawings.

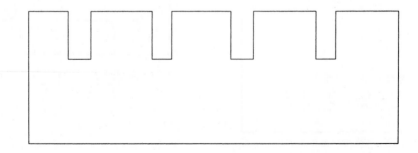

EX8-8

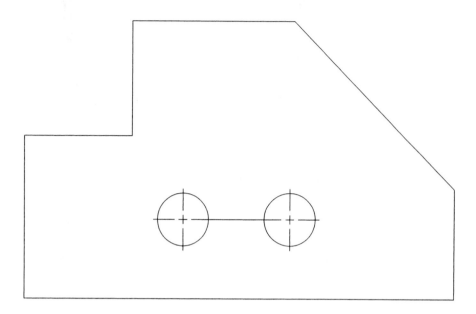

EX8-9

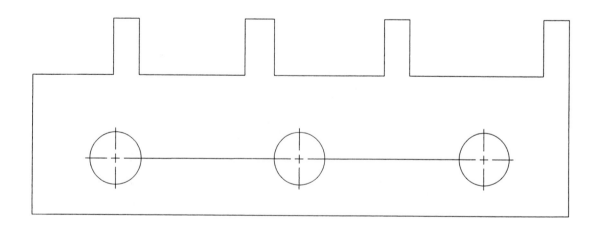

EX8-10

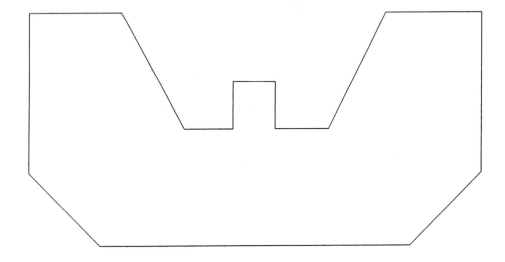

EX8-11

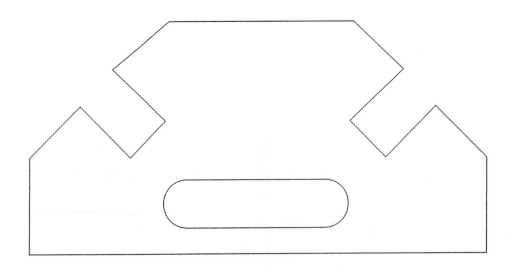

EX8-12

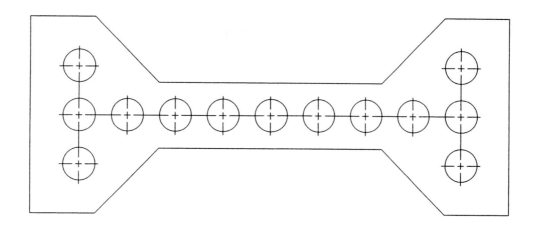

EX8-13

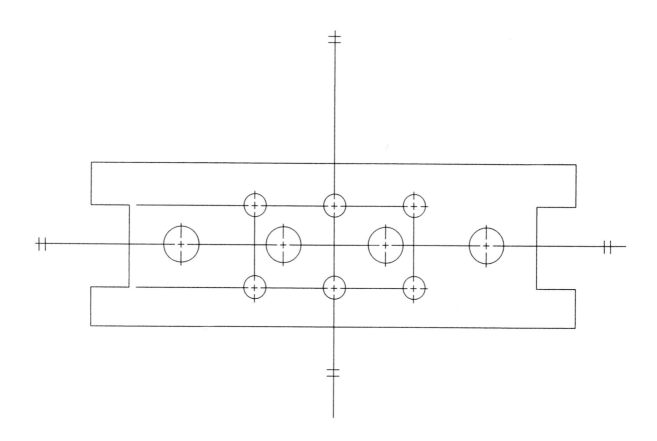

EX8-14

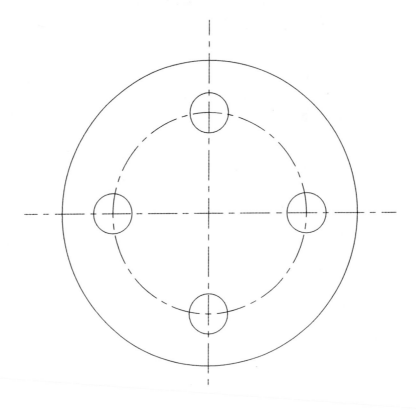

EX8-15

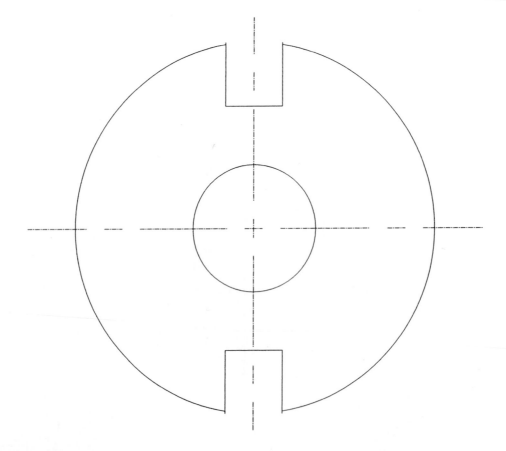

EX8-16

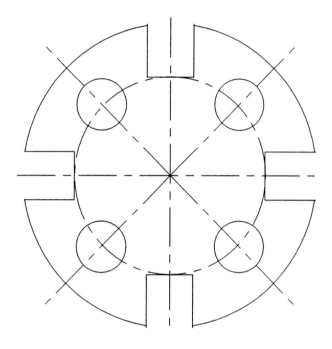

EX8-17

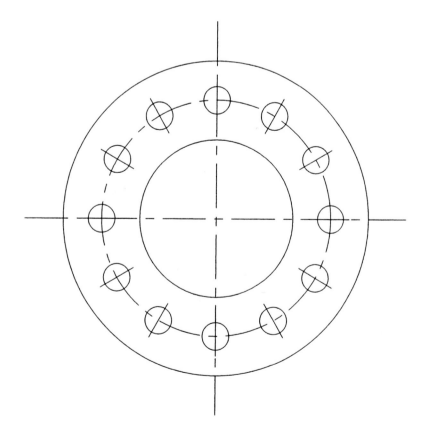

EX8-18

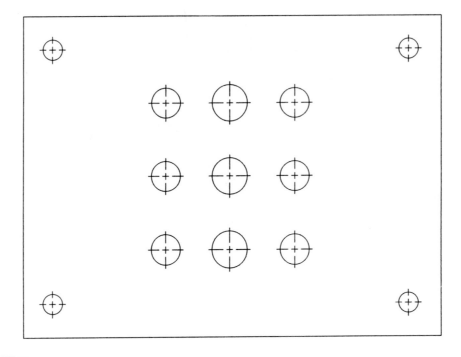

EX8-19

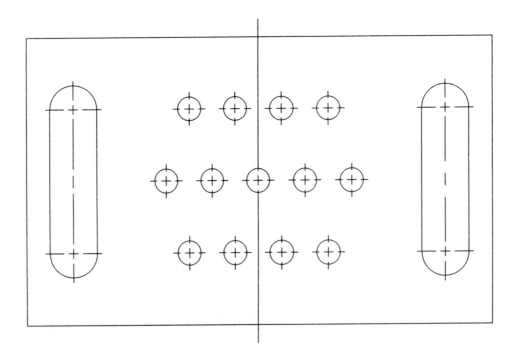

EX8-20

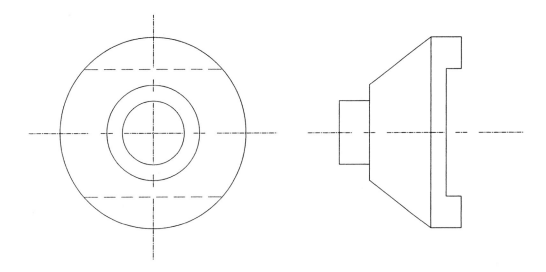

EX8-21

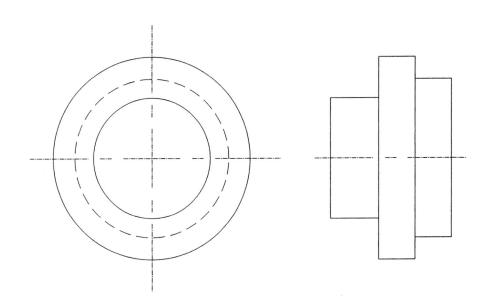

EX8-22

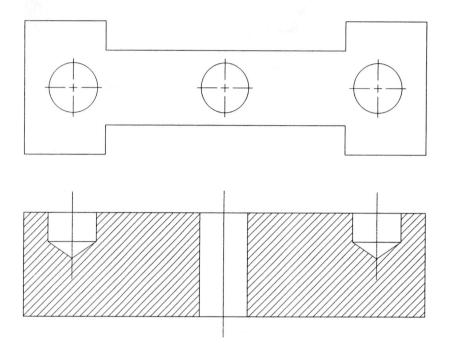

EX8-23

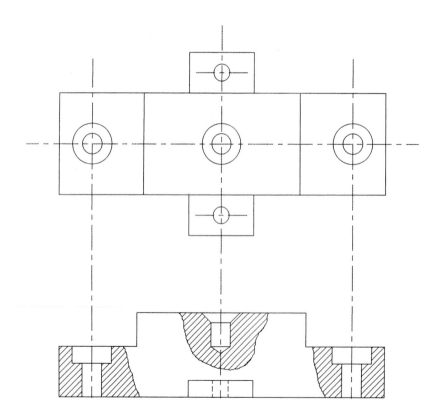

CHAPTER 9

Attributes, Layers, and Blocks

9-1 INTRODUCTION

This chapter introduces the BLOCKS, ATTRIBUTES, and LAYERS commands. The chapter explains how each command is used, and shows how the commands are applied when preparing drawings.

9-2 BLOCKS

Blocks are groups of entities saved as a single unit. Blocks are used to save shapes and groups of shapes that are used frequently when creating drawings. Once created, blocks can be inserted into the drawings, thereby saving drawing time.

AutoCAD offers the AutoDESK Mechanical Library which contains more than 200,000 drawn blocks of standard manufacturers parts and material specifications. These blocks can be inserted directly into a drawing.

The Make Block tool is located in the Draw toolbar. See Figure 9-1. The Insert Block tool is associated with the Make Block tool.

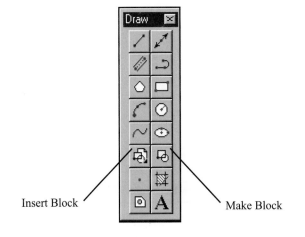

Figure 9-1

To create a BLOCK

Figure 9-2 shows an object. A block can be made from this existing drawing as follows.

1. Select the Make Block tool from the Draw tool-bar.

The Block Definition dialog box will appear. See Figure 9-3.

2. Type the name SHAPE in the Block name box.

Block names may contain up to 31 characters.

3. Select the Select Objects box.

Select objects:

4. Select the object.
5. Select the Select point box in the Base Point box.

Insertion base point:

6. Select the object's center point.

The Block Definition dialog box will appear.

7. Select the OK box.

The object will disappear from the screen. It will be saved as part of the existing drawing.

To insert a Block

1. Select the Insert Block tool from the Draw tool-bar.

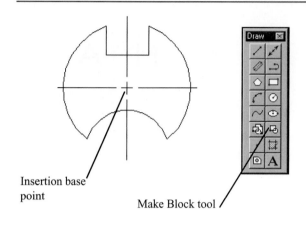

Insertion base point

Make Block tool

Figure 9-2

The Insert dialog box will appear. See Figure 9-4.

2. Select the Block box.

The Defined Blocks dialog box will appear. See Figure 9-5.

3. Select SHAPE.

The word SHAPE will appear in the Selection box.

4. Select OK.

The Insert dialog box will reappear with the word SHAPE in the box next to the Block box.

5. Select OK.

Type block name here.

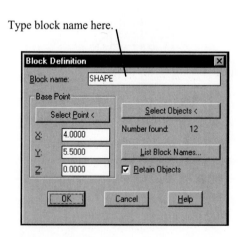

Figure 9-3

Block name will appear here.

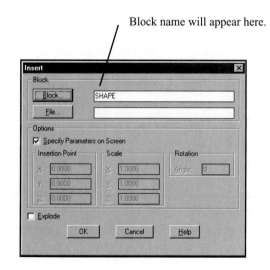

Figure 9-4

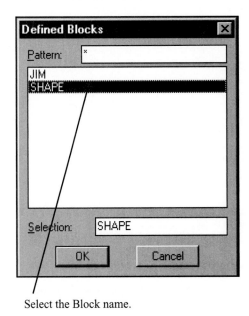

Select the Block name.

Figure 9-5

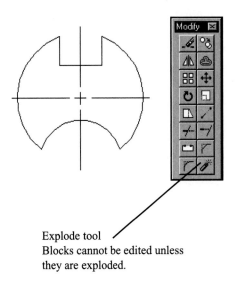

Explode tool
Blocks cannot be edited unless
they are exploded.

Figure 9-6

The SHAPE block will appear on the screen with its insertion point aligned with the cursor. The shape will move as you move the cross hairs.

Command: _ddinsert
Insertion point:

6. Select an insertion point for the Shape.

Insertion point: X scale factor <1>/ Corner/ XYZ:

The default scale factor is 1. This means that if you accept the default value by pressing Enter, the shape will be redrawn at its original size. The shape of the object may be changed by moving the cross hairs. Verify that the shape can be changed dynamically by moving the cross hairs around the drawing and observing how the shape changes.

7. Press Enter.

Y scale factor (default =X):

The shape will temporarily disappear from the screen. The Y scale factor will automatically be made equal to the X scale factor unless a different value is defined.

8. Press Enter.

Rotation angle <0.00>:

9. Press Enter.

The block is now part of the drawing. However, the block in its present form may not be edited. Blocks are treated as a single entity and not as individual lines. You can verify this by trying to erase any one of the lines in the block. The entire object will be erased. The object can be returned to the screen by clicking the Undo tool.

A block must first be exploded before it can be edited.

To explode a block

See Figure 9-6.

1. Select the Explode tool from the Modify toolbar.

Command: _explode
Select objects:

2. Window the entire object.

Select objects:

3. Press Enter.

The object is now exploded and can be edited. There is no visible change in the screen, but there will be a short blink after the EXPLODE command is executed.

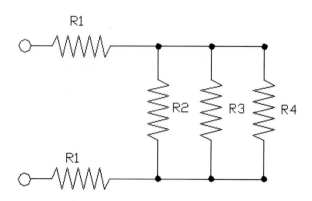

Figure 9-7

9-3 WORKING WITH BLOCKS

Figure 9-7 shows a resistor circuit. It was created from an existing block called RESISTOR. The drawing uses Decimal units with a Grid set to .5 and a Snap set to .25. The default Drawing Limits were accepted. The procedure is as follows.

To insert blocks at different angles

1. Select the Insert Block tool from the Draw toolbar.

   The Insert dialog box will appear.

2. Select Block.

   The Defined Blocks dialog box will appear.

3. Select the block called RESISTOR.

Figure 9-8

Figure 9-9

See Figure 9-8.

4. Select OK, OK to return to the drawing screen.

   The RESISTOR block will appear attached to the cross hairs. The block will attach to the cross hairs at the predefined insertion point. See Figure 9-9.

   The resistor is too large for the drawing, so a reduced scale will be used to generate the appropriate size.

   Insertion point:

5. Select an insertion point.

   Insertion point: X scale factor <1> /Corner/XYZ:

6. Type .50; press Enter.

   Y scale factor (default=X):

7. Press Enter.

   Rotation angle <0.00>:

8. Press Enter.

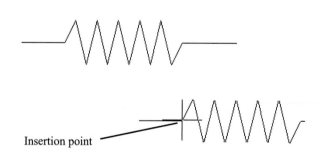

Insertion point

Figure 9-10

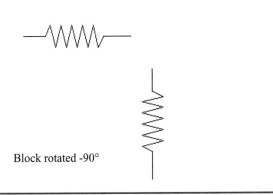

Block rotated -90°

Figure 9-11

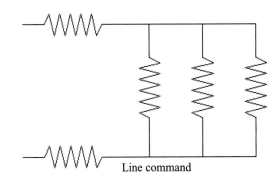

Line command

Figure 9-13

This will set the resistor in place.

9. Press Enter.

This will reactivate the INSERT BLOCK command. The Insert dialog box will appear on the screen with the word RESISTOR in the box next to the block button.

10. Select OK.

A second resistor will appear on the screen.

11. Select an insertion point as shown in Figure 9-10.

Insertion point: X scale factor <1> /Corner/XYZ:

12. Type .50; press Enter.

Y scale factor (default=X):

13. Press Enter.

Rotation angle <0.00>:

14. Type -90; press Enter.

Your screen should look like Figure 9-11.

15. Use the COPY command (Modify toolbar) to add the additionally required resistors.

See Figure 9-12.

16. Use the LINE command (Draw toolbar) to draw the required lines.

See Figure 9-13.

17. Type donut; press Enter.

Command: _donut
Inside diameter <0.5000>:

18. Type 0.0; press Enter.

Outside diameter <1.0000>:

19. Type .125; press Enter.

Center of doughnut:

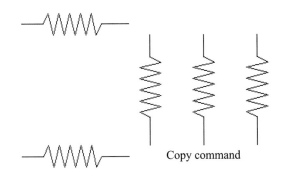

Copy command

Figure 9-12

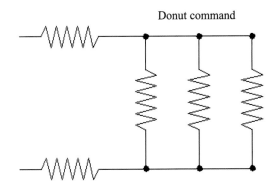

Donut command

Figure 9-14

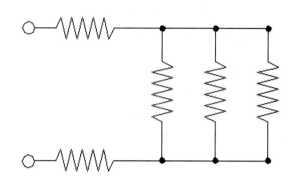

Figure 9-15

20. Locate a solid donut on each of the connection points.

   See Figure 9-14.

21. Select the CIRCLE command (Draw toolbar) and draw a circle of diameter .375.
22. Use the COPY command (Modify toolbar) to create a second circle.
23. Use the EXTEND command to draw the lines that touch the edges of the two open circles.

   See Figure 9-15.

24. Use the TEXT command (Draw toolbar) to add the appropriate text. See Figure 9-7.

To insert blocks with different scale factors

Figure 9-16 shows four different-sized threads, all created from the same block. The block labeled A used the default scale factor of 1, so it is exactly the same size as the original drawing used to create the block. The block labeled B was created with an X scale factor equal to 1, and a Y scale factor equal to 2. The procedure is as follows.

1. Select the Insert Block tool from the Draw toolbar.
2. Select the Thread block.

The Thread block is not an AutoCAD creation. Thread was created specifically for this example.

   Insertion point:

3. Select an insertion point.

   Insertion point: X scale factor <1> /Corner/XYZ:

4. Press Enter.

   Y scale factor (default=X):

5. Type 2.

   Rotation angle <0.00>:

6. Press Enter.

The thread labeled C has an X scale factor of .75 and a Y scale factor of 1.25. The thread labeled D has an X scale factor of .5, a Y scale factor of .75, and a rotation angle of 180 degrees.

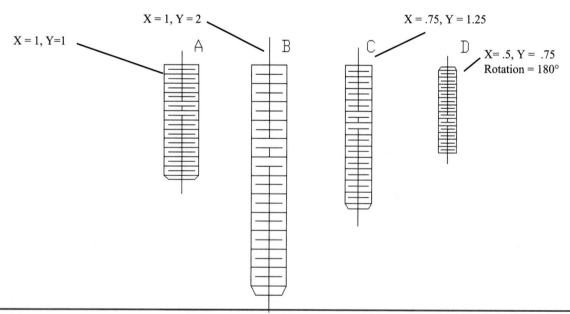

Figure 9-16

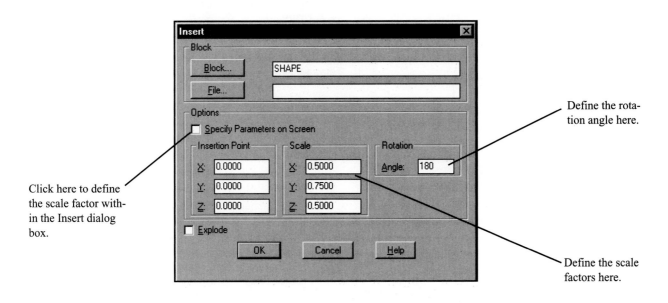

Click here to define the scale factor within the Insert dialog box.

Define the rotation angle here.

Define the scale factors here.

Figure 9-17

To use the Insert dialog box to change the shape of a block

The D thread scale factors and rotation angle were defined using the Insert dialog box. The procedure is as follows.

1. Select the Insert Block.

The Insert dialog box will appear. See Figure 9-17.

2. Click the box to the left of the words Specify Parameters on Screen.

The click should remove the X in the box, indicating that the option has been turned off. When the X is removed, the options labeled Insertion Point, Scale, and Rotation should change from gray to black in the dialog box.

3. Locate the cursor arrow within the X Scale box, backspace to remove the existing value, and type .50.
4. Change the Y scale factor to .75.
5. Change the Angle to 180.

The insertion point could also have been defined using the Insert dialog box. In this example, the thread will appear on the screen with its insertion point at the 0.0000,0.0000 point. The thread could then be moved using the Move command (Modify toolbar). When the prompt

Base point or displacement:

appears when using the Move command, respond by typing 0,0 and pressing Enter; then locate the thread by moving the cross hairs and selecting a point.

To combine blocks

Figure 9-18 shows a hex head screw that was created from two blocks. The threaded portion of the screw was created first, then the head portion was added. Note in Figure 9-18 that the original proportions of the head were changed. When the head block was created, the insertion point was deliberately selected so that it could easily be aligned with the centerline and top surface of the Thread block.

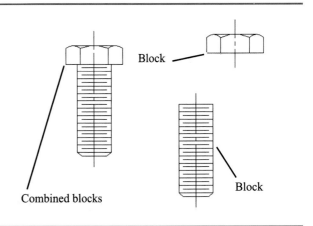

Block

Block

Combined blocks

Figure 9-18

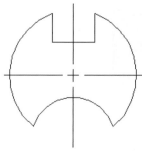

This block is on the drawing screen.
A WBlock must be created from an
exisiting block.

Figure 9-19

9-4 WBLOCK

WBlocks are blocks that can be entered into any drawing. When a block is created, it is unique to the drawing on which it was defined. This means that if you create a block on a drawing, then Save and Exit the drawing, the block is saved with the drawing but cannot be used on another drawing. If you start a new drawing, the saved blocks will not be available.

Any block can be defined and saved as a WBlock. WBlocks are saved as individual drawing files and can be inserted into any drawing.

To create a WBlock saved on a disk in the A drive

Figure 9-19 shows the block shape inserted into a drawing screen. WBlocks can be created only from existing blocks.

1. At a command prompt, type wblock and press Enter.

The Create Drawing File dialog box will appear. See Figure 9-20. There is no WBlock tool.

2. Select the arrow to the right of the Save in box.

The available drives list will cascade.

3. Select the A drive.
4. Locate the cursor in the File name box and backspace to remove the current drawing name, then type the new WBlock name.

See Figure 9-21. In this example the WBlock was given the same name, shape, as the block.

5. Select OK.

The original drawing will appear with the following prompt.

Command: _wblock
block name:

6. Type shape; press Enter.

The WBlock must be created from an existing block.

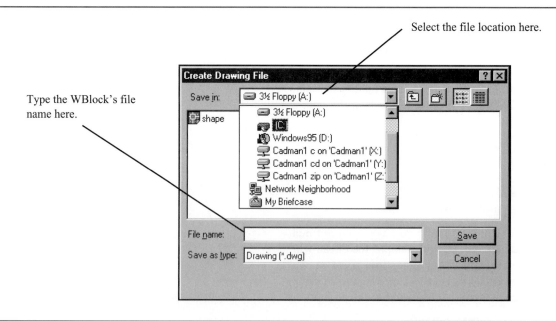

Select the file location here.

Type the WBlock's file name here.

Figure 9-20

Selected drive

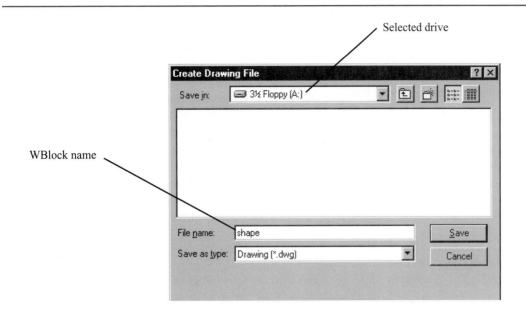

WBlock name

Figure 9-21

In this example the block and WBlock used the same name, but different names may be used. The original drawing will return.

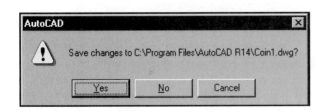

Figure 9-22

To verify that a WBlock has been created

1. Select the Open tool from the Standard toolbar.

   The Save changes box will appear. See Figure 9-22.

2. Select the appropriate response.

   The Select File dialog box will appear.

3. Select the A drive from the Look in box.

   A listing of all drawings on the A drive will appear under the File name box. This list will include all WBlocks. Note in Figure 9-23 that the WBlock named shape is listed.

A listing of existing WBlocks located on the A drive

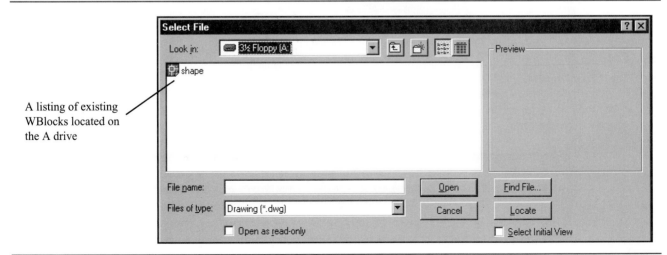

Figure 9-23

To insert a WBlock into a drawing

This example assumes that you have started a new drawing, or are working on a drawing other than the one on which the original block was created.

1. Select the Insert Block tool from the Draw toolbar.

The Insert dialog box will appear. Blocks unique to the drawing are accessed by clicking the Blocks box. WBlocks are accessed by clicking the File box.

2. Select the File box.

The Select Drawing File dialog box will appear.

3. Select the shape WBlock file.

Shape.dwg will appear in the File name box.

4. Select OK.

The WBlock will appear on the drawing screen. If the WBlock appears at an awkward location, or only partially appears on the screen, use the MOVE or PAN commands to position the WBlock in an appropriate location.

9-5 ATTRIBUTES

Attributes are sections of text added to a block that prompt the drawer to add information to the drawing. For example, a title block could have an attribute that prompts the illustrator to add the date to the title block as it is inserted into a drawing. Attributes have their own toolbar.

To access the Attribute toolbar

1. Select the Draw pull-down menu.
2. Select Blocks.
3. Select Define Attributes.

See Figure 9-23.

To add an attribute to a block

Figure 9-25 shows a block. This example will add attributes that request information about the product's part number, material, quantity, and finish.

1. Select the Define Attributes command from the

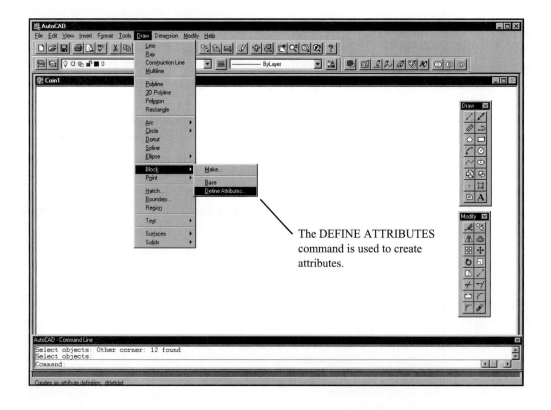

The DEFINE ATTRIBUTES command is used to create attributes.

Figure 9-24

Draw pull-down menu.

The Attribute Definition dialog box will appear. See Figure 9-26.

2. Select the Tag box and type NUMBER.

A tag is the name of an attribute and is used for filing and reference purposes. Tag names must be one word with no spaces.

3. Select the Prompt box and type: Define a part number.

The prompt line will eventually appear at the bottom of the screen when a block containing an attribute is inserted into a drawing. If you do not define a prompt, the Tag name will be used as a prompt.

4. Leave the Value box empty.

The Value box entry will be the default value if none is entered when the prompt appears. Based on the defined prompt and value inputs, the following prompt line would appear when the attribute is combined with a block.

Define the part number <>:

5. Select Pick Point.

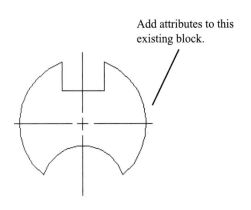

Add attributes to this existing block.

Figure 9-25

The original drawing will appear with cursor. Select a location for the Attribute tag by moving the cross hairs, then press the left mouse button.

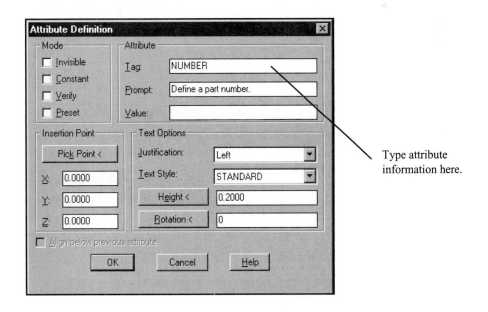

Type attribute information here.

Figure 9-26

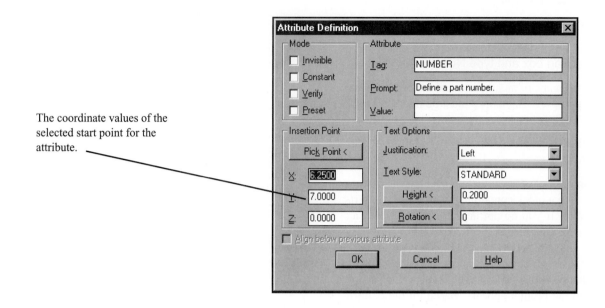

The coordinate values of the selected start point for the attribute.

Figure 9-27

The Attribute Definition dialog box will reappear with the X,Y coordinate values of the selected point listed in the Insertion Point box. See Figure 9-27.

6. Select OK.

Figure 9-28 shows the Attribute tag applied to a drawing. Figures 9-29, 9-30, and 9-31 show three more Attribute Definition dialog boxes. Note that the Align

below previous attribute box has an X indicating it has been turned on. The option is turned on by clicking the box. Figure 9-32 shows the resulting drawing.

To create a new block that includes attributes

1. Select the Make Block tool from the Draw toolbar.

   The Block Definition dialog box will appear.

2. Enter the block name Coin-A; select the Select object box.

   Select object:

3. Window the block and all the attribute tags shown in Figure 9-32.

   The Block Definition dialog box will reappear.

4. Select the Select point box within the Base point box.

   Insertion base point:

5. Select the center point of the block.

   The Block definition dialog box will reappear.

6. Select the OK box.

The Attribute tag from Figure 9-26 applied to a Block.

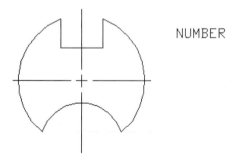

NUMBER

Figure 9-28

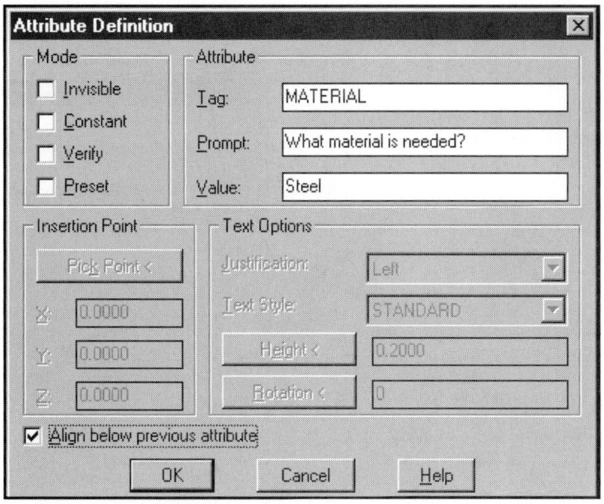

Figure 9-29

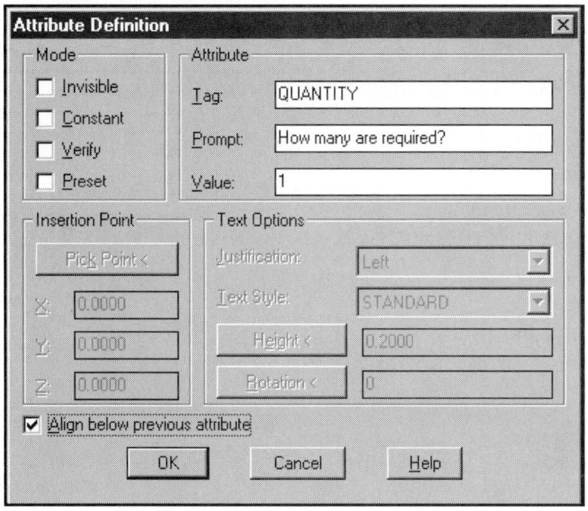

Figure 9-30

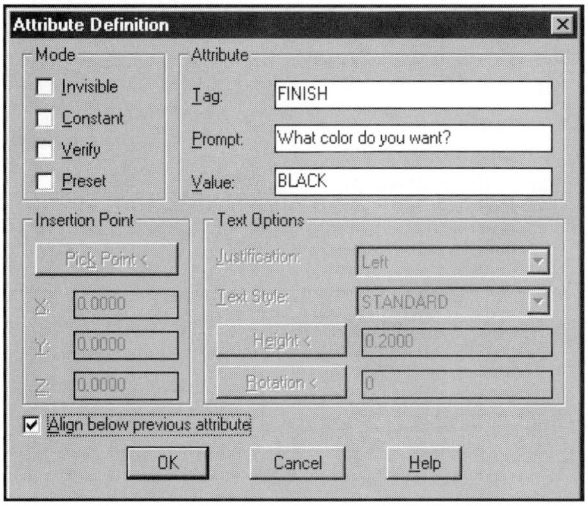

Figure 9-31

The attributes created in Figures 9-29, 9-30, and 9-31 added to a block

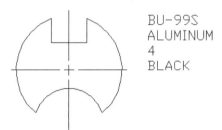

Create a new block, COIN-A, that includes the existing block, SHAPE, and the attributes.

Figure 9-32

To insert an existing block with attributes

1. Select the Insert Block tool on the Draw toolbar.

   The Insert dialog box will appear. See Figure 9-33.

2. Select Block.

   The Defined Blocks dialog box will appear. See Figure 9-34.

3. Select COIN-A, then OK, OK.

   Insertion point:

4. Select an insertion point.

   Insertion point: X scale factor <1> / Corner / XYZ:

5. Press Enter.

   Y scale factor (default=X):

6. Press Enter.

   Rotation angle <0.00>:

   The block will appear on the screen and the attribute prompt values you assigned will appear at the bottom of the screen in the prompt line. See Figure 9-35.

Select Block.

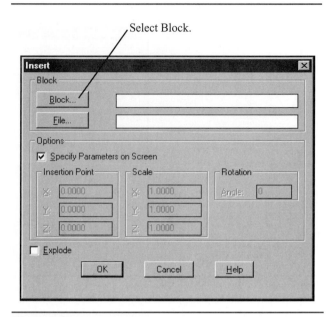

Figure 9-33

New block that includes attributes

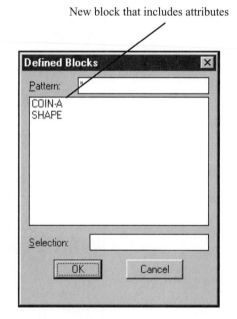

Figure 9-34

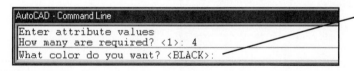

The words entered in the Attribute Definition prompt box will appear on the prompt line when the block is inserted. The value input will be the default value.

Figure 9-35

What color do you want? <BLACK>:

7. Accept the default value BLACK by pressing Enter.

    How many are required<1>:

8. Type 4; press Enter.

    What material is required <STEEL>:

9. Type Aluminum.

    Define the part number <None>:

10. Type BU-99S; press Enter.

    Figure 9-36 shows the resulting drawing.

To edit an existing attribute

Once a block has been created that includes attributes, it may be edited using the Edit Attributes command. The procedure is as follows. The block to be edited must be on the drawing screen.

1. Select the Edit Attributes tool from the Modify II toolbar.

    Command: _ddate
    Select block:

2. Select block COIN-A.

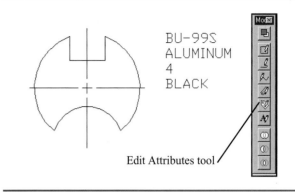

Edit Attributes tool

Figure 9-36

The Edit Attributes dialog box will appear. See Figure 9-37. In this example, the color will be changed from BLACK to GREEN. Note that the attribute prompt lines originally entered in the Attribute Definition dialog box are listed on the left side of the Edit Asttributes dialog box.

3. Locate the cursor to the right of the word BLACK, backspace to remove BLACK, then type GREEN.

    Figure 9-38 shows the resulting changes.

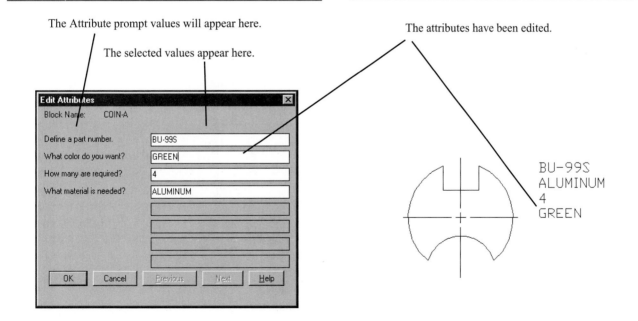

The Attribute prompt values will appear here.

The selected values appear here.

The attributes have been edited.

Figure 9-37

Figure 9-38

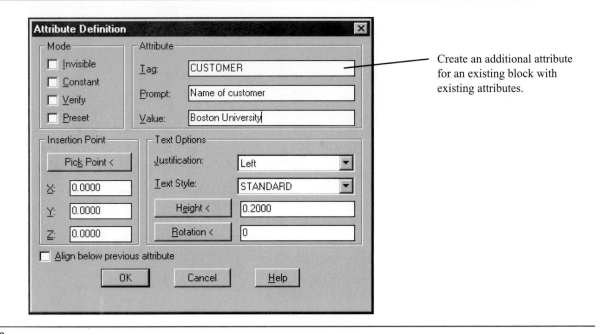

Create an additional attribute for an existing block with existing attributes.

Figure 9-39

To use the REDEFINE ATTRIBUTE command

The REDEFINE ATTRIBUTE command is used to add additional attributes to an existing block with attributes.

1. Insert block COIN-A (a block with attributes) onto the drawing screen.
2. Explode the block.
3. Select the Draw pull-down menu, then Block, then Define Attributes.

The Attribute Definition dialog box will appear.

4. Create an additional attribute tag as shown in Figure 9-39.
5. Select Pick Point.

Start point:

6. Select an insertion point for the new tag.

Select a point aligned with the existing attributes. The Attribute Definition dialog box will reappear with the coordinate values of the selected start point listed in the Insertion Point box.

7. Select OK.

The new attribute tag will appear on the screen as shown in Figure 9-40.

8. Select the Make Block tool.

The Block Definition dialog box will appear.

9. Name the block COIN-A, select the entire shape including the attributes, and select a base point.

This is the same name as the old block so a Warning box will appear.

10. Select Redefine.

The new attribute has been added to the block.

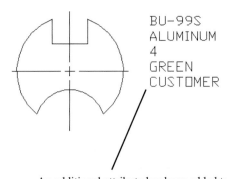

BU-99S
ALUMINUM
4
GREEN
CUSTOMER

An additional attribute has been added to the block.

Figure 9-40

To edit an existing attribute

The Edit Attribute command is used to change information in a block independently of the block's definitions. The block to be edited must be on the screen.

1. Type attedit; press Enter.

   Command: _attedit
   Edit attributes one at a time? <Y>:

2. Press Enter.

   block name specification<>:*

3. Type COIN-A; press Enter.

   Attribute tag specification:

4. Type QUANTITY; press Enter.

   Attribute value specification:

5. Type 4; press Enter.

   Select attributes:

6. Select the number 4 associated with the COIN-A block on the screen; press Enter.

   Value/Position/Height/Angle/Style/Layer/Color/ Next <N>:

   The value 4 will disappear from the screen and a large X will appear on the screen. See Figure 9-41.

7. Type V; press Enter.

   Change or replace? <R>:

   The change option is used to modify a few characters of the existing value. The replace option is used to create an entirely new value. The replace option is the default option.

8. Press Enter.

   New attribute value:

9. Type 12; press Enter.

   Value/Position/Height/Angle/Style/Layer/Color/ Next <N>:

10. Press Enter.

    The new value will replace the former value. See Figure 9-42.

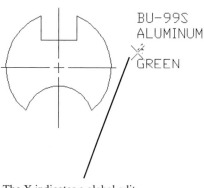

The X indicates a global edit.

Figure 9-41

> **NOTE:**
>
> It is possible to create a block that contains only attributes. A drawing is not required. Use the Define Attribute command and define as many tags as desired, locate the tags on the drawing screen, then save them as a block.

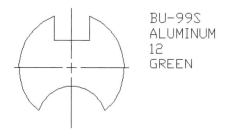

The attribute has been edited.

Figure 9-42

9-6 TITLE BLOCKS WITH ATTRIBUTES

Figure 9-43 shows a title block that has been saved as a block. It would be helpful to add attributes to the block so that when it is called up to a drawing, the drawer will be prompted to enter the required information.

Figures 9-44, 9-45, 9-46, 9-47, and 9-48 show the five Attribute Definition dialog boxes used to define the title block attributes. Note that the text height in Figures 9-46 and 9-47 was changed to 0.250. Also note that no attribute values were assigned. This means that the space on the drawing will be left blank if no prompt value is entered. Figure 9-49 shows the resulting title block with the attribute tags in place.

The new block with attributes was saved using the block command as TITLE-A. It can now be saved as a WBlock and used on future drawings. Figure 9-50 shows the block-generated prompt lines with responses, and Figure 9-51 shows a possible title block created from the TITLE-A block.

Title block saved as a block.

Figure 9-43

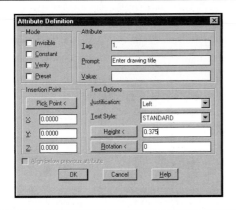

Figure 9-44

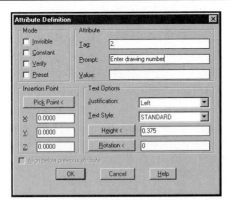

Figure 9-45

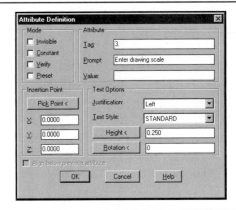

Figure 9-46

Figure 9-47

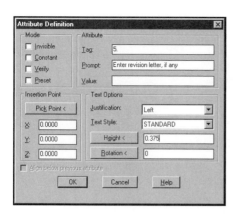

Figure 9-48

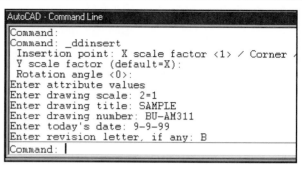

Prompt lines created in the Attribute Definition dialog boxes

Figure 9-50

Title block with attribute tags in place

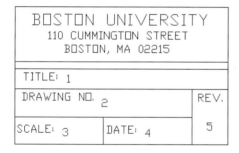

Figure 9-49

Some possible responses

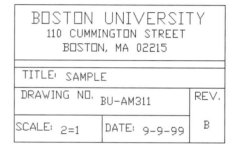

Figure 9-51

9-7 LAYERS

A layer is like a clear piece of paper you can lay directly over the drawing. You can draw on the layer and see through it to the original drawing. Layers can be made invisible and information can be transferred between layers.

In the example shown, a series of layers will be created, then a group of lines will be moved from the initial 0 layer to the other layers.

The Layer tool is located just below the Standard toolbar and includes six associated tools and headings. See Figure 9-52. The associated tools are not command tools but indicators of the status of the Layer options. For example, the open padlock shown indicates that the Layer is not locked. If it were locked, the Padlock tool would change to the closed position.

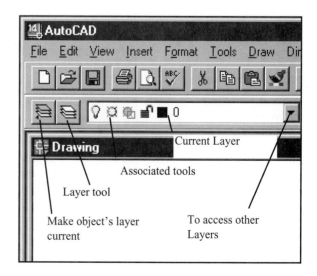

Figure 9-52

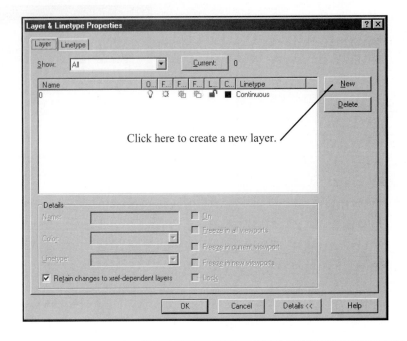

Figure 9-53

To create new layers

This exercise will create two new layers: Hidden and Center.

1. Select the Layer tool located on the left of the screen just below the Sandard toolbar.

The Layer and Linetype Properties dialog box will appear. See Figure 9-53.

2. Click the New box.

A new line will be added to the name list. The line will read Layer1.

3. Backspace out the name Layer1 and type in HIDDEN.

See Figure 9-54.

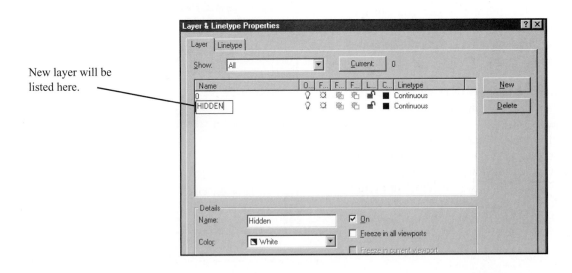

Figure 9-54

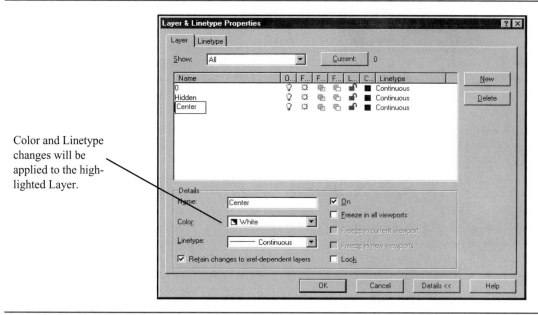

Figure 9-55

4. Click the New box again and name the next new layer Center

The name Center will appear in the Layer Name box. There are now three layers associated with the drawing. The 0 layer is the current layer as indicated by the 0 to the right of the Current Layer heading.

To change the color of a layer

1. Move the arrow cursor to the Center layer line and press the left mouse button.

The word Center will be highlighted as shown in Figure 9-55.

2. Select the Color box.

The color options will cascade down. See Figure 9-56.

3. Select the red color patch at the top of the dialog box.

The word Red should appear in the Color box along with a red patch.

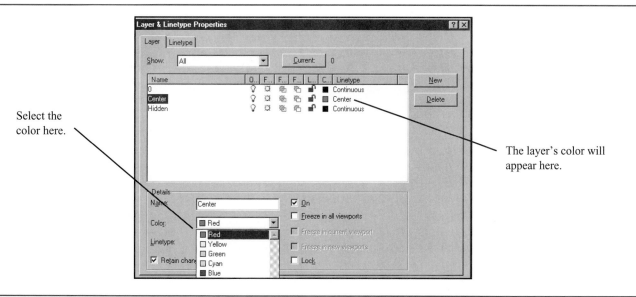

Figure 9-56

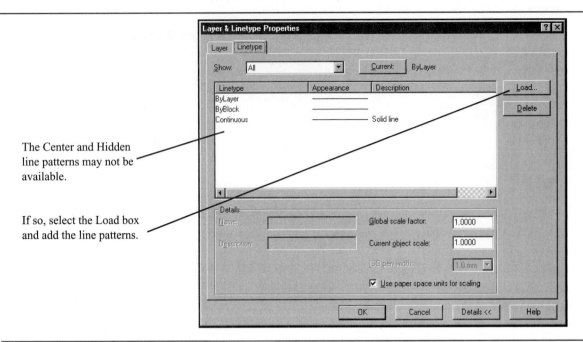

The Center and Hidden line patterns may not be available.

If so, select the Load box and add the line patterns.

Figure 9-57

4. Select the Hidden layer and change its color to blue.

5. Select OK.

The Layer and Linetypes Properties dialog box will disappear. Click the Layer Control tool to return the Layer Control dialog box to the screen.

To change the linetype of a layer

1. Select the Linetype tab in the Layer & Linetypes Properties dialog box..

The Linetype folder will appear. See Figure 9-57. The Hidden and Center line patterns may not be displayed under the Linetype heading. If not, select the Load box, then select the Hidden and Center lines from the Load or Reload Linetypes dialog box shown in Figure 9-58.

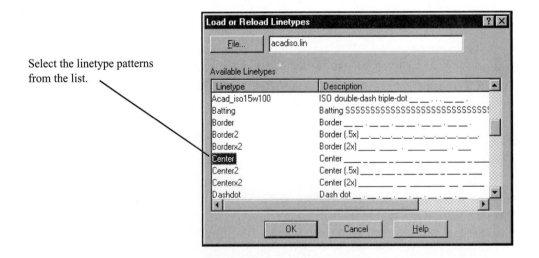

Select the linetype patterns from the list.

Figure 9-58

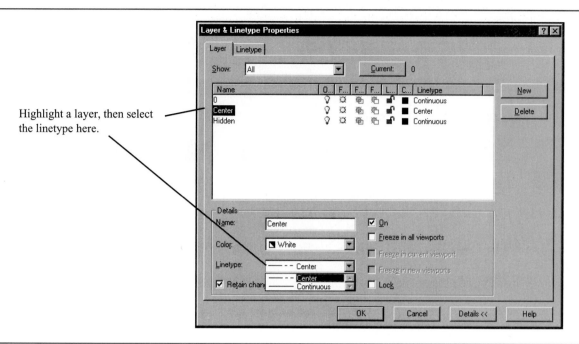

Highlight a layer, then select the linetype here.

Figure 9-59

2. Select OK, OK to return to the Layers & Linetypes Properties dialog box.
3. Highlight the Center layer.
4. Select the arrow to the right of the Linetype box.

A listing of avaible linetypes will cascade down. See Figure 9-59.

5. Select the Hidden layer, then the Hidden linetype pattern.

Your Layer & Linetype Properties dialog box should look like Figure 9-60.

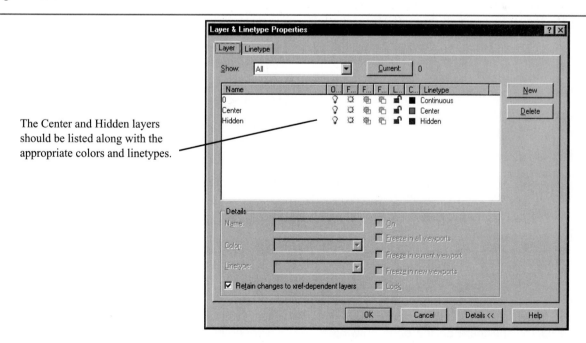

The Center and Hidden layers should be listed along with the appropriate colors and linetypes.

Figure 9-60

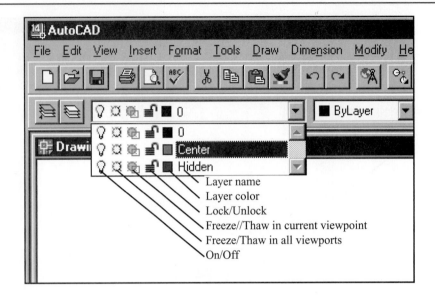

Figure 9-61

To access the defined layers

It is not necessary to use the layer tool to access the Layer Control dialog box in order to make changes to the layers. The Layer Control box, see Figure 9-60, can be used to access the layers and their option directly.

1. Select the arrow box on the right side of the Layer Control box.

A listing of defined layers will cascade from the box. See Figure 9-61. The tools shown on the same line as the layer name can be used to change the layer's options. The tools are toggle switches that allow you to turn the various options on or off.

2. Select the Open Lock tool on the 0 layer line.

The tool will change to a closed lock. The layer is now locked.

3. Select the Sun tool on the Center layer line.

The Sun tool will be replaced with a snowflake, and the word Center will become lighter, indicating the layer is now frozen.

4. Select the lightbulb on the Hidden layer line.

The yellow color will be replaced by a dark color indicating that the layer is off.

5. Click all the tools again, returning them to their original settings.
6. Click the cursor anywhere on the open drawing screen to return to a complete drawing screen.

To draw in different layers

The default layer is the 0 layer. This is indicated on the screen by the 0 in the Layer Control box.

1. Select the Line tool in the Draw toolbar and draw a line on the screen.

See Figure 9-62. This line is on the 0 layer. You can draw only on the current layer. You can see other layers, but you can work only on the current layer.

2. Select the arrow box to the right of the layer Control box, then click the Center layer line.

The current layer is now the Center layer. The color patch in the Layer Control box will be red and the word Center will be next to the red color patch.

3. Select Circle from the Draw toolbar and draw a circle on the screen.

The circle will appear in red and in the centerline pattern.

To change the scale of a linetype

The circle in Figure 9-62 is drawn using the centerline pattern. The spacing of the centerline pattern is based on default values which may not be acceptable for all drawings.

1. Select the Properties tool from the Object Properties toolbar.

Select objects:

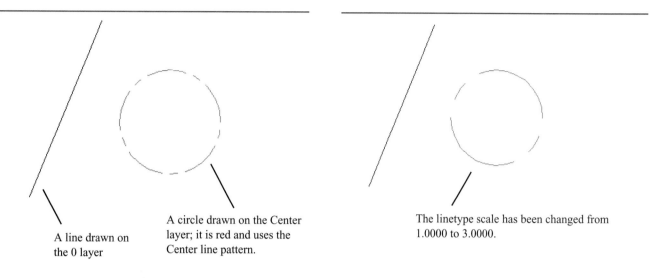

A line drawn on
the 0 layer

A circle drawn on the Center
layer; it is red and uses the
Center line pattern.

The linetype scale has been changed from
1.0000 to 3.0000.

Figure 9-62

Figure 9-64

2. Select the circle.

The Modify Circle dialog box will appear. See Figure 9-63. Remeber, you can only work on the current layer. The Center layer should be the current layer.

3. Select the Linetype scale box.
4. Change the scale factor from 1.0000 to 3.0000.

See Figure 9-64. Note the difference between the centerline patterns shown in Figures 9-63 and 9-64.

To change layers

An object may be drawn on one layer, then moved to another layer. This feature is helpful when designing; an original layout may be created in a single layer, then lines may be transferred to different layers as needed. As a line changes layers, it assumes the color and linetype of the new layer.

1. Return to layer 0 and draw a rectangle.

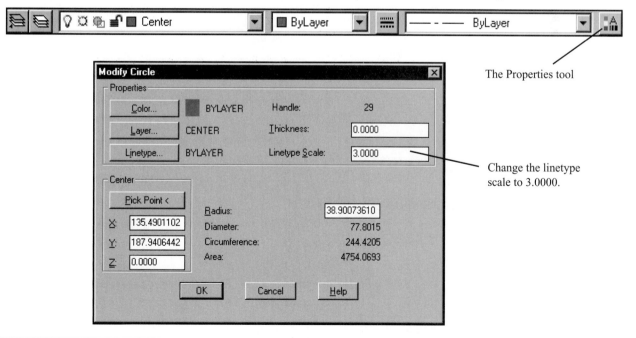

The Properties tool

Change the linetype
scale to 3.0000.

Figure 9-63

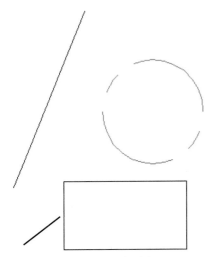

The rectangle was drawn on the 0 layer.

Figure 9-65

Select the Hidden layer.

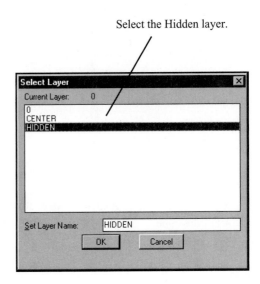

Figure 9-67

See Figure 9-65.

2. Select the Properties tool from the Objects Properties toolbar.

Select objects:

3. Select the rectangle.

Select objects:

4. Press Enter.

The Modify Polyline dialog box will appear. See Figure 9-66.

5. Select the layer box.

The Select Layer dialog box will appear. See Figure 9-68.

6. Select the HIDDEN layer.

The highlight line will switch from the 0 layer to the Hidden layer.

7. Select OK, OK to return to the original drawing.

The rectangle will now be drawn using the Hidden line pattern, and will be blue. See Figure 9-68.

Select the Layer box

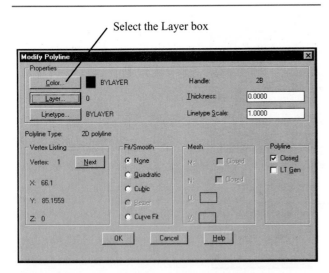

Figure 9-66

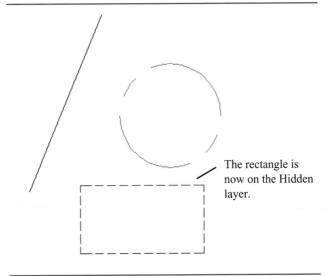

The rectangle is now on the Hidden layer.

Figure 9-68

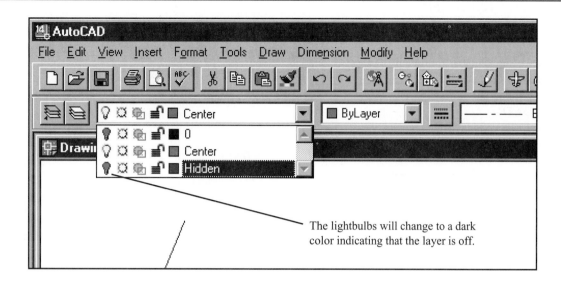

Figure 9-69

To work with layers

When preparing a drawing, it is sometimes desirable not to see a layer; or to see a layer, but not be able to accidentally edit it. For example, you want to work in the Center layer, but not see either the 0 or Hidden layers. You need to make the Center layer the current layer and turn off the other layers.

To turn layers off

1. Select the Layer Control tool.

   A listing of current layers will cascade from the box.

2. Select the word Center.

   The Center layer is now the current layer.

3. Select the Layer Control tool again.
4. Select the lightbulb icon for both the 0 and Hidden layers.

The lightbulb changes to a dark color, indicating that the layer is off. See Figure 9-69. The resulting drawing will show only the circle. See Figure 9-70. The line and the rectangle are still part of the drawing, but because their layers are turned off, they are not visible.

Other layer options

The Lock/Unlock option allows you to lock a layer. A locked layer cannot be edited, but you can add shapes to a locked layer.

The Freeze/Thaw option makes layers invisible. Frozen layers will not appear in a plot of the drawing.

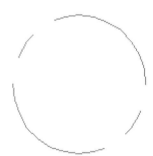

Only the circle appears because it is drawn on the Center layer and the other layers have been turned off.

Figure 9-70

9-8 EXERCISE PROBLEMS

EX9-1

A. Create a block of the following drawing layout. All dimensions are in inches.

B. Add the following attributes.

Tag = DATE
Prompt = Enter today's date
Value = (leave blank)
Tag = DRAWING
Prompt = Enter the drawing number
Value = (leave blank)

Tag = NAME
Prompt = Enter your name
Value = (leave blank)

Align the attribute tags on the drawing.

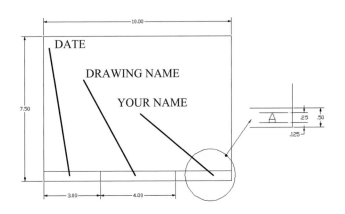

EX9-2

A. Create a block of the following drawing layout. All dimensions are in millimeters.

B. Add the following attributes

Tag = DATE
Prompt = Enter today's date
Value = (leave blank)

NOTE: MAKE ALL LETTERING 0.25 HIGH
UNLESS OTHERWISE STATED.

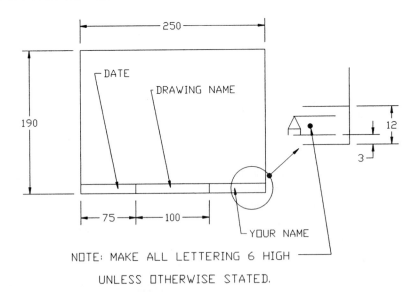

DRAWING LAYOUT − 2(MILLIMETERS)

NOTE: MAKE ALL LETTERING 6 HIGH
UNLESS OTHERWISE STATED.

EX9-3

Create a block for the standard tolerance block shown below.

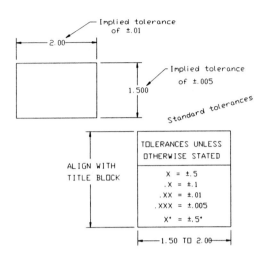

EX9-4

A. Create a block for one of the title blocks shown below. A sample completed title block is also shown.

B. Add attributes that will help the user complete the title block satisfactorily.

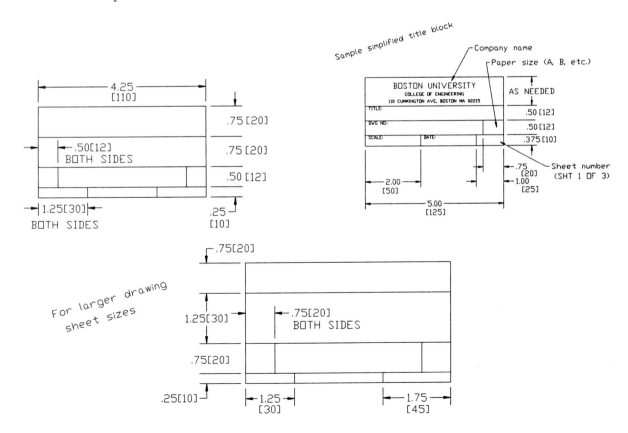

EX9-5

Create a block of the following release block and add the appropriate attributes. All dimensions are in inches.

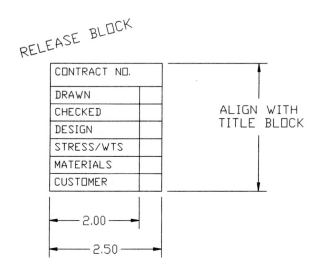

EX9-6

Create blocks for the following parts lists formats. All dimensions are in inches.

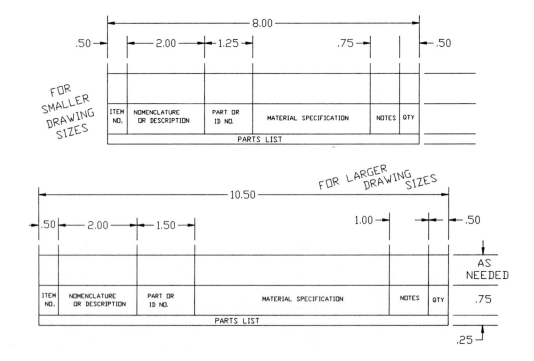

EX9-7

Create blocks for the following revision block formats. The dimensions within the dimension lines are inches; the dimensions within the brackets are millimeters.

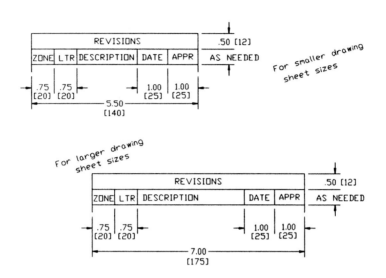

EX9-8

Use the blocks created in the previous exercises to create the drawing format shown below. Scale the blocks to align with the following drawing limits.

A. 8 x 11.5
B. 297 x 210
C. 18 x 24
D. 11 x 17

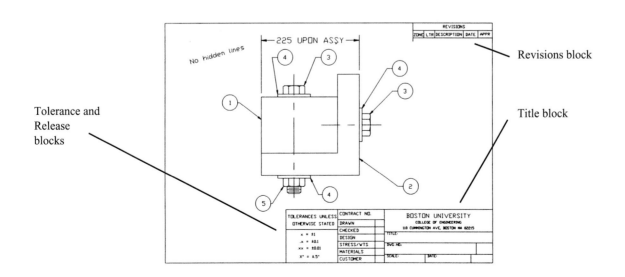

Tolerance and Release blocks

No hidden lines

225 UPON ASSY

Revisions block

Title block

EX9-9

A. Create blocks for the following screws. Use a diametric value of Ø 1.00. This will make it simpler to calculate scale factors when adding screws to the drawings.

B. Add attributes that call for the set screws' size specifications.

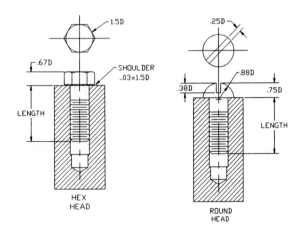

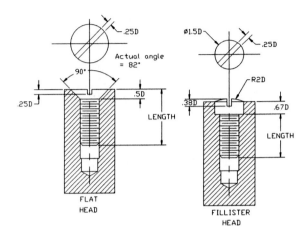

EX9-10

A. Create blocks for the following set screws. Assume a diametric value of Ø 1.00.

B. Add attributes that call for the set screws' size specifications.

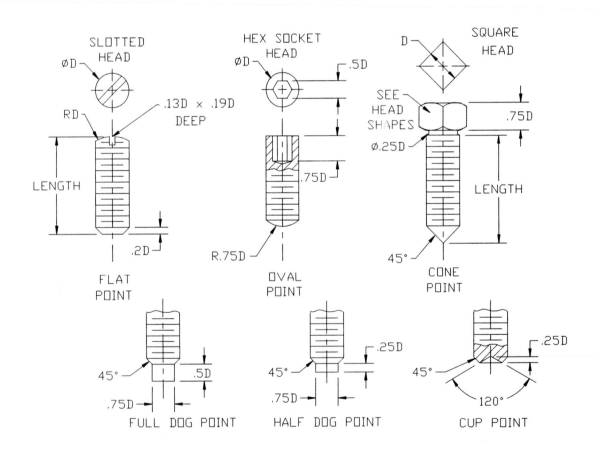

C H A P T E R   **10**

Fundamentals of 3D Drawing

10-1 INTRODUCTION

This chapter introduces the fundamental concepts needed to produce 3D drawings using AutoCAD. The chapter shows how to change viewpoints and how to create, save, and work with user-defined coordinate systems called user coordinate systems, or UCSs.

The chapter will demonstrate how to use both the Viewpoint and UCS toolbars. It will also show how to create orthographic views from given 3D objects using the View toolbar.

10-2 THE WORLD COORDINATE SYSTEM

AutoCAD's absolute coordinate system is called the World Coordinate System (WCS). The default setting for the WCS is a viewing position located so that you are looking at the system 90 degrees to its XY plane. See Figure 10-1. The Z axis is also perpendicular to the XY plane, or directly aligned with your viewpoint. This setup is ideal for 2D drawings and is called a plan view. All your drawings up to now have been done in this orientation.

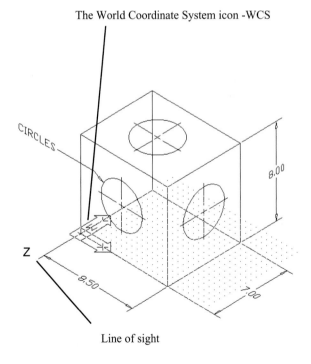

Figure 10-1

255

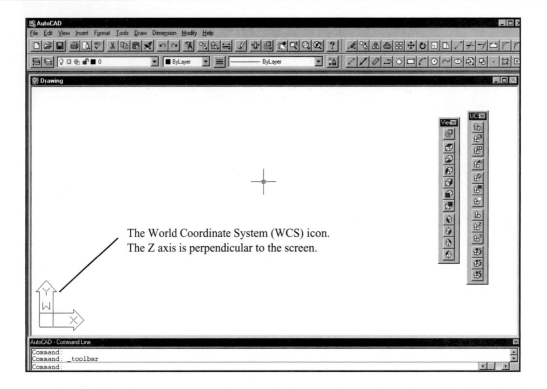

The World Coordinate System (WCS) icon.
The Z axis is perpendicular to the screen.

Figure 10-2

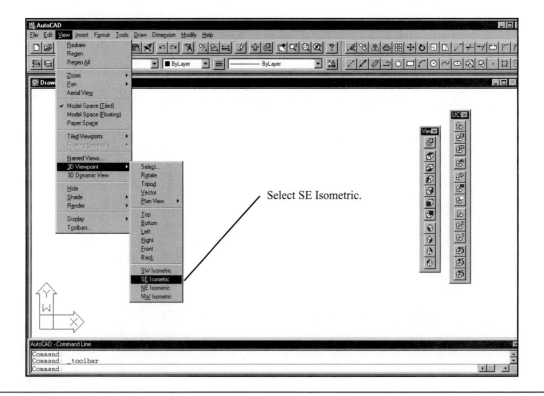

Select SE Isometric.

Figure 10-3

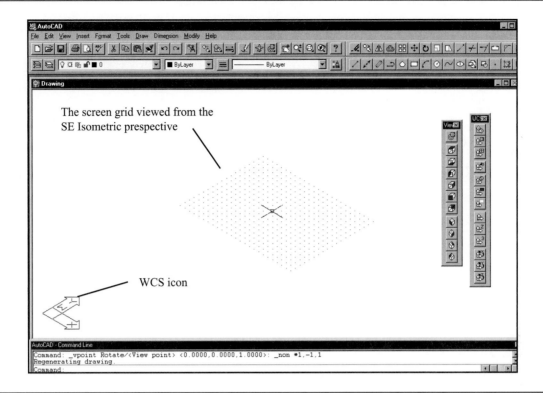

Figure 10-4

Figure 10-2 shows a standard drawing screen setup oriented so that you are looking directly down on the XY plane of the World Coordinate System. Note the icon in the lower left corner of the screen. The W indicates that it is the WCS, and the X and Y indicate the orientation of these axes. The WCS icon will always be present when you are in the WCS.

10-3 VIEWPOINTS

The orientation of the WCS may be changed by changing the drawing's viewpoint. There are two ways to change the viewpoint: use the View pull-down menu or the View toolbar.

To change the viewpoint using the View pull-down menu

This exercise assumes that the screen includes a grid. The grid serves to help define a visual orientation.

1. Select the View pull-down menu.
2. Select 3D Viewpoint.

See Figure 10-3. The current WCS orientation is the Plan View setting.

3. Select SE Isometric.

The drawing will now be oriented as shown in Figure 10-4. Note the change in the WCS icon. It is important to remember that you are still in the WCS, but are merely looking at it from a different perspective. This concept may be verified by drawing some simple shapes. Figure 10-5 shows a rectangular shape created using the LINE command, and a circle created using the CIRCLE command. These shapes are 2D shapes drawn on the WCS.

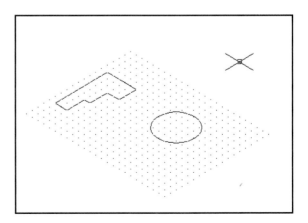

Figure 10-5

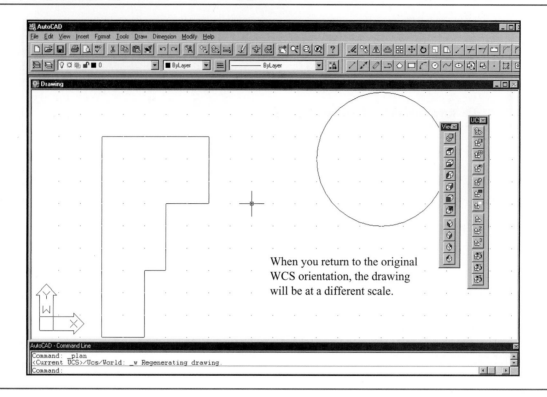

When you return to the original WCS orientation, the drawing will be at a different scale.

Figure 10-6

To return to the original WCS orientation

1. Select the View pull-down menu.
2. Select 3D Viewpoint, Plan View, World UCS.

See Figure 10-6. The drawing will reappear on the screen at a different scale than was originally defined.

3. Type zoom; press Enter.

All/Center/Dynamic/Extents/Previous/Scale/(X/X P)/Window/<Realtime>:

4. Type .75; press Enter.

Figure 10-7 shows the resulting screen scale.

To change the drawing's viewpoint using the View toolbar

1. Select the View pull-down menu.
2. Select Toolbars, then View.

The View toolbar will appear on the screen. It can be moved and reshaped as described in Chapter 1. Figure 10-8 shows both the View toolbar and the View, 3D Viewpoint menus. The same commands are included in both.

3. Select the SE Isometric tool.

The screen will return to the same orientation as was achieved using the SE Isometric command listed on the

View pull-down menu, Figure 10-4. The scale will be larger, so use the ZOOM command to achieve a smaller scale.

4. Type zoom; press Enter.

All/Center/Dynamic/Extents/Left/Previous/Vmax/ Window/<Scale(X/XP)>:

5. Type .75; press Enter.

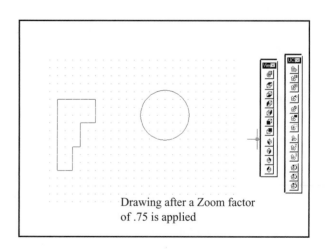

Drawing after a Zoom factor of .75 is applied

Figure 10-7

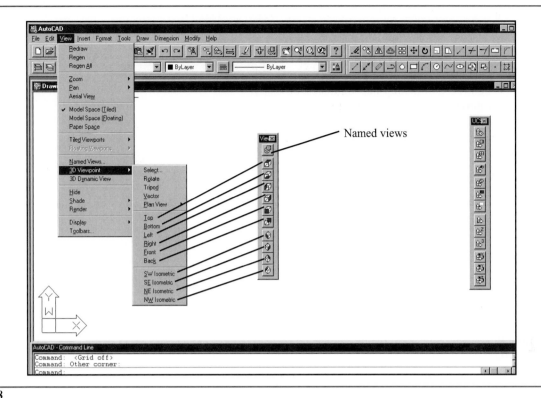

Named views

Figure 10-8

To return to the original WCS orientation

1. Select the Top tool from the View toolbar.

The original plan view orientation will return to the screen. Use the ZOOM command to reduce the drawing's scale size if necessary. In Figure 10-9 a Zoom scale factor of .50 was used.

It is suggested that you try several of the View toolbar commands and watch how the WCS icon and grid pattern change. Remember that the TOP VIEW command will return the Plan View in WCS orientation.

10-4 USER COORDINATE SYSTEMS (UCSs)

User coordinate systems are coordinate systems that you define relative to the WCS. Drawings often contain several UCSs. UCSs can be saved and recalled.

Figure 10-10 shows a wedge oriented using the SE Isometric View icon. This section assumes that a wedge shape already exists on the screen and will create a UCS aligned with the slanted surface of the wedge.

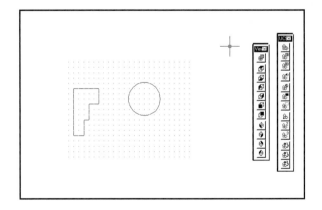

Drawing with a Zoom scale factor of .50

Figure 10-9

To create a UCS

1. Select the View pull-down menu.
2. Select toolbars, then UCS.

The UCS toolbar will appear on the screen. As with viewpoints, UCS-related commands can be accessed through a toolbar or through pull-down menus. Figure 10-11 shows both the UCS toolbar and the UCS pull-down menus. The UCS commands are accessible using the Tools pull-down menu.

3. Select the 3 Point option.

   Origin/Zaxis/3point/OBject/View/X/Y/ZPrev/
   Restore/Save/Del/?/<World>: _3
   Origin point <0,0,0>:

4. Select the lower corner of the wedge (Use OSNAP if necessary).

   Point on positive portion of the X-axis
   <7.0000,0.0000,0.0000>:

The given coordinate values are the coordinate values of the new origin point relative to the WCS. See Figure 10-12.

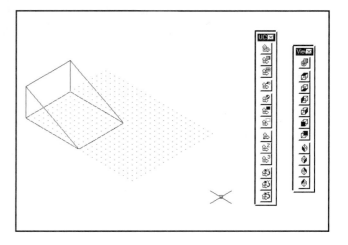

Figure 10-10

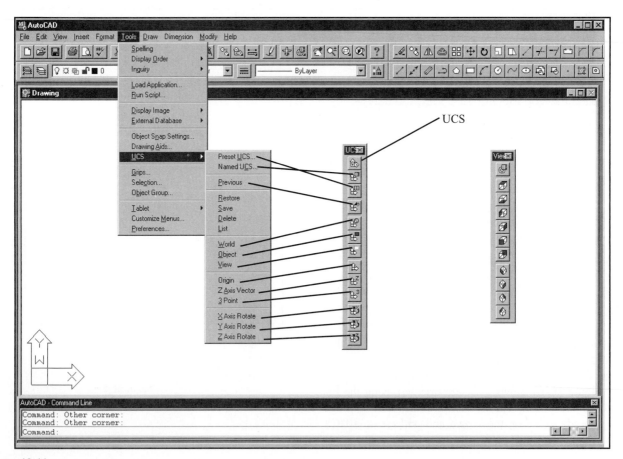

Figure 10-11

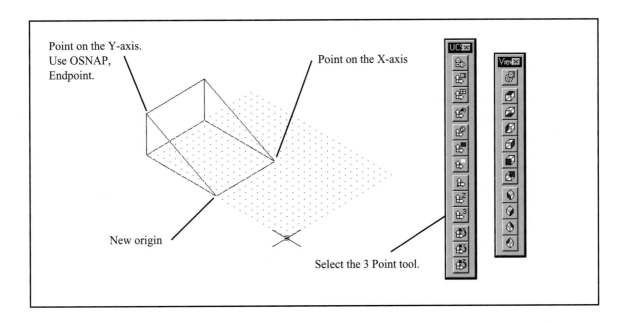

Point on the Y-axis.
Use OSNAP,
Endpoint.

Point on the X-axis

New origin

Select the 3 Point tool.

Figure 10-12

5. Select the lower right corner of the wedge.

Point on the positive-Y portion of the UCS XY plane <5.0000,0.0000,0.00000>:

6. Use OSNAP, Endpoint option and select the upper left corner of the wedge.

It is imperative that the Endpoint option of the OSNAP command be used to locate the point. The crosshairs will move only in the current coordinate plane. In this exercise we are still in the WCS because the new UCS has not been completely defined yet. If you were to try to select the upper corner of the wedge visually, that is, by moving the crosshairs so that they appeared to be located on the corner, you would get an incorrect result. The selected point would actually be a point on the WCS directly behind the corner point. Figure 10-13 shows the reoriented coordinate system.

To save a UCS

It is helpful to save defined UCSs so that if they are needed again during the preparation of the drawing, they will not have to be redefined.

1. Select the Tools pull-down menu, then UCS, then Save.

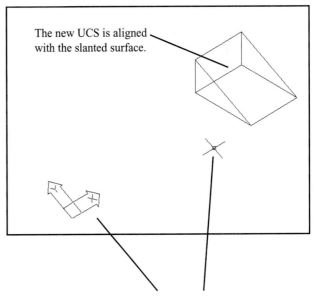

The new UCS is aligned with the slanted surface.

The UCS icon and cursor will change orientation to align with the new UCS.

Figure 10-13

Origin/Zaxis/3point/OBject/View/X/Y/ZPrev/
Restore/Save/Del/?/<World>: _s
?/Desired UCS name:

2. Type Slant; press Enter.

Command:

The new UCS named Slant has been saved and can be retrieved using the Named UCS tool on the UCS toolbar.

To return to the WCS using named UCSs

1. Select the Named UCS tool.

The UCS Control dialog box will appear. See Figure 10-14. Note that the Slant UCS is the current UCS.

2. Select *WORLD* in the UCS Control dialog box.
3. Select Current.

The word Current will move from the SLANT line to the *WORLD* line.

4. Select OK.

The original WCS will return to the screen.

To return to the WCS directly

1. Select the World UCS icon or the World command on the Set UCS menu.

10-5 WORKING WITH UCSs

Drawing with AutoCAD in 3D is limited to the UCS orientation. You can draw only in the current UCS, as shown in Figure 10-15. You cannot draw a circle on the slanted surface of the wedge or on the side or back surfaces of the wedge using the current UCS orientation. You must draw on a surface oriented to the surface you wish to work on.

To draw on a slanted surface

To draw on the slanted surface shown in Figure 10-15, we must first move to a UCS oriented to the slanted surface. A UCS named Slant was created in the previous section, so we will return to that UCS and then draw a circle. The procedure is as follows.

1. Select the Named UCS tool on the UCS toolbar.

The UCS Control dialog box will appear. See Figure 10-16.

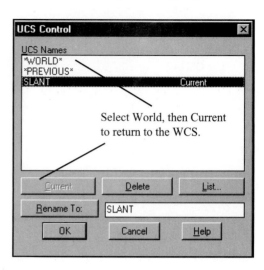

Select World, then Current to return to the WCS.

Figure 10-14

2. Select the Slant UCS and make it Current.

3. Select OK.

The screen will be reoriented to the Slant UCS settings.

4. Use the Circle tool to draw the circle.

See Figure 10-17.

5. Select the World UCS icon to return to the original drawing orientation.

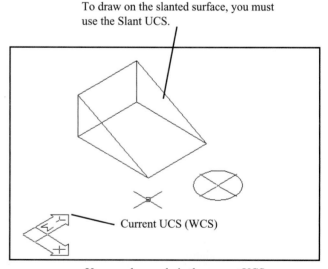

To draw on the slanted surface, you must use the Slant UCS.

Current UCS (WCS)

You can draw only in the current UCS, in this example, the WCS.

Figure 10-15

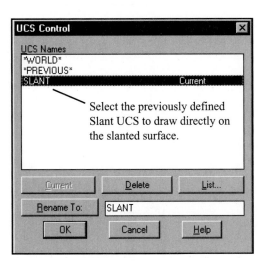

Figure 10-16

To draw on a surface perpendicular to the WCS XY axis

The side surface of the wedge is perpendicular to the WCS XY axis. To draw on the side surface we must reorient the current surface UCS to be aligned with the side surface. In addition, we must also align the origin of the axis with the side surface.

1. Select the X Axis Rotate tool from the UCS toolbar.

   *Origin/Zaxis/3point/OBject/View/X/Y/ZPrev/
   Restore/Save/Del/?/<World>: _x
   Rotation angle about X axis <0>:*

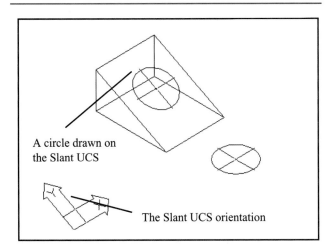

A circle drawn on the Slant UCS

The Slant UCS orientation

Figure 10-17

2. Type 90; press Enter.

The drawing icon in the lower left corner of the screen will shift positions, and the letter W will disappear from the icon because the drawing is no longer in the WCS.

3. Select the Origin tool from the UCS toolbar.

   *Origin/Zaxis/3point/OBject/View/X/Y/ZPrev/
   Restore/Save/Del/?/<World>: _o
   Origin point<0,0,0>:*

4. Use OSNAP, Endpoint (shift key, right mouse button), and select the lower left corner of the wedge.

The origin has now been shifted. This can be verified by locating the cross hairs on the lower left corner of the wedge and checking the coordinate display at the bottom of the screen.

5. Use the Circle tool to draw a circle on the side surface of the wedge.

   See Figure 10-18.

6. Select the World tool on the UCS toolbar to return the drawing to its original WCS orientation.

A circle drawn in the current UCS

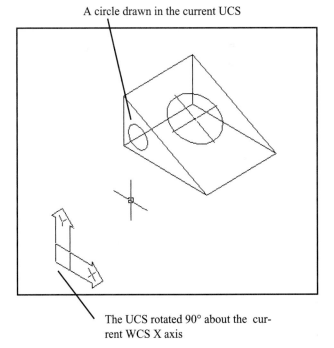

The UCS rotated 90° about the current WCS X axis

Figure 10-18

To draw a circle on the back surface of the wedge

1. Select the Y Axis Rotate tool from the UCS tool-bar.

   Origin/Zaxis/3point/OBject/View/X/Y/ZPrev/
   Restore/Save/Del/?/<World>: _y
   Rotation angle about the Y axis <0.00>:

2. Type 90; press Enter.

The drawing icon in the lower left corner of the screen will shift positions, and the letter W will disappear from the icon because the drawing is no longer in the WCS.

3. Select the ORIGIN command.

   Origin/Zaxis/3point/OBject/View/X/Y/ZPrev/
   Restore/Save/Del/?/<World>: _o
   Origin point<0,0,0>:

4. Use OSNAP, Endpoint (shift key, right mouse button), and select the lower left corner of the wedge.
5. Use the Circle tool to draw a circle on the side surface of the wedge.

   See Figure 10-19.

10-6 PRESET UCSs

AutoCAD includes a group of preset UCS orientations. Figure 10-20 shows the UCS Orientation dialog box. The preset orientations are set up to help generate standard orthographic views. The icons in the dialog box can also be used to return to the last UCS you were working on, or to return to the WCS.

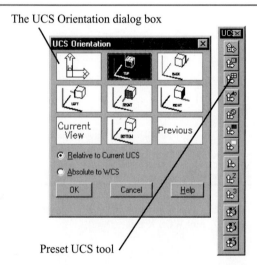

The UCS Orientation dialog box

Preset UCS tool

Figure 10-20

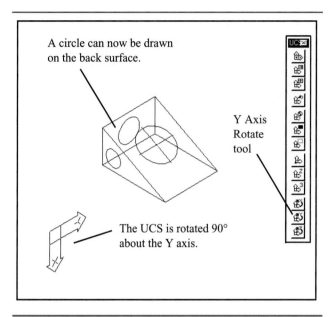

A circle can now be drawn on the back surface.

Y Axis Rotate tool

The UCS is rotated 90° about the Y axis.

Figure 10-19

Figure 10-21 shows an L-shaped object. This section assumes that a 3D L-shaped object, as shown in Figure 10-21, already exists on the screen. The toolbars shown in Figures 10-8 and 10-11 show that the commands listed in the View and Set UCS menus under the View pull-down menu are directly equivalent to the View and UCS toolbars.

To avoid visual errors

When working in 3D, it is important to remember that you cannot rely on visual inputs to locate shapes. What you see may be misleading. For example, Figure 10-22

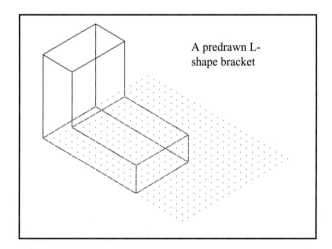

A predrawn L-shape bracket

Figure 10-21

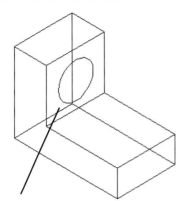

This circle appears to be on the
upper right surface, but it is not.

Figure 10-22

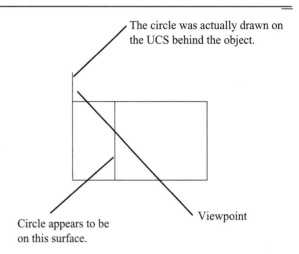

The circle was actually drawn on
the UCS behind the object.

Circle appears to be
on this surface.

Viewpoint

Figure 10-23

shows a circle that appears to be drawn on the upper right
surface of an L-shaped object. In fact, the circle is not
located on the surface; it only appears to be on the surface.
Figure 10-23 shows a top view of the same drawing. The
circle is actually drawn behind the surface and off the
object. It appears to be located on the surface because of
the line of sight of the current viewpoint orientation. This
visual distortion can be avoided by ensuring that the origin
for the current UCS is located on the plane where you are
working.

To locate the screen icon on the current origin

To help avoid visual errors based on an incorrectly
located origin, it is best to set the screen icon so that it is
located directly on the current origin.

1. Select the View pull-down menu.

   See Figure 10-24.

2. Select Display, then UCS Icon.
3. Select Origin.

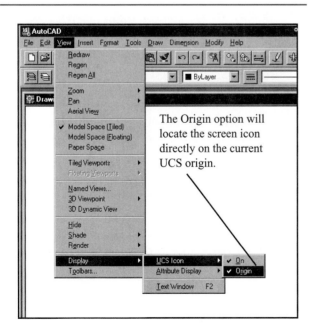

The Origin option will
locate the screen icon
directly on the current
UCS origin.

Figure 10-24

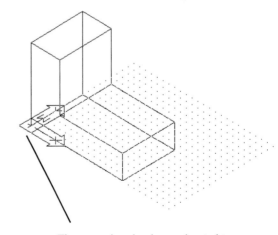

The screen icon has been relocated to
the current UCS origin.

Figure 10-25

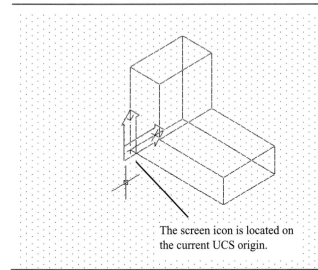

The screen icon is located on the current UCS origin.

Figure 10-26

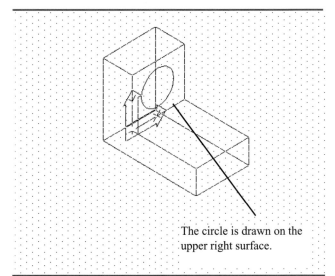

The circle is drawn on the upper right surface.

Figure 10-28

There should be a check mark to the right of the word Icon. After Icon Origin has been selected, there should be check marks next to both the words Icon and Origin. The screen icon should move to the current origin: 0,0,0 on the WCS. See Figure 10-25.

To draw in the right plane of an object using pre-set UCSs

This section will draw a circle on the inside right plane of the object shown in Figure 10-21.

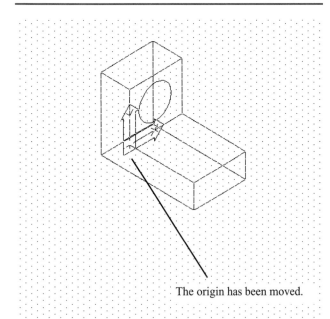

The origin has been moved.

Figure 10-27

1. Select the Preset UCS tool from the UCS tool-bar.

   The UCS Orientation dialog box will appear.

2. Select the Right tool, then OK.

   The screen icon and grid pattern will change as shown in Figure 10-26. It is important to note that the origin of the current Right UCS is still located at the 0,0,0 point on the WCS. If you try to draw now, the shapes will be located on a plane aligned with the 0,0,0 origin. The origin must be moved before you can draw on the inside right surface of the object.

3. Select the ORIGIN command.

   Origin/Zaxis/3point/OBject/View/X/Y/ZPrev/
   Restore/Save/Del/?/<World>: _o
   Origin point<0,0,0>:

4. Use OSNAP, Endpoint (shift key, right mouse button) and select the lower left corner of the inside right surface.

   The origin of the Right UCS is now located at the lower left corner of the inside right surface of the object. The screen icon will move to this new origin location. See Figure 10-27.

5. Use the Circle tool to draw a circle as shown in Figure 10-28.

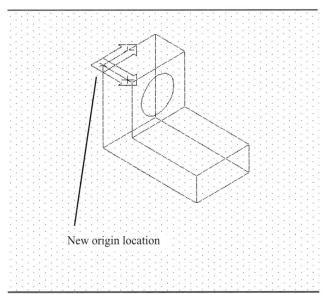

New origin location

Figure 10-29

To draw more complex shapes on UCSs

In this section, a slot shape will be drawn on the upper top surface of the object shown in Figure 10-28. The shape can be drawn directly on the surface once a UCS has been established, but in this exercise we will change the view orientation to create a 2D drawing surface.

1. Select the World tool from the UCS toolbar.
2. Select the Origin tool.

   *Origin/Zaxis/3point/OBject/View/X/Y/ZPrev/
   Restore/Save/Del/?/<World>: _o
   Origin point<0,0,0>:*

3. Use OSNAP, Endpoint (shift key, right mouse button) and select the lower left corner of the upper top surface.

The screen icon will shift to the designated origin corner. See Figure 10-29.

4. Select the TOP VIEW command.
5. Type Zoom; press Enter.

   *All/Center/Dynamic/Extents/Left/Previous/Vmax/
   Window/<Scale(X/XP)>:*

6. Type .75; press Enter.

A top view of the object will appear on the screen as shown in Figure 10-30.

7. Use the Line, Circle, and Trim tools to create the slot shape shown in Figure 10-31.
8. Select the SE Isometric View tool.

The object will appear oversized. Use the ZOOM command to reduce its appearance and size.

9. Type Zoom; press Enter.

   *All/Center/Dynamic/Extents/Left/Previous/Vmax/
   Window/<Scale(X/XP)>:*

10. Type .75; press Enter.

Your screen should look like Figure 10-32.

11. Select the WORLD command to return to the original WCS orientation.

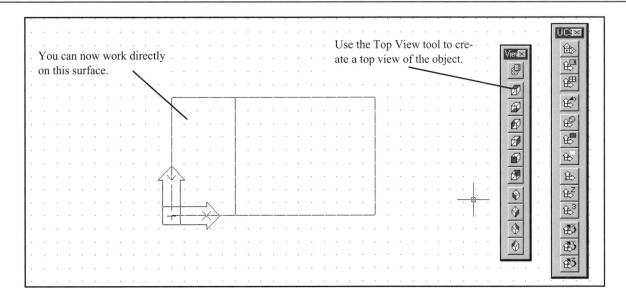

You can now work directly on this surface.

Use the Top View tool to create a top view of the object.

Figure 10-30

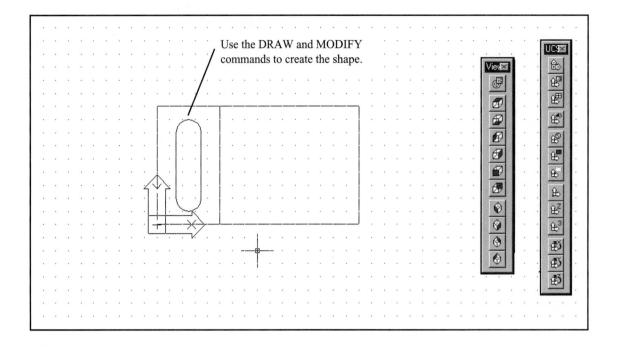

Figure 10-31

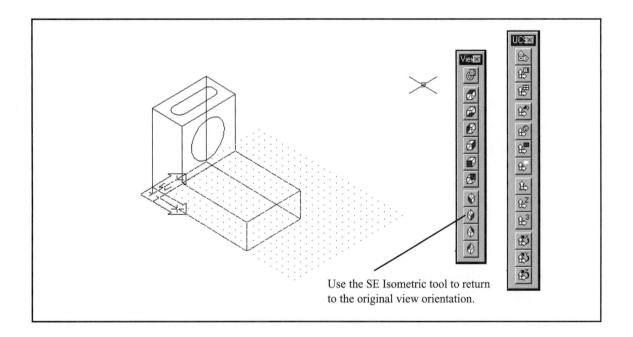

Figure 10-32

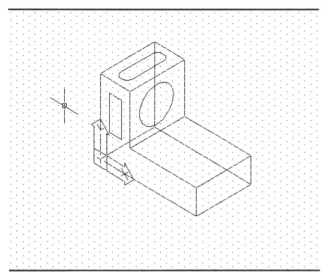

Figure 10-33

To draw on the Front surface

1. Select the Preset UCS tool.

   The UCS Orientation dialog box will appear.

2. Select the Front option, OK.

   The screen icon will appear at the lower left corner of

the front surface because this corner of the object was originally located on the origin of the WCS. It is not necessary to relocate the origin.

3. Use the Line tool to draw a rectangular shape on the Front surface as shown in Figure 10-33.

   The rectangular shape will appear to be out of visual perspective as you draw it, but it will be drawn in the correct orientation after the corner points are defined.

4. Use the WORLD command to return to the original WCS orientation.

10-7 ORTHOGRAPHIC VIEWS

Once a 3D object has been created, orthographic views may be taken directly from the object. The screen is first split into four ports, each showing the 3D object. The viewpoint of three of the ports will be changed to create the front, top, and right-side views of the object.

1. Select the View pull-down menu.

   See Figure 10-34.

2. Select Tiled Viewports, 4 Viewports.

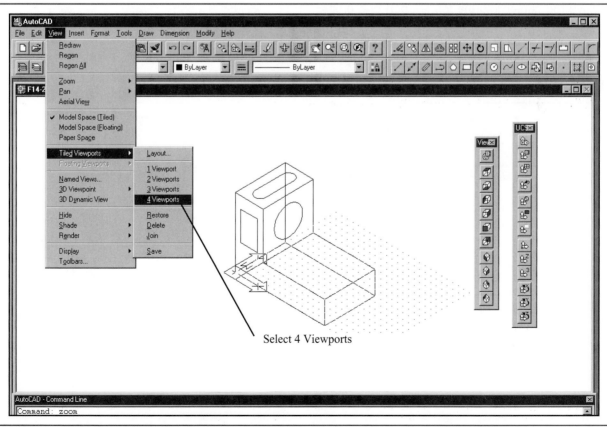

Select 4 Viewports

Figure 10-34

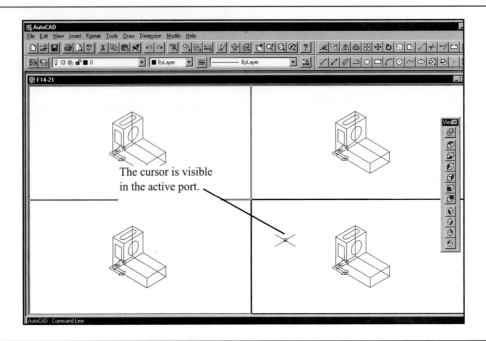

Figure 10-35

See Figure 10-35. Note that cross hairs are visible in only the lower right viewport. An arrow will appear if the cursor is moved into any other port.

3. Move the cursor into the top left port and press the left mouse button.

The cross hairs will appear in the port.

4. Select the Top tool.

An oversized top view of the object will appear in the port.

5. Type Zoom; press Enter.

All/Center/Dynamic/Extents/Left/Previous/Vmax/Window/<Scale(X/XP)>:

6. Type .75; press Enter.

See Figure 10-36.

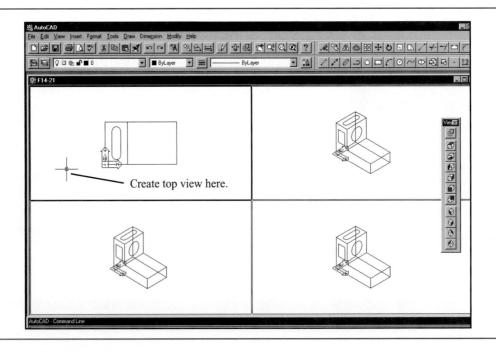

Figure 10-36

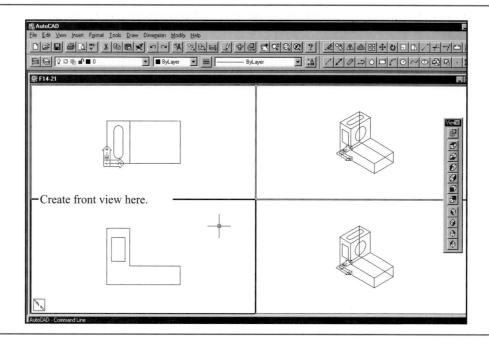

Figure 10-37

7. Move the cursor into the lower left port and press the left mouse button.

   The cross hairs will appear in the port.

8. Select the Front View tool.
9. Type Zoom; press Enter, then type .75; press Enter.

   See Figure 10-37.

10. Move the cursor to the lower right port and press the left mouse button.
11. Select the Right View command.
12. Type Zoom; press Enter, then type .75; press Enter.

   See Figure 10-38.

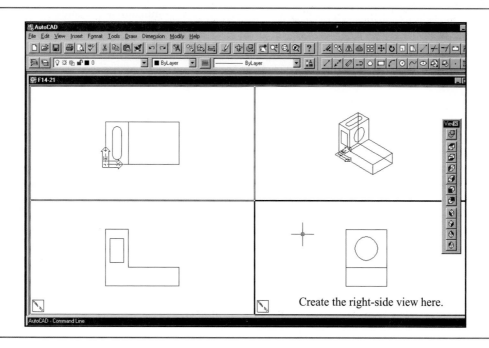

Figure 10-38

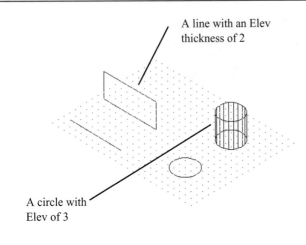

A line with an Elev
thickness of 2

A circle with
Elev of 3

Figure 10-39

10-8 ELEV

The ELEV command (elevation) command is used to create surfaces from 2D drawing commands. Figure 10-39 shows a line and a circle drawn as normal 2D entities and then drawn a second time with the ELEV command thickness set for 2. The LINE command generates a plane perpendicular to the plane of the original line, and the CIRCLE command generates a cylinder perpendicular to the plane of the base circle.

To use the ELEV command

There is no icon for the ELEV command. It is accessed by typing ELEV in response to a command prompt. In the example shown below, Grid and Snap have

.5 spacing using decimal units, and there is an SE Isometric viewpoint.

Command:

1. Type Elev; press Enter.

New current elevation <0.0000>:

The current elevation is the base plane for the constructions. In this example the base plane is the current XY (WCS) plane, that is, the plane with the grid.

2. Press Enter.

New current thickness <0.0000>:

The thickness defines the height of the generated planes. It is currently set at 0.0000, so any shapes drawn have no thickness. They are two-dimensional.

3. Type 2; press Enter.

The height of the elevation is now set for 2 units.

4. Select the Line tool.

Command: _line From point:

5. Draw the box shape shown in Figure 10-40.

The shape shown in the figure is not a box, but is four perpendicular planes. There is no top or bottom surface. This can be seen more clearly in Figure 10-41, which shows a rendering of the surfaces. Chapter 15 shows how to add a 3D Face to objects, closing in the top and bottom, and Chapter 16 shows how to create a solid object of the same shape.

Figure 10-42 shows a hexagon and an arc drawn with the ELEV command set at thickness values of 2 and 4, respectively.

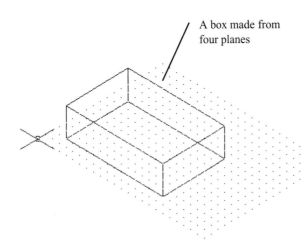

A box made from
four planes

Figure 10-40

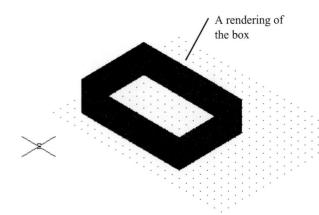

A rendering of
the box

Figure 10-41

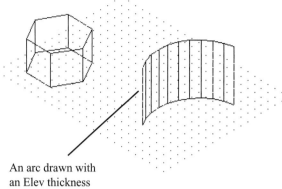

An arc drawn with
an Elev thickness

Figure 10-42

To draw a curve using ELEV

See Figure 10-43.

1. Type Elev; press Enter.

   New current elevation <0.0000>:

2. Press Enter.

   New current thickness <0.0000>:

3. Type 3.5; press Enter.
4. Select the Polyline tool from the Draw toolbar.

   Command: _pline
   From point:

5. Draw a polyline approximately like the one shown in Figure 10-43.
6. Select the Edit Polyline tool from the Modify II toolbar.

   Command: _pedit Select polyline:

7. Select the polyline.

   Close/ Join/ Width/ Edit vertex/ Fit/ Spline Decurve/ Ltype gen/ Undo/ eXit/ <X>:

8. Type F; press Enter.

   See Figure 10-44.

10-9 USING THE ELEV COMMAND TO CREATE OBJECTS

The ELEV command can be used with different UCSs to create 3D objects. The objects will actually be open-ended plane structures. The following procedure shows how to create the object shown in Figure 10-45. The drawing was created with Grid and Snap set to .5 using decimal units and with an SE Isometric viewpoint.

To draw the box

1. Type Elev; press Enter.

   New current elevation <0.0000>:

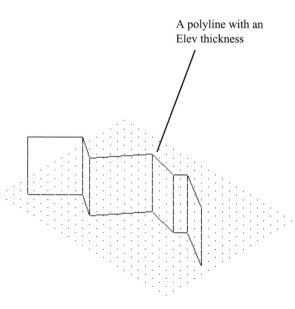

A polyline with an
Elev thickness

Figure 10-43

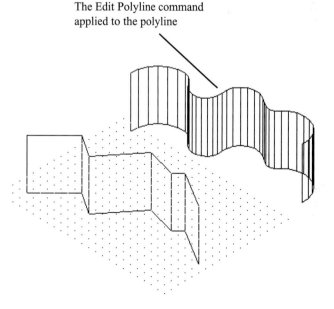

The Edit Polyline command
applied to the polyline

Figure 10-44

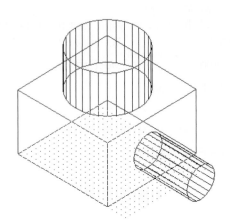

Figure 10-45

The box is drawn so that its corner is on the origin of the WCS.

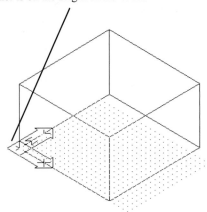

Figure 10-46

2. Press Enter.

   New current thickness <0.0000>:

3. Type 6; press Enter.

   Command:

4. Select the Line tool and draw a 10 x 10 box as shown in Figure 10-46.

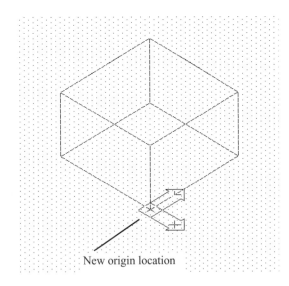

New origin location

Figure 10-47

To create a new UCS

1. Select the ORIGIN command.

   Origin point <0,0,0>:

2. Select the lower right corner as shown in Figure 10-47.

3. Press Enter.

   Origin/ZAxis/3point/OBject/View/X/Y/Z/Prev/ Restore/Save/Del/?/<World>:

4. Type 3; press Enter.

   Origin point:

5. Press Enter.

   Point on positive portion of the X-axis <1.0000,0.0000,0.0000>:

6. Use OSNAP, Endpoint and select the right corner of the box.

   Point on positive-Y portion of the UCS XY plane <−1.0000,0.0000,0.0000>:

7. Use OSNAP, Endpoint and select the end of the line directly above the origin.

   Your screen should look like Figure 10-48.

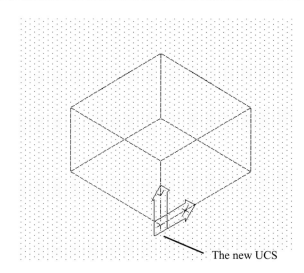

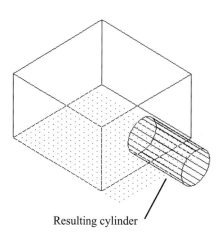

Resulting cylinder

Figure 10-48

Figure 10-49

To draw the right cylinder

This cylinder will be 6 units long, so there is no need to change the Elev thickness setting.

1. Select the Circle tool and draw a cylinder centered on the right face of the box.

The coordinate values for the cylinder's center point are 5,3 and the radius value is 2.00.

To draw the top cylinder

First return the drawing to the original WCS XY axis, then change the location of the XY plane so the top cylinder can be drawn in the correct position.

1. Select the Preset UCS tool and select the WCS box, then OK.

Your screen should look like Figure 10-49.

2. Type Elev; press Enter.

New current elevation <0.0000>:

3. Type 6; press Enter.

The value 6 is used because this is the thickness of the box.

New current thickness <0.0000>:

4. Type 4; press Enter.

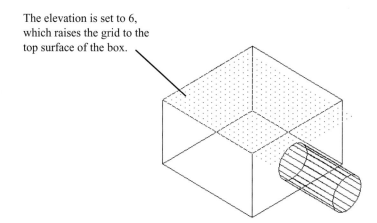

The elevation is set to 6, which raises the grid to the top surface of the box.

Figure 10-50

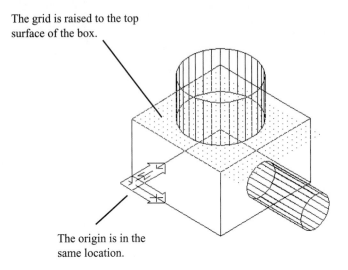

The grid is raised to the top surface of the box.

The origin is in the same location.

Figure 10-51

Notice the shift in the Grid pattern. The WCS origin icon is still located at the same place, but the grid origin is on the top surface. See Figure 10-50. Positions on the top surface may be located using the displayed coordinate values.

5. Select the Circle tool and draw the cylinder as shown in Figure 10-51.

The cylinder's center point is at 5,5 and its radius = 4.

6. Type Elev; press Enter.

   New current elevation <6.0000>:

7. Type 0; press Enter.

   New current thickness <4.0000>:

8. Type 0; press Enter.

The object should look like the one shown in Figure 10-51.

10-10 EXERCISE PROBLEMS

Exercise problems EX10-1 to EX10-4 require you to draw 2D shapes on various surfaces of 3D objects created using the ELEV command. All 2D shapes should be drawn at the center of the surfaces on which they appear.

A. Draw the 2D shapes as shown.

B. Divide the screen into 4 viewports, and create front, top, and right-side orthographic views for each object.

EX10-1 INCHES

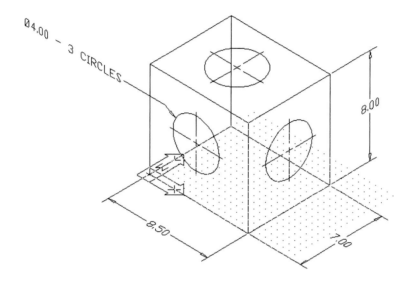

EX10-2 MILLIMETERS

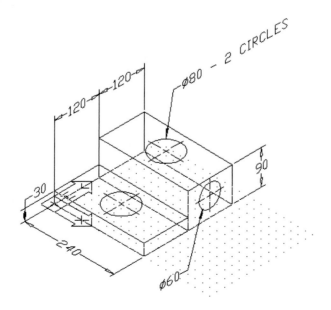

EX10-3 MILLIMETERS

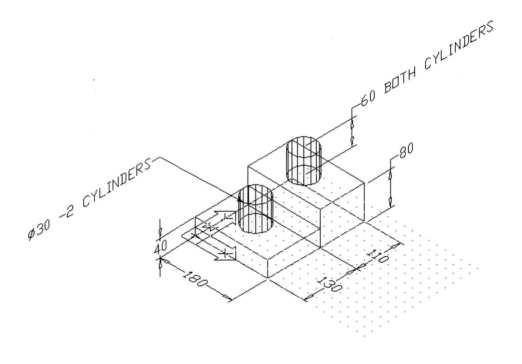

EX10-4 MILLIMETERS

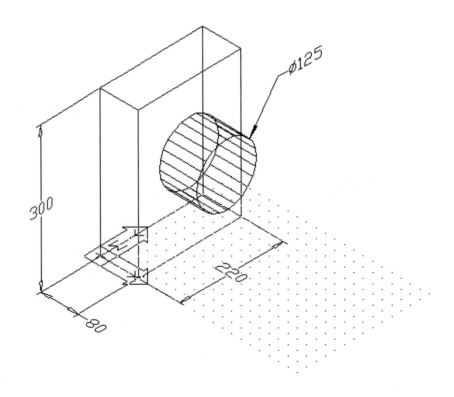

Redraw the figures presented in exercise problems EX10-5 to EX10-17 as wire frame models using the ELEV command. Divide the screen into four viewports and create front, top, and right-side orthographic views and one isometric view as shown in Figure 10-38.

EX10-5 INCHES

Box a; X = 6, Y = 5, Z = 2
Box b: X = 4, Y = 4, Z = 4
Box c: X = 5, Y = 2, Z = 1

       Hint: Consider a modified version of the NE Isometric viewpoint.

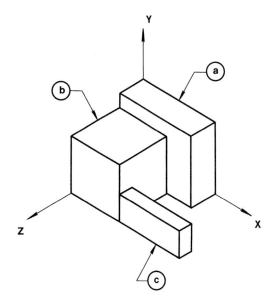

EX10-6 INCHES

Box a: X = 8, Y = 8, Z = 1
Box b: X = 6, Y = 6, Z = 2
Box c: X = 2, Y = 2, Z = 6

EX10-7 MILLIMETERS

Each box is 2 x 2 x 5.

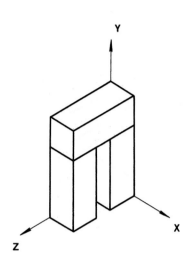

EX10-9 MILLIMETERS

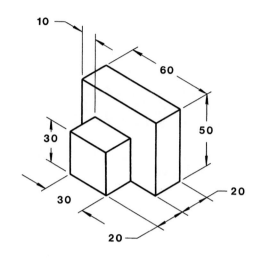

EX10-8 MILLIMETERS

Cylinder a: Ø10 x 30 LONG
Cylinder b: Ø20 x 8 LONG
Cylinder c: Ø35 x 18 LONG

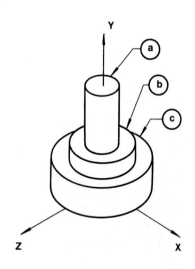

EX10-10 MILLIMETERS

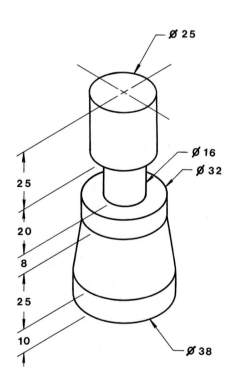

EX10-11 MILLIMETERS

EX10-12 MILLIMETERS

Cylinders are 30 LONG.

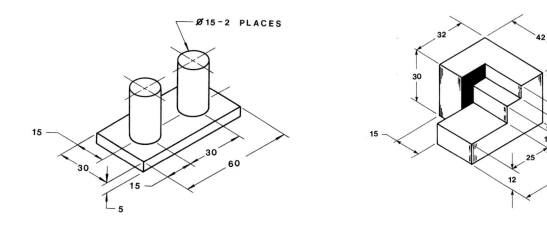

EX10-13 MILLIMETERS

The Ø12 cylinder is 20 LONG.
The Ø20 cylinder is 12 LONG.

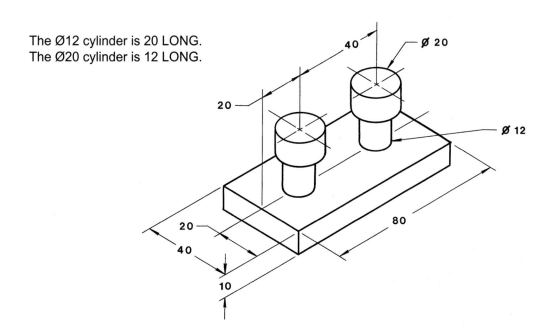

EX10-14 MILLIMETERS

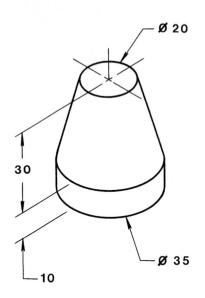

EX10-16 MILLIMETERS

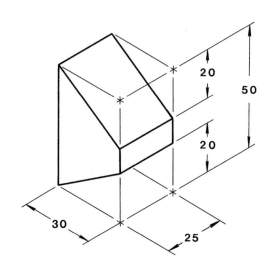

EX10-15 MILLIMETERS

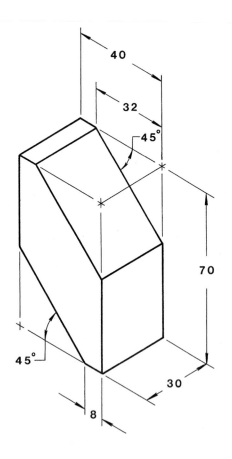

EX10-17 MILLIMETERS

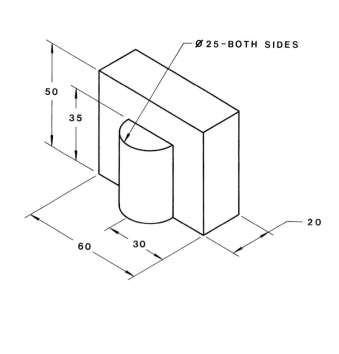

CHAPTER 11

The Surface Toolbar

11-1 INTRODUCTION

This chapter introduces surface modeling. Surface modeling creates 3D objects by joining surfaces together. The SURFACE commands are accessed using the Surface toolbar (see Figure 11-1) or by using the Draw pull-down menu, then selecting Surfaces, then 3D Surfaces. See Figure 11-2. AutoCAD's surface commands generate what AutoCAD calls "faceted surfaces using a polygonal mesh." This means that the curved surfaces generated are only mathematical approximations and not true, smooth surfaces. The approximations are, however, very accurate and present minimal visual distortion.

Surface modeling is different from solid modeling. Surface modeling deals only with individual surfaces, whereas solid modeling deals with entire solid shapes. A box created as a surface model comprises six surfaces. A box created using solid modeling is a solid object with six sides. Surface models cannot be unioned or subtracted as can solid models.

Surface modeling is particularly well suited for drawing complex 3D meshes, such as might be found on a surface profile map, or for the transition area between an airfoil and the fuselage on an aircraft.

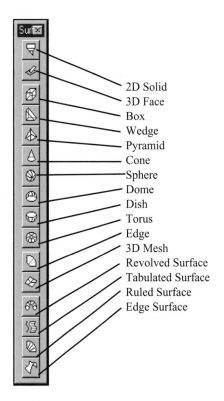

2D Solid
3D Face
Box
Wedge
Pyramid
Cone
Sphere
Dome
Dish
Torus
Edge
3D Mesh
Revolved Surface
Tabulated Surface
Ruled Surface
Edge Surface

Figure 11-1

283

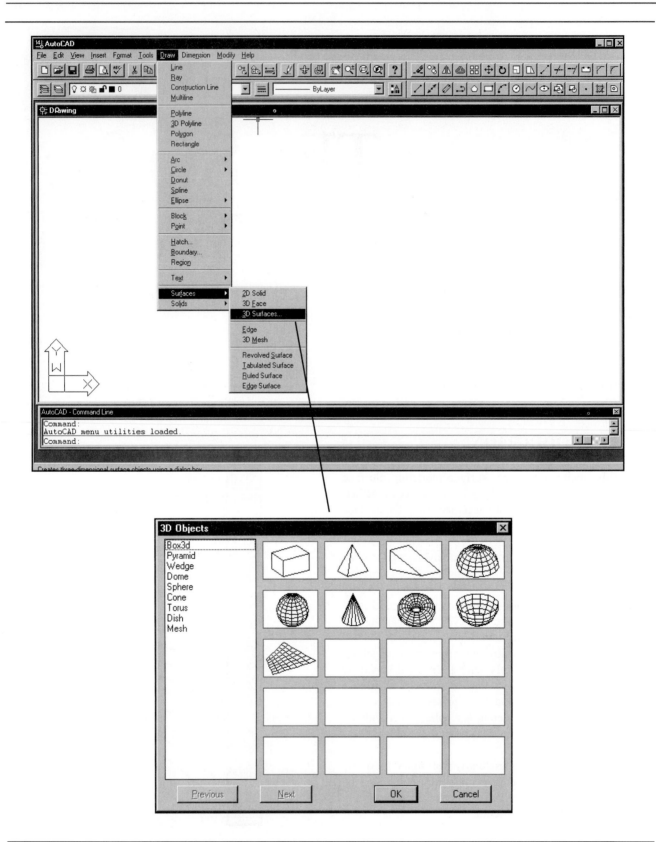

Figure 11-2

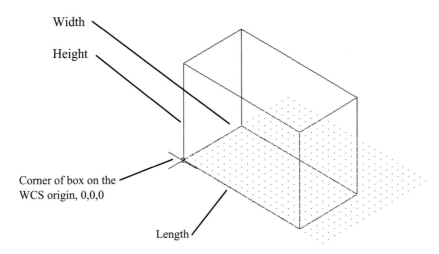

Width

Height

Corner of box on the
WCS origin, 0,0,0

Length

Figure 11-3

11-2 BOX

The BOX command is used to draw a box comprised of six surfaces. See Figure 11-3. The command sequence is as follows.

To draw a box

1. Select the Box tool from the Surfaces toolbar.

   Command: ai_box
   Corner of box:

2. Select a corner location for the box.

The example shown, the coordinate display was used to locate the corner point on 0,0,0 WCS origin. The corner could have been defined using a 3D coordinate value (X,Y,Z).

3. Move the cursor until the cursor display reads 0,0,0 then press the left mouse button.

   Length:

4. Type 10; press Enter.

The length could also be determined by moving the crosshairs and then picking a random point, using OSNAP to align with an existing entity, or using the values displayed in the coordinate display box.

   Cube/<Width>:

5. Type 5; press Enter.

If the box is a cube, enter a C response instead of a width value. AutoCAD will automatically take the length value and apply it to the width and height, skipping the width and height prompts in the sequence.

   Height:

6. Type 7; press Enter.
7. Rotation angle about the Z axis:

AutoCAD is now in a dynamic mode, meaning that as you move the cross hairs, the box will rotate about a Z axis projected out of the original corner point.

8. Type 0; press Enter.

11-3 WEDGE

The wedge in this example will be drawn aligned with the box drawn in Section 11-2. See Figure 11-4.

To draw a wedge

1. Select the Wedge tool from the Surfaces toolbar.

   Command: ai_wedge
   Corner of wedge:

2. Use OSNAP, Endpoint (press the shift key and right mouse button simultaneously) and select the lower right corner of the box.

   Length:

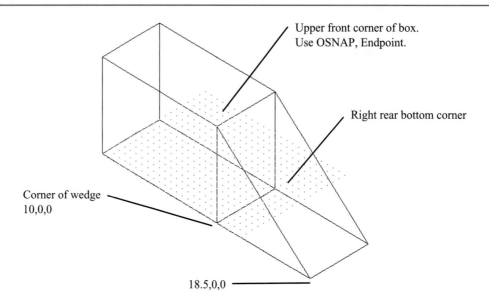

Upper front corner of box.
Use OSNAP, Endpoint.

Right rear bottom corner

Corner of wedge
10,0,0

18.5,0,0

Figure 11-4

3. Move the cross hairs away from the corner until the coordinate display reads 18.5,0,0, then press the left mouse button.

   Width:

4. Use OSNAP, Endpoint and select the right rear bottom corner of the box.

   Height:

5. Use ONAP, Endpoint and select the upper front corner of the box as shown in Figure 11-4.

   Rotation angle about the Z axis:

6. Type 0; press Enter.

The box and the wedge are not joined objects. They each have a surface that occupies the same space, but they are still two individual entities.

7. Clear the screen using the Erase tool.

11-4 PYRAMID

There are several different types of pyramids. Pyramids may have triangular, rectangular, or square bases. Any polygon can be used as the base of a pyramid, but the Surfaces toolbar commands are limited to rectangular bases. Polygons with more points can be used with the solid modeling options.

A tetrahedron is a pyramid made from three triangles. Pyramids that have their apexes located directly over the center point of their base polygons are called right pyramids. Pyramids whose apexes are not located directly over the center point of their base polygons are called oblique pyramids. Pyramids with their top sections removed are called truncated pyramids, and pyramids whose tops are edges and not point apexes are called ridged pyramids.

To draw a rectangular pyramid

See Figure 11-4.

1. Select the Pyramid tool from the Surfaces toolbar.

   Command: ai_pyramid
   First base point:

2. Type 0,0; press Enter.

   Second base point

3. Press F8 to turn the ORTHO command on.
4. Move the cross hairs to the 8,0 point, and press the left mouse button.

   Third base point:

5. Move the cross hairs to the 8,10 point and press the left mouse button.

   Tetrahedron/<Fourth base point>:

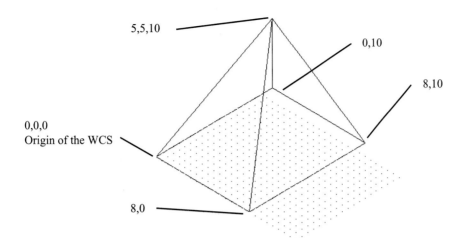

Figure 11-5

6. Move the cross hairs to the 0,10 point and press the left mouse button.

   Ridge/Top/<Apex point>:

   The apex point location requires a Z component. All of the previous points have been in the XY plane, so only X and Y coordinates were necessary. The apex point can be located randomly; that is, you can simply move the crosshairs and select a point, but this procedure can be visually misleading. It is better to use either coordinate values or the OSNAP command to locate a pyramid's apex.

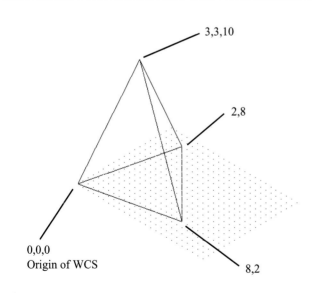

Figure 11-6

7. Type 5,5,10; press Enter.

   A negative Z value could have been entered.

To draw a tetrahedron

   See Figure 11-6.

1. Select the Pyramid tool from the Surfaces toolbar.

   Command: ai_pyramid
   First base point:

2. Type 0,0; press Enter.

   Second base point:

3. Type 8,2; press Enter.

   Third base point:

4. Type 2,8; press Enter.

   Tetrahedron/<Fourth base point>:

5. Type T; press Enter.

   Top/<Apex point>:

6. Type 3,3,10; press Enter.

To draw a truncated pyramid

   See Figure 11-7.

1. Select the PYRAMID command.

   Command: ai_pyramid
   First base point:

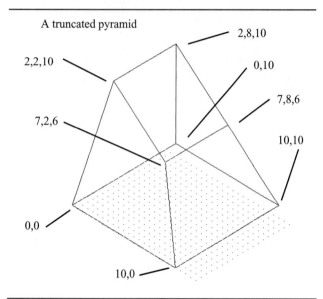

A truncated pyramid

2,8,10
2,2,10
0,10
7,8,6
7,2,6
10,10
0,0
10,0

Figure 11-7

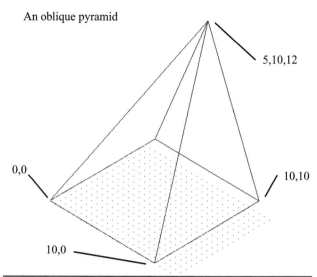

An oblique pyramid

5,10,12
0,0
10,10
10,0

Figure 11-8

2. Type 0,0; press Enter.

   Second base point:

3. Type 10,0; press Enter.

   Third base point:

4. Type 10,10; press Enter.

   Tetrahedron/<Fourth base point>:

5. Type 0,10; press Enter.

   Ridge/Top/<Apex point>:

6. Type T; press Enter.

   First top point:

7. Type 2,2,10; press Enter.

   Remember, top points are not in the base plane and therefore require a Z component in their coordinate values. Also note that a line will appear from the cross hairs to the corner point whose top point is being defined.

   Second top point:

8. Type 7,2,6; press Enter.

   Third top point:

9. Type 7,8,6; press Enter.

   Fourth top point:

10. Type 2,8,10; press Enter.

To draw an oblique pyramid

   See Figure 11-8.

1. Select the Pyramid tool from the Surfaces toolbar.

   Command: ai_pyramid
   First base point:

2. Type 0,0; press Enter.

   Second base point:

3. Type 10,0; press Enter.

   Third base point:

4. Type 10,10; press Enter.

   Tetrahedron/<Fourth base point>:

5. Type 0,10; press Enter.

   Ridge/Top/<Apex point>:

6. Type 5,10,12; press Enter.

   This apex point is located over the center point of the rear line in the base polygon. Any apex not located over the base polygon's center point is classified as an oblique pyramid.

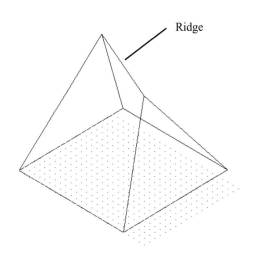

Ridge

Figure 11-9

To draw a ridged pyramid

See Figure 11-9.

1. Select the Pyramid tool from the Surfaces toolbar.

   Command: ai_pyramid
   First base point:

2. Type 0,0; press Enter.

   Second base point:

3. Type 10,0; press Enter.

   Third base point:

4. Type 10,10; press Enter.

   Tetrahedron/<Fourth base point>:

5. Type 0,10; press Enter.

   Ridge/Top/<Apex point>:

6. Type R; press Enter.

   First ridge point:

7. Type 3,5,10; press Enter.

   Second ridge point:

8. Type 7,5,7; press Enter.

11-5 CONE

See Figure 11-10.

To draw a cone

1. Select the CONE command.

   Command: ai_cone
   Base center point:

2. Type 5,5; press Enter.

   The cone's center point could also have been selected by locating a point using the cross hairs, then pressing the left mouse button.

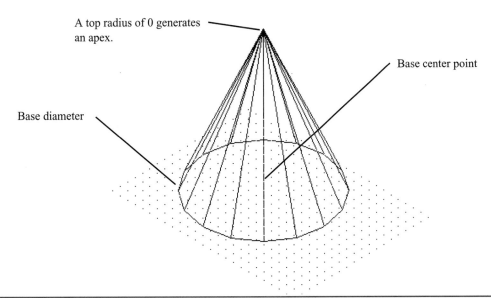

A top radius of 0 generates an apex.

Base center point

Base diameter

Figure 11-10

Diameter/<radius> of base:

3. Type 4; press Enter.

Diameter/<radius> of top <0>:

A top radius of 0 will create a cone with an apex. A radius value other than 0 will create a truncated cone.

4. Press Enter.

Height:

5. Type 9; press Enter.

Number of segments <16>:

The number of segments defines the number of facets used to show the cone. The more segments, the smaller each segment will be, and the resulting cone will appear smoother. However, a large number of segments will use more memory and will require more time to process. The default value of 16 is a good compromise between visual accuracy and drawing speed.

6. Press Enter.

To draw a truncated cone

See Figure 11-11.

1. Select the Cone tool from the Surfaces toolbar.

   Command: ai_cone
   Base center point:

2. Type 5,5; press Enter.

   Diameter/<radius> of base:

3. Type 6; press Enter.

   Diameter/<radius> of top <0>:

4. Type 3; press Enter.

   Height:

5. Type 8; press Enter.

   Number of segments <16>:

6. Press Enter.

11-6 SPHERE

See Figure 11-12.

To draw a sphere

In this example, the sphere will be drawn above the XY plane.

1. Select the Sphere tool from the Surfaces toolbar.

   Command: ai_sphere
   Center of sphere:

2. Type 5,5,5; press Enter.

   Diameter/<radius>:

3. Type 5; press Enter.

   Number of longitudinal segments <16>:

4. Press Enter.

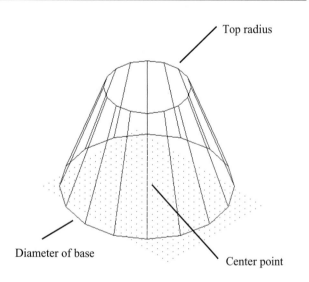

Diameter of base   Center point

Figure 11-11

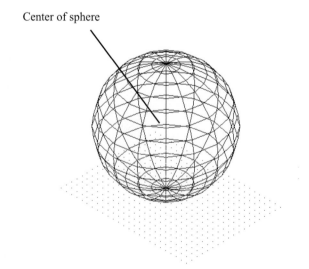

Center of sphere

Figure 11-12

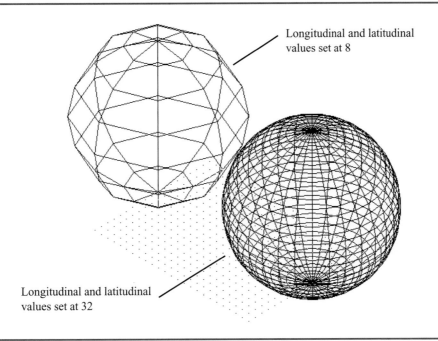

Longitudinal and latitudinal values set at 8

Longitudinal and latitudinal values set at 32

Figure 11-13

Number of latitudinal segments <16>:

5. Press Enter.

The number of longitudinal and latitudinal segments controls the number of facets that will show on the sphere. Longitudinal lines are equivalent to vertical lines. Latitudinal lines are equivalent to horizontal lines. The more segments defined, the smoother the resulting sphere, but more line segments will increase the drawing time needed to generate the sphere.

Figure 11-13 shows two spheres; one drawn with 8 longitudinal and latitudinal segments, and one drawn with 32 longitudinal and latitudinal segments. Note the differences in the overall smoothness of the visual presentation.

11-7 DOME

A dome is a semisphere, or the top half of a sphere. See Figure 11-14.

To draw a dome

1. Select the Dome tool from the Surfaces toolbar.

   Command: ai_dome
   Center of dome:

2. Type 5,5; press Enter.

Diameter/<radius>:

3. Type 8; press Enter.

   Number of longitudinal segments <16>:

4. Press Enter.

   Number of latitudinal segments <8>:

5. Press Enter.

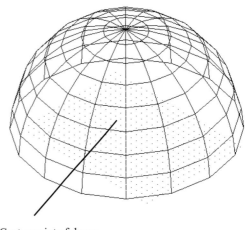

Center point of dome

Figure 11-14

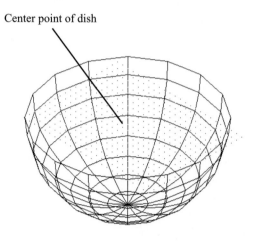

Center point of dish

Figure 11-15

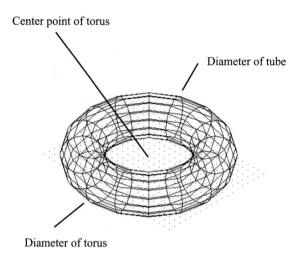

Center point of torus

Diameter of tube

Diameter of torus

Figure 11-16

11-8 DISH

A dish is a semisphere, or the bottom half of a sphere. See Figure 11-15.

To draw a dish

1. Select the Dish tool from the Surfaces toolbar.

   Command: ai_dome
   Center of dome:

2. Type 5,5; press Enter.

   Diameter/<radius>:

3. Type 7; press Enter.

   Number of longitudinal segments <16>:

4. Press Enter.

   Number of latitudinal segments <8>:

5. Press Enter.

11-9 TORUS

A torus is a donutlike shape. See Figure 11-16.

To draw a torus

1. Select the Torus tool from the Surfaces toolbar.

   Command: ai_torus
   Center of torus:

2. Type 5,5; press Enter.

   Diameter/<radius> of torus:

3. Type 6; press Enter.

   Diameter/<radius> of tube:

4. Type 1.5; press Enter.

   The radius of a torus is the distance from the center point to the outside edge of the torus, as measured along the center plane.

   Segments around the tube circumference <16>:

5. Press Enter.

   Segments around the torus circumference <16>:

6. Press Enter.

A surface created using 3D Face

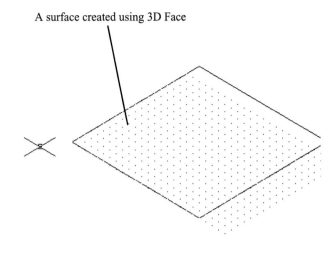

Figure 11-17

Adjoining surfaces created using 3D Face

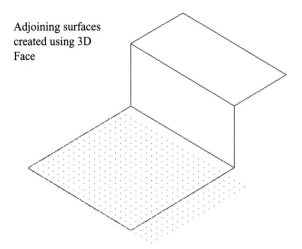

Figure 11-18

11-10 3D FACE

A 3D face is a plane. 3D faces may be created as individual planes or as adjoining groups of planes. A 3D face can be added to an existing 3D object to close an open area, or to create a transition between existing objects.

To draw a single 3D face

In this example we will select points on a grid. Coordinate values can also be used to define the 3D face's corner points. See Figure 11-17.

1. Select the 3D Face tool from the Surfaces tool-bar.

   Command: _3dface First point:

2. Select the 0,0 point on the grid.

   Use the coordinate display at the lower left of the screen to verify the point location.

   Second point:

3. Select 10,0.

   Third point:

4. Select 10,10.

   Fourth point:

5. Select 0,10.

   Third point:

   This point assumes that you will also be creating an adjacent plane, but because you are not, enter a null value.

6. Press Enter.

To draw adjoining 3D faces

In this example we will draw three adjoining planes, each 90 degrees to the others. See Figure 11-18.

1. Select the 3D Face tool from the Surfaces tool-bar.

   Command: _3dface First point:

2. Type 0,0; press Enter.

   Second point:

3. Type 10,0; press Enter.

   Third point:

4. Type 10,10; press Enter.

   Fourth point:

5. Type 0,10; press Enter.

   Third point:

6. Type 0,10,5; press Enter.

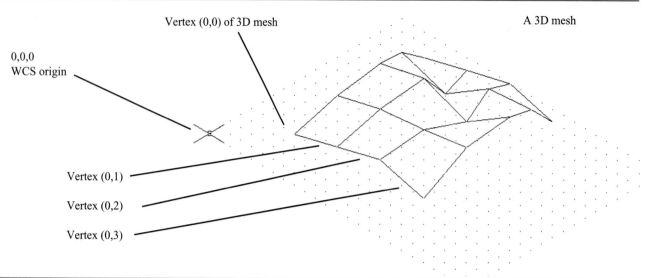

Vertex (0,0) of 3D mesh

A 3D mesh

0,0,0
WCS origin

Vertex (0,1)

Vertex (0,2)

Vertex (0,3)

Figure 11-19

The adjoining plane will be perpendicular to the line on the Y axis.

Fourth point:

7. Type 10,10,5; press Enter.

Third point:

8. Type 10,15,5; press Enter.

Fourth point:

9. Type 0,15,5; press Enter.
10. Press Enter.

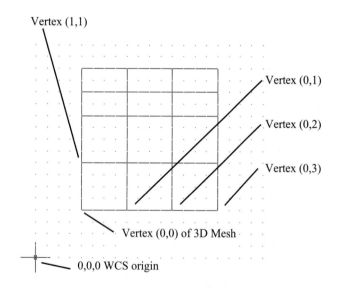

Vertex (1,1)

Vertex (0,1)

Vertex (0,2)

Vertex (0,3)

Vertex (0,0) of 3D Mesh

0,0,0 WCS origin

Figure 11-20

11-11 3D MESH

The 3D MESH command defines a surface by defining a series of points on the surface. Mesh surfaces are usually irregular surfaces, that is, not flat. See Figure 11-19.

The points used to define a mesh surface are located in terms of an M,N axis. First the number of points in both the M and N directions is defined, then each point is automatically assigned a vertex notation. The vertex notation 0,2 means first row, that is, the first line in the M direction, and the third column, third line in the N direction. The first line in both the M and N directions is labeled 0. Figure 11-20 is a top view of the finished 3D mesh.

Coordinate values for each mesh point are expressed in terms of their X,Y,Z values relative to the existing XY plane.

To draw a 3D mesh

1. Select the 3D MESH command.

   Command: _3dmesh
   Mesh M size:

2. Type 5; press Enter.

   Mesh N size:

3. Type 4; press Enter.

   Vertex (0,0):

This prompt is asking for the first point value on the M,N axis. It is not a numerical input, but rather a point location definition relative to the M,N axis. Note that the numbers are enclosed by ( ), which means that they are not default values.

4. Type 2,2,0; press Enter.

This input establishes the corner of the mesh surface at the 2,2,0 point on the current XYZ axis.

Vertex (0,1):

5. Type 4,2,.5; press Enter.

Vertex (0,2):

6. Type 6,2,1; press Enter.

Vertex (0,3):

7. Type 8,2,.5; press Enter.

Vertex (1,0):

8. Type 2,4,.5; press Enter.

Vertex (1,1):

9. Type 4,4,1; press Enter.

Vertex (1,2):

10. Type 6,4,1.2; press Enter.

Vertex (1,3):

11. Type 8,4,1.5; press Enter.

Vertex (2,0):

12. Type 2,6,.75; press Enter.

Vertex (2,1):

13. Type 4,6,1.2; press Enter.

Vertex (2,2):

14. Type 6,6,.5; press Enter.

Vertex (2,3):

15. Type 8,6,1.7; press Enter.

Vertex (3,0):

16. Type 2,7,.7; press Enter.

Vertex (3,1):

17. Type 4,7,.1; press Enter.

Vertex (3,2):

18. Type 6,7,1.3; press Enter.

Vertex (3,3):

19. Type 8,7,1.5; press Enter.

Vertex (4,0):

20. Type 2,8,0; press Enter.

Vertex (4,1):

21. Type 4,8,.5; press Enter.

Vertex (4,2):

22. Type 6,8,1; press Enter.

Vertex (4,3):

23. Type 8,8,.5; press Enter.

11-12 REVOLVED SURFACE

A revolved surface is created by revolving a given shape about an axis of revolution. Cones, cylinders, spheres, toruses, and ellipsoids are all shapes that can be created as revolved surfaces. AutoCAD can also create unusual shapes by revolving polylines.

To create a cylinder

A cylinder can be created by revolving a line around another line. The resulting cylinder is not a solid cylinder, but a cylindrical shape with open ends. It is analogous to a rolled sheet of paper. Figure 11-21 shows two lines that will be used to construct a cylinder.

1. Select the Revolved Surface tool from the Surface toolbar.

Command: _revsurf
Select curve path:

The curve path is the line that will be revolved. In this example it is the shorter of the two lines.

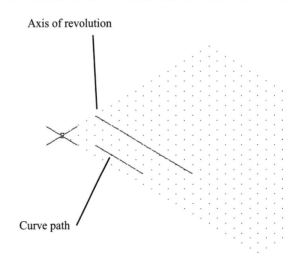

Axis of revolution

Curve path

Figure 11-21

A revolved surface with
SURFTAB1 and SURFTAB2 set
to a value of 6

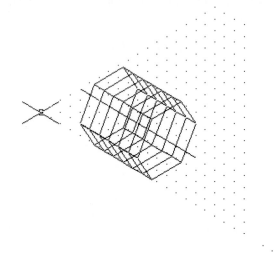

Figure 11-22

A revolved surface with
SURFTAB1 and SURFTAB2
set to a value of 18

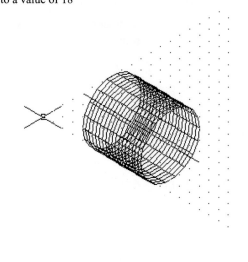

Figure 11-23

2. Select the shorter line.

   Select the axis of revolution:

3. Select the longer line.

   Start angle <0>:

Revolved shapes need not be 360 degrees. They can be started at and ended at any angle.

4. Press Enter.

   Included angle (+=ccw, −=cw) <Full circle>:

This prompt reminds you that the counterclockwise direction is the positive direction. Either positive or negative values may be entered. The default value is a full circle, or 360 degrees.

5. Press Enter.

See Figure 11-22. Note that the resulting cylinder doesn't look like a cylinder. This is because the default system variables SURFTAB1 and SURFTAB2 are set too small for this example. Their values must be increased to produce a smooth looking cylinder.

To change SURFTAB1 and SURFTAB2

1. Type SURFTAB1 and press Enter in response to a Command: prompt.

   Command: surftab1

New value for SURFTAB1 <6>:

2. Type 18; press Enter.

   Command:

3. Type SURFTAB2; press Enter.

   Command: surftab2
   New value for SURFTAB2 <6>:

4. Type 18; press Enter.

The values may also be changed by using the System Variables option in the Options pull-down menu.

The surftab variables are now reset from a value of 6 to a value of 18. Figure 11-23 shows the resulting cylinder created using the Revolved Surface command.

Figure 11-24 shows a cylinder drawn perpendicular to the XY plane. The lines used to define the surface of revolution were defined using the XYZ values of the line's endpoints. For example, the shorter vertical line was drawn between points 0,0,0 and 0,0,5.

Figure 11-24 also shows an example of a partial cylinder that was created by not revolving the line a full 360 degrees. The partial cylinder shown was created by revolving a line 120 degrees.

Figure 11-25 shows an object that was created by first drawing a polyline, then revolving the polyline around an axis of revolution. Any curved shape can be used to create a surface of revolution. A torus, for example, can be created by revolving a circle around an axis of revolution.

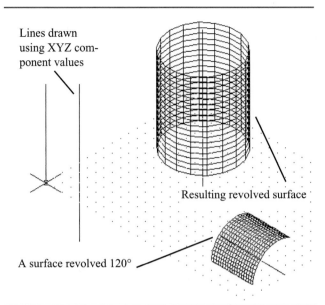

Lines drawn using XYZ component values

Resulting revolved surface

A surface revolved 120°

Figure 11-24

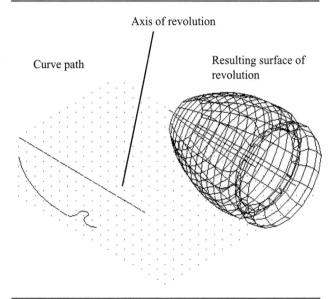

Axis of revolution

Curve path

Resulting surface of revolution

Figure 11-25

11-13 TABULATED SURFACE

A tabulated surface creates an object by tracking a given shape along a direction vector. Only straight lines can be used as direction vectors. A curved line may be specified as a direction vector, but AutoCAD will draw a straight line from the first point to the last point on the curve and use that straight line as the direction vector.

Tabulated surfaces often involve objects drawn in two different planes. The following example will create a cylinder by extruding a circle along a straight line. The circle will be drawn in a plane perpendicular to the plane of the line.

Select the Right UCS.

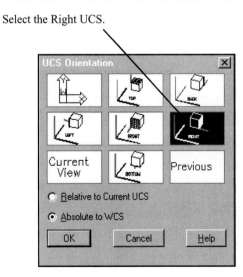

Figure 11-26

To draw a tabulated surface

1. Use the Line tool from the Draw toolbar to draw a line in the current XY plane (WCS).
2. Select the Preset UCS tool from the UCS toolbar.

The UCS Orientation dialog box will appear. See Figure 11-26.

3. Select the Right UCS option.

The screen grid pattern will shift to the new UCS orientation.

4. Select the Origin tool from the UCS toolbar.
5. Use the OSNAP, Endpoint command to define the start point of the line as the new origin for the Right UCS.

See Figure 11-27.

6. Use the Circle tool from the Draw toolbar to draw a circle about the start point of the line.

See Figure 11-28.

7. Return to the WCS by selecting the World tool from the UCS toolbar.
8. Select the Tabulated Surface tool from the Surface toolbar.

Command: _tabsurf
Select curve path:

9. Select the circle.

Select direction vector:

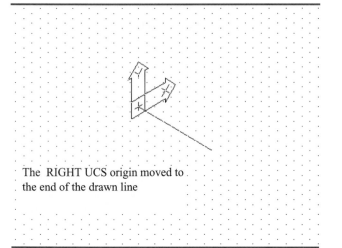

The RIGHT UCS origin moved to
the end of the drawn line

Figure 11-27

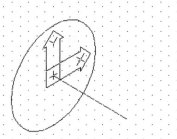

A circle drawn around
the origin

Figure 11-28

10. Select the line.

Figure 11-29 shows the resulting extruded surface. The circle has been extended in the direction and length of the line. The cylinder is drawn perpendicular to the base plane, which was established as the Right UCS with its origin at the start point of the line. Had the same procedure been followed, but the other end of the line selected as the origin, the cylinder would have covered the line.

Figure 11-30 shows a polyline that has been edited to form a curve. The TABULATED SURFACE command was then applied using a line perpendicular to the plane of the polyline as a direction vector.

11-14 RULED SURFACE

A ruled surface is created between two existing curves or between a point and a curve.

To draw a ruled surface between two curves

See Figure 11-31.

1. Select the RULED SURFACE command.

Command: _rulesurf
Select the defining curve:

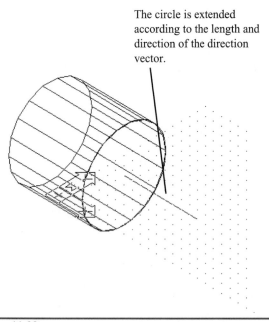

The circle is extended according to the length and direction of the direction vector.

Figure 11-29

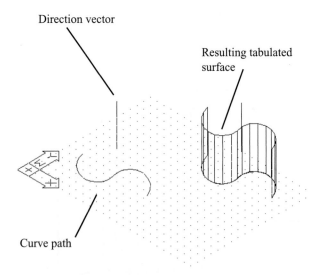

Direction vector

Resulting tabulated surface

Curve path

Figure 11-30

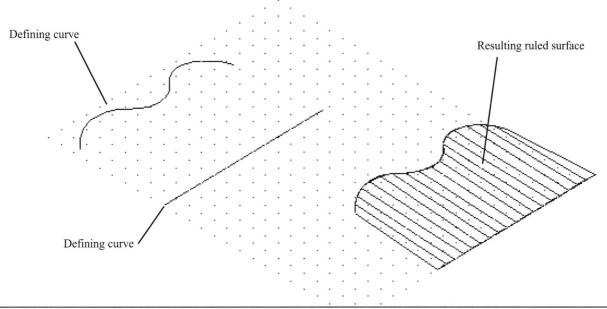

Defining curve

Defining curve

Resulting ruled surface

Figure 11-31

2. Select the straight line.

Select second defining line:

Ruled surfaces are selection sequence dependent. Figure 11-32 shows two ruled surfaces created from the same defining curves. The differences between the resulting surfaces are the result of different selection point locations. The selection points for the defining curves for the lower left surface were the left end of each line. The selection points for the upper surface were the left end of the curved line and the right end of the straight line.

Figure 11-33 shows a ruled surface drawn between curves drawn in different planes. The left curve was drawn in the preset Right UCS and the right curve was drawn in the WCS.

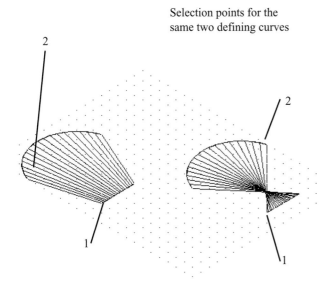

Selection points for the same two defining curves

2

2

1

1

Figure 11-32

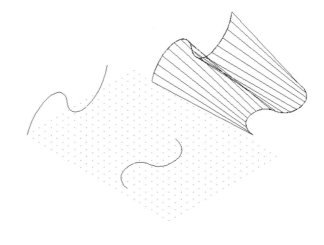

Figure 11-33

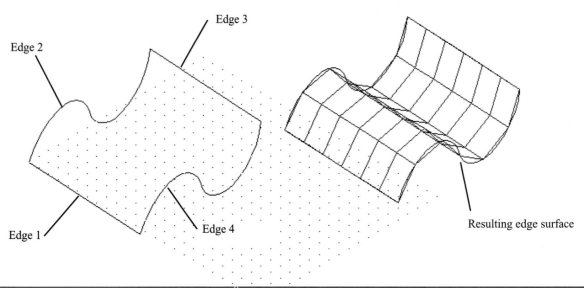

Figure 11-34

11-15 EDGE SURFACE

The EDGE SURFACE command is used to create a surface among four existing adjoining lines. See Figure 11-34.

1. Select the Edge Surface tool from the Surface toolbar.

   Command: _edgesurf
   Select edge 1:

2. Select one of the lines.

   Select edge 2:

3. Select a second line.

Select edge 3:

4. Select a third line.

Select edge 4:

5. Select a fourth line.

The edge surface shown in Figure 11-34 was created with a SURFTAB setting of 6. Note how the surface contains a 6 by 6 pattern of rectangles. Figure 11-35 shows an edge surface created on the same four lines, but with SURFTAB1 and SURFTAB2 set to 18. Note the difference in visual appearance.

Figure 11-36 shows an edge surface created among three 3D polylines edited to form splines and a single straight line. Both SURFTABs were set at 18.

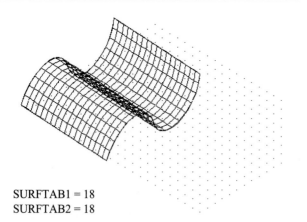

SURFTAB1 = 18
SURFTAB2 = 18

Figure 11-35

An example of an edge surface

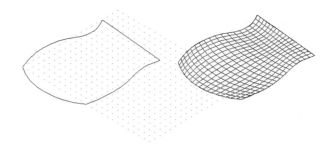

Figure 11-36

11-16 COMBINING SURFACES

Figure 11-37 shows an object created by combining two surfaces: a cylinder and a truncated cone. The procedure is as follows. The object will be centered around the 0,0 point on the XY axis.

To draw the truncated cone

1. Set the Grid and Snap spacing for .5, and select Decimal Units.
2. Select the SE Isometric tool from the Views toolbar.
3. Select the Zoom Window tool and create a zoom window around the 0,0,0 point.
4. Select the Cone tool from the Surfaces toolbar.

   Base center point:

5. Type 0,0,0; press Enter.

   Diameter/<radius> of base:

6. Type D; press Enter.

   Diameter of base:

7. Type 1.38; press Enter.

   Diameter/<radius> of top <0>:

8. Type D; press Enter.

   Diameter of top <0>:

9. Type .62; press Enter.

   Height:

10. Type .75; press Enter.

    Number of segments <16>:

11. Press Enter.

See Figure 11-38. Use the Zoom Window tool again, if necessary, to approximate the screen size shown.

The problem now requires that a cylinder be joined to the top of the truncated cone. The ELEV command can be used to create a cylinder located on top of the cone, but in this example, two circles will be drawn, then joined, using the RULED SURFACE command. The cone was created using a SURFACE command and has no top surface. The RULED SURFACE command requires an edge, and the top of the cone does not constitute an edge, so a circle must be added at the top of the cone.

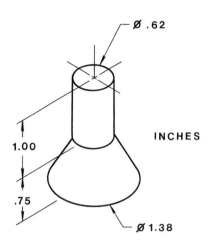

Figure 11-37

A cone drawn using the Cone tool from the Surfaces toolbar

Base = Ø1.38
Top = Ø0.62

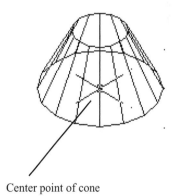

Center point of cone

Figure 11-38

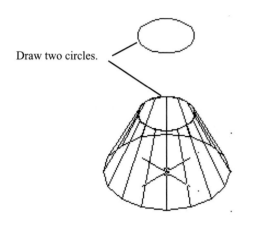

Draw two circles.

Figure 11-39

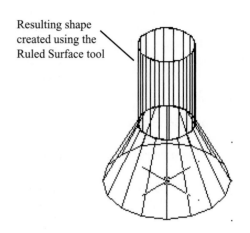

Resulting shape created using the Ruled Surface tool

Figure 11-40

To add the cylinder to the top of the cone

1. Select the Circle tool.

   Command: _circle 3P/2P/TTR/<Center point>:

2. Type 0,0,.75; press Enter.

   Diameter/Radius <0.3100>: _d diameter <0.6200>:

3. Press Enter.

   There is no visual change in the object, although a tick mark may appear at the center point of the circle. The circle and the top of the cone occupy the same space.

4. Press Enter again to restart the CIRCLE command sequence.

   CIRCLE 3P/2P/TTR/<Center>:

5. Type 0,0,1.75; press Enter.

   Diameter/<Radius>: <0.3100>:

   See Figure 11-39.

To create the cylinder

1. Select the Ruled Surface tool from the Surface toolbar.

   Select the first defining curve:

2. Select the top circle.

   Select the second defining curve:

3. Select the circle drawn on the top surface of the cone.

   See Figure 11-40. Use the Zoom Window tool to enlarge the top surface of the cone, if necessary, to select the Ø.62 circle. The density of line segments in the cylinder occurs because the SURFTAB settings are at 18.

11-17 USING SURFACES WITH A UCS

Figure 11-41 shows a dimensioned object. The following section shows how the object was created as a surface model.

To draw the box portion

1. Set the Grid and Snap spacing for .5, and select Decimal Units.
2. Select the SE Isometric tool from the View toolbar.
3. Select the Zoom tool and create a zoom window around the 0,0,0 point.
4. Select the Box tool.

   Corner of box:

5. Type 0,0,0.

   Length:

6. Type 1.50; press Enter.

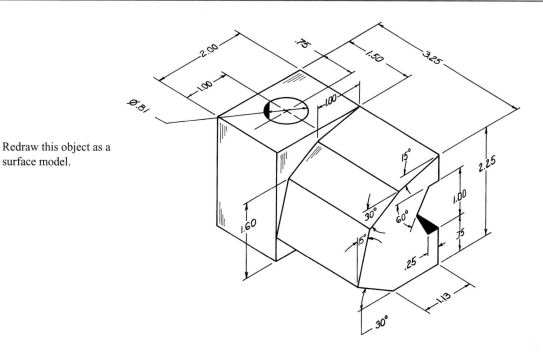

Redraw this object as a surface model.

Figure 11-41

Cube/<Width>:

7. Type 2.00; press Enter.

   Height:

8. Type 2.25; press Enter.

   Rotation angle about Z axis:

9. Type 0; press Enter.

See Figure 11-42. The size values for the box were taken from the dimensions given in Figure 11-41. Use the ZOOM command again, if necessary, to approximate the size shown in Figure 11-41.

To add the hole to the object

The hole could be created using the ELEV command, but in this example, the RULED SURFACE command will be used.

1. Select the Circle tool.

   Command: _circle 3P/2P/TTR/<Center point>:

2. Type .75,1.00,0; press Enter.

   The hole location came from the given dimensions.

A surface model created using the Box tool

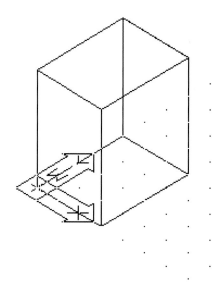

Figure 11-42

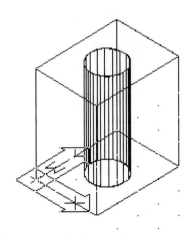

The hole was created using two circles and the Ruled Surface tool.

Figure 11-43

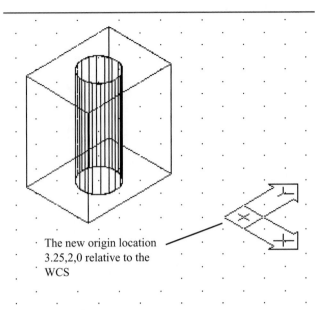

The new origin location 3.25,2,0 relative to the WCS

Figure 11-44

Diameter/Radius <0.3100>:

3. Type .405; press Enter.
4. Press Enter again to restart the CIRCLE command sequence.

CIRCLE 3P/2P/TTR/<Center point>:

5. Type .75,1.00, 2.25; press Enter.

This will locate the second circle on the top surface of the box.

Diameter/<Radius> <0.4050>:

6. Press Enter.
7. Select the Ruled Surface tool.

Select first defining curve:

8. Select the lower circle.

Select the second defining curve:

9. Select the upper circle.

See Figure 11-43.

To create the extended portion of the object

This construction requires that a new UCS be established with its origin at the 3.25,2.00,0 point on the WCS (current UCS) which is the right back corner of the object.

1. Select the ORIGIN command.

Origin point:

2. Type 3.25,2,0; press Enter.

Command:

See Figure 11-44. The screen icon is on the new origin.

3. Press Enter again.

Origin/ZAxis/3point/OBject/View/Z/Y/Z/Prev/ Restore/Save/Del/?/<World>:

4. Type 3; press Enter.

Origin point <0.0.0>:

5. Press Enter.

Point on the positive portion of the x-axis <1.0000,0.0000,0.0000>:

6. Type 0,–2,0; press Enter.

A Y value is entered because of the new origin location. This value will position the new UCS so that the construction of the object's extended portion will be in the positive quadrant of the UCS.

Point on the positive-Y portion of the UCS XY plane <1.0000,0.0000,0.0000>:

7. Type 0,0,2.25; press Enter.

The new UCS is now defined as indicated by the screen icon shift. See Figure 11-44. A value of 1 could have been used rather than 2.25. Any value in the positive-Y direction is acceptable.

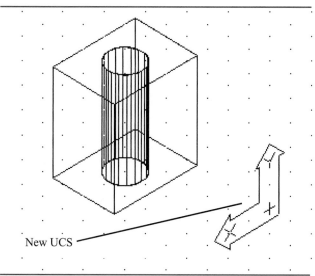

New UCS

Figure 11-45

To draw the end shape of the object

1. Select the Right View tool from the View toolbar or from the UCS Orientation dialog box (see Figure 15-26).

   Grid too dense to display
   Command:

2. Type Zoom; press Enter.

   All/Center/Dynamic/Extents/Left/Previous/Vmax/ Window/<Scale(X/XP)>:

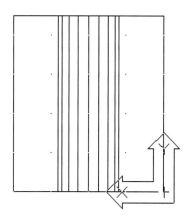

The RIGHT view of the new UCS.

Figure 11-46

3. Type 1 (or whatever factor presents a clear picture of the object); press Enter.

   See Figure 11-46.

4. Use the DRAW and MODIFY commands to create the needed shape.

Figure 11-47 shows the untrimmed construction of the extension shape. Remember, you are working in the new UCS, and even though the box lines appear to be in the same plane, they are not. You cannot use them as part of the construction process. They are visible, but not usable

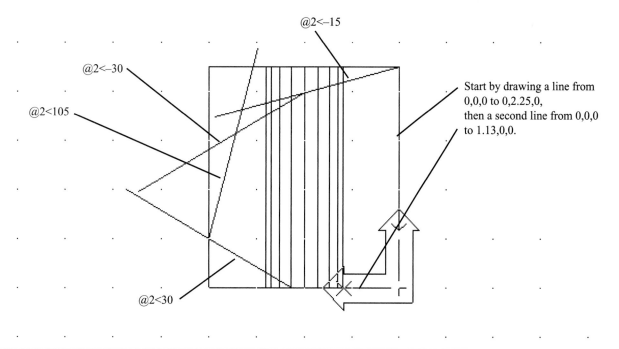

Figure 11-47

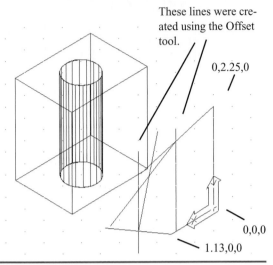

These lines were created using the Offset tool.

0,2.25,0

0,0,0

1.13,0,0

Figure 11-48

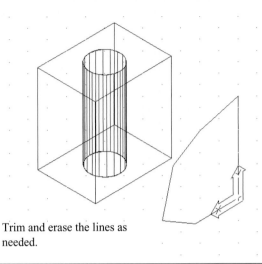

Trim and erase the lines as needed.

Figure 11-49

in this plane. Figure 11-48 shows the same construction from the SE Isometric viewpoint. Horizontal and vertical construction lines were added to the UCS plane so that the OSNAP and LINE commands could be used. Lines were also drawn using relative coordinate values. Figure 11-49 shows the final trimmed shape with the construction lines erased. Figure 11-50 shows the completed shape with the internal cutout added.

5. Use the Edit Polyline tool from the Modify II toolbar and change the shape into a polyline.

6. Draw a line from the 0,0,0 point of the new UCS to 0,0,–1.75.
7. Select the Tabulated surface tool from the Surface toolbar and use the polyline as the path curve and the line created in step 6 as the direction vector.

If necessary, set the SURFTABs to a value of 6. Also use the ZOOM command to help select the correct lines. Figure 11-51 shows the finished drawing of the object oriented to the WCS.

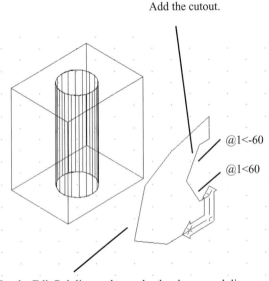

Add the cutout.

@1<-60

@1<60

Use the Edit Polyline tool to make the shape a polyline.

Figure 11-50

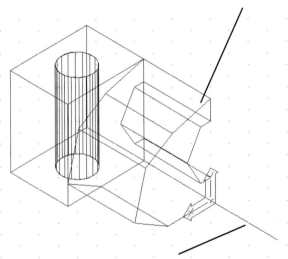

Use the polyline as the path curve.

Draw a line from 0,0,0 to 0,0,–1.75 and use it as the direction vector for the Tabulated surface tool.

Figure 11-51

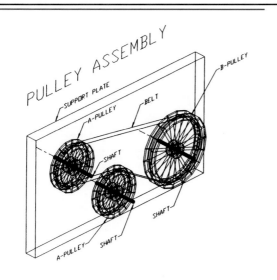

Figure 11-52

11-18 SAMPLE PROBLEM

The following section explains in detail how to draw the pulley assembly shown in Figure 11-52. When creating a complex design or drawing of an object, it is best to start with a freehand sketch and locate all the important assembly and insertion points. A sketch is not required, but it provides a quick and easy way to become familiar with the drawing requirements, resulting in a more efficient drawing approach. Figure 11-53 shows a design layout for the pulley assembly. Note how the sizes and shapes are defined. Center point locations for the pulleys are given in terms of coordinate values, shaft sizes and locations are defined, and assembly points use coordinate values.

It is also good practice to draw each part on its own Layer. This allows individual parts to be shown or not shown as needed during the construction. The following Layers are recommended. See Chapter 3 for an explanation of the Layer command.

Select the Layers tool to access the Layer &

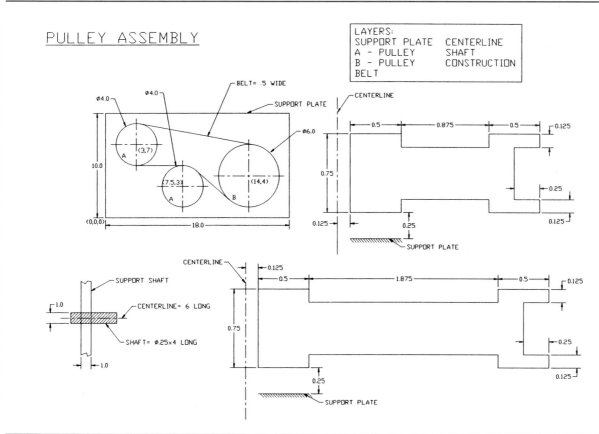

Figure 11-53

Linetype Properties dialog box and create the following Layers. See Figure 11-54.

LAYER - COLOR - LINETYPE
CONSTRUCTION - white - continuous
CENTERLINES - white - centerlines
PLATE - white - continuous
A-PULLEY - blue - continuous
B-PULLEY - cyan - continuous
SHAFT - red - continuous
BELT - yellow - continuous

To draw the support plate

See Figure 11-55. The drawing uses decimal units.

1. Set Grid and Snap = 1, and the viewpoint for SE Isometric.
2. Select the Layer Control tool from the Standard toolbar.
3. Select the Plate layer and make it the current layer.
4. Select Box from the Surfaces toolbar and draw a 1 x 18 x 10 box with its lower front corner on the 0,0,0 origin of the WCS.

To draw the pulley centerlines

See Figure 11-56.

1. Select the Layer Control tool.
2. Select the Centerlines layer.
3. Select the Line tool from the Draw toolbar.

Draw the three pulley centerlines using coordinate values based on the dimensions given in Figure 11-52. Each shaft is to be 4″ long, so the centerlines should be greater than 4″ long. A length of 6″ was chosen. The centerlines should start at a point 1″ beyond the end of the shaft and end 1″ beyond the other end.

The centerline coordinate points are as follows.

1. From point: –2,3,7 to point: 4,3,7
2. From point: –2,7.5,3 to point: 4,7.5,3
3. From point: –2,11,4 to point: 4,11,4

To draw the A and B pulleys' profiles

The profiles of the pulleys will be drawn on the Construction layer and from a Top viewpoint. The Top view will give the drawing a 2D appearance and make it visually easier to draw the profiles. The Elev command will be used to position the working planes for the profiles at the correct height above the WCS X,Y plane.

1. Select the Top tool from the View toolbar.

Define the needed layers.

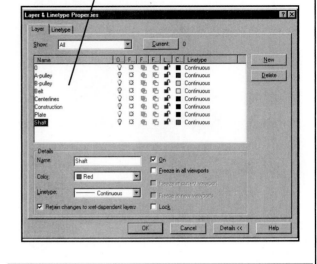

Figure 11-54

A box of size 1,18,10

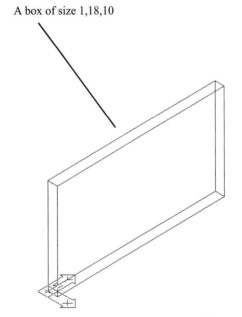

Figure 11-55

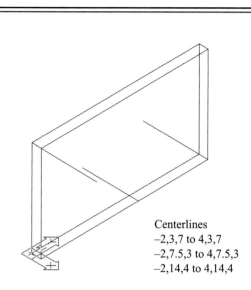

Centerlines
–2,3,7 to 4,3,7
–2,7.5,3 to 4,7.5,3
–2,14,4 to 4,14,4

Figure 11-56

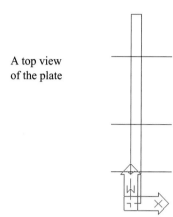

A top view
of the plate

Figure 11-57

The view will change to a top view of the plate. See Figure 11-57.

2. Select the Tools pull-down menu, then Drawing Aids.

The Drawing Aids dialog box will appear.

3. Change the Snap setting to .125.

The .125 value was chosen because it is the length of the smallest distance on the pulley's profile.

4. Select the Zoom Window tool from the Standard toolbar and zoom the plate and pulley centerlines.
5. Select the Layer Control tool, then select the Construction Layer.

At this time you can also turn off any layers that you find visually confusing.

The profiles of the pulleys will be drawn using Polyline, then the REVOLVED SURFACE command will be used to generate the final pulley shapes. The Polyline construction will be done on the Construction layer, and the REVOLVED SURFACE command will be applied to the profiles on the A-pulley and B-pulley layers.

6. Type Elev and set the new current elevation to 7 and the new current thickness to 0.0000.

The value of 7 was derived from the centerline location of the first A-pulley which, according to the sketch in Figure 11-52, has a Y component of 7. The chosen orientation on the WCS changes the component value to the Z

axis.

7. Select Polyline from the Draw toolbar and draw the A-pulley profile.

The new current elevation of 7 has raised the working plane of the drawing so that it is aligned with the pulley's centerline. This means you can draw the pulley directly. Turn on the ORTHO command (F8) if desired to help produce the required horizontal and vertical lines of the profile. See Figure 11-57.

8. Type Elev and change the new current elevation to 4.
9. Use the Zoom All tool to return the drawing to its original size, then use the Zoom Window tool to enlarge the area for the B-pulley.
10. Select the Layer Control tool and select the B-pulley layer.
11. Use Polyline to draw the B-pulley's profile.
12. Type Elev and change the new current elevation to 0.

Turn off the ORTHO command, if it was turned on. See Figure 11-58.

To copy the A-pulley's profile

The assembly requires two A-pulleys, so it is easier to copy the existing profile than to redraw it. The pulleys are located at two different Z planes so you cannot make the copy directly. The displacement points must be defined in terms of their X,Y,Z component values relative to the

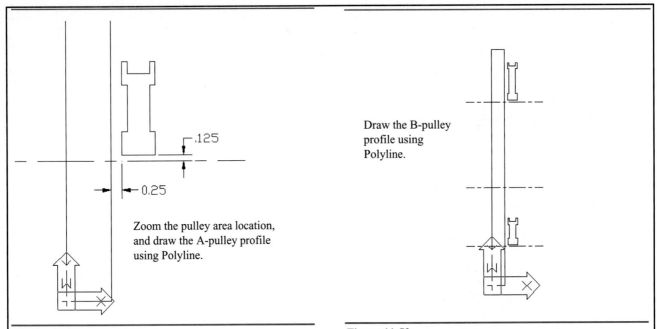

Zoom the pulley area location, and draw the A-pulley profile using Polyline.

Figure 11-58

Draw the B-pulley profile using Polyline.

Figure 11-59

WCS.

1. Select the Copy tool from the Modify toolbar.

    Select objects:

2. Select the A-pulley profile.

    <Base point of displacement>/Multiple:

3. Type 1.25,3,7; press Enter.

    Second point of displacement:

4. Type 1.25,7.5,3; Press Enter.

    See Figure 11-60.

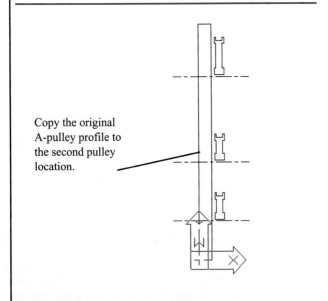

Copy the original A-pulley profile to the second pulley location.

Figure 11-60

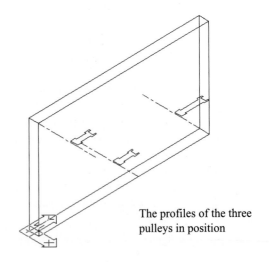

The profiles of the three pulleys in position

Figure 11-61

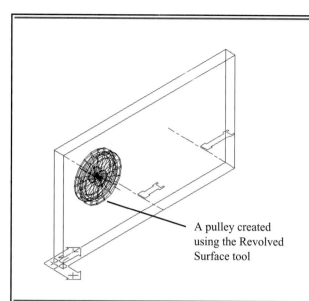

A pulley created
using the Revolved
Surface tool

Figure 11-62

To create the A- and B-pulleys

1. Select the SE Isometric viewpoint from the View toolbar.

   See Figure 11-61.

2. Select the Layer Control tool, then select the A-pulley layer.

Turn off any layers that are not needed or that cause visual confusion.

3. Type Surftab1 and set the value to 18.
4. Type Surftab2 and set the value to 18.
5. Select the Revolved Surface tool from the Surfaces toolbar.
6. Select the first A-pulley profile, use the centerline as an axis of revolution, and rotate the profile 360 degrees.

   See Figure 11-62.

7. Press Enter and repeat the sequence for the second A-pulley.
8. Select the Layer tool, then select the B-pulley layer.
9. Again use the Revolved Surface tool on the Surfaces toolbar to rotate the B-pulley profile 360 degrees around its centerline.

   See Figure 11-63.

To draw the shafts

The three shafts are each 4″ long, perpendicular to the face of the support plate. The Preset, Right view will be used to create a viewpoint perpendicular to the face of the support plate; then the ELEV command can be used to draw the shafts.

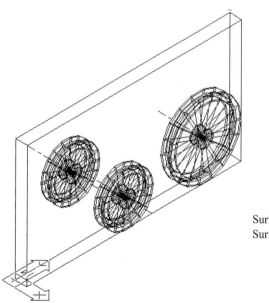

Surftab1 = 18
Surftab2 = 18

Figure 11-63

1. Select the Preset tool from the UCS toolbar, then select the Right UCS.
2. Select Right from the View toolbar.

See Figure 11-64. The origin is located at the lower left corner of the support plate and will not move. Remember that the working plane of this view is the back face of the support plate, even though you are looking at the front face.

3. Type Elev; press Enter.

New current elevation <0.0000>:

4. Type –1; press Enter.

This value moves the working plane 1″ from the back face of the support plate.

New current thickness <0.0000>:

5. Type 4; press Enter.

The value 4 is the length of the shafts.

6. Select the Diameter Circle tool from the Draw toolbar.

Command: _Circle 3P/2P/TTR/<Centerpoint>:

7. Type 3,7,0; press Enter.

The 3,7 X,Y component values were obtained from the design sketch, Figure 11-53. The Z component is 0 because the working plane has been set to align with the end of the shaft.

Radius/<Diameter>:

8. Type .25; press Enter.
9. Press Enter.

Command: _Circle 3P/2P/TTR/<Centerpoint>:

10. Type 7.5,3,0; press Enter.

Radius/<Diameter>:

11. Type .25; press Enter.
12. Press Enter.

Command: _Circle 3P/2P/TTR/<Centerpoint>:

13. Type 11,4,0; press Enter.

Radius/<Diameter>:

14. Type .25; press Enter.

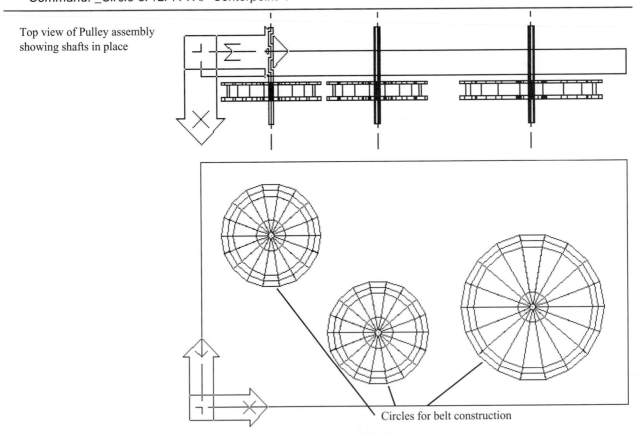

Top view of Pulley assembly showing shafts in place

Circles for belt construction

Figure 11-64

All layers turned off, except for the belt layer.
Drawing is still using the Right UCS orientation.

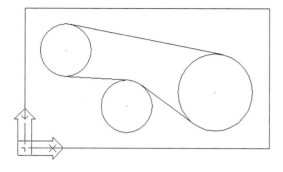

Figure 11-65

Finished drawing

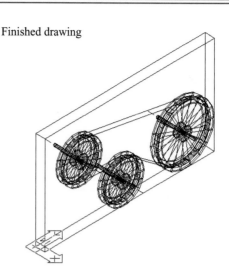

Figure 11-66

To draw the belt

The belt is .50 wide and can be created by constructing the required shape using the CIRCLE, OSNAP TANGENT, LINE, and TRIM commands. The working plane will be moved to a plane .625 from the face of the support plate, and the current elevation thickness set to .50. The .625 value was derived by adding the distance from the front face of the support plate to the edge of the pulleys (.50), and the .125 distance from the edge of the pulleys to the edge of the belt.

The drawing should still be in the Right view orientation. See Figure 11-65. Turn off all layers except for the Belt layer.

1. Type Elev.

New current elevation <-1.0000>:

2. Type 1.625; press Enter.

The additional 1″ is in the thickness of the plate. The origin that you are working from is still located at the original 0,0,0 point of the WCS.

New current thickness <4.0000>:

3. Type .5; press Enter.
4. Select the Layer Control tool, then the Belt layer.

Turn off any layers that you find visually confusing.

5. Select the Options pull-down menu, then Drawing Aids.
6. Set the Grid value for 1.00 and the Snap value for .50.

7. Select the Circle Radius tool from the Draw toolbar and draw three circles as shown in Figure 11-65.

The center point coordinate values are 3,7,0; 7.5,3,0; and 11,4,0. The same points were used as the center points for the shafts. Again the Z component is 0 because the working plane has been moved using the ELEV command.

The radius values of 1.625 and 2.625 were derived from the design sketch in Figure 11-53 and define the location on the pulley surfaces where the belt interfaces.

8. Draw the belt path by drawing lines between the circles using the OSNAP and TANGENT commands.
9. Use the TRIM command to remove the excess portions of the circles.

To create the final assembly drawing

1. Select the Layer tool and turn on all layers except the Construction layer.
2. Select the Preset tool from the UCS dialog box, then select the WCS box, then OK.
3. Select the SE Isometric tool from the View toolbar.

The final drawing should appear and look similar to Figure 11-66.

4. Select the File pull-down menu, then Save As.
5. Save the drawing as PULLEYS.

The drawing will be used again in Chapter 16 when rendering is discussed.

11-19 EXERCISE PROBLEMS

Draw surface models of the objects in exercise problems EX11-1 to EX11-3.

EX11-1 INCHES

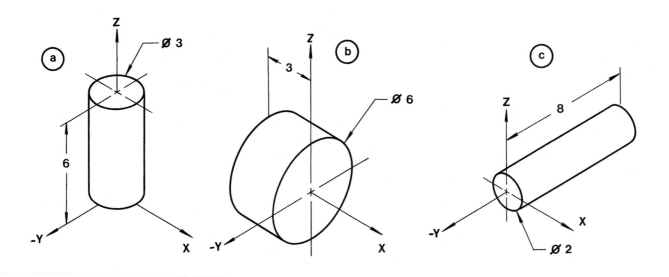

EX11-2 INCHES

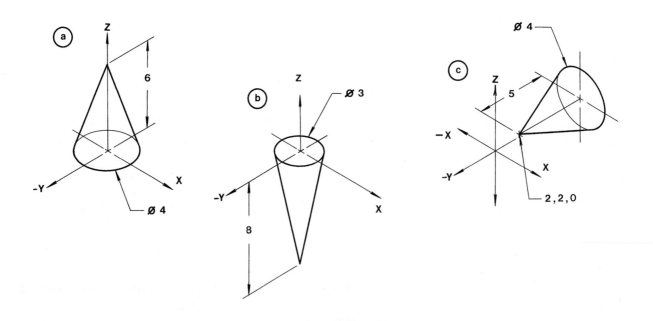

EX11-3 INCHES

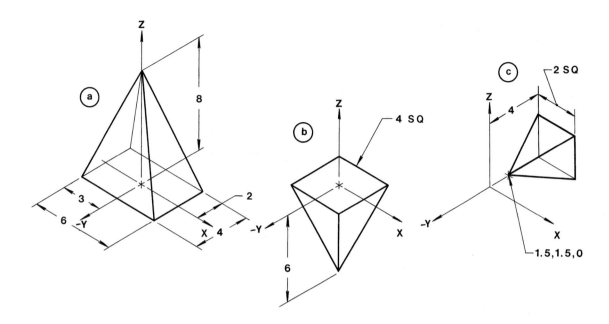

EX11-4 INCHES

Draw the three boxes positioned as shown. The length, width, and height for each box are as follows.

    a. X = 6, Y = 5, Z = 2
    b. X = 4, Y = 4, Z = 4
    c. X = 5, Y = 2, Z = 1

EX11-5 INCHES

Draw the three boxes as shown centered about the Z axis. The length, width, and height for each box are as follows.

    a. X = 8, Y = 8, Z = 1
    b. X = 6, Y = 6, Z = 2
    c. X = 2, Y = 2, Z = 6

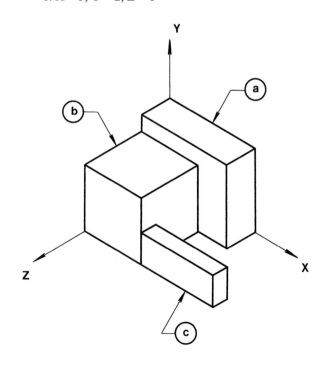

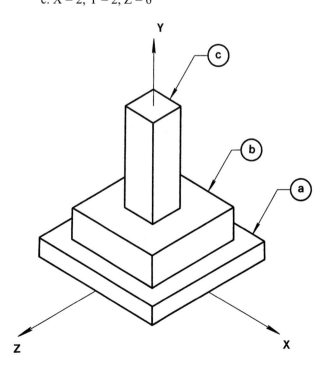

EX11-6 INCHES

Draw the three cylinders shown below centered about the Y axis positioned as shown. The sizes of the cylinders are as follows. Use either the REVOLVED SURFACE or RULED SURFACE command.

    a. Diameter = 2, height = 4
    b. Diameter = 4, height = 1
    c. Diameter = 6, height = 2

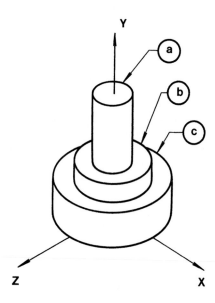

EX11-7 INCHES

Draw the three boxes positioned as shown. The length, width, and height for each box is 2 x 2 x 5.

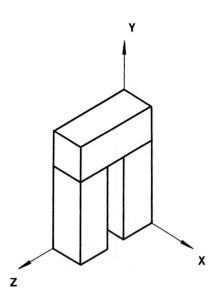

Draw surface models of the objects shown in exercise problems EX11-8 to EX11-27.

EX11-8 FEET

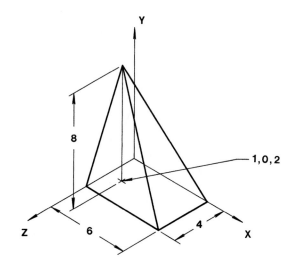

EX11-9 MILLIMETERS

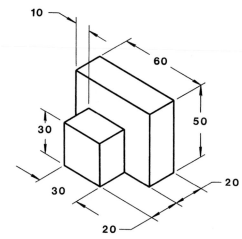

EX11-10 MILLIMETERS

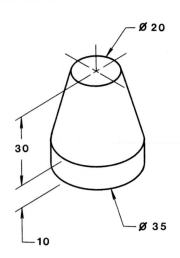

EX11-11 MILLIMETERS

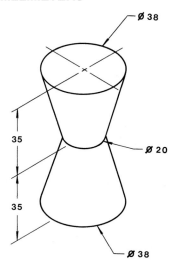

EX11-12 MILLIMETERS

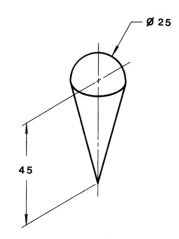

EX11-13 INCHES

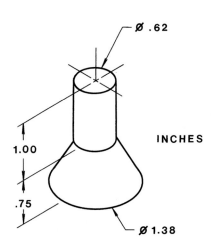

INCHES

EX11-14 MILLIMETERS

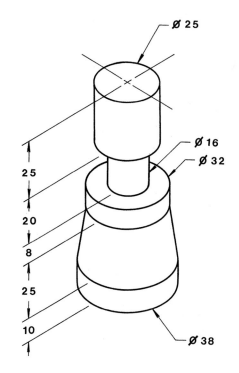

EX11-15 MILLIMETERS

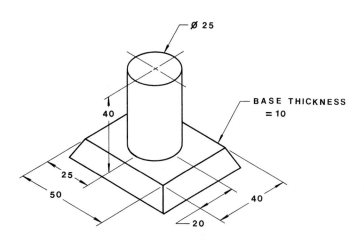

NOTE: LOWER SURFACE OF BASE IS +5
GREATER THAN THE TOP SURFACE
ALL AROUND.

EX11-16 MILLIMETERS

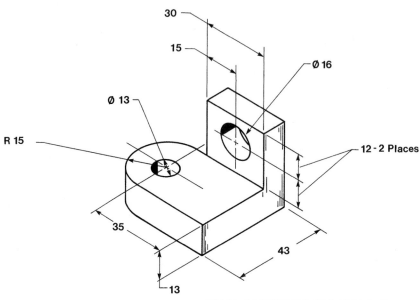

EX11-17 MILLIMETERS

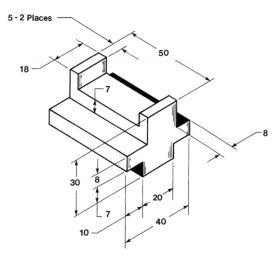

EX11-19 INCHES (SCALE 4:1)

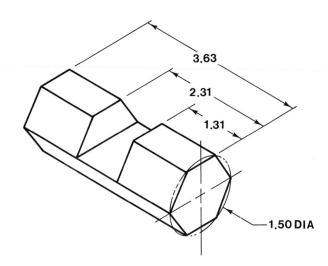

EX11-18 MILLIMETERS

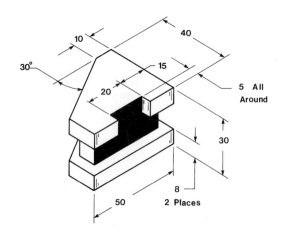

EX11-20 INCHES

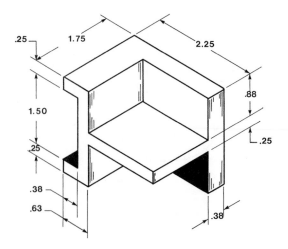

EX11-21 MILLIMETERS

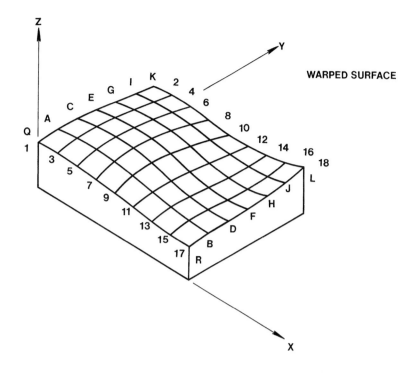

WARPED SURFACE

	LINE Q R		
N	X	Z	Y
1	0	20.0	0
3	10	20.4	0
5	20	20.0	0
7	30	19.6	0
9	40	17.8	0
11	50	16.8	0
13	60	16.0	0
15	70	15.0	0
17	80	14.8	0

	LINE A B		
N	X	Z	Y
1	0	20.5	10
3	10	21.5	10
5	20	22.0	10
7	30	21.5	10
9	40	20.5	10
11	50	18.0	10
13	60	16.5	10
15	70	15.8	10
17	80	15.2	10

	LINE CD		
N	X	Z	Y
1	0	20.0	20
3	10	21.2	20
5	20	22.2	20
7	30	22.7	20
9	40	21.3	20
11	50	17.8	20
13	60	16.0	20
15	70	15.5	20
17	80	15.8	20

	LINE E F		
N	X	Z	Y
1	0	19.3	30
3	10	20.8	30
5	20	22.0	30
7	30	22.5	30
9	40	20.5	30
11	50	17.5	30
13	60	15.8	30
15	70	15.0	30
17	80	16.0	30

	LINE G H		
N	X	Z	Y
1	0	17.5	40
3	10	19.0	40
5	20	20.3	40
7	30	20.0	40
9	40	18.8	40
11	50	16.3	40
13	60	15.2	40
15	70	15.4	40
17	80	16.8	40

	LINE I J		
N	X	Z	Y
1	0	16.5	50
3	10	18.0	50
5	20	18.8	50
7	30	17.8	50
9	40	16.6	50
11	50	15.5	50
13	60	15.3	50
15	70	16.0	50
17	80	18.0	50

	LINE K L		
N	X	Z	Y
1	0	15.0	60
3	10	16.5	60
5	20	17.0	60
7	30	14.5	60
9	40	13.8	60
11	50	14.3	60
13	60	15.6	60
15	70	17.5	60
17	80	20.0	60

EX11-22 MILLIMETERS

EX11-24 MILLIMETERS

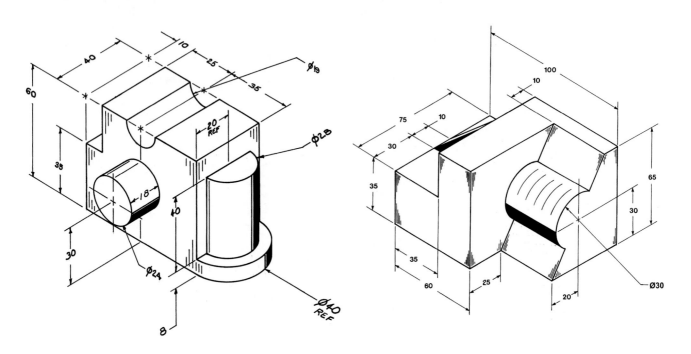

EX11-23 MILLIMETERS

EX11-25 MILLIMETERS

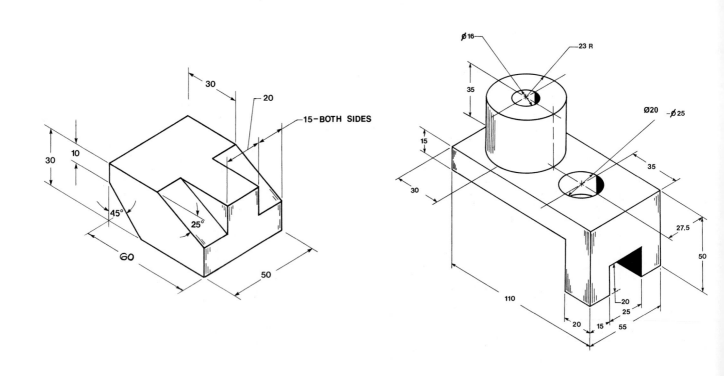

EX11-26 INCHES

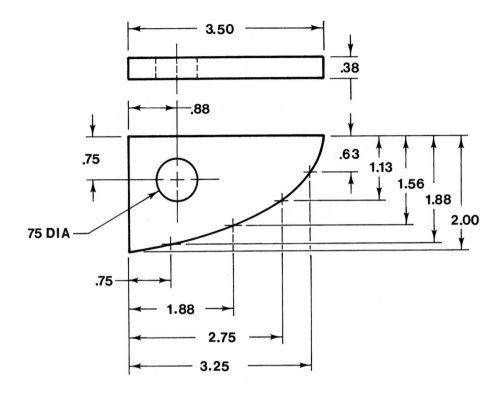

EX11-27 INCHES

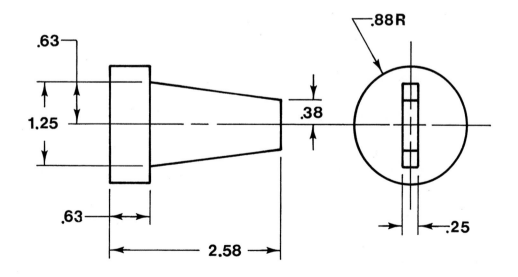

EX11-28 MILLIMETERS

Study the sample problem presented in Section 11-18, then create a similar surface model for the pulley assembly described below.

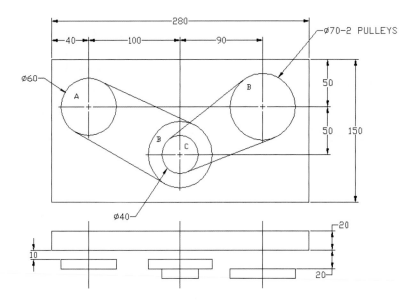

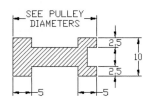

DETAIL A
SCALE:4=1

If the pulley in the upper left of the plate is turning at 1750 rpm, what is the speed of the other pulleys?

EX11-29 MILLIMETERS

Study the sample problem presented in Section 11-18, then create a similar surface model for the pulley assembly described below.

If the pulley in the upper left of the plate is turning at 3550 rpm, what is the speed of the other pulleys?

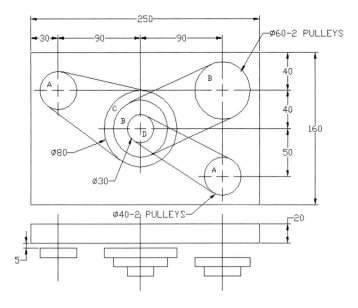

DETAIL A
SCALE:4=1

DESIGN

EX11-30 MILLIMETERS

Design a plate to support the following gear pattern. Each gear is mounted on a Ø5.00 shaft with a tolerance of +0.02,– 0.00. The pitch diameters of the gears are as follows.

1. Ø20.00±.01
2. Ø36.00±.01
3. Ø40.00±.01
4. Ø18.00±.01

DESIGN PARAMETERS:

1. The edge of any gear must be at least 20mm from the edge of the support plate.
2. The plate must be at least 20mm thick.
3. The gears must be at least 3mm from the support plate.

REQUIREMENTS:

1. A surface model of the assembly including the gears and the shafts.
2. A dimensioned and toleranced drawing of the support plate.

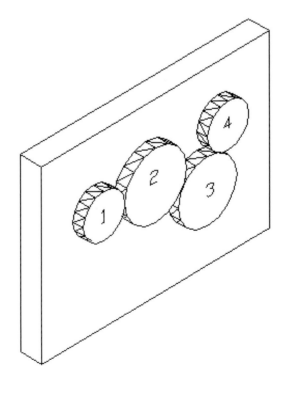

The Solids Toolbar

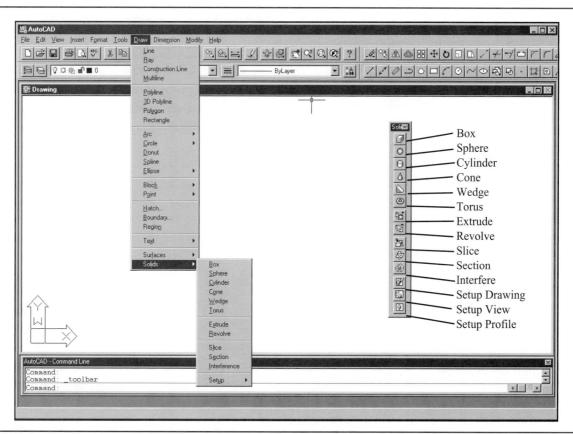

Figure 12-1

12-1 INTRODUCTION

This chapter introduces solid modeling. The solid modeling commands can be accessed using the Solids toolbar or by using the Draw pull-down menu. See Figure 12-1.

Solid modeling allows you to create objects as solid entities. Solid models differ from the surface models created in the last chapter because solid models have density and are not merely joined surfaces.

Solid models are created by joining together, or unioning, basic primitive shapes: boxes, cylinders, wedges, etc., or by defining a shape as a polyline and extruding it into a solid shape. Solid primitives may also be subtracted from one another. For example, to create a hole in a solid box, draw a solid cylinder, then subtract the cylinder from the box. The result will be an open volume in the shape of a hole.

The first part of the chapter deals with the individual tools of the Solids toolbar. The second part gives examples of how to create solid objects by joining primitive shapes and changing UCSs.

12-2 BOX

The Box tool on the Solids toolbar has two options; Center and Corner. Corner is the default option. The Center tool is used to draw a box by first locating its center point. The Corner tool is used to draw a box by first locating one of its corner points.

To draw a box using the Center option

See Figure 12-2.

1. Select the Box tool from the Solids toolbar.

    Command: _box
    Center/<Corner of box> <0,0,0>:

2. Type C; press Enter.

    Center of box <0,0,0>:

3. Type 5,5,0; press Enter or locate the center point using the coordinate display at the lower left of the screen.

    Cube/Length/<corner of box>:

4. Type 10,10,0; press Enter.

    Height:

5. Type 3; Press Enter.

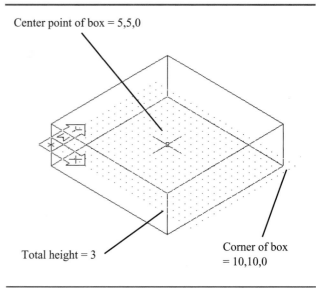

Figure 12-2

The box is centered about the 5,5,0 point. The size of the box along the XY plane is defined by the distance from the center point (5,5,0) to the right corner of the box (10,10,0). The height input of 3 represents half the total height of 6 centered about the XY plane.

To draw a box using the Corner option

See Figure 12-3.

1. Select the Box tool from the Solids toolbar.

    Command: _box
    Center/<Corner of the box> <0,0,0>:

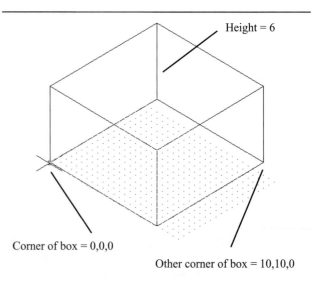

Figure 12-3

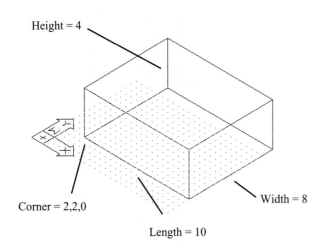

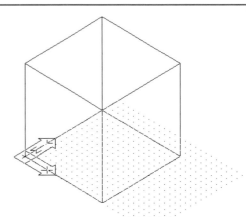

The cube option makes the length, width, and height of box of equal length.

Figure 12-4

Figure 12-5

2. Press Enter.

This means that the corner of the box will be located on the 0,0,0 point of the XY plane.

Cube/Length/<corner of box>:

3. Type 10,10,0; press Enter.

Height:

4. Type 6; press Enter.

The box shown in Figure 12-3 has its bottom surface aligned with the XY plane, whereas the XY plane passes through the middle of the box shown in Figure 12-2.

To draw a box from given dimensions

See Figure 12-4. Draw a box with a length of 10, a width of 8, and a height of 4, with its corner at the 2,2,0 point.

1. Select the Box tool from the Surfaces toolbar.

Command: _box
Center/<Corner of the box> <0,0,0>:

2. Type 2,2,0; press Enter.

Cube/Length/<corner of box>:

3. Type L; press Enter.

Length:

4. Type 10; press Enter.

Width:

5. Type 8; press Enter.

Height:

6. Type 4; press Enter.

To draw a cube

See Figure 12-5.

1. Select the Corner tool from the Solids toolbar.

Command: _box
Center/<Corner of the box> <0,0,0>:

2. Press Enter.

Cube/Length/<corner of box>:

3. Type C; press Enter.

Length:

4. Type 7.5; press Enter.

AutoCAD will automatically make all three edges of the box equal lengths.

12-3 SPHERE

The visual accuracy of solids depends on the settings of the ISOLINES command. The sphere shown in Figure 12-6 was drawn with an Isolines setting of 4. For this section the Isolines setting will be changed to 18.

To change the Isolines settings

1. Type Isolines; press Enter.

Center point = 5,5,0

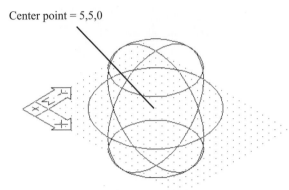

A sphere drawn with Isolines set at 4

Figure 12-6

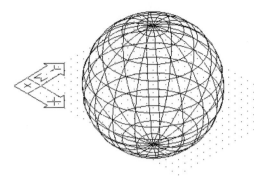

A sphere drawn with Isolines set at 18

Figure 12-7

There is no tool for the ISOLINES command.

New value for Isolines <4>;

2. Type 18; press Enter.

Command:

The new value is now in place and will be used for all solids drawn until it is changed or a new drawing is started.

To draw a sphere

See Figure 12-7.

1. Select the Sphere tool from the Solids toolbar.

Center of sphere <0,0,0>:

2. Type 5,5,0; press Enter.

Diameter/<Radius> of sphere:

3. Type 4; press Enter.

12-4 CYLINDER

There are two options associated with the CYLINDER command: circular and elliptical. This means cylinders may be drawn with either elliptical or circular base planes. The base elliptical shape is drawn using the same procedure as was outlined for the ELLIPSE command in Chapter 3.

To draw a cylinder with an elliptical base

See Figure 12-8. The sequence described below uses coordinate value input. The same points could have been selected by moving the cross hairs and pressing the left mouse button.

1. Select the Cylinder tool from the Surfaces toolbar.

Command: _cylinder
Elliptical/<center point> <0,0,0>:

2. Type e; press Enter.

Center/<Axis endpoint>:

3. Type 0,0,0; press Enter.

First point = 0,0,0

Third point = 5,3.5,0

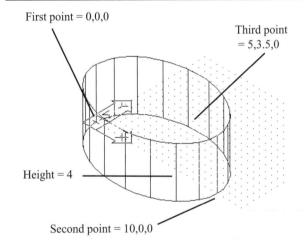

Height = 4

Second point = 10,0,0

Figure 12-8

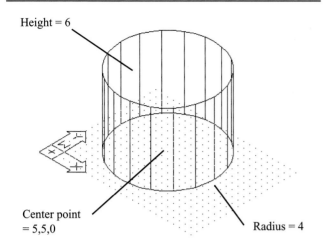

Figure 12-9

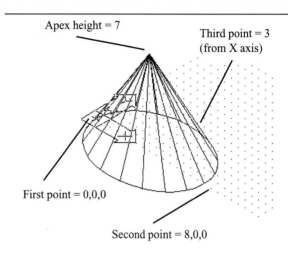

Figure 12-10

This input will locate one end of the base axis on the 0,0,0 point of the XY plane.

Axis endpoint 2:

4. Type 10,0,0.

This input will locate the axis line along the X axis.

Other axis distance:

Note that the line dragging from the cross hairs has one end centered on the axis just defined.

5. Type 5,3.5,0; press Enter.

Center of other end/<Height>:

6. Type 4; press Enter.

An elliptical base can also be defined by first defining a center point for the ellipse, then defining the length of the radii of the major and minor axes.

To draw a cylinder with a circular base

See Figure 12-9.

1. Select the Cylinder tool from the Solids toolbar.

Command: _cylinder
Elliptical/<center point> <0,0,0>:

2. Type 5,5,0; press Enter.

Diameter/<Radius>:

3. Type 4; press Enter.

Center of other end/<Height>:

4. Type 6; press Enter.

12-5 CONE

There are two options associated with the Cone tool; circular and elliptical. This means that cones can be drawn with either an elliptical or circular base plane. The base elliptical shape is drawn using the same procedure as was outlined for the ELLIPSE command in Chapter 3. The CONE command cannot be used to draw truncated cones as could the Cone tool associated with the Surfaces toolbar. Solid truncated cones are created by subtracting the top portion of the cone.

To draw a cone with an elliptical base

See Figure 12-10.

1. Select the Cone tool from the Solids toolbar.

Command: _cone
Elliptical/<center point> <0,0,0>:

2. Type e; press Enter.

Center/<Axis endpoint>:

3. Type 0,0,0; press Enter.

This input will locate one end of the ellipse axis at the 0,0,0 point of the XY plane.

Axis point 2:

4. Type 8,0,0.

Other axis distance:

Note that the line dragging from the cross hairs has one end centered on the axis just defined.

5. Type 3; press Enter.

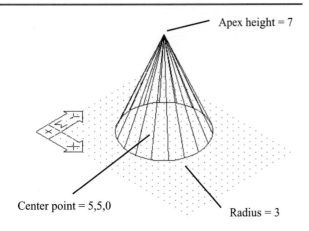

Figure 12-11

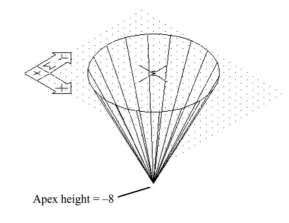

Figure 12-12

Apex/<Height>:

6. Type 6; press Enter.

A response of A to the Apex/<Height>: prompt allows you to select the height of the cone using the cross hairs.

To draw a cone with a circular base

See Figure 12-11.

1. Select the Cone tool from the Solids toolbar.

   Command: _cone
   Elliptical <center point> <0,0,0>:

2. Type 5,5,0; press Enter.

   Diameter/<Radius>:

3. Type 3; press Enter.

   Apex/<Height>:

4. Type 7; press Enter.

A response of A to the Apex/<Height>: prompt allows you to select the height of the cone using the cross hairs. Figure 12-12 shows a cone drawn using a negative height input.

12-6 WEDGE

There are two options associated with the WEDGE command; Center and Corner.

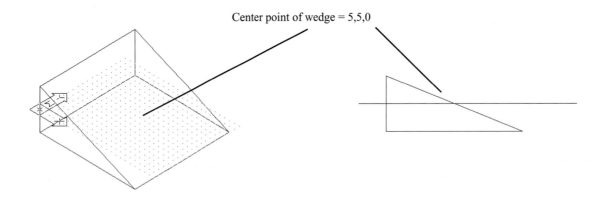

Figure 12-13

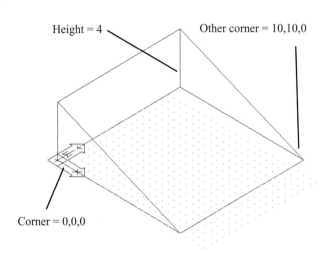

Height = 4

Other corner = 10,10,0

Corner = 0,0,0

Figure 12-14

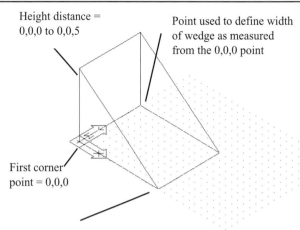

Height distance =
0,0,0 to 0,0,5

Point used to define width
of wedge as measured
from the 0,0,0 point

First corner
point = 0,0,0

Use this point to define the length of
wedge as measured from the 0,0,0 point.

Figure 12-15

To draw a wedge by defining its center point

See Figure 12-13.

1. Select the Wedge tool from the Solids toolbar.

   Command: _Wedge
   Center/<Center of wedge> <0,0,0>:

2. Type c; press Enter.

   Center of wedge <0,0,0>:

3. Type 5,5,0; press Enter.

   Cube/Length/<Corner of wedge>:

4. Type 10,10,0 ; press Enter.

   Height:

5. Type 4; press Enter.

The wedge shown in Figure 12-13 is centered about the XY plane; that is, part of the wedge is above the plane, and part is below. This is not easy to see even with the grid shown. The far right corner of the wedge is actually located below the grid at the 10,10 point. Figure 12-13 also shows a side view of the same wedge with a line drawn on the XY plane added. Note how the line bisects the height line of the wedge.

To draw a wedge by defining its corner point

See Figure 12-14.

1. Select the Wedge tool from the Solids toolbar.

Command: _wedge
Center/<Corner> <0,0,0>:

2. Press Enter.

This input will locate the corner of the wedge on the 0,0,0 point of the XY plane.

Cube/Length/<other corner>:

The default response to this command defines the diagonal corner of the wedge's base.

3. Type 10,10,0; press Enter.

   Height:

4. Type 4; press Enter.

To draw a wedge using selected points

See Figure 12-15.

1. Select the Wedge tool from the Solids toolbar.

   Command: _wedge
   Center/<Corner> <0,0,0>:

2. Press Enter.

This input will locate the corner of the wedge on the 0,0,0 point of the XY plane.

Cube/Length/<other corner>:

3. Type L; press Enter.

   Length:

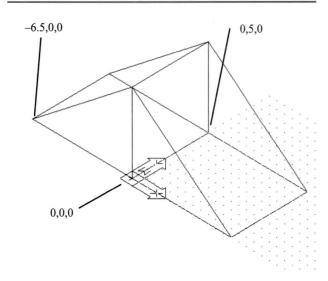

Figure 12-16

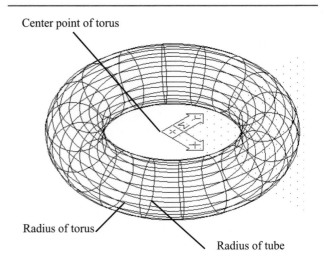

Figure 12-17

4. Select the 0,0 point on the XY plane by moving the cross hairs to that point and pressing the left mouse button.

   Length: Second point:

5. Select a point on the X axis.

The length input values are measured only along the X axis. If you had selected a point within the XY plane, the resulting length distance would have been the X component of the selected point relative to the first point selected. The width distances are measured only in the Y direction, and the height distances in the Z direction.

   Width:

6. Select the 0,0 point.

   Width: Second point:

7. Select a point on the Y axis.

   Height:

8. Select the 0,0,5; press Enter.

The technique of drawing a wedge by selecting points is best used when another object already exists and the OSNAP command can be used to align the selection points with the existing object.

To align a wedge with an existing wedge

This example serves to illustrate how you can use different inputs to position a wedge. Figure 12-15 shows a

wedge. The problem is to draw another wedge with its back surface aligned with the back surface of the existing wedge.

1. Select the Wedge tool from the Solids toolbar.

   Command: _wedge
   Center/<Corner of wedge> <0,0,0>:

2. Press Enter.

   Cube/Length/<other corner>:

3. Type L; press Enter.

   Length:

4. Type –6.5; press Enter.

The negative value will orient the wedge in the direction opposite that of the previously drawn wedges.

   Width:

5. Type 5; press Enter.

   Height:

6. Type 5; press Enter.

See Figure 12-16. The construction could also have been achieved by using the COPY command to create a second wedge, the ROTATE command to rotate the new wedge 180 degrees, and the MOVE command to align the wedge with the existing wedge. Use OSNAP, Endpoint to ensure exact alignment.

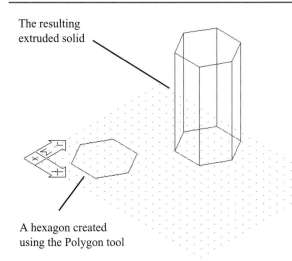

The resulting extruded solid

A hexagon created using the Polygon tool

Figure 12-18

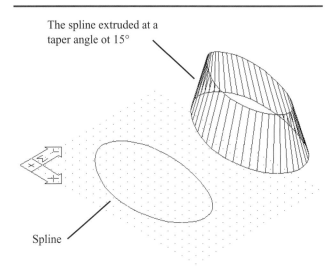

The spline extruded at a taper angle ot 15°

Spline

Figure 12-19

12-7 TORUS

A torus is a donutlike shape. See Figure 12-17.

To draw a torus

1. Select the TORUS command.

   Command: _torus
   Center of torus <0,0,0,>:

2. Press Enter.

   This input will locate the center of the torus at the 0,0,0 point on the XY plane.

   Diameter/<Radius> of torus:

3. Type 5; press Enter.

   Diameter/<Radius> of tube:

4. Type 1.5; press Enter.

   The torus shown in Figure 12-17 was created with Isolines set at 18.

12-8 EXTRUDE

The EXTRUDE command is used to extend existing 2D shapes into 3D shapes.

> The Extrude tool can only be applied to a polyline.

To extrude a 2D Polyline

Figure 12-18 shows a hexagon drawn using the POLYGON command. All shapes drawn using the POLYGON command are automatically drawn as a polyline, so the hexagon can be extruded. How to draw a polygon is discussed in Chapter 3.

1. Select the Extrude tool from the Solids toolbar.

   Command: _extrude
   Select objects:

2. Select the hexagon.

   Select objects:

3. Press Enter.

   Path/<Height of Extrusions>:

4. Type 6; press Enter.

   Extrusion taper angle <0>:

5. Press Enter.

   Figure 12-19 shows an irregular spline and an extrusion created from the spline. The procedure is as outlined above with a 15 response to the Extrusion taper angle prompt.

To create a polyline from line segments

Figure 12-20 shows a 2D shape that was created using the Line and Circle tools from the Draw toolbar. The object must be converted to a polyline before it can be extruded.

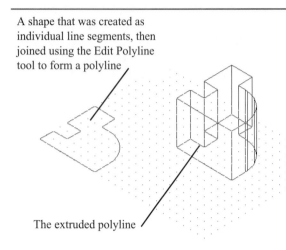

A shape that was created as individual line segments, then joined using the Edit Polyline tool to form a polyline

The extruded polyline

Figure 12-20

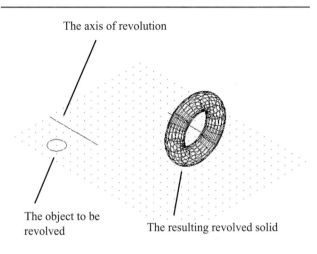

The axis of revolution

The object to be revolved

The resulting revolved solid

Figure 12-21

1. Select the Edit Polyline tool from the Modify II toolbar.

   Command: _pedit Select polyline:

2. Select any one of the lines in the 2D shape.

   Object selected is not a polyline
   Do you want to turn it into one? <Y>:

3. Press Enter.

   Close/ Join/ Width/ Edit Vertex/ Fit/ Spline/ Decurve /Ltype gen/ Undo/ eXit <X>:

4. Type J (for Join); press Enter.

   The polyline will be defined by joining together all the line segments to form a polyline. A curve is considered to be a line segment.

   Select objects:

5. Window the entire object.

   Select objects: Other corner: 11 found
   Select objects:

6. Press Enter.

   10 Segments added to polyline
   Close/ Join/ Width /Edit Vertex/ Fit/ Spline/ Decurve/ Ltype gen/ Undo/ eXit <X>:

7. Select the EXTRUDE command and create an extrusion 3 units high with a 0 degree taper angle.

12-9 REVOLVE

The Revolve tool on the Solids toolbar is used to create a solid 3D object by rotating a 2D shape around an axis of revolution. Figure 12-21 shows a torus created by rotating a circle around a straight line. The density of the resulting object is controlled by the ISOLINES command. In this example Isolines is set at 18.

To create a revolved solid object

This procedure assumes that the curve path (2D shape) and the line that will be used as the axis of revolution already exist on the drawing.

1. Select the Revolve tool from the Solids toolbar.

   Command: _revolve
   Select objects:

2. Select the circle.

   Select objects:

3. Press Enter.

   Axis of revolution — Object/X/Y/<Start point axis>:

4. Select one end of the line to be used as the axis of revolution.

   Use OSNAP, Endpoint to ensure accuracy.

   <End point of Axis>:

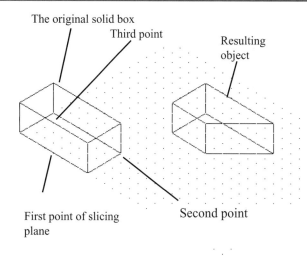

The original solid box

Third point

Resulting object

First point of slicing plane

Second point

Figure 12-22

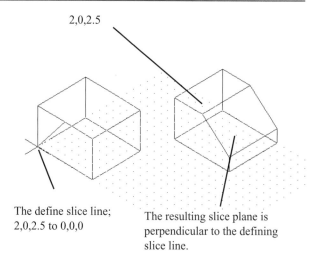

2,0,2.5

The define slice line; 2,0,2.5 to 0,0,0

The resulting slice plane is perpendicular to the defining slice line.

Figure 12-23

5. Select the other end of the axis line.

Angle of revolution <full circle>:

6. Press Enter.

12-10 SLICE

The Slice tool is used to remove part of an existing solid object. The plane of the slice can be defined using several different methods.

To slice an object — 3 points

See Figure 12-22.

1. Select the Slice tool from the Solids toolbar.

Command: _slice
Select objects:

2. Select the box.

Select objects:

3. Press Enter.

Slicing plane by object/Zaxis/View/XY/YZ/ZX/ <3 points>:

4. Press Enter.

This input means that the cutting plane for the slice will be defined using three points. This is a process similar to that used to define a UCS in Chapter 13.

1st point on plane:

5. Select a point.

In this example a first point was selected along the X axis of the box.

2nd point on plane:

6. Select the far right corner of the box.

The OSNAP, Endpoint tool can be used to ensure accuracy.

3rd point on plane:

7. Select the third point on the plane.

In this example a coordinate value (X,Y,Z) was used to define the third point. Remember, the cross hairs move only on the XY plane, so either OSNAP commands or an X,Y,Z coordinate value must be used to define the point. It cannot be defined by moving the cross hairs and selecting a point.

Both sides/<Point on desired side of the plane>:

8. Select a point on the part of the object you want to remain.

In this example the far left corner was selected. Figure 12-22 shows the resulting 3D shape.

To slice an object — Z axis

The Z axis option allows you to specify a slicing plane by defining a line with its origin on the Z axis and its other end in an XY plane. The resulting slice will be generated normal to the defining line. See Figure 12-23.

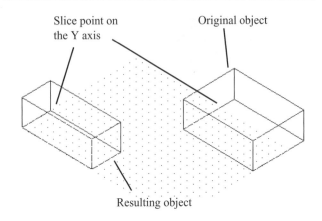

Slice point on the Y axis

Original object

Resulting object

Figure 12-24

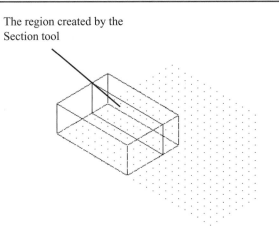

The region created by the Section tool

Figure 12-25

1. Select the Slice tool from the Solids toolbar.

   Command: _slice
   Select objects:

2. Select the box.

   Select objects:

3. Press Enter.

   Slicing plane by object/Zaxis/View/XY/YZ/ZX/ <3 points>:

4. Type Z; press Enter.

   Point on plane:

5. Type 5,0,4; press Enter.

   This point could also have been selected using the crosshairs.

   Point on Z-axis (normal) of the plane:

6. Select a point on the Z axis.

   In this example the point 0,0,2 was selected. Again the point could have been selected using the cross hairs.

   Both sides/<Point on desired side of the plane>:

7. Select a point on the object on the side of the plane you want to keep.

   In this example, a point on the Y axis was selected. Figure 12-23 shows the resulting sliced object.

To Slice an object — ZX

The Slice command lists three plane options: XY, YZ, and ZX. Each of these options allows you to take slices parallel to these principal planes. See Figure 12-24.

1. Select the Slice tool from the Solids toolbar.

   Command: _slice
   Select objects:

2. Select the box.

   Select objects:

3. Press Enter.

   Slicing plane by object/Zaxis/View/XY/YZ/ZX/<3 points>:

4. Type ZX; press Enter.

   Point on the zx plane <0,0,0>:

   This prompt is asking for the location of the ZX slicing plane. The plane will be drawn parallel to the ZX plane based at the 0,0,0 origin, so select a point along the Y axis.

5. Select a point on the Y axis.

   Both sides/<Point on desired side of the plane>:

6. Select a point on the object on the side of the plane you want to keep.

12-11 SECTION

The SECTION command is used to define a region within an existing solid. Regions are defined in the same manner as the slicing planes discussed in the previous section.

1. Select the Section tool from the Solids toolbar.

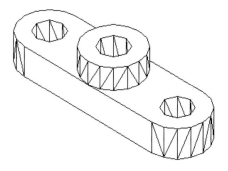

Figure 12-26

Command: _section
Select objects:

2. Select the box.

Select objects:

3. Press Enter.

Slicing plane by object/Zaxis/View/XY/YZ/ZX/<3 points>:

4. Type ZX; press Enter.

Point on the zx plane <0,0,0>:

This prompt is asking for a location parallel to the ZX plane based at the 0,0,0 origin. The defining point for the region will be a point along the Y axis.

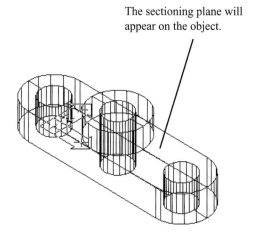

The sectioning plane will appear on the object.

Figure 12-28

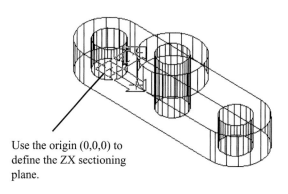

Use the origin (0,0,0) to define the ZX sectioning plane.

Figure 12-27

5. Select a point on the Y axis.

Figure 12-25 shows the resulting region. The region is an independent object or plane. This can be verified by erasing the box. After the box is erased, the region will remain.

To create a sectional view from a solid object

Figure 12-26 shows a solid object. Use the SECTION command to create a sectional view along the longitudinal center line of the object.

1. Select the Section tool from the Solids toolbar.

Command:_section
Select objects

2. Select the object.

Select objects:

3. Press Enter.

Slicing plane by object/Zaxis/View/XY/YZ/ZX/<3 points>:

4. Type ZX; press Enter.

Point on the zx plane <0.00>:

5. Locate a point on the object for the sectional cutting plane.

See Figure 12-27. The example shown was drawn centered about the X axis, so the origin (0,0,0) can be used to define the location of the ZX sectioning plane. See Chapter 6 for section view terminology. The section cutting plane will appear on the object. See Figure 12-28.

6. Select the Scommand and move the cutting

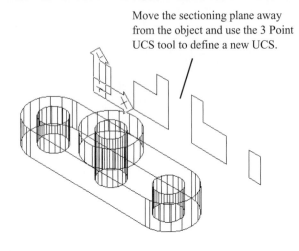

Move the sectioning plane away from the object and use the 3 Point UCS tool to define a new UCS.

Figure 12-29

plane away from the solid object.

See Figure 12-29. In order to apply Hatch lines to the sectional view just created, a UCS aligned to the sectional view must be created.

7. Select the 3 Point UCS tool from the UCS toolbar.
8. Select the lower left corner of the sectional view as the origin, the lower right corner as the second point, and the upper left corner as the third point.
9. Use the Line tool to add the lines needed to complete the sectional view including centerlines.

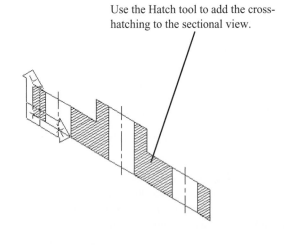

Use the Hatch tool to add the crosshatching to the sectional view.

Figure 12-31

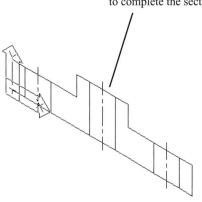

Use the Line tool to add lines to complete the sectional view.

Figure 12-30

Use OSNAP, Endpoint to ensure accuracy of the point selections. See Figure 12-30.

To add hatching to the sectional view

1. Select the Hatch tool from the Draw toolbar and select the appropriate areas on the sectional view.

See Figure 12-31. See Chapter 6 for instructions on the Hatch command.

2. Create a layer called SECTION and use the Change Properties (Modify pull-down menu) command to move the sectional view to the SECTION layer.
3. Turn off the 0 layer, and turn on the SECTION layer and make it the current layer.

Only the sectional view should be seen on the screen.

4. Select the Front tool from the View toolbar.

If the sectional view appears too large for the screen, use the ZOOM command to reduce the view's size. Figure 12-32 shows the resulting sectional view.

12-12 INTERFERE

The INTERFERE command is used to define a volume common to two or more existing solid objects. Figure 12-33 shows a solid cylinder and box that intersect each other. They were both drawn on the XY plane. The following procedure will define the volume common to both of them.

The resulting sectional view taken
from a solid model

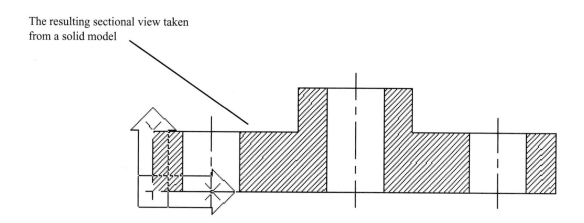

Figure 12-32

1. Select the Interfere tool from the Solids toolbar.

   Command: -interfere Select the first set of solids:
   Select objects:

2. Select the box.

   Select objects:

3. Press Enter.

   Select the second set of solids:

Select objects:

4. Select the cylinder.

   Select objects:

5. Press Enter.

   Create interference solids ? <N>:

6. Type Y; press Enter.

   The common volume is now defined but difficult to see.

The cylinder's center
point is on the corner
of the box.

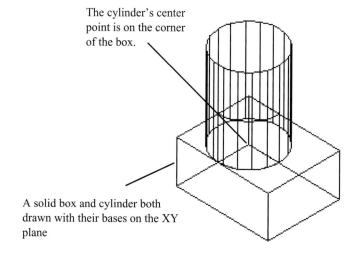

A solid box and cylinder both
drawn with their bases on the XY
plane

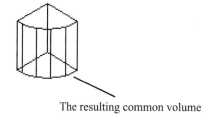

The resulting common volume

Figure 12-33

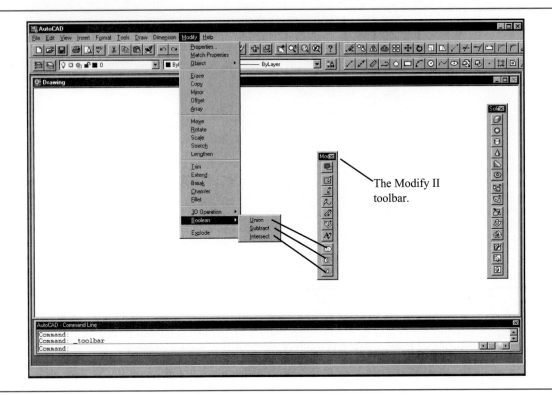

Figure 12-34

7. Select the Erase tool from the Modify toolbar and erase the box and cylinder.

Figure 12-33 shows the resulting common volume.

12-13 UNION AND SUBTRACTION

Solid objects may be combined to form more complex objects. Objects can be added together using the UNION command and subtracted from each other using the SUBTRACT command. A volume common to two or more objects may be defined using the INTERSECT command. The tools for these three commands are located on the Modify II toolbar or under the BOOLEAN command located on the Modify pull-down menu. See Figure 12-34.

To union two objects

Figure 12-35 shows two solid boxes drawn with adjoining surfaces.

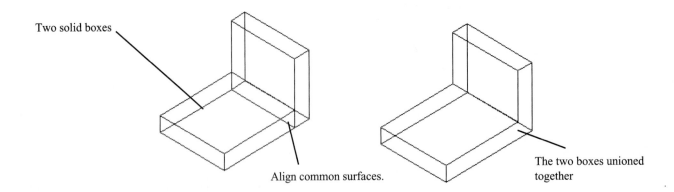

Two solid boxes

Align common surfaces.

The two boxes unioned together

Figure 12-35

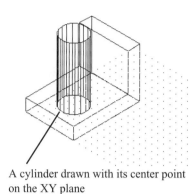

A cylinder drawn with its center point on the XY plane

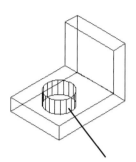
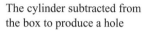

The cylinder subtracted from the box to produce a hole

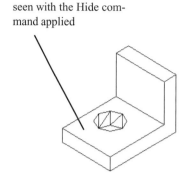

The resulting shape as seen with the Hide command applied

Figure 12-36

1. Select the Union tool from the Modify II toolbar.

   Command: _union
   Select objects:

2. Select the two boxes.

   Note the changes in the boxes after they have been unioned. The solid object is no longer two boxes but an L-shaped object.

To subtract an object

Figure 12-36 shows the L-shaped bracket formed above. This exercise will add a hole to the front surface. Holes are created in solid objects by subtracting solid cylinders from the existing objects.

1. Select the Cylinder tool from the Solids toolbar.

   Command: _cylinder
   Elliptical/<center point> <0,0,0>:

2. Select a point on the XY plane approximately in the middle of the front surface.

   Remember, the cross hairs move only in the XY plane, so you can select a point only on the bottom surface of the box.

   Diameter/<Radius>:

3. Select a radius for the cylinder.

   Center of other end/<Height>:

4. Type 5; press Enter.

   The height of the cylinder was deliberately drawn

higher than the top surface of the bracket to illustrate the fact that the two heights need not be equal for the SUBTRACT command. The only requirement is that the cylinder be equal to or greater than the height of the box surface. See Figure 12-36.

5. Select the Subtract tool from the Solids toolbar.

   Command: _subtract Select solids and regions to subtract from...
   Select objects:

   This prompt is asking you to define the main object, that is, the object you want to remain after the subtraction.

6. Select the L-shaped bracket.

   Select objects:

7. Press Enter.

   Select solids and regions to subtract...
   Select objects:

   This prompt is asking you to define the object you want removed by the subtraction.

8. Select the cylinder.

   Select objects:

9. Press Enter.

12-14 SOLID MODELING AND UCSs

In this section we will again work with the L-shaped bracket and add a hole to the upper surface. The procedure

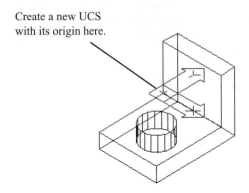

Create a new UCS with its origin here.

Figure 12-37

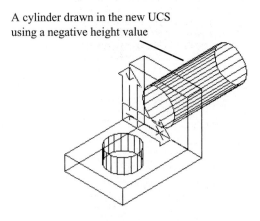

A cylinder drawn in the new UCS using a negative height value

Figure 12-38

is to create a new UCS with its origin at the left intersection of the two perpendicular surfaces, then create and subtract a cylinder. See Figure 12-37.

1. Select the Origin UCS tool from the UCS toolbar.

   Command: _ucs
   Origin/Zaxis/3point/OBject/View/X/Y/Z/Prev/
   Restore/Save/Del/?/<World>: _o
   Origin point <0,0,0>:

2. Use OSNAP, Endpoint and select the new origin.
3. Select the Preset UCS tool from the UCS toolbar.
4. Select the FRONT preset Ucs option.
5. Select the Cylinder tool from the Solids toolbar.

   Command: _cylinder
   Elliptical/<center point>: <0,0,0>:

6. Select a center point in the approximate center of the surface.

   Diameter/<Radius>:

7. Select a radius for the cylinder.

   Center of other end/<Height>:

8. Type –5; press Enter.

   See Figure 12-38. The negative value is required to project the cylinder into the back surface. The direction on the Z-axis is determined using the right-hand rule.

   Figure 12-39 shows the right-hand rule. The positive direction of a Z axis relative to a given XY plane is determined by aligning your thumb with the X axis so that the

end of your thumb is pointing in the positive X direction. Align your first finger with the Y axis so that it is pointing in the positive Y direction. Extend your second finger so that it is perpendicular to the plane formed by your thumb and first finger. This is the positive Z direction.

9. Select the SE Isometric tool from the View toolbar.
10. Type Zoom and enter a .5 scale factor.
11. Select the Subtract tool from the Modify II toolbar.

    Command: _subtract Select solids and regions to subtract from...
    Select objects:

12. Select the L-shaped bracket.

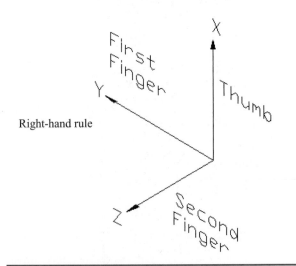

Right-hand rule

Figure 12-39

The finished solid model

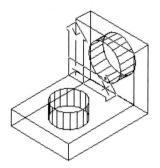

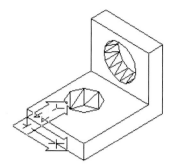

The solid model with the HIDE command applied

Figure 12-40

Select objects:

13. Press Enter.

   Select solids and regions to subtract...
   Select objects:

14. Select the cylinder.

   Select objects:

15. Press Enter.

   Figure 12-40 shows the resulting solid object.

12-15 COMBINING SOLID OBJECTS

Figure 12-41 shows a dimensioned object. The following section explains how to create the object as a solid model. There are many different ways to create a solid model. The sequence presented here was selected to demonstrate several different input options.

To set up the drawing

Set up the drawing as follows:

Units = decimal (millimeters)
Drawing Limits = 297,210
Grid = 10
Snap = 10
View = SE Isometric
Toolbars = Draw, Modify, Solids, View, and UCS

The object is relatively small so use the Zoom tool to create a size that you find visually comfortable.

To draw the first box

The size specifications for this box are based on the given dimensions.

1. Select the Box tool from the Solids toolbar.

   Command: _box
   Center/<Corner of box> <0,0,0>:

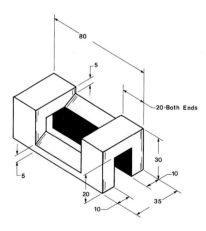

Figure 12-41

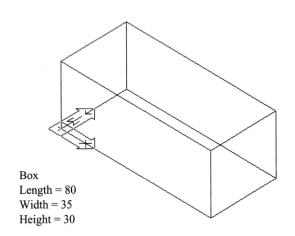

Box
Length = 80
Width = 35
Height = 30

Figure 12-42

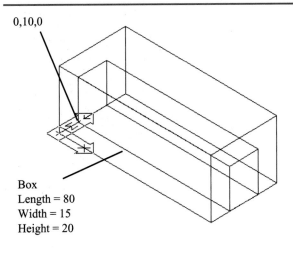

0,10,0

Box
Length = 80
Width = 15
Height = 20

Figure 12-43

2. Select the 0,0,0 point on the WCS.

   Cube/Length/<other corner>:

3. Type L; press Enter.

   Length:

4. Type 80; press Enter.

   Width:

5. Type 35; press Enter.

   Height:

6. Type 30; press Enter.

   See Figure 12-42.

To create the internal open volume

The volume will be created by subtracting a second box from the first box.

1. Select the Box tool from the Solids toolbar.

   Command: _box
   Center/<Corner of box> <0,0,0>:

2. Select the 0,10,0 point on the WCS.

The point 0,10,0 was selected based on the given 10mm dimension. The point can be selected using the cross hairs because it is on the grid located on the XY plane.

   Cube/Length/<other corner>:

3. Type L; press Enter.

   Length:

4. Type 80; press Enter.

   Width:

5. Type 15; press Enter.

   Height:

6. Type 20; press Enter.

   Figure 12-43 shows the second box within the first box.

7. Select the Subtract tool from the Modify II toolbar.

   Command: _subtract Select solids and regions to subtract from...
   Select objects:

8. Select the first box.

   Select objects:

9. Press Enter.

   Select solids and regions to subtract...
   Select objects:

10. Press Enter.

   See Figure 12-44.

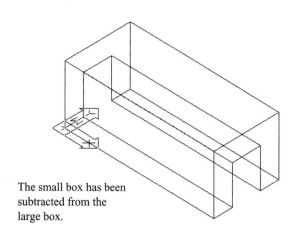

The small box has been subtracted from the large box.

Figure 12-44

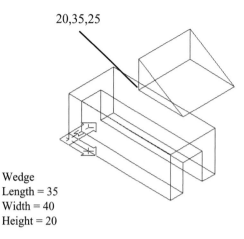

20,35,25

Wedge
Length = 35
Width = 40
Height = 20

Figure 12-45

To create the wedge-shaped cutout

The wedge-shaped cutout will be created by first drawing a wedge based on the given dimensions, then moving the wedge into the correct location. A small box will be unioned to the wedge and both will be subtracted from the first box.

1. Select the Wedge tool from the Solids toolbar.

   Command: _wedge
   Center/<Center of wedge> <0,0,0>:

2. Type 20,35,25; press Enter.

The corner point locates the corner of the wedge on the back surface of the first box, 5mm below the top surface as specified by the given dimensions.

   Cube/Length/<other corner>:

3. Type L; press Enter.

   Length:

4. Type 35; press Enter.

The WEDGE command always interprets a length dimension as being along the current X axis. In this example, the wedge will be rotated 90 degrees into place, so the 35mm length must be equal to the width of the object.

   Width:

5. Type 40; press Enter.

   Height:

6. Type 20; press Enter.

See Figure 12-45.

To rotate the wedge in the XY plane

1. Select the Rotate tool from the Modify toolbar.

   Command: _rotate
   Select objects:

2. Select the wedge.

   Select objects:

3. Press Enter.

   Base point:

4. Use OSNAP, Endpoint and select the corner point of the wedge.

You may have to turn the SNAP command off temporarily to select the point.

   <Rotation angle>/Reference:

5. Type –90; press Enter.

See Figure 12-46. The counterclockwise direction is the positive direction.

To rotate the wedge along the X axis

The ROTATE command will rotate only in the current XY plane, so a new UCS is required.

1. Select the Origin UCS tool from the UCS toolbar.

   Command: _ucs

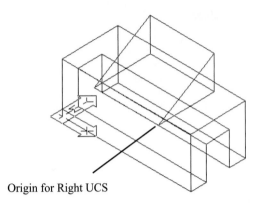

Origin for Right UCS

Figure 12-46

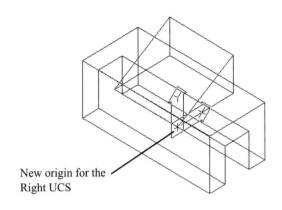

New origin for the
Right UCS

Figure 12-47

*Origin/Zaxis/3point/OBject/View/X/Y/Z/Prev/
Restore/Save/Del/?/<World>: _o
Origin point <0,0,0>:*

2. Use OSNAP, Endpoint and select the new origin.

   See Figure 12-47.

3. Select the Preset UCS tool from the UCS tool-bar.
4. Select the Right UCS option.
5. Select the Rotate tool from the Modify toolbar.

   *Command: _rotate
   Select objects:*

6. Select the wedge.

   Select objects:

7. Press Enter.

   Base point:

8. Use OSNAP, Endpoint and select the new origin, or type 0,0,0; press Enter.

   <Rotation angle>/Reference:

9. Type 180; press Enter.

   See Figure 12-48.

To relocate the wedge

1. Select the Move tool from the Modify toolbar.

   *Command: _move
   Select objects:*

2. Select the wedge.

Select objects:

3. Press Enter.

   Base point or displacement:

4. Type 0,0,0; press Enter.

   Second point of displacement:

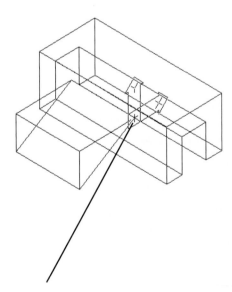

The wedge rotated 180° about the Right UCS

Figure 12-48

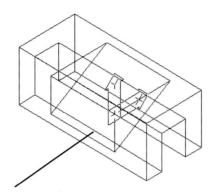

The wedge moved 35 along
the current X azis

Figure 12-49

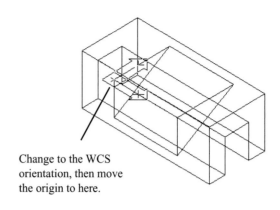

Change to the WCS
orientation, then move
the origin to here.

Figure 12-50

5. Type 35,0,0; press Enter.

The displacement inputs serve to move the wedge 35mm along the current X axis. Remember, you are in the Right UCS.

6. Select the World tool from the UCS toolbar.

See Figure 12-49.

To add a small box to the wedge

The wedge is located 5mm from the top surface of the first box. A small box will be added to reach the top surface of the first box.

1. Select the Origin tool from the UCS toolbar.

Command: _ucs
Origin/Zaxis/3point/OBject/View/X/Y/Z/Prev/
Restore/Save/Del/?/<World>: _o
Origin point <0,0,0>:

2. Use OSNAP, Endpoint and select the new origin.
See Figure 12-50.

3. Select the Box tool from the Solids toolbar.

Command: _box
Center/<Corner of box> <0,0,0>:

4. Press Enter.

The corner of the box could have been defined directly relative to the WCS as point 20,0,25.

Cube/Length/<other corner>:

5. Use OSNAP, Endpoint and select the diagonal corner from the origin.

Height:

6. Type 5; press Enter.

See Figure 12-51.

7. Select the WCS tool from the UCS toolbar and return to the World Coordinate System.

To subtract the wedge and small box

1. Select the Subtract tool from the Modify II toolbar.

Command: _subtract Select solids and regions to subtract from...

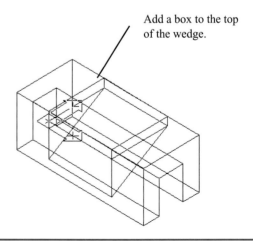

Add a box to the top
of the wedge.

Figure 12-51

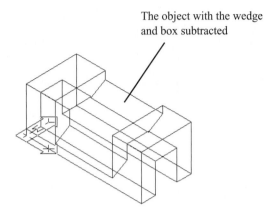

The object with the wedge and box subtracted

Figure 12-52

Select objects:

2. Select the first box.

Select objects:

3. Press Enter.

   Select solids and regions to subtract...
   Select objects:

4. Select the small box.

Select objects:

5. Select the wedge.

Select objects:

6. Press Enter.

   See Figure 12-52.

7. Type Hide; press Enter.

   See Figure 12-53.

8. Type Regen; press Enter.

   The Regen command will return the screen to a wire frame type of object display.

9. Save the drawing if desired.

12-16 INTERSECTING SOLIDS

Figure 12-54 shows an incomplete 3D drawing of a cone and a cylinder. The problem is to complete the drawing in 3D and show the front, top, and right-side orthographic views of the intersecting objects. If this problem

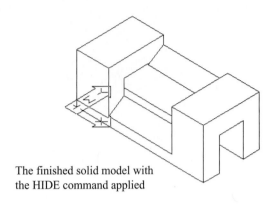

The finished solid model with the HIDE command applied

Figure 12-53

were to be done by hand on a drawing board, it would require extensive projection between views, as well as a high degree of precision in the line work. Done as a solid model, the problem is much simpler and serves to show the strength of solid modeling as a design tool.

To set up the drawing

Set up the drawing screen as follows.

Grid = 0.50
Snap = 0.50
Units = decimal
View = SE Isometric

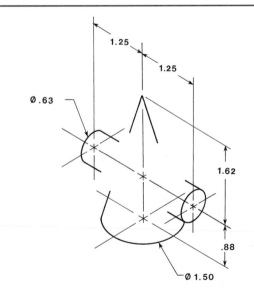

Figure 12-54

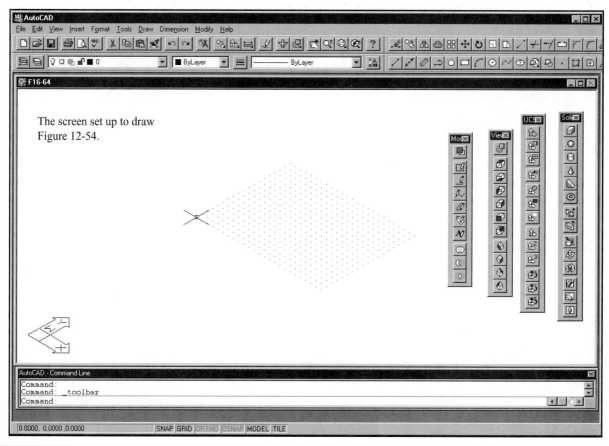

The screen set up to draw Figure 12-54.

Figure 12-55

Toolbars = Solids, View, UCS, Modify, and Modify II
Isolines = 18

See Figure 12-55. The objects are small, so use the Zoom Window command to create a comfortable visual size.

To draw the cone

1. Select the Cone tool from the Solids toolbar.

   Command: _cone
   Elliptical/<center point> <0,0,0>:

2. Press Enter.

   The center point of the cone will be located on the origin of the WCS.

   Diameter/<Radius>:

3. Type D; press Enter.

   Diameter:

4. Type 1.50.

   Apex/<Height>:

5. Type 2.50; press Enter.

   See Figure 12-56.

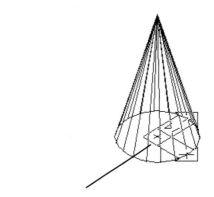

The cone drawn about the WCS origin

Figure 12-56

To draw the cylinder

1. Select the Origin UCS tool from the UCS toolbar.

   Command: _ucs
   Origin/Zaxis/3point/OBject/View/X/Y/Z/Prev/
   Restore/Save/Del/?/<World>: _o
   Origin point <0,0,0>:

2. Type 1.25,0,0; press Enter.

   This input locates the origin in the same plane as the end of the cylinder.

3. Select the Preset UCS tool from the UCS tool-bar, then the Right UCS.

   See Figure 12-57.

4. Select the Cylinder tool from the Solids toolbar.

   Command: _cylinder
   Elliptical/<center point> <0,0,0>:

5. Type 0,.88,0.

   Diameter/<Radius>:

6. Type D; press Enter.

   Diameter:

7. Type .63; press Enter.

   Center of other end/<Height>:

8. Type –2.50; press Enter.

   The negative value is used based on the right-hand rule applied to the Right UCS. See Figure 12-58.

To complete the 3D drawing

1. Select the Union tool from the Modify II toolbar.

   Command: _union
   Select objects:

2. Select the cone.

   Select objects:

3. Select the cylinder.

   Select objects:

4. Press Enter.

   Command:

5. Select the WCS command.

   See Figure 12-59.

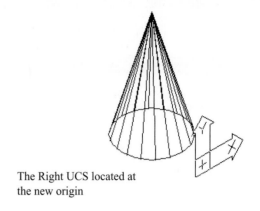

The Right UCS located at the new origin

Figure 12-57

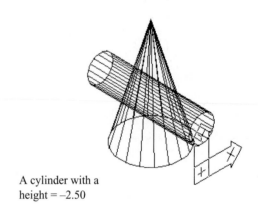

A cylinder with a height = –2.50

Figure 12-58

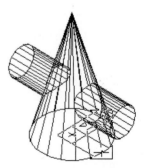

The cone and cylinder unioned together

Figure 12-59

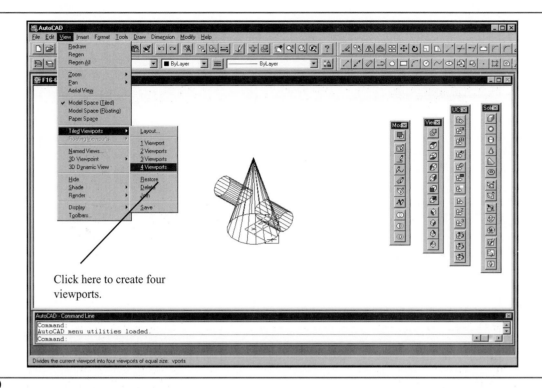

Click here to create four viewports.

Figure 12-60

To create the viewports for the orthographic views

1. Select the View pull-down menu.

2. Select Tiled Viewports, then 4 Viewports.

See Figure 12-60 and Figure 12-61.

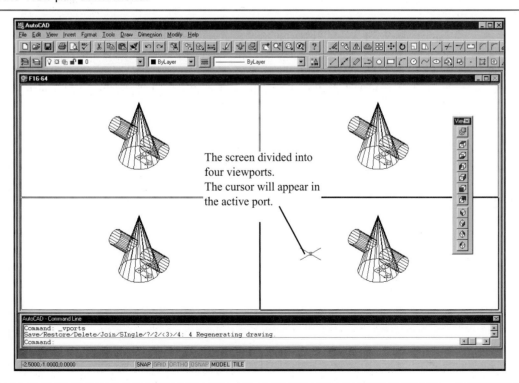

The screen divided into four viewports.
The cursor will appear in the active port.

Figure 12-61

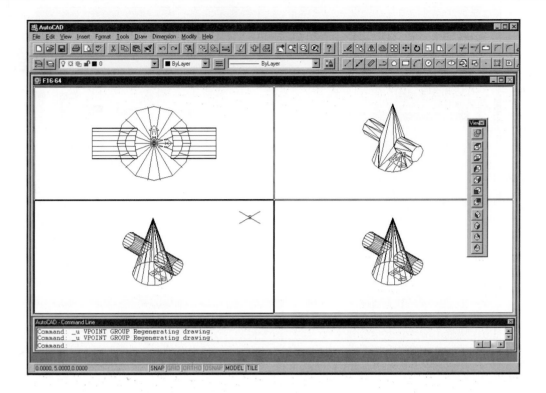

Figure 12-62

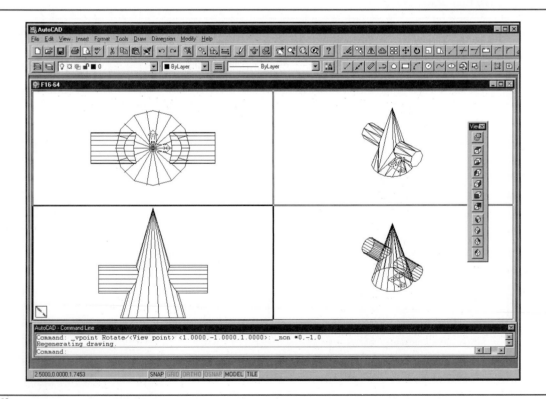

Figure 12-63

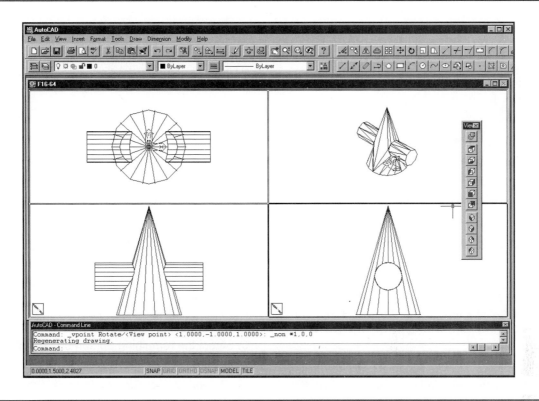

Figure 12-64

To create a top orthographic view

1. Move the cursor into the top left port and press the left mouse button.

   The cross hairs will appear in the port.

2. Select the Top View tool from the View toolbar.

   An oversized top view of the objects will appear.

3. Type Zoom; press Enter.

   All/Center/Dynamic/Extents/Left/Previous/Vmax/ Window/<Scale(X/XP)>:

4. Type 4; press Enter.

   See Figure 12-62.

To create the front orthographic view

1. Move the cursor into the lower left port and press the left mouse button.

   The cross hairs will appear in the port.

2. Select the Front View tool from the View toolbar.

   See Figure 12-63.

To create the right-side orthographic view

1. Move the cursor to the lower right port and press the left mouse button.

2. Select the Right View tool from the View toolbar.

   See Figure 12-64. The model shown in the upper right port has the HIDE command applied.

12-17 SOLID MODELS OF CASTINGS

Figure 12-65 shows a casting. Note that the object includes rounded edges. These rounded edges can be created on a solid model using the FILLET command found on the Modify toolbar. The FILLET command was explained in Chapter 3.

This example will be presented without specific dimensions and will use a generalized approach to creating the model.

To draw the basic shape

The basic shape will first be drawn in 2D, then extruded into the 3D solid model.

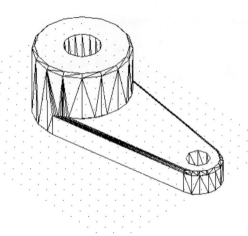

Figure 12-65

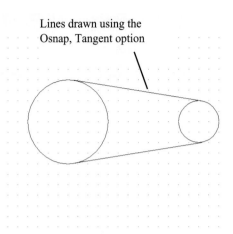

Lines drawn using the
Osnap, Tangent option

Figure 12-66

1. Set up the drawing screen as needed.
2. Draw the basic shape using the Circle tool and then the Line tool along with the OSNAP, Tangent option.

See Figure 12-66. It is important to know the center point locations for the two circles in terms of their XY components. In this example the center point location for the large circle is 4.5,5, and the location for the small circle is 11,5.

To create a polyline from the basic shape

Only polylines can be extruded; and so some of the lines in the basic shape must be formed into a polyline. The large circle can be extruded, so it need not be included as part of the polyline. However, the polyline must be a closed area, and so it will need part of the circle. The needed circular segment can be created by drawing a second large circle directly over the existing circle and then using the TRIM command to remove the excess portion. Remember that two lines can occupy the same space in AutoCAD drawings. If there is difficulty working with the two large circles, trim the circles, then add another larger circle if needed.

1. Use the Circle tool and draw a second large circle directly over the first circle.
2. Use the Trim tool to remove the excess portion of both the large and small circles.

Figure 12-67 shows the resulting shape that will be joined to form a polyline.

3. Select View (pull-down menu), Redraw to return the original large circle to the screen.
4. Select the Edit Polyline tool from the Modify II

toolbar.

Command: _pedit Select polyline:

5. Select the remaining portion of the small circle.

Object is not a polyline
Do you want to turn it into one? <Y>

6. Press Enter.

Close/ Join/ Width/ Edit vertex/ Fit/ Spline/ Decurve /Ltype gen/ Undo/ eXit/ <X>:

7. Type J; press Enter.

Select objects:

8. Select the line segments for the polyline.

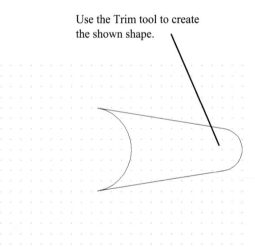

Use the Trim tool to create
the shown shape.

Figure 12-67

Change to a SE Isometric viewpoint.

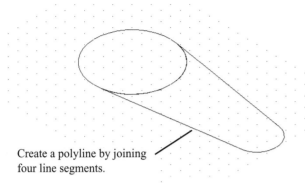

Create a polyline by joining
four line segments.

Figure 12-68

The arc portion of the large circle created in step 2 can be selected just like the other line segments. AutoCAD will select the last entity created if two or more entities occupy the same space.

Close/ Join/ Width/ Edit vertex/ Fit/ Spline/ Decurve/ Ltype gen/ Undo/ eXit/ <X>:

9. Press Enter.

To extrude the shape

1. Select the SE Isometric tool from the View toolbar.
2. Use the Zoom tool if needed to present the figure at a comfortable visual size.

   See Figure 12-68.

3. Select the Extrude tool from the Solids toolbar.

   Command: _extrude
   Select objects:

4. Select the polyline, and assign a height and 0 degree taper.

   In this example a height of 1 was assigned. See Figure 12-69.

   Command:

5. Press Enter.

   Select objects:

6. Select the large circle and assign a height and a 0 degree taper angle.

   In this example a height of 3 was assigned. See Figure 12-70.

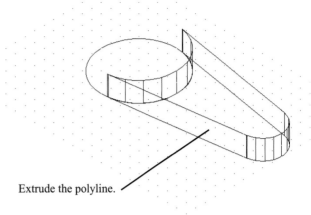

Extrude the polyline.

Figure 12-69

To add the holes

Create the holes by subtracting cylinders from the object.

1. Select the Cylinder tool from the Solids toolbar.

   Command: _cylinder
   Elliptical/<center point> <0,0,0>:

2. Type 4.5,5,0; press Enter.

   This value came from the original circle's center point location.

3. Enter the appropriate diameter and height values.

   Command:

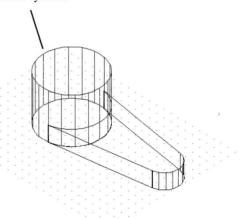

The large circle was extruded
to form a cylinder.

Figure 12-70

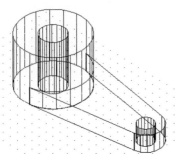

Use the Cylinder tool to create
two holes in the model.

Figure 12-71

4. Press Enter.

   Elliptical/<center point> <0,0,0>:

5. Type 11,5,0.
6. Enter the appropriate diameter and height values.

   See Figure 12-71.

7. Select the Union tool from the Modify II toolbar and join the large cylinder portion of the object to the polyline portion.
8. Select the Subtract tool from the Modify II toolbar and subtract the cylinders from the basic shape.

   See Figure 12-72.

To create the rounded edges

1. Select the Fillet tool from the Modify toolbar.

   Command: _fillet

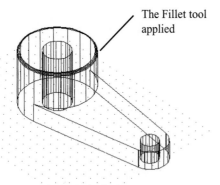

The Fillet tool
applied

Figure 12-73

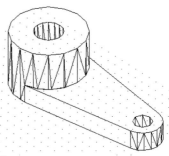

Subtract the cylinders
from the model.

Figure 12-72

   *(TRIM mode) Current fillet radius = 0.2500
   Polyline/Radius/Trim/<Select first object>:*

2. Type R and enter the appropriate radius value if necessary.

   In this example a radius of 0.125 was used.

3. Select the outside edge of the top surface of the large cylindrical portion of the object.
4. Press Enter.

   Chain/Radius/<Select edge>:

5. Press Enter.

   See Figure 12-73.

6. Use the Fillet tool to create a fillet along the top edges of the object as shown in Figure 12-74.

   Note that the arc edge line between the large cylindrical portion of the object and the extended flat area cannot be filleted.

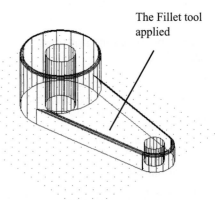

The Fillet tool
applied

Figure 12-74

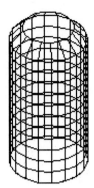

A representation of a thread using a solid model

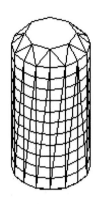

A solid model thread representation with the HIDE command applied

Figure 12-75

12-18 THREAD REPRESENTATIONS IN SOLID MODELS

This section explains how to draw thread representations for solid models. The procedure presented represents only a thread. It is not an actual detailed solid drawing of a thread. As with the thread representations presented in Chapter 11 for 2D drawings, 3D representations are acceptable for most applications.

1. Select the Cylinder tool from the Solids toolbar and draw a cylinder.

In the example shown, a cylinder of diameter 3 and a height of 6 was drawn centered about the 0,0,0 point of the WCS. See Figure 12-75.

2. Draw a circle with a diameter equal to the diameter of the cylinder using the same center point that was originally used to create the cylinder.
3. Select the 3D Array command from the Modify pull-down menu, located under the 3D Operation heading.

   Select object:

4. Select the circle. You may have to use Zoom to make the object large enough to select the 2D circle.

   Rectangular or Polar array:

5. Type R; press Enter.

   Number of rows (___)<1>:

6. Press Enter.

   Number of columns: (|||)<1>:

7. Press Enter.

   Number of levels:

8. Type 11; press Enter.

The number 11 is used because the cylinder is 6 units high, and in this example circles representing threads will be spaced .5 apart. The top edge of the thread will be chamfered.

   Distance between levels: (...):

9. Type .5; Press Enter.
10. Select the Chamfer tool from the Modify toolbar and draw a .5 x .5 chamfer around the top edge of the cylinder.

12-19 LIST

The LIST command is used to display database information for a drawn solid object. Figure 12-76 shows a listing for the thread model created for Figure 12-75. There is no icon for the LIST command. Type List in response to a command prompt and select the object.

12-20 MASSPROP

The MASSPROP command is used to display information about the structural characteristics of an object. Figure 12-77 shows the Massprop information for the object shown in Figure 12-75. To access the Massprop command, type massprop in response to a command prompt and select the object.

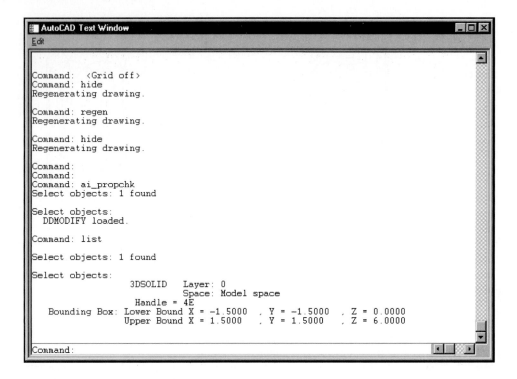

Figure 12-76

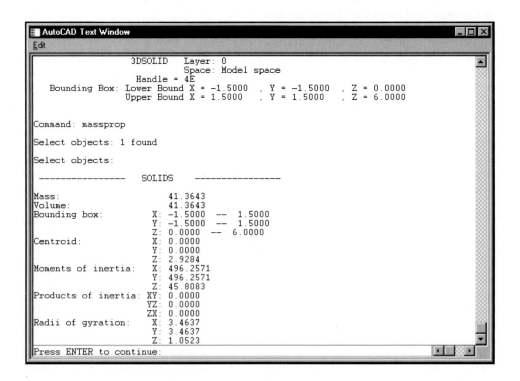

Figure 12-77

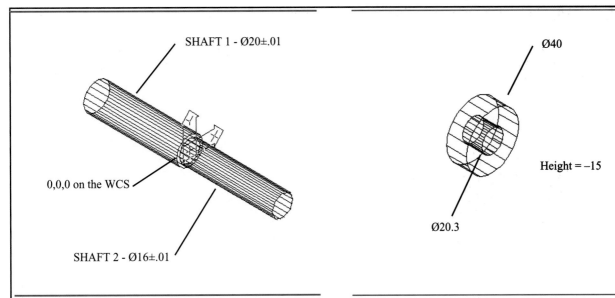

SHAFT 1 - Ø20±.01

0,0,0 on the WCS

SHAFT 2 - Ø16±.01

Figure 12-78

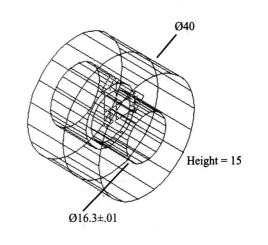

Ø40

Height = –15

Ø20.3

Figure 12-79

12-21 DESIGN PROBLEM

Design a coupling that will hold together the two shafts shown in Figure 12-78. The coupling is to be cylindrical and have an outside diameter of 40. It should also include two threaded holes for M4 setscrews to hold the shafts in place. The final design was created as follows.

The two shafts are positioned on the screen so that the intersection of their centerlines is at the 0,0,0 point on the WCS. This was done to make it easier to determine any required coordinate points.

Assume that analysis has determined that the inside diameters of the coupling should be Ø20.3±.05 and Ø12.3±.05.

1. Select the Preset UCS tool from the UCS toolbar.
2. Select the Right option from the UCS orientation dialog box.

This step will create a new UCS perpendicular to the longitudinal axis of the two shafts. The origin for the UCS is located on the 0,0,0 point of the WCS.

3. Select the Cylinder tool from the Solids toolbar and draw two cylinders with their center points located at the 0,0,0 point.

The diameter of the smaller cylinder is Ø20.3 and the diameter of the larger cylinder is Ø40, and both have a height of –15. See Figure 12-79. The two shafts were moved to a new layer, shafts, and the layer turned off.

4. Subtract the smaller cylinder from the larger cylinder.
5. Draw a second set of solid cylinders.

The diameter of the smaller cylinder is Ø12.3 and the diameter of the larger cylinder is Ø40, and both have a height of 15. See Figure 12-80.

6. Subtract the smaller cylinder from the larger cylinder.
7. Union the two resulting hollow cylinders to form the overall shape of the coupling.

Ø40

Height = 15

Ø16.3±.01

Figure 12-80

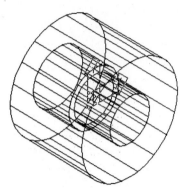

The two solid cylinders are
unioned to form the coupling.

Figure 12-81

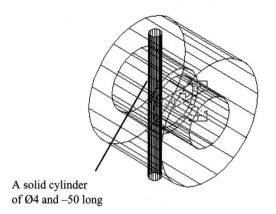

A solid cylinder
of Ø4 and –50 long

Figure 12-82

See Figure 12-81.

8. Create a new layer called Coupling and move
   the unioned cylinder to the new layer.

To draw the holes for the setscrews

1. Return to the WCS using the World UCS tool
   from the UCS toolbar.
2. Draw a solid cylinder centered about center point
   –7.5,0,20 of Ø4 with a height of –50.

   See Figure 12-82.

3. Subtract the Ø4 cylinder from the coupling.

   See Figure 12-83.

4. Draw a second Ø4 solid cylinder centered about

7.5,0,20 with a height of –50.
5. Subtract the Ø4 cylinder from the coupling.

   See Figure 12-84.

To create a sectional view of the coupling

1. Select the Preset UCS tool from the UCS toolbar, then the Front option.
2. Select the Origin tool from the UCS toolbar and
   move the origin to –15,–20,0.

   The given coordinate values are based on the fact
that the new Front UCS has its initial location at the WCS
0,0,0 point.

3. Select the Section tool from the Solids toolbar.

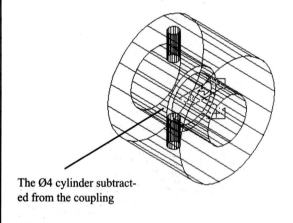

The Ø4 cylinder subtract-
ed from the coupling

Figure 12-83

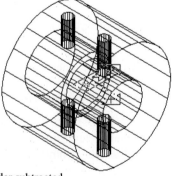

A second cylinder subtracted
from the coupling

Figure 12-84

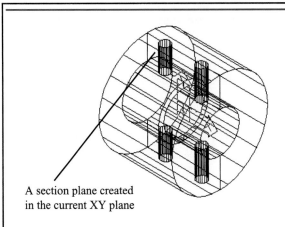

A section plane created
in the current XY plane

Figure 12-85

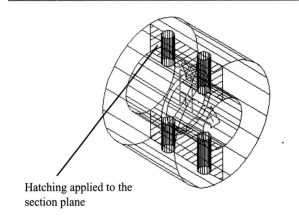

Hatching applied to the
section plane

Figure 12-86

Select object:

4. Select the coupling.

Section plane by Object/ Zaxis/ View/ XY/ YZ/ ZY/ <3 points>:

5. Type XY; press Enter.

Point on XY plane <0,0,0>:

6. Press Enter.

See Figure 12-85. The default 0,0,0 point can be accepted because the origin for the Front UCS was moved from its original location.

To hatch the sectional plane

1. Select the Hatch tool from the Draw toolbar.

The Boundary Hatch dialog box will appear. See Chapter 6 for an explanation of the HATCH command. Use the ANSI 31 pattern.

2. Set the Scale factor equal to .5, then pick points.

Select internal points:

3. Select the points within all six areas of the section plane; press Enter.

The Boundary Hatch dialog box will reappear.

4. Select the Apply box.

The section plane will be hatched. See Figure 12-86. In Figure 12-86 the UCS icon is located on the UCS's origin in the Front orientation.

To dimension the coupling

The coupling will be dimensioned by using a sectional view and a 3D view. The screen must first be divided into two viewports in Paper space. Model space will then be used to manipulate the objects individually. The drawing will be returned to Paper space and the dimensions applied.

Paper space treats the screen as one sheet of paper and does not acknowledge individual screen ports. Model space treats each port as an individual drawing, allowing each port to be manipulated independently of the others.

1. Move the drawing to Paper space by double-clicking the word MODEL at the bottom of the screen.
2. Select the View pull-down menu, then Floating Viewports, then 2 Viewports.

ON/OFF/Hideplot/Fit/2/3/4/Restore/<First point>:_2
Horizontal/<Vertical>:

3. Press Enter.

Fit/<First point>:

4. Type f; press Enter.

The screen will display two identical objects. See Figure 12-87. The left port will now be changed to a sectional view by using the Front UCS orientation.

5. Move the drawing to Model space by double-clicking the word PAPER at the bottom of the screen.

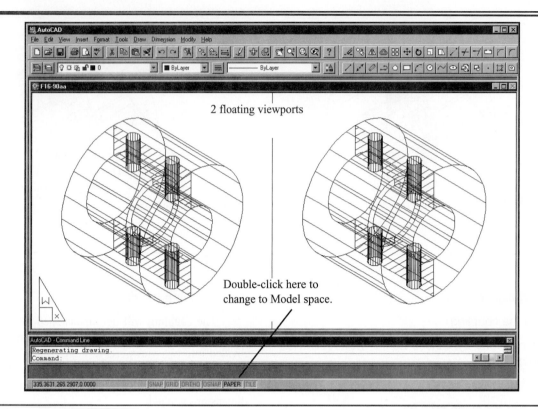

Figure 12-87

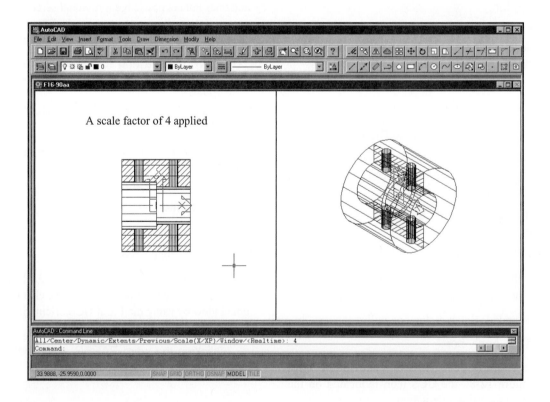

Figure 12-88

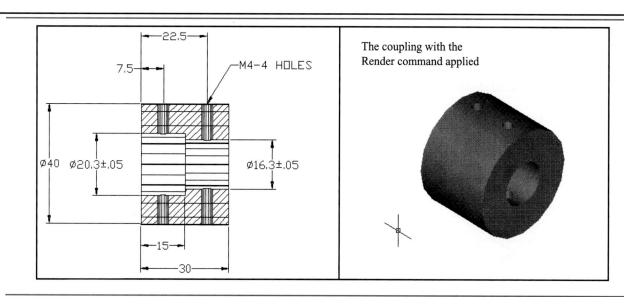

Figure 12-89

6. Make the left port active by locating the cursor or arrow within the port and pressing the left mouse button.

7. Select the Front tool from the View toolbar.

The coupling will appear as a sectional view in the left port. Use the Zoom Window tool to create a visually comfortable size. Don't forget to leave room for the dimensions. In this example both ports were redrawn using a Zoom scale factor of 4. See Figure 12-88.

To dimension the sectional view

The port views of the coupling may be dimensioned directly by using the DIMLFAC command. There is no tool for the DIMLFAC command.

1. Move the drawing to Paper space by double-clicking the word MODEL at the bottom of the screen.

The word MODEL will change to PAPER indicating that the drawing is now in Paper space.

2. Type Dim in response to a command prompt; press Enter.

Dim:

3. Type dimlfac; Press Enter.

Current value <1.0000>: New Value (Viewport):

4. Type v: press Enter.

Select viewport to set to scale:

5. Select the left viewport by clicking one of its boundary lines.

DIMLFAC set to –0.3344
Dim:

6. Dimension the coupling using the Dimensions toolbar.

Figure 12-89 shows the final dimensioned coupling. AutoCAD's ability to combine 3D and 2D views of any object allows both types of views to be used within the same drawing, giving a clearer understanding of both an object's size and shape.

12-22 EXERCISE PROBLEMS

Draw the objects in exercise problems EX16-1 to EX16-45 as follows:

A. Draw each as a solid model.

B. Create front, top, and right-side orthographic views from the solid model.

C. Dimension the orthographic views.

EX12-1 INCHES

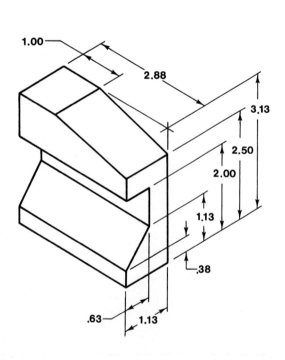

EX12-2 INCHES

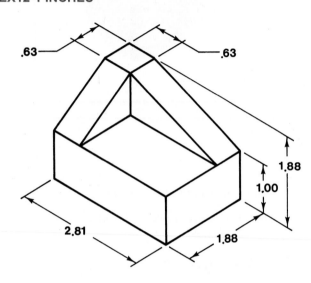

EX12-3 MILLIMETERS

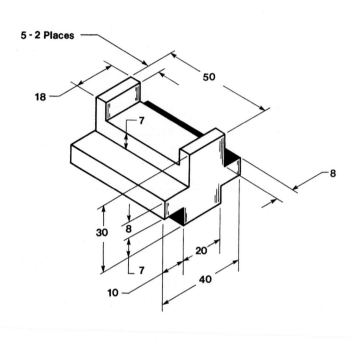

EX12-4 MILLIMETERS

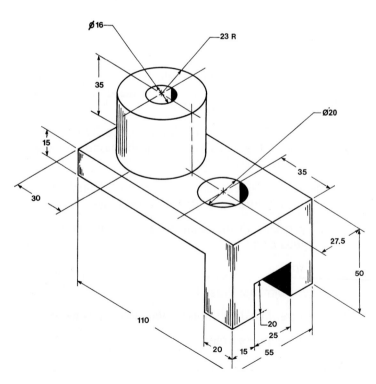

EX12-5 INCHES

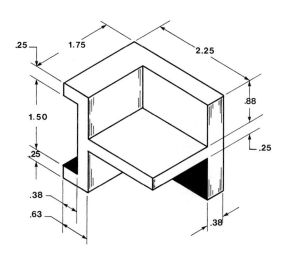

EX12-7 MILLIMETERS

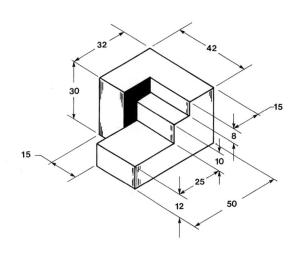

EX12-6 MILLIMETERS

Slot is 15 DEEP.

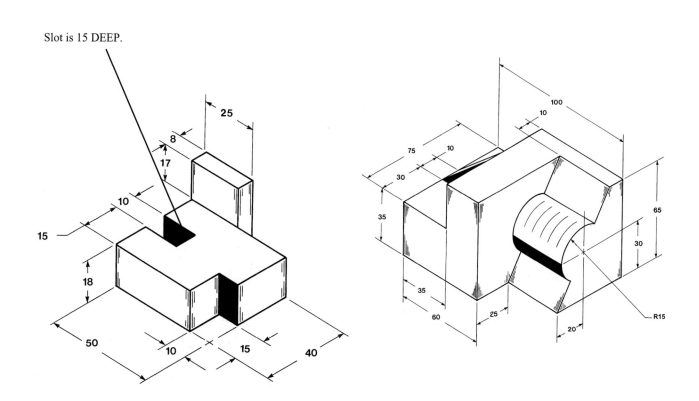

EX12-8 MILLIMETERS

EX12-9 INCHES

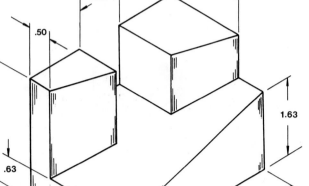

EX12-11 MILLIMETERS

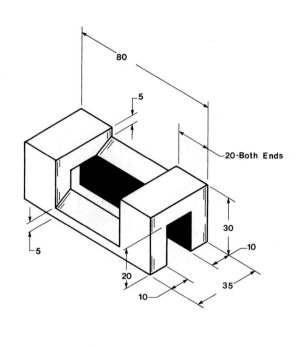

EX12-10 MILLIMETERS

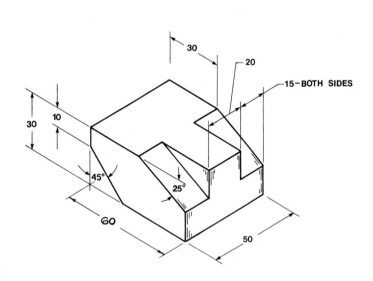

EX12-12 INCHES

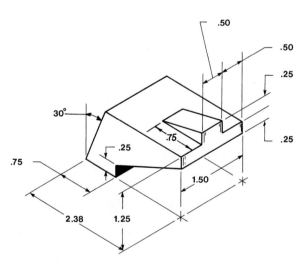

EX12-13 MILLIMETERS

EX12-15 MILLIMETERS

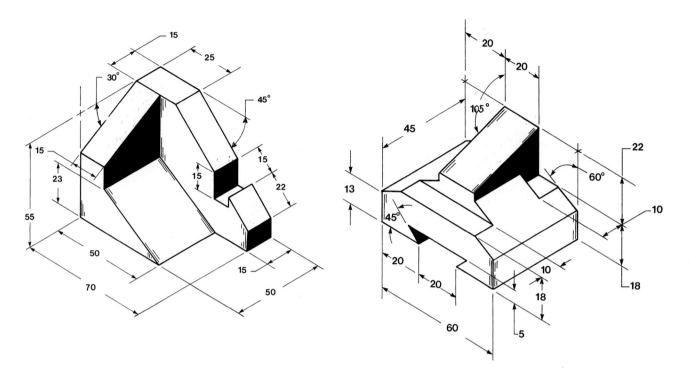

EX12-14 INCHES

EX12-16 MILLIMETERS

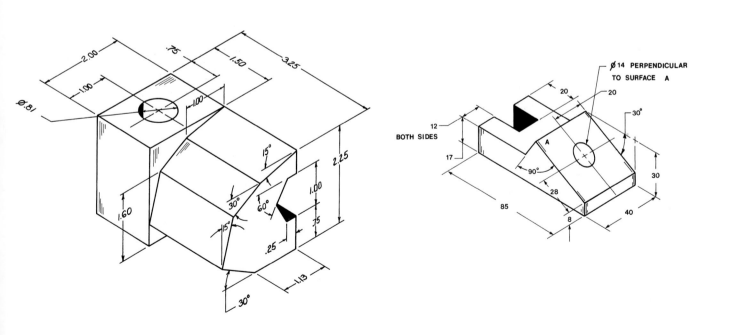

EX12-17 INCHES

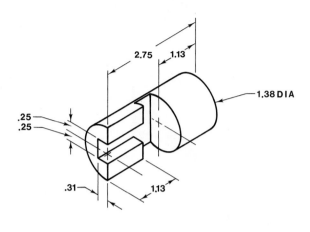

EX12-19 MILLIMETERS

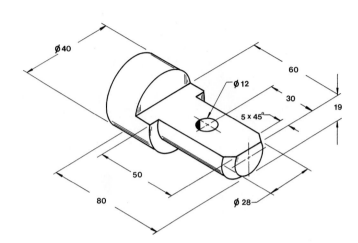

EX12-18 MILLIMETERS

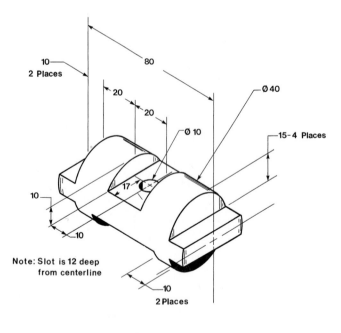

EX12-20 MILLIMETERS

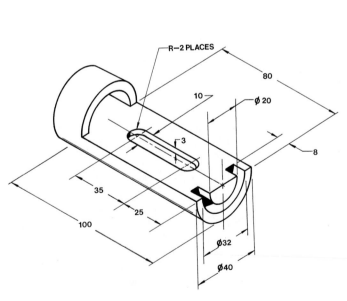

EX12-21 INCHES

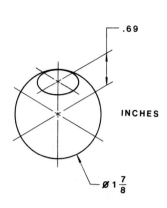

.69

INCHES

Ø 1 $\frac{7}{8}$

EX12-22 MILLIMETERS

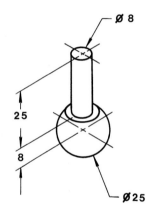

Ø 8

25

8

Ø25

EX12-23 MILLIMETERS

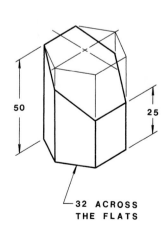

50

25

32 ACROSS
THE FLATS

EX 12-24 MILLIMETERS

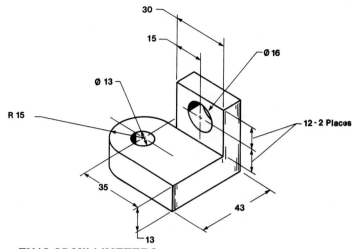

30

15

Ø 16

Ø 13

R 15

12 - 2 Places

35

43

13

EX12-25 MILLIMETERS

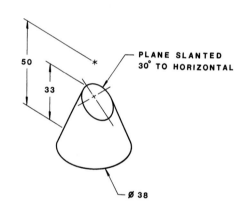

50

33

PLANE SLANTED
30° TO HORIZONTAL

Ø 38

EX12-26 MILLIMETERS

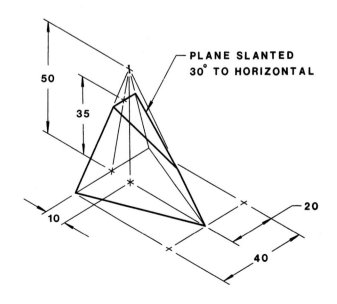

PLANE SLANTED
30° TO HORIZONTAL

50

35

10

20

40

EX12-27 MILLIMETERS

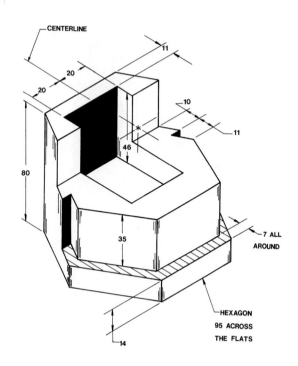

EX12-29 MILLIMETERS

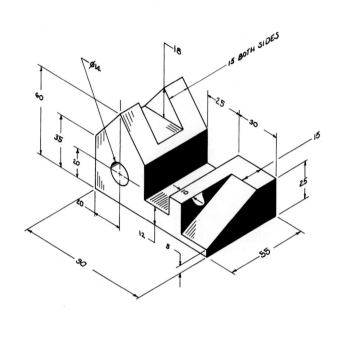

EX12-28 INCHES

EX12-30 INCHES (SCALE 2:1)

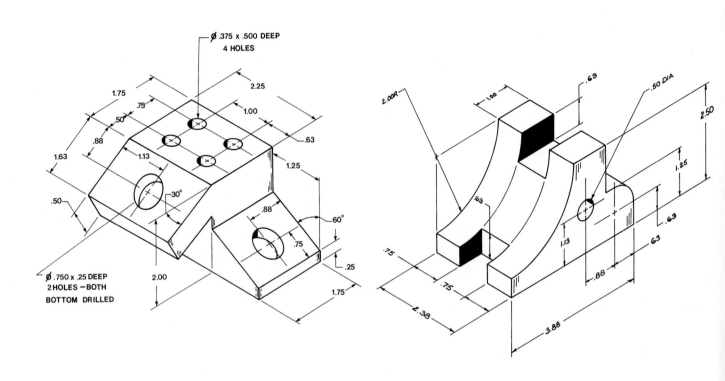

EX12-31 MILLIMETERS

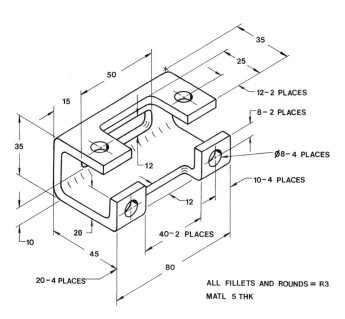

12- 2 PLACES
8- 2 PLACES
Ø8- 4 PLACES
10- 4 PLACES
40- 2 PLACES
20- 4 PLACES

ALL FILLETS AND ROUNDS = R3
MATL 5 THK

EX12-33 MILLIMETERS

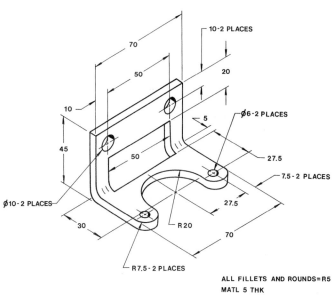

10- 2 PLACES
Ø6- 2 PLACES
27.5
7.5 - 2 PLACES
27.5
R20
Ø10 - 2 PLACES
R7.5 - 2 PLACES

ALL FILLETS AND ROUNDS=R5
MATL 5 THK

EX12-32 MILLIMETERS

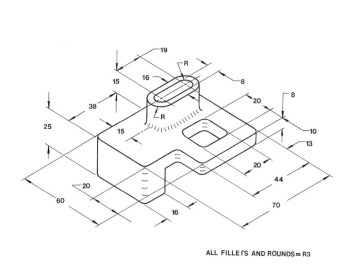

ALL FILLETS AND ROUNDS= R3

EX12-34 MILLIMETERS

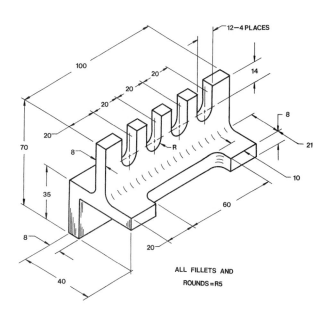

12—4 PLACES

ALL FILLETS AND
ROUNDS=R5

EX12-35 INCHES (SCALE 2:1)

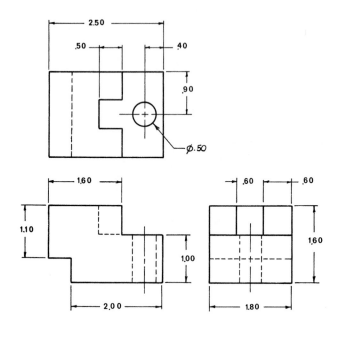

EX12-37 INCHES

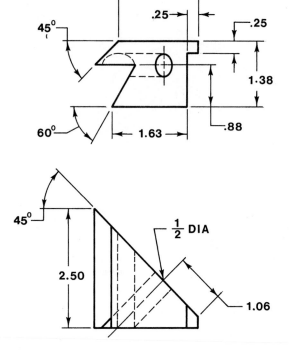

EX12-36 INCHES (SCALE 4:1)

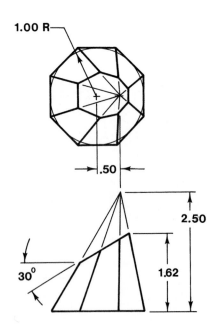

EX12-38 INCHES

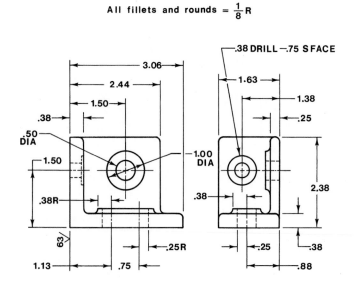

EX12-39 MILLIMETERS

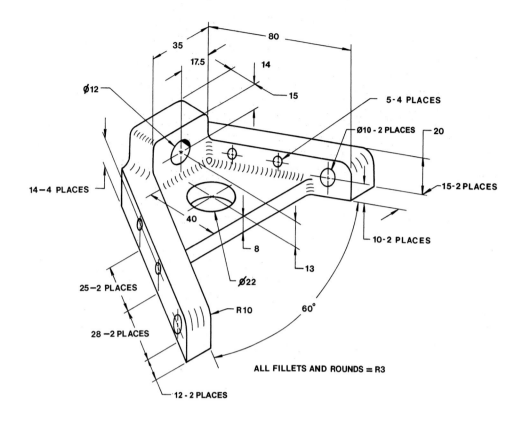

ALL FILLETS AND ROUNDS = R3

EX12-40 MILLIMETERS

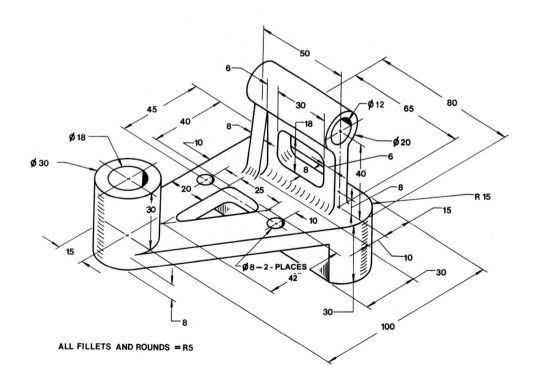

ALL FILLETS AND ROUNDS = R5

EX12-41 MILLIMETERS

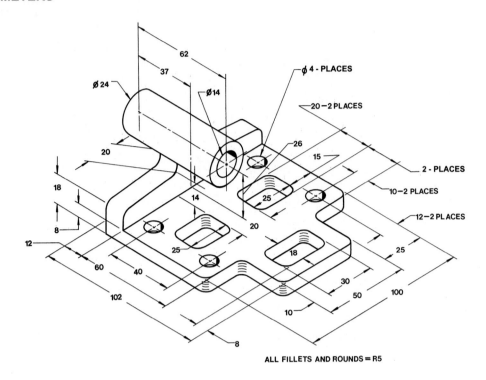

ALL FILLETS AND ROUNDS = R5

EX12-42 INCHES

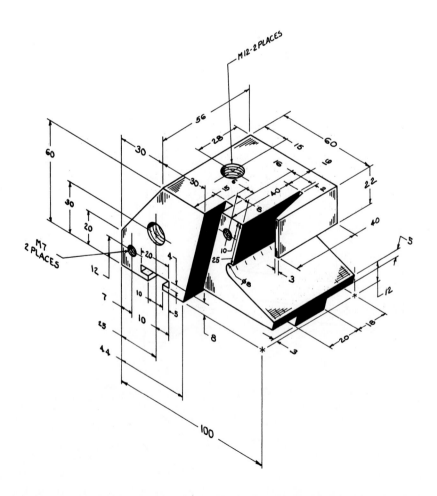

EX12-43 MILLIMETERS

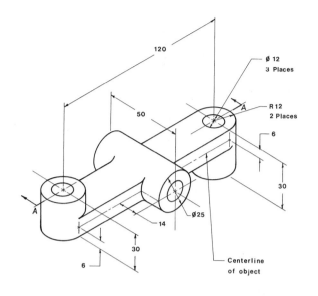

EX12-45 MILLIMETERS

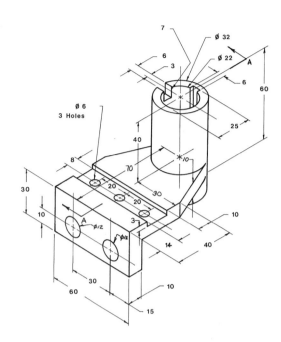

EX12-44 INCHES

Draw a solid model of the object, then create the three indicated sectional views from the model.

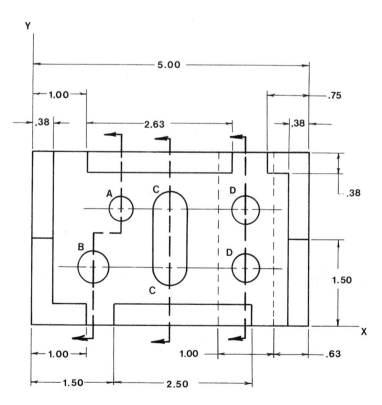

HOLE	X	Y	DIA
A	1.63	2.00	.44
B	1.13	1.00	.56
C	2.50	2.00 1.00	.63
D	3.88	2.00 1.00	.50

Redraw the assemblies in exercise problems EX12-46 to EX12-48 as solid models with the individual parts located in approximately the positions shown.

EX12-46

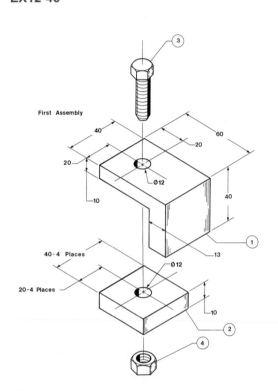

EX12-47

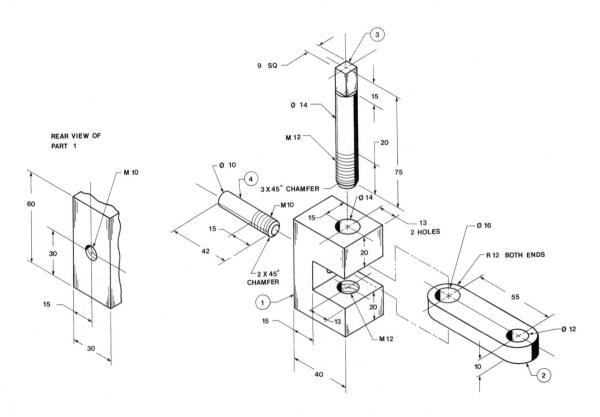

EX12-48

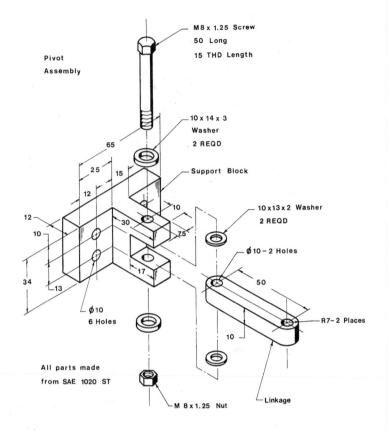

Prepare solid models and three-dimension orthographic views of the intersecting objects in exercise problems EX16-49 to EX16-55.

EX12-49 INCHES

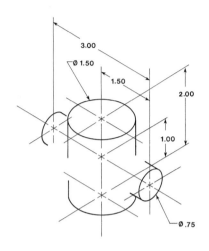

EX12-50 MILLIMETERS

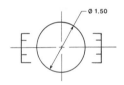

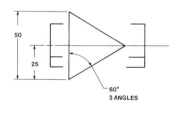

EX12-51 MILLIMETERS

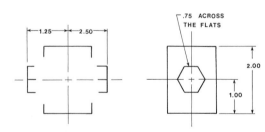

EX12-52 MILLIMETERS

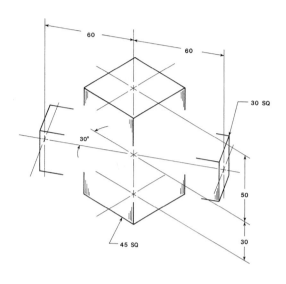

EX12-53 MILLIMETERS

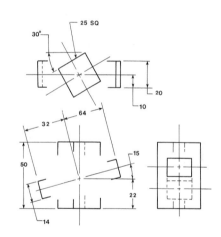

EX12-54 INCHES

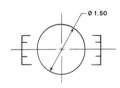

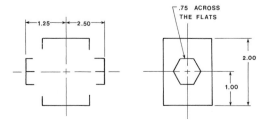

EX12-55

Redraw the given objects as solid models and add the bolts with the appropriate nuts at the L and H holes. Add the appropriate drawing callouts. Specify standard bolt lengths.

A. Use the inch values.

B. Use the millimeter values.

C. Draw the front assembly view using a sectional view.

D. Prepare orthographic assembly views from the solid model.

E. Prepare a parts list.

F. Prepare detail drawings of each part.

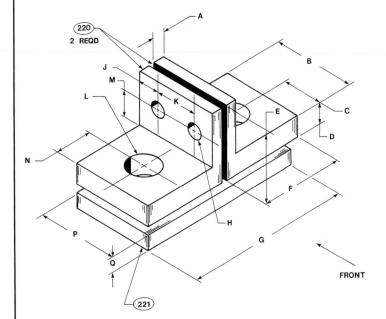

DIMENSION	INCHES	mm
A	.25	6
B	2.00	50
C	1.00	25
D	.50	13
E	1.75	45
F	2.00	50
G	4.00	100
H	Ø.438	Ø11
J	.50	12.5
K	1.00	25
L	Ø.781	Ø19
M	.63	16
N	.88	22
P	2.00	50
Q	.25	6

EX12-56

Redraw the given objects as solid models and add the hex head machine screws at M and N. Use standard length screws and allow at least two unused threads at the bottom of each threaded hole. Add a bolt and the appropriate nut at hole P.

A. Use the inch values.

B. Use the millimeter values.

C. Draw the front assembly view using a sectional view.

D. Prepare orthographic assembly views from the solid model.

E. Prepare a parts list.

F. Prepare detail drawings of each part.

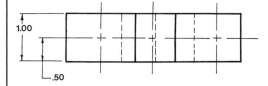

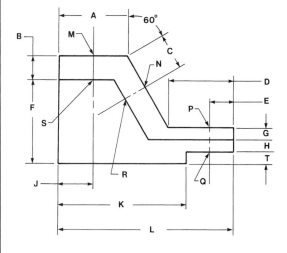

DIMENSION	INCHES	mm
A	1.50	38
B	.50	13
C	.75	I9
D	1.38	35
E	.50	13
F	1.75	44
G	.25	6
H	.25	6
J	.75	I9
K	2.75	70
L	3.75	96
M	Ø.31	Ø8
N	Ø.25	Ø6
P	Ø.41	Ø12
Q	Ø.41	Ø12
R	.164–32 UNF X .50 DEEP	M4 X 14 DEEP
S	.250–20 UNC X 1.63 DEEP	M6 X 14 DEEP
T	.25	6

EX12-57

Redraw the given objects as solid models and add the appropriate hex head bolts and nuts. Use only standard length bolts and include callouts for the bolts and nuts on the drawing.

A. Use the inch values.

B. Use the millimeter values.

C. Draw the front assembly view using a sectional view.

D. Prepare orthographic assembly views from the solid model.

E. Prepare a parts list.

F. Prepare detail drawings of each part.

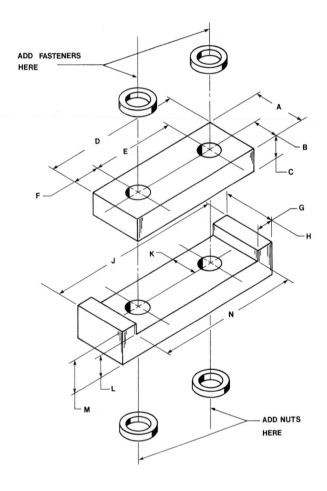

DIMENSION	INCHES	mm
A	1.25	32
B	.63	16
C	.50	13
D	3.25	82
E	2.00	50
F	.63	16
G	.38	10
H	1.25	32
J	4.13	106
K	.63	16
L	.50	13
M	.75	10
N	3.38	86

EX12-58

The objects below are to be assembled as shown. Select sizes for the parts that make the assembly possible. (Choose dimensions for the top and bottom blocks and determine the screw and stud lengths.) The hex head screws (5) have a major diameter of either .375 or M10. The studs (3) are to have the same thread size as the screws and are to be screwed into the top part (2). The holes in the lower part (1) that accept the studs are to be clearance holes.

A. Use the inch values.

B. Use the millimeter values.

C. Draw the front assembly view using a sectional view.

D. Prepare orthographic assembly views from the solid model.

E. Prepare a parts list.

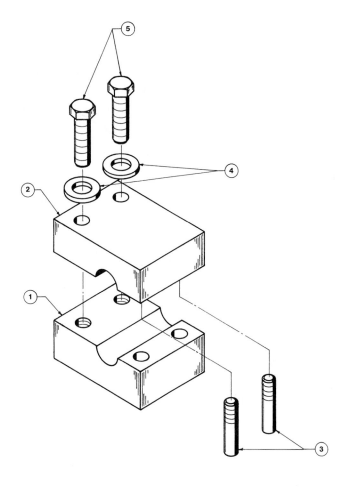

EX12-59

Select values for the dimension indicated in A through J. Use either inches or millimeters. Use either .375 or M10 as the major diameter of the screws. Assemble the parts using three identical screws.

The holes in part 1 are clearance holes and are threaded in part 2. Allow at least two unused threads in the holes beyond the ends of their assembled screws.

A. Use the inch values.

B. Use the millimeter values.

C. Draw the front assembly view using a sectional view.

D. Prepare orthographic assembly views from the solid model.

E. Prepare a parts list.

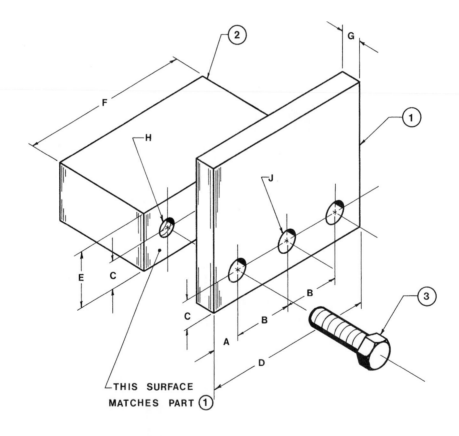

THIS SURFACE MATCHES PART ①

EX12-60

Design an access controller based on the information given below. The controller works by moving an internal cylinder up and down within the base to align with output holes A and B. Liquids will enter the internal cylinder from the top, then exit the base through holes A and B. Include as many holes in the internal cylinder as necessary to create the following liquid exit combinations.

1. A open, B closed
2. A open, B open
3. A closed, B open

The internal cylinder is held in place by an alignment key and a stop button. The stop button is to be spring loaded so that it will always be held in place. The internal cylinder will be moved by pulling out the stop button, repositioning the cylinder, then reinserting the stop button.

Prepare the following drawings.

A. Draw the objects as solid models.

B. Draw an assembly drawing.

C. Draw detail drawings of each nonstandard part. Include positional tolerances for all holes.

D. Prepare a parts list.

INTERNAL CYLINDER

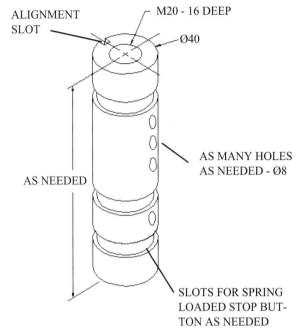

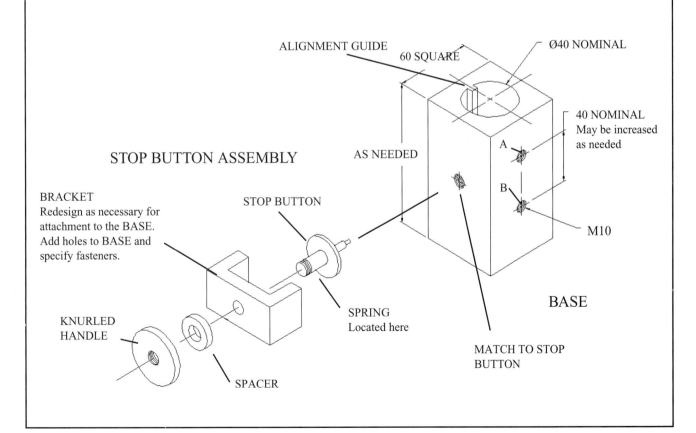

STOP BUTTON ASSEMBLY

BASE

EX12-61

Design a hand-operated grinding wheel specifically for sharpening a chisel. The chisel is to be located on an adjustable rest while it is being sharpened. The mechanism should be able to be clamped to a table during operation using two thumb screws.

A standard grinding is Ø6.00″, is 1/2 inch thick, and has an internal mounting nole with a 50.00±0.3 millimeter bore.

Prepare the following drawings.

A. Draw the objects as solid models.

B. Draw an assembly drawing.

C. Draw detail drawings of each nonstandard part. Include positional tolerances for all holes.

D. Prepare a parts list.

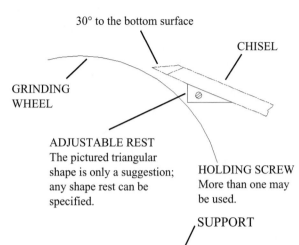

30° to the bottom surface

CHISEL

GRINDING WHEEL

ADJUSTABLE REST
The pictured triangular shape is only a suggestion; any shape rest can be specified.

HOLDING SCREW
More than one may be used.

SUPPORT

GRINDING WHEEL
1/2″ Thick, Ø6″, 50.00±0.3mm bore

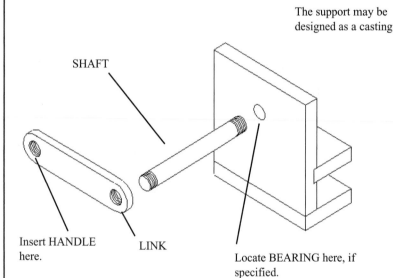

SHAFT

Insert HANDLE here.

LINK

Locate BEARING here, if specified.

The support may be designed as a casting

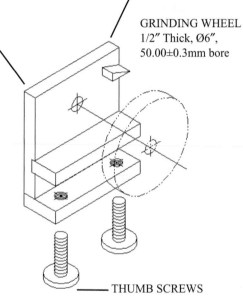

THUMB SCREWS

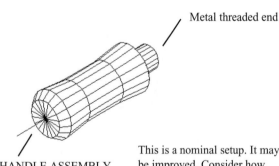

Metal threaded end

HANDLE ASSEMBLY
wooden with threaded metal insert

This is a nominal setup. It may be improved. Consider how the SPACERs rub against the stationary SUPPORT, and consider double NUTs at each end of the shaft.

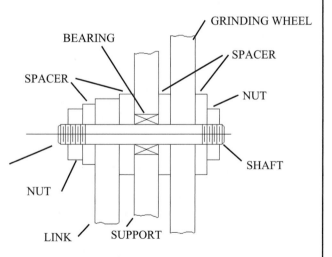

GRINDING WHEEL

BEARING

SPACER

SPACER

NUT

SHAFT

NUT

LINK

SUPPORT

The Render Toolbar

13-1 INTRODUCTION

This chapter explains the Render toolbar. See Figure 13-1. Some of the tools are better suited to architecural drawings and will not be covered in this chapter.

13-2 SAMPLE FIGURE

Figure 13-2 shows three 3D objects. This group of objects will be used throughout the chapter to demonstrate the various Render options. Draw the objects as follows.

To set up the drawing

1. Select the Tools pull-down menu, then Drawing Aids.

   The Drawing Aids dialog box will appear.

2. Turn on both Grid and Snap, set their X and Y spacing for 0.500.
3. Select the View pull-down menu, 3D Viewpoint Presets, and SE Isometric.
4. Use the Pan command to move the grid to the center of the drawing screen if needed.

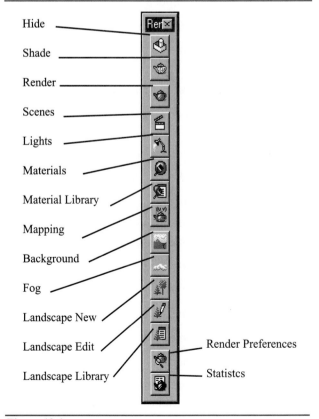

Hide
Shade
Render
Scenes
Lights
Materials
Material Library
Mapping
Background
Fog
Landscape New
Landscape Edit
Landscape Library
Render Preferences
Statistcs

Figure 13-1

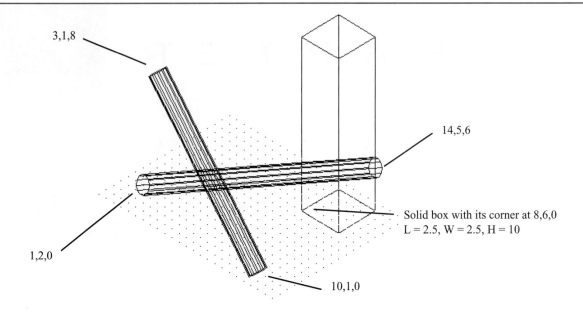

3,1,8

14,5,6

1,2,0

Solid box with its corner at 8,6,0
L = 2.5, W = 2.5, H = 10

10,1,0

Figure 13-2

To change the number of lines used to represent cylinders

1. Type Isolines in response to a command prompt.

   New value for ISOLINES <4>:

2. Type 12; press Enter.

To draw the objects shown in Figure 13-2

1. Select the Tools pull-down menu, Toolbars, and Solids.

   The Solids toolbar will appear on the screen.

2. Select the Center option from the Cylinder tools.

   Command: _cylinder
   Elliptical/<center point> <0,0,0>:

3. Type 1,2,0; press Enter.

   Diamter/<Radius>:

4. Type .5; press Enter.

   Center of other end/<Height>:

5. Type C; press Enter.

   Center of other end:

6. Type 14,5,6; press Enter.
7. Draw a second cylinder of radius .5 from point

   10,1,0 to 3,1,8.

8. Select the Corner option from the Box tools.

   Cube/Length/<other corner>:

9. Select the other corner so that the length and width of the base both equal 2.5.

   Height:

10. Type 10; press Enter.
11. Save the drawing as F13-2.

13-3 HIDE

The Hide command is used to determine the relative current placement of an object within a drawing. In Figure 13-2, it is difficult to determine the relative locations of the three objects. Which is in front of which?

To use the HIDE command

1. Select the Hide tool from the Render toolbar.

   Command: _hide Regenerating drawing.
   Hiding lines 100% done

The Hide command will automatically hide the appropriate lines for all objects in the drawing. See Figure 13-3.

Drawing with the Hide command applied

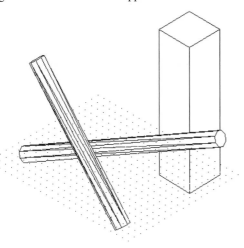

Figure 13-3

To return the drawing to its original form

1. Type Regen in response to a command prompt, then press Enter.

13-4 SHADE

The SHADE command is used to add shading to the drawing. The HIDE command shows only the relative positions of the object. The SHADE command first hides the objects, then adds shading to give a more realistic picture.

To use the SHADE command

1. Select the Shade tool from the Render toolbar.

   Command: _shade Regenerating drawing
   Shading complete

The SHADE command will automatically shade all the objects in the drawing. See Figure 13-4.

To return the drawing to its original form

1. Type Regen in response to a Command prompt, then press Enter.

13-5 SHADEDGE

The intensity of the SHADE command is controlled using the SHADEDGE command. SHADEDGE has four

Drawing with the Shade command applied

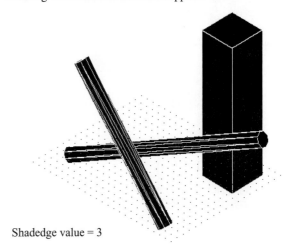

Shadedge value = 3

Figure 13-4

settings 0 through 3. Figure 13-5 shows a drawing shaded using a shadedge value of 2.

To use the SHADEDGE command

1. Type Shadedge.

   New value for shadedge <3>:

2. Type 2; press Enter.
3. Click the Shade tool and the Render toolbar.

Shadedge value = 2

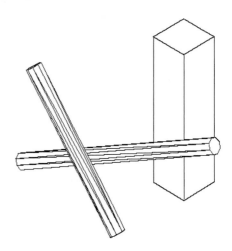

Figure 13-5

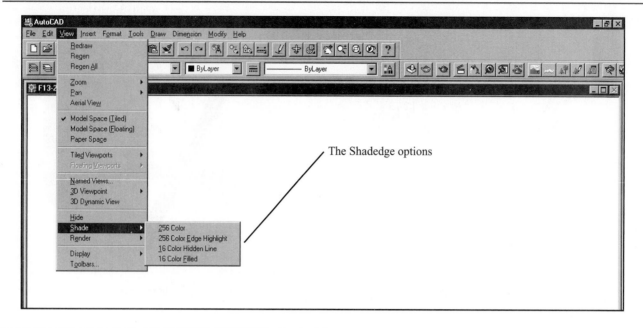

The Shadedge options

Figure 13-6

Shadedge = 0 = 256 Color

Shadedge = 1 = 256 Color Edge Highlight

Shadedge = 2 = 16 Color Hidden Line

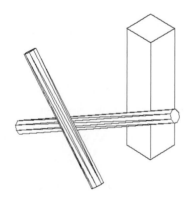

Shadedge = 3 = 16 Color Filled

Figure 13-7

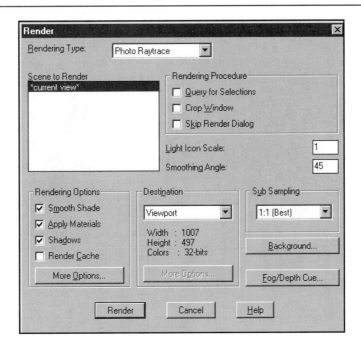

Figure 13-8

The Shadedge option can also be accessed using the View pull down menu as shown in Figure 13-6. The four options listed are the same as the 0 through three settings specified previously for Shadedge. The results for the four options are as follows and are displayed in Figure 13-7.

- 0 = 256 Color — The object's edge lines are not visible.
- 1 = 256 Color Edge Highlight — The object's edges are visible.
- 2 = 16 Color Hidden Line — This option is very similar to the Hide command.
- 3 = 16 Color Filled — The object is filled completely with edges showing.

13-6 RENDER

The RENDER command is used to create true images of objects including various colors and textures. Rendering can be a time-consuming process. This will probably be the first time you will have to wait as the computer processes the rendering. The speed of the rendering depends on the number of surfaces involved, the quality of the rendering desired, and any special texture requirements specified. During a rendering operation, status listing will appear on the command line. The status values indicate that the computer is processing the information.

To render a drawing using the default settings

1. Select the Render tool from the Render toolbar.

The Render dialog box will appear. See Figure 13-8. For this example, all default values were accepted.

2. Select the Render Scene box.

The drawing will be rendered. See Figure 13-9. AutoCAD assumes that the light falling on a drawing is coming from over the shoulder, that is, the light source is behind you, the viewer, and is coming from behind you, over your shoulder, to the drawing. The light source may be changed.

Drawing with the RENDER command applied

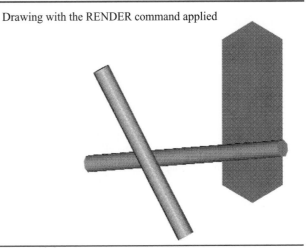

Figure 13-9

Smooth shading applied

Smooth shading not applied

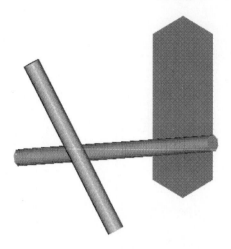

Figure 13-10

Smooth Shading

The Smooth Shading option in the Render dialog box affects the quality of the rendering. Figure 13-10 shows a drawing rendered with Smooth Shading both on and off. Figure 13-11 shows a solid sphere (the isolines values are set at 12) rendered with and without the smooth shading option.

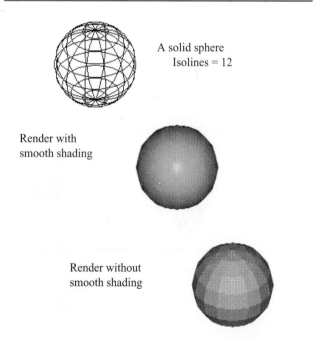

A solid sphere
Isolines = 12

Render with
smooth shading

Render without
smooth shading

Figure 13-11

To render just one object at a time

1. Select for Query for Selections box in the Rendering Procedures area of the Render dialog box.

    Select objects:

2. Select one of the cylinders; press Enter.

    Figure 13-12 shows the resulting rendered view. The Crop Window option is used to render a specific area of the drawing. Figure 13-13 shows a drawing with a window defined and the resulting rendered cropped area.

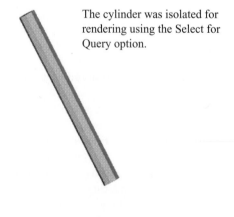

The cylinder was isolated for rendering using the Select for Query option.

Figure 13-12

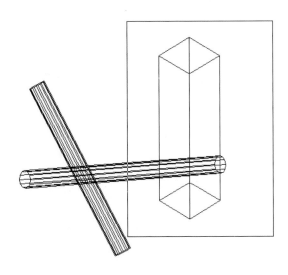

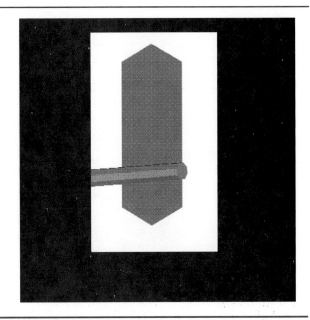

Figure 13-13

To use the Background command

1. Select the Background box on the Render dialog box or the Background tool on the Render toolbar.

The Background dialog box will appear. See figure 13-14.

2. Select the Solid radio button, then turn off the AutoCAD Background default setting.
3. Set the Blue color value to 1.0000 and the Red and Green values to 0.0000.

This will create an all blue background.

4. Select OK.

The Render dialog box will reappear.

5. Select the Render box.

Figure 13-15 shows the resulting rendered screen.

To use the Gradient option

1. Select the Gradient option on the Background dialog box.

Select Solid.

Set the Blue color value for 1.0000 and the Green and Red values for 0.0000.

Turn off the default AutoCAD Background.

Preview will appear here.

Click here to see a preview.

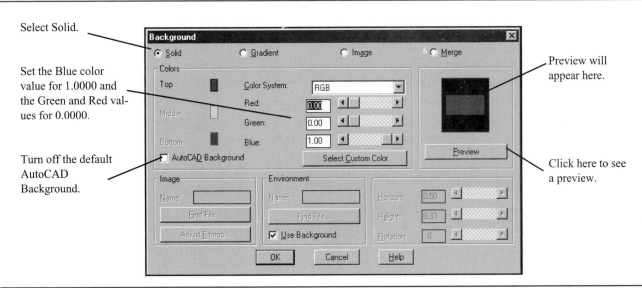

Figure 13-14

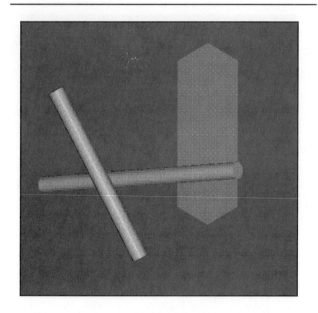

Figure 13-15

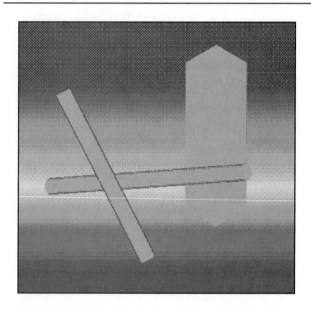

Figure 13-17

2. Set the color values for Red, Green, and Blue for 0.23.

See Figure 13-16.

3. Select OK, then Render on the Render dialog box.

Figure 13-17 shows the resulting background.

To use the Image option

Any image that uses .bmp, tga, .tif, .gif, .or .pcx format files can be pulled into AutoCAD to create a rendered

background. Figure 13-18 shows an example of a photograph pull from the World Wide Web and used as a background for a rendered drawing. The drawing must be on file before it can be used as a background. In the example shown, the drawing was pulled from www.bu.edu and saved as a .gif file. It was incorperated as a background as follows.

1. Select the Image option on the Background dialog box.

The Image area of the Background dialog box will activate. See Figure 13-19.

Select the Gradient option.

Set the Red, Green, and Blue color values for 0.23.

Preview

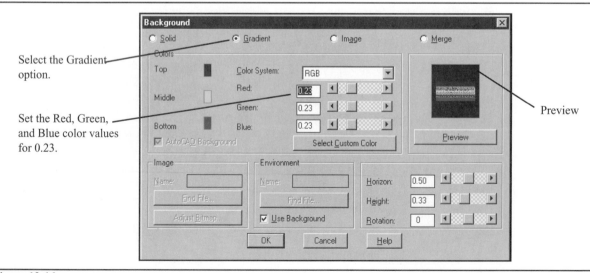

Figure 13-16

Figure 13-18

Select the Image option.

Use the Find File box to access the computer's files.

In this example file C:\TEMP\bu.edu was used.

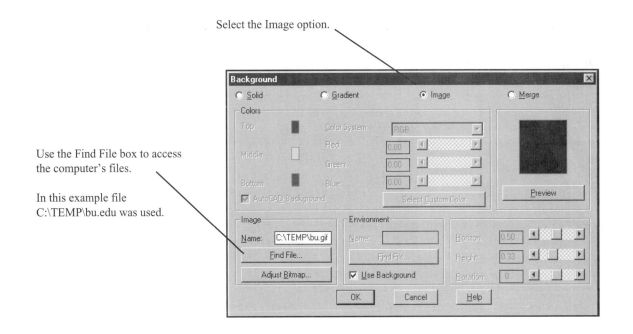

Figure 13-19

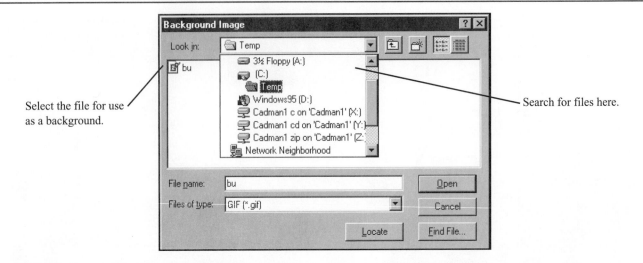

Select the file for use as a background.

Search for files here.

Figure 13-20

2. Select the Find File box.

The Background Image dialog box will appear. See Figure 13-20. In this example the file was saved in a folder called Temp located on the C drive.

3. Select the File, then the Open box.

The Background Image dialog box will reappear.

4. Select OK.

The Render dialog box will reappear.

5. Select the Render box.

To use the FOG command

The Fog/Depth Cue option is used to add the appearance of depth to the drawing. Fog adds white to the drawing, Depth Cue adds black.

1. Select the Fog/Depth Cue option from the Render dialog box or select the Fog tool from the Render toolbar.

The Fog/Depth Cue dialog box will appear. See Figure 13-21.

2. Click the Enable Fog box.
3. Select OK, then render the scene.

Figure 13-22 shows the resulting scene using the default values.

Click here to Enable Fog.

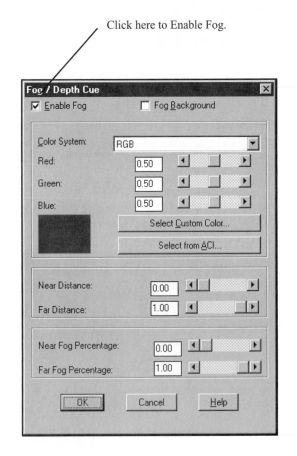

Figure 13-21

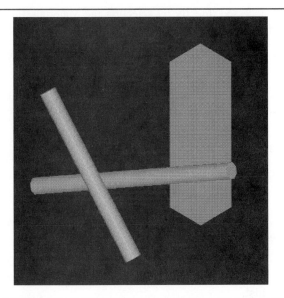

Figure 13-22

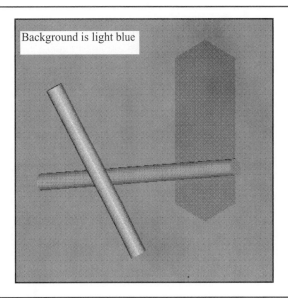

Background is light blue

Figure 13-24

Click Fog Background.

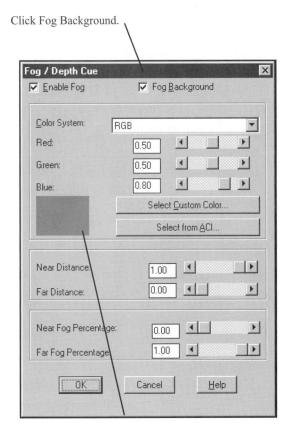

Preview screen will be light blue.

Figure 13-23

To use the Fog Background option

1. Select the Fog/Depth Cue option from the Render dialog box or select the Fog tool from the Render toolbar.

   The Fog/Depth Cue dialog box will appear. See Figure 13-23.

2. Enable the Fog Background option
3. Set the Various values as shown in Figure 13-23; Blue = .80, Near Distance = 1.00, Far Distance = 0.00, Near Fog Percentage = 0.00, Far Fog Percentage = 1.00.

   Figure 13-24 shows the resulting screen. The background color is light blue.

13-7 MATERIALS AND MATERIALS LIBRARY

The Materials and Materials Library commands are closely related, so they will be discussed together. These commands allow you to add texture to your drawings that simulate various materials. The Materials Library command contains a listing of the materials available on the standard Release 14 version. In addition, AutoCAD has a "Materialspec" CD available as part of its Mechanical Library series that includes extensive design information about 25,000 different materials. The materials presented in this section will deal only with the rendering properties of the materials.

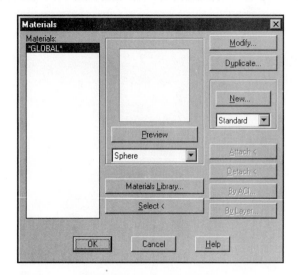

Figure 13-25

To define an object's material

1. Select the Materials tool from the Render toolbar.

The Materials dialog box will appear. See Figure 13-25.

2. Select the Materials Library box.

The Materials Library dialog box will appear. See Figure 13-26. The Materials Library dialog box could have been accessed directly by selecting the Materials Library tool from the Render toolbar.

3. Select CHECKER TEXTURE from the Library List.

The words CHECKER TEXTURE will be highlighted.

4. Select the Preview box.

A rendering of a sphere using the CHECKER TEXTURE pattern will appear in the Preview box.

5. Select the Import box.

The words CHECKER TEXTURE will appear in the Materials List box.

6. Repeat the above procedure and add AQUA GLAZE and BEIGE MATTE to the Materials List.

7. Select the Save box.

The Library File dialog box will appear. This box is used to save the defined materials for the current drawing as .mln files. See Figure 13-27.

8. Type in a file name and click OK twice to return to the Materials dialog box.

In this example the file name F13-2 was selected. The Library file dialog box can be accessed from the Materials dialog box using the Open box.

9. Select from the Materials box, then select the

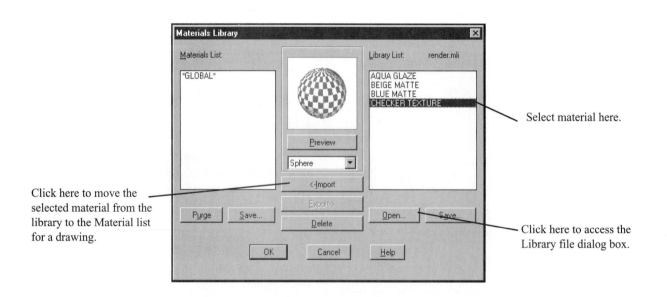

Click here to move the selected material from the library to the Material list for a drawing.

Select material here.

Click here to access the Library file dialog box.

Figure 13-26

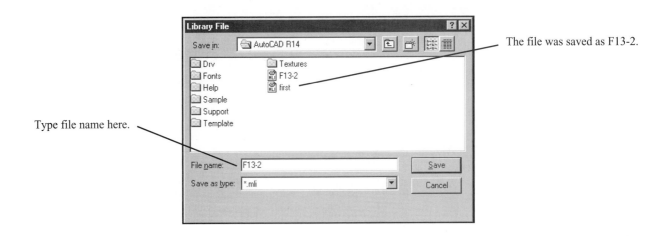

Type file name here.

The file was saved as F13-2.

Figure 13-27

Attach box.

Doneand: _rmat
Gathering objects...0 found
Select objects to attach "CHECKER TEXTURE"
to:

10. Select the front cylinders in the drawing.

    Select objects:

11. Press Enter.

    The Materials dialog box will appear on the screen.

12. Repeat the above procedure and assign AQUA GLAZE to the box and BEIGE MATTE to the other cylinder, then select OK to return to the drawing.
13. Select the Render tool from the Render toolbar and render the drawing.

    Figure 13-28 shows the resulting scene. In the example shown, the default settings were used,

13-8 LIGHTS

There are four different light sources that can be used to change the appearance of a rendered drawing: ambient light, distant light, point lights, and spotlights.

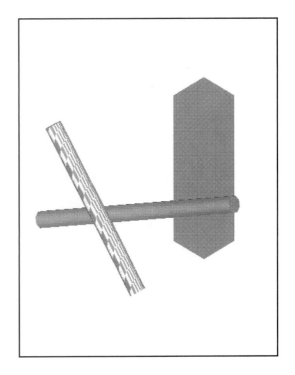

Figure 13-28

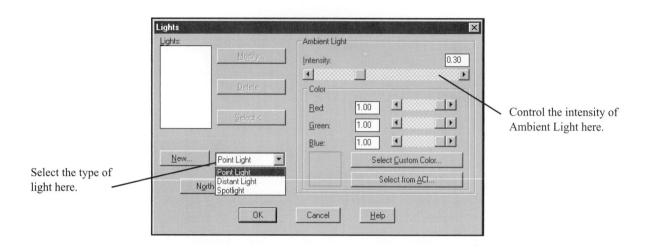

Figure 13-29

To access the Lights option

1. Select the Lights tool from the Render toolbar.

   The Lights dialog box will appear. See Figure 13-29.

2. Select the desired settings, then OK.

   The drawing will return to the screen. The RENDER command may now be used to create a rendered drawing with the new Lights settings.

Ambient light

Ambient light is light that comes from no single or specific source. The intensity of ambient light is controlled using the Intensity scroll bar under the Ambient Light heading in the Lights dialog box. See Figure 13-30. The default setting is 0.30. Figure 13-30 shows the same drawing rendered with two different ambient light intensities: .15 and .85. AutoCAD recommends that ambient light intensity be kept at the lower intensity levels; otherwise, drawings may have a washed-out look.

Ambient Light Intensity set at 0.15   Ambient Light Intensity set at 0.85

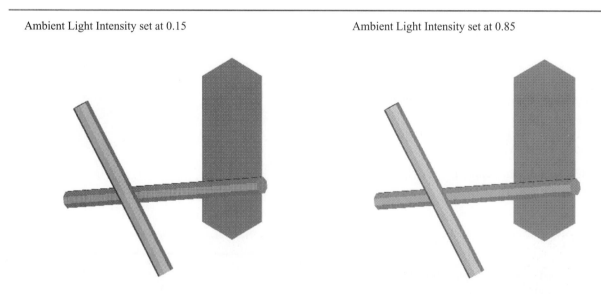

Figure 13-30

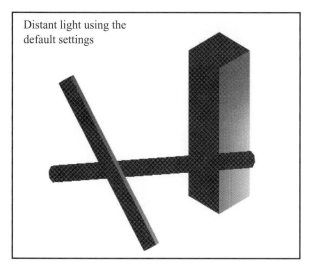

Distant light using the default settings

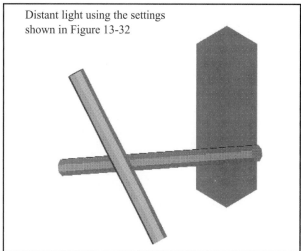

Distant light using the settings shown in Figure 13-32

Figure 13-31

Distant light

A distant light source emits light rays in only one direction. The source of the distant light source may be moved to create different rendering effects on the drawing. Figure 13-31 shows the same drawing with two different distant light sources and two different intensities. The differences were created as follows.

1. Select the Lights tool from the Render toolbar.

   The Lights dialog box will appear. See Figure 13-32.

2. Set the Ambient Light Intensity for .90 and the light source for Distant Light.

3. Set the Blue intensity for 1.00 and the Red and Green intensities for 0.00.

4. Select OK to return to the drawing screen.

5. Select the Render tool and render the objects.

   The Lights dialog box contains many options.

To change the distant light source location

1. Select the Lights tool from the Render toolbar.

   The Lights dialog box will appear.

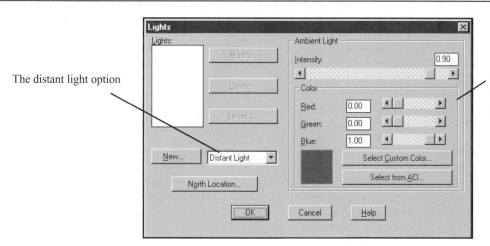

The distant light option

The settings used to create the shading for the right object in Figure 13-31

Figure 13-32

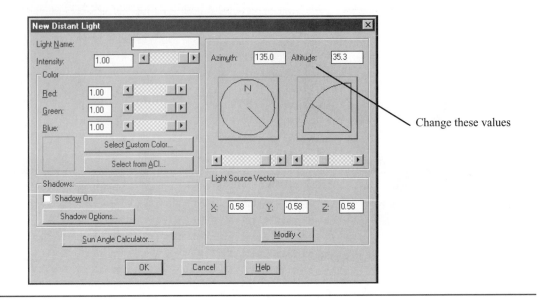

Change these values

Figure 13-33

2. Set the Ambient Light Intensity for .90 and the light source for Distant Light.
3. Select the New box.

The New Distant Light dialog box will appear. See Figure 13-33.

4. Name the new distant light source A1.
5. Change the Azimuth to -42 by using the scroll bar located under the Azimuth circle.

The values may be changed by using the scroll bar or by clicking the arrows at the ends of the scroll bar.

6. Change the Altitude to 80.
7. Select OK.

The Lights dialog box will reappear with the source A1 listed.

8. Select OK.
9. Render the drawing using the Render tool.

Figure 13-34 shows a possible resulting rendering.

The Sun Angle Calculator option

The Sun Angle Calculator option located on the New Distant Light dialog box is used to create shadows based on the day of the year, the time of day, and the latitude and longitude of the object. It is very helpful to architects who wish to show clients rendered views of a future house. Figure 13-35 shows the Sun Angle Calculator dialog box. The dialog box also includes a Geographic location dialog box that can be used to approximately define a location

based on the location of major cities. See figure 13-36.

Figure 13-37 shows the shadow created by sunlight using the values specified in Figure 13-36. It is suggested that a variety of light sources and intensities be defined and viewed, so that you can get a feel for what settings will generate what rendering results. Any excess light sources can be removed using the Delete option in the New Distant Light dialog box.

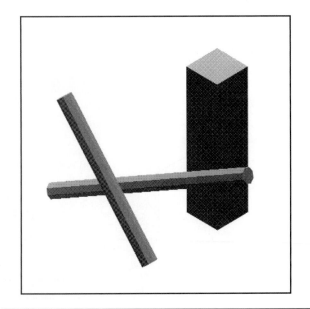

Figure 13-34

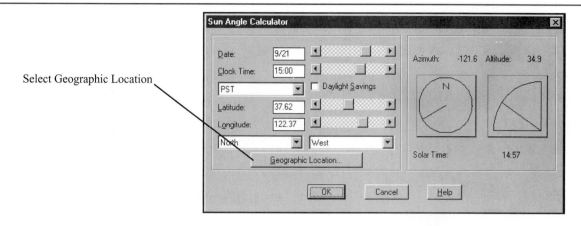

Select Geographic Location

Figure 13-35

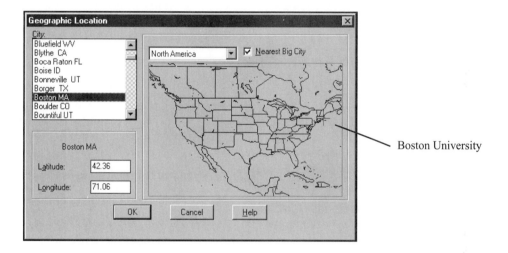

Boston University

Figure 13-36

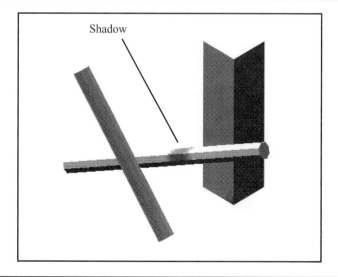

Shadow

Figure 13-37

The default Point Light location

 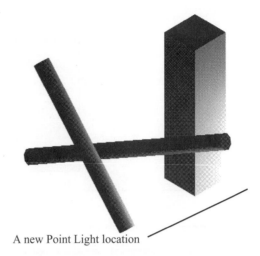

A new Point Light location

Figure 13-38

Point Lights

A point light is a single light source that radiates light in all directions. Think of it as a single light bulb located on the drawing. Point light source locations are defined as follows. This example assumes that all settings are in their default values.

1. Select the Lights tool from the Render toolbar.
2. Select the Point Light option (the Distant Light option should still be in place).
3. Select OK.

4. Render the scene.

The left scene in Figure 13-38 was created using the default Point Light settings.

To move the Point Light location

1. Select the Lights tool from the Render toolbar.
2. Select the New box.

The New Point Light dialog box will appear.

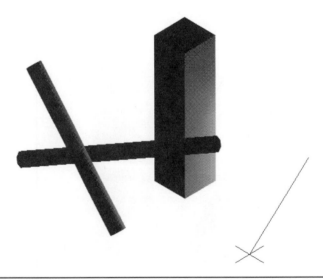

Figure 13-39

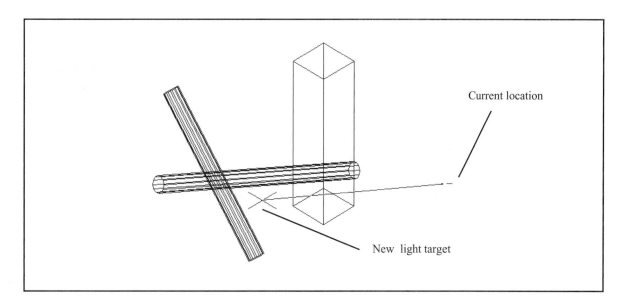

Current location

New light target

Figure 13-40

3. Name the new Point Light location A3 and set the intensity value for 7.00.
4. Select the Modify box.

The cursor will be attached to the original light source by a line that will move as a new location is selected. See Figure 13-39.

The right scene in Figure 13-38 shows the resulting lighting from the new Point Light location.

Spotlight

Spotlights are single source lights that emit light in a cone-shaped trajectory and in a single direction.

1. Select the Light tool from the Render toolbar.
2. Select the Spotlight option and change the intensity to 0.90.
3. Select the New box.
4. Name the Spotlinght A5 and set the intensity for 7.00.
5. Select the Modify box.

Enter light target <current>:

Select a location in the approximate center of the objects. See Figure 13-40.

Enter light location <current>:

Locate the light using coordinate values. Use the coordinate display at the lower left corner of the screen to determine the X and Y values for the new target location.

The Z value should be 0.0000. Change the Z value to 12. Figure 13-41 shows the new spotlight light location.

6. Select OK and then render the scene.

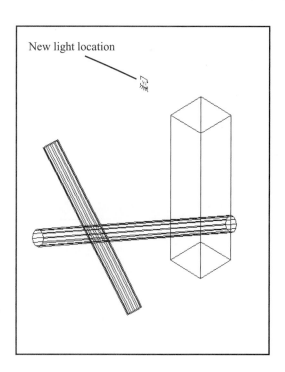

New light location

Figure 13-41

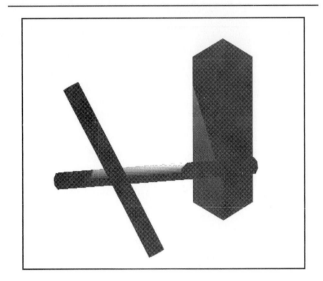

Figure 13-42

Figure 13-42 shows the resulting lighting.

13-9 SCENE

A scene is a saved view and light source. In the example presented here the two dome light sources L1 and L2 were defined as scenes.

To create a scene

1. Select the Scenes tool from the Render toolbar.

The Scenes dialog box will appear. See Figure 13-43.

2. Select the New box.

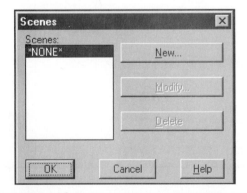

Figure 13-43

Type scene name here.

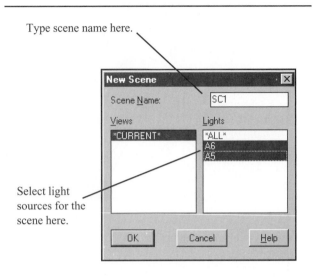

Select light sources for the scene here.

Figure 13-44

The New Scene dialog box will appear. See Figure 13-44.

3. Type the name of the scene, SC1, in the Scene Name box.
4. Highlight the A5 and A6 light source.

The A5 light source was created for the new spotlight location in the last section. The A6 light source was created specifically for this example. Light sources must be previously defined before they can be incorporated into a scene.

5. Select OK.

The Scenes dialog box will appear with SC1 listed in the Scenes box.

6. Select OK.

The drawing screen will reappear with the two scene light sources indicated. See Figure 13-45.

7. Select the Render tool from the Render toolbar.

Figure 12-46 shows the Render dialog box with the scene SC1 listed.

8. Render scene SC1

Figure 13-47 shows the rendered scene.

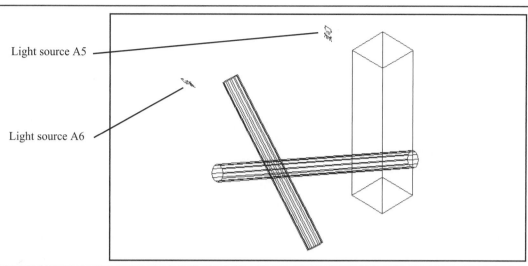

Light source A5

Light source A6

Figure 13-45

Figure 13-46

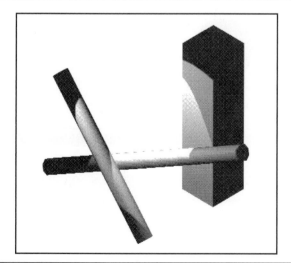

Figure 13-47

13-10 STATISTICS

The STATISTICS command provides information about the current rendering. The information cannot be changed, but it can be saved to an ASCII file. Figure 13-48 shows the Statistics dialog box. Note that the Render + Display Time line specifies the time to render the SC1 scene.

The total time needed to render the SC1 scene

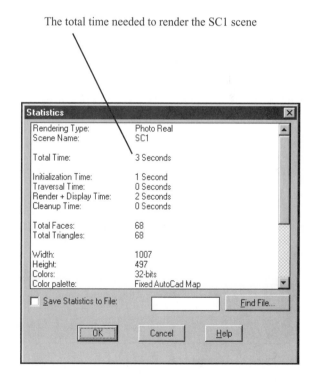

Figure 13-48

13-11 EXERCISE PROBLEMS

EX13-1 INCHES

Set the drawing at the SE Isometric viewpoint and use the default drawing limits. Set isolines for 16. Draw a sphere centered at the 0,0,0 point with a radius of 4.00. Create the following light conditions and save them as scenes. Vary the intensities of the light as desired.

Create four ports as shown below and define each port using a different scene. (Hint: Use View, Titled Viewports, 4 Viewports). Your renderings may look slightly different than those shown below.

Port 1. Define the default Distant ambient light as A1, scene SC1.

Port 2. Define a Distant Light with an Azimuth of 180 degrees and an Altitude of 0.80 degrees. Save the light source as A2, scene SC2. Set the Red, Blue, and Green values for 1.00. Set the ambient light intensity for 80. Change the light source vector so that the Direction from point is 10 inches above (0,0,10) the current direction to point (0,0,0).

Port 3. Define a Point Light source with a light location of 0,-8,0, and an ambient light intensity of 0.80. Select color settings as desired. Save the light source as A3, scene SC3.

Port 4. Define a Spotlight with a target point of 0,0,0 and a light location of 8,0,0 and maximum intensity. Save the light source as SP1, scene SC4.

Draw the following figures as solid models. Do not

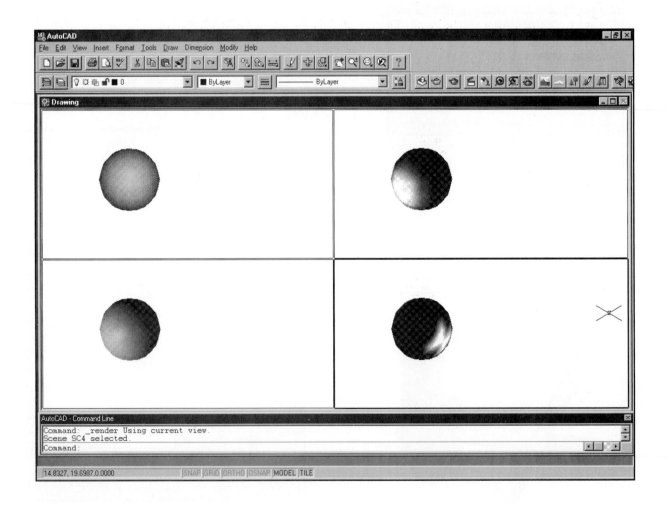

include dimensions. Create scenes as needed. Define light sources, light intensity, and materials. Divide the screen into the number of ports needed to display all the assigned renderings. Render the solid models as follows.

A. Use ambient light.
B. Use distant light.
C. Use a point light.
D. Use a spotlight.

EX13-2 MILLIMETERS

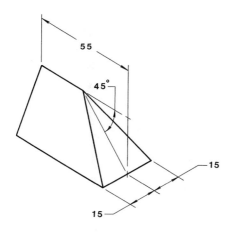

EX13-3 INCHES (SCALE: 3 =1)

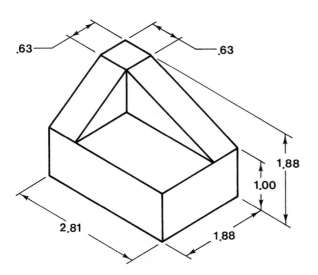

EX13-4 MILLIMETERS

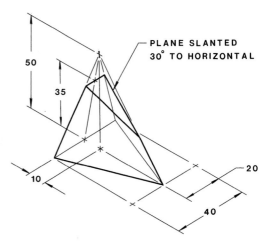

EX13-5 MILLIMETERS

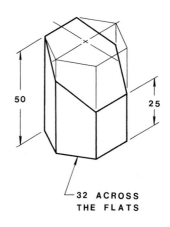

EX13-6 INCHES (SCALE: 4=1)

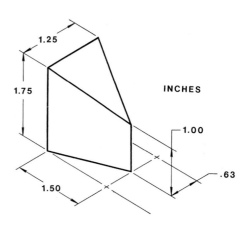

EX13-7 INCHES

EX13-9 MILLIMETERS

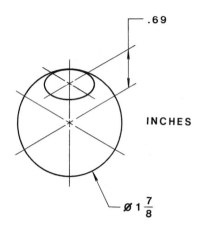

INCHES

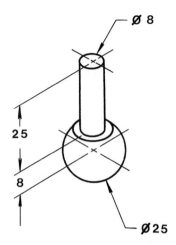

EX13-8 MILLIMETERS

Draw the assembly in the exploded position as shown. Estimate the distances between the parts. Create three different lighting scenes for the drawing; one with the lighting above the objects, one with the lighting below the objects, and a third with the lighting in front of the objects.

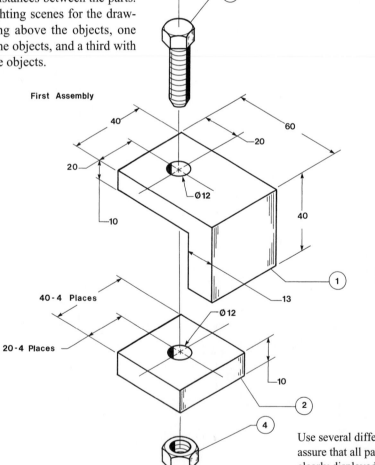

Use several different light sources to assure that all parts of the assembly are clearly displayed.

EX13-10 MILLIMETERS

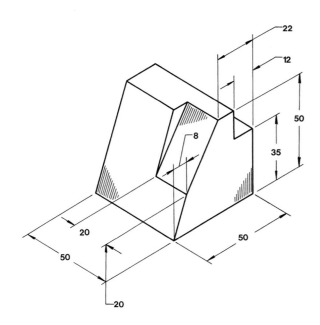

EX13-12 INCHES

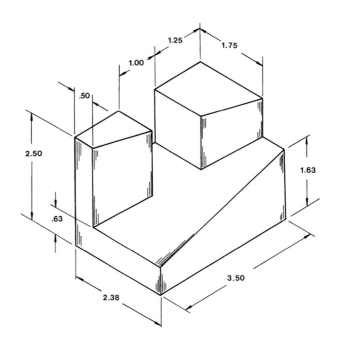

EX13-11 MILLIMETERS

Position light source so that the internal surfaces are illuminated.

EX13-13 MILLIMETERS

Locate one light source inside the object and at least two others outside the object.

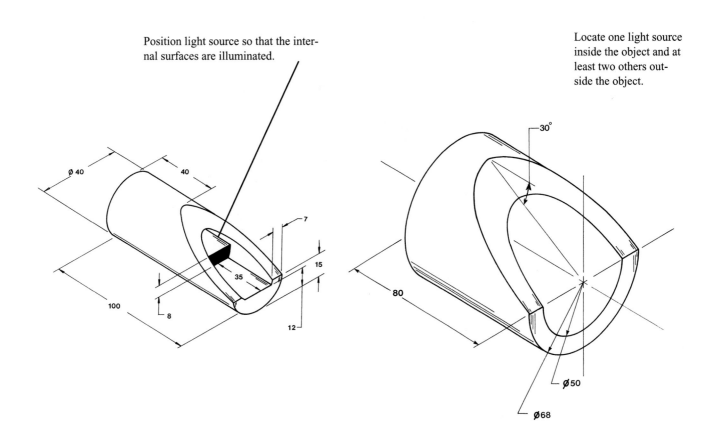

EX13-14 MILLIMETERS

Draw the assembly in the exploded position as shown. Estimate the distances between the parts. Create three different lighting scenes for the drawing; one with the lighting above the objects, one with the lighting below the objects, and a third with the lighting in front of the objects.

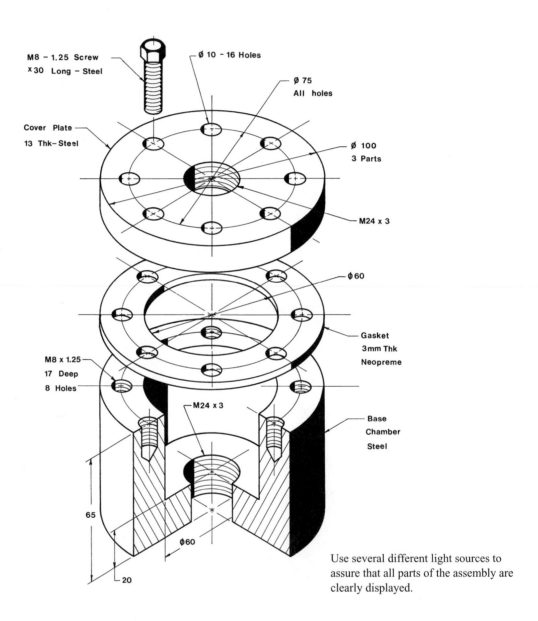

M8 – 1.25 Screw
× 30 Long – Steel

Ø 10 - 16 Holes

Ø 75
All holes

Cover Plate
13 Thk–Steel

Ø 100
3 Parts

M24 x 3

Ø 60

Gasket
3mm Thk
Neopreme

M8 x 1.25
17 Deep
8 Holes

M24 x 3

Base
Chamber
Steel

65

ϕ60

20

Use several different light sources to assure that all parts of the assembly are clearly displayed.

C H A P T E R

Viewpoints

14-1 INTRODUCTION

This chapter presents the fundamentals of working in three dimensions with AutoCAD Release 14 for Windows. AutoCAD allows you to manipulate three-dimensional objects rapidly, making it easier to create and modify them. The first part of the chapter explains the various commands used to view and manipulate 3D objects. The second part demonstrates how to create an object using three-dimensional applications.

14-2 3D VIEWPOINT PRESETS — ISOMETRICS

AutoCAD includes a set of preset 3D viewpoints. The preset viewpoints are accessed using either the Viewpoint toolbar or through the View pull-down menu. Figure 14-1 shows the Viewpoint toolbar and the 3D Viewpoint Presets commands. The Presets commands are used initially to define a 3D viewpoint; they can then be used to view the object from different points.

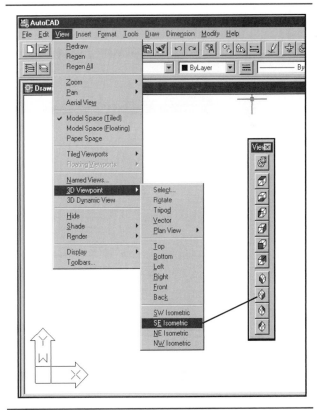

Figure 14-1

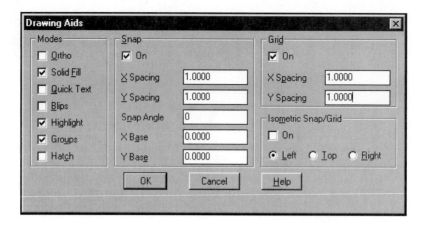

Figure 14-2

To create a 3D object

In this example, an object will be created using the SE Isometric viewpoint.

1. Select the Tools pull-down menu, then Drawing Aids. Turn on both SNAP and GRID commands and set them both for 1.0000 X and Y spacing as shown in Figure 14-2.

2. Select the SE Isometric View tool from the Viewpoint toolbar.

   See Figure 14-3.

3. Select the View-pull down menu, then Display, UCS icon, and click the Icon option to locate the XY icon on the origin of the grid.

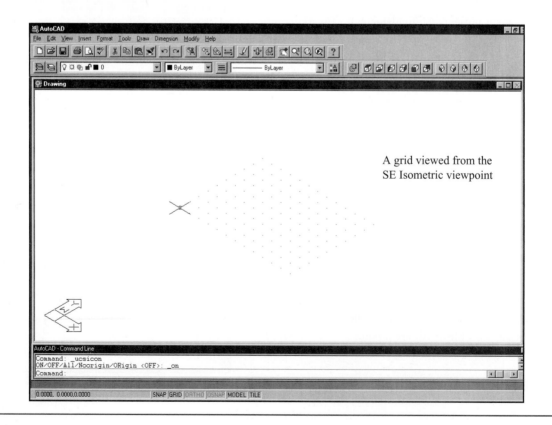

A grid viewed from the SE Isometric viewpoint

Figure 14-3

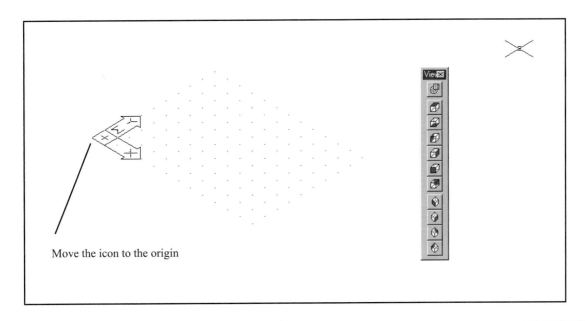

Move the icon to the origin

Figure 14-4

Locating the XY icon on the origin will help create a better visual understanding of the grid in 3D space. See Figure 14-4.

4. Use the Box tool from the Solids toolbar and create two boxes both with their corners located on the origin (0,0,0) and with dimensional X,Y,Z values of 2,8,8 and 6,8,1.5.

5. Union the boxes using the Union tool on the Modify II toolbar to create a solid 3D object. See Figure 14-5.

To change to a different preset viewpoint

1. Select the SW Isometric tool from the Viewpoint toolbar.

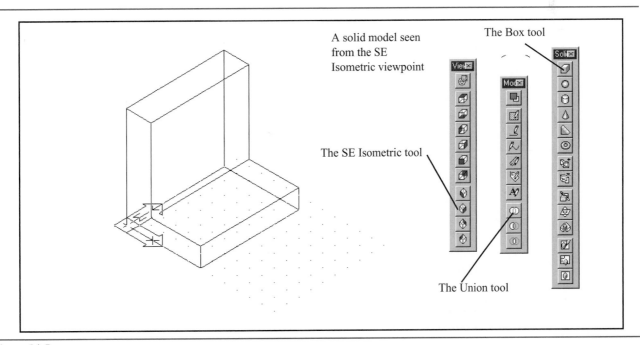

A solid model seen from the SE Isometric viewpoint

The Box tool

The SE Isometric tool

The Union tool

Figure 14-5

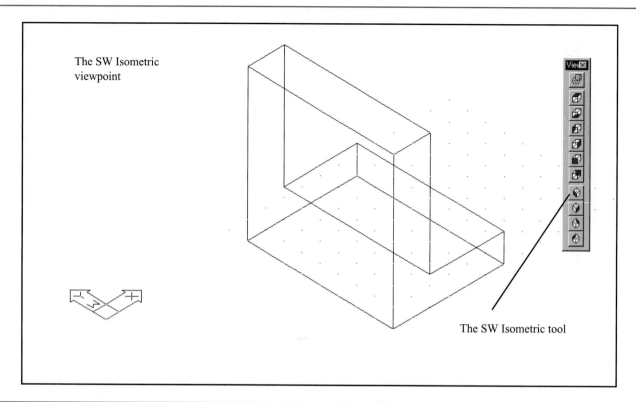

The SW Isometric
viewpoint

The SW Isometric tool

Figure 14-6

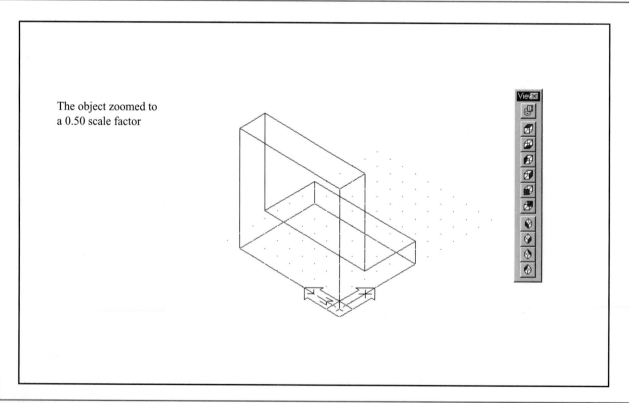

The object zoomed to
a 0.50 scale factor

Figure 14-7

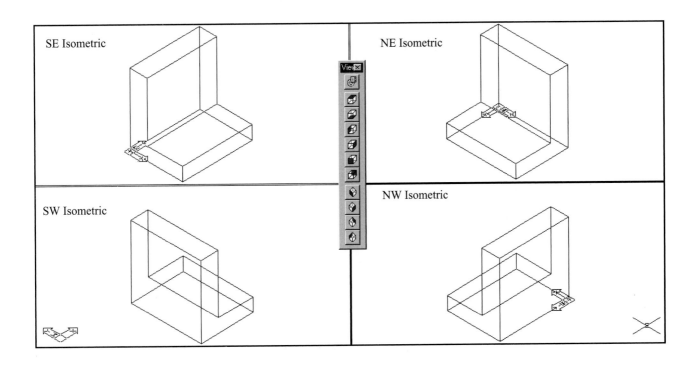

Figure 14-8

The viewpoint of the object will change as shown in Figure 14-6. The rotated object will be slightly larger than the original.

2. Type Zoom in response to a Command prompt.

All/Center/Dynamic/Extents/Left/Previous/Vmax/ Window/<Scale(X/XP)>:

3. Type .5; press Enter.

The object will appear at approximately the original size. See Figure 14-7. The object can be returned to the SE Isometric viewpoint by selecting the SE Isometric preset viewpoint.

Figure 14-8 shows the object in the four possible preset viewpoints.

14-3 3D VIEWPOINTS PRESETS — ORTHOGRAPHIC

The Viewpoint toolbar also includes six preset orthographic viewpoints: top, bottom, left, right, front, and back. The views are taken relative to the WCS regardless of the object's isometric orientation.

To create a front viewpoint

1. Select the Front tool from the Viewpoint toolbar.

A front view from the X,Z plane will appear and be displayed across the entire screen. See Figure 14-9.

2. Type Zoom in response to a Command prompt and define a factor of .5.

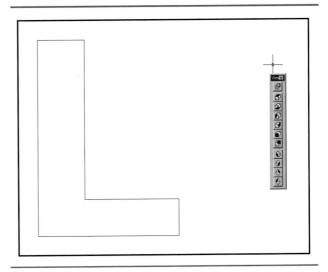

Figure 14-9

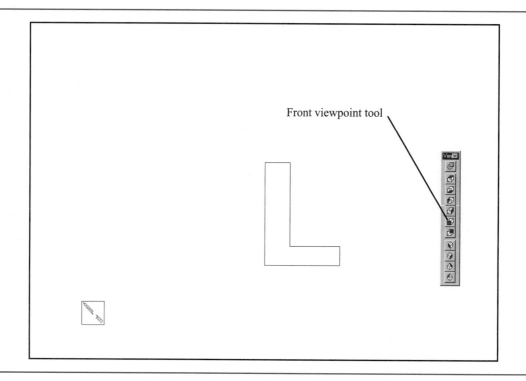

Front viewpoint tool

Figure 14-10

Figure 14-10 shows the resulting object.

The other five possible orthographic views can be created in a similar manner. As all views are relative to the WCS axis, it is not necessary to return to the original draw-ing to create other views. Figure 14-11 shows the 3D object and front, top, and right orthographic views created using 3D Viewpoint Preset commands. A zoom factor of .5 was used for each view.

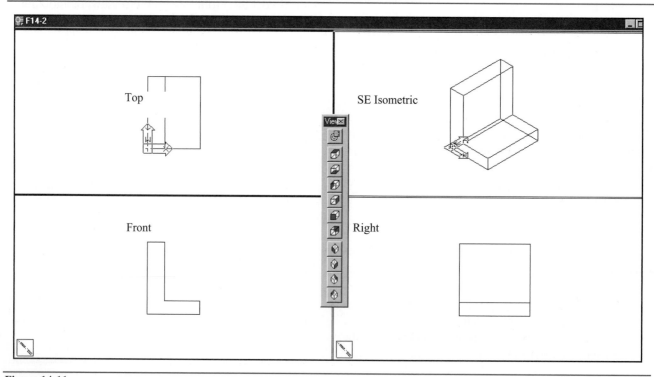

Figure 14-11

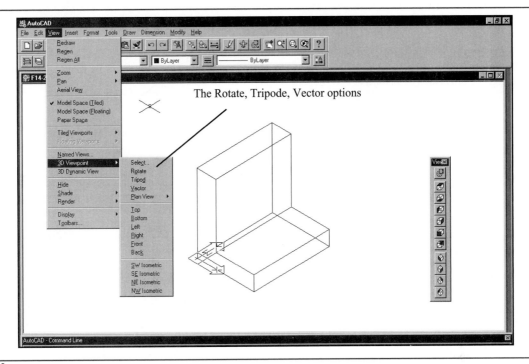

The Rotate, Tripode, Vector options

Figure 14-12

14-4 3D VIEWPOINT

The 3D VIEWPOINT command has three related commands, ROTATE, TRIPOD, and VECTOR. See Figure 14-12.

To use the ROTATE command

1. Select the View pull-down menu, then 3D Viewpoint, then Rotate.

The Rotate option does not actually rotate the object, but rather changes the viewing angle relative to the WCS and current viewing setting. In this example, the SE Isometric viewpoint generated values of 315.0 and 35.3. Changing these values will change the viewpoint.

Command:_vpoint Rotate/ <View point> <1.0000,-1.0000,1.0000>:_r Enter angle in XY plane from X axis <315>:

2. Type 135.0; Press Enter

Enter angle from XY plane <35>:

3. Type -30.0; press Enter

Figure 14-13 shows the resulting new orientation. The revised object is automatically fitted to the total screen, so a zoom factor is needed to return the object to its original size.

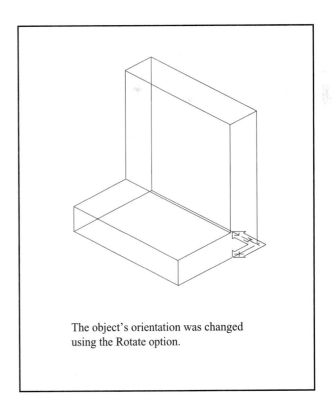

The object's orientation was changed using the Rotate option.

Figure 14-13

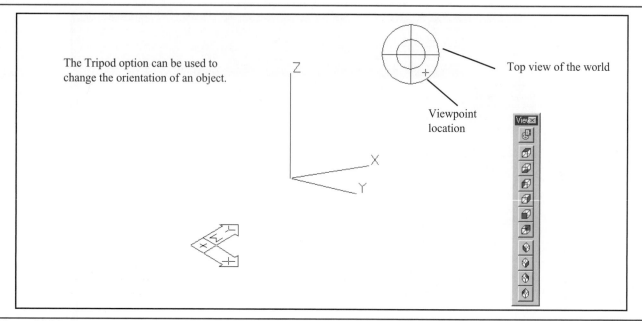

The Tripod option can be used to change the orientation of an object.

Top view of the world

Viewpoint location

Figure 14-14

To use the TRIPOD command

1. Select the View pull-down menu, then 3D Viewpoint, then Tripod.

The object will disappear and a coordinate reference system will appear as shown in Figure 14-14. In addition, a circular symbol will appear. This symbol represents a top view of the world (WCS). The small cross represents the viewpoint. The viewpoint can be moved by moving the mouse.

2. Once a new position has been established, click the mouse; the object will reappear on the

screen in its new position.

See Figure 14-16. The object can always be returned to its previous position using the UNDO command.

The tripod option presents an easy way to view an object from different orientations, but it is difficult to create views at specific orientations because it has no angular inputs.

To use the VECTOR command

1. Select the View pull-down menu, then 3D Viewpoint, then Vector.

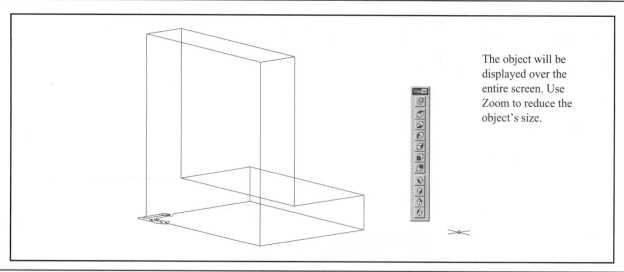

The object will be displayed over the entire screen. Use Zoom to reduce the object's size.

Figure 14-15

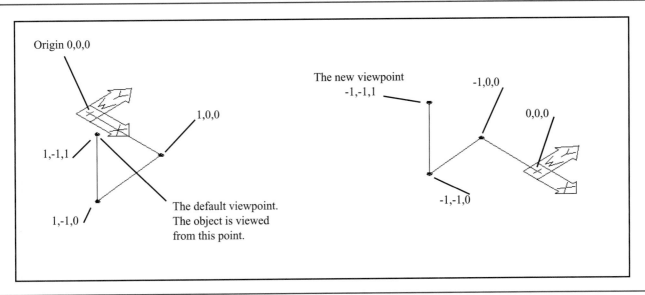

Figure 14-16

The VECTOR command defines a viewpoint using X,Y,Z coordinate values. The object shown from the SE Isometric viewpoint has a default vector coordinate value of 1.0000,-1.0000,1.0000. See Figure 14-16. This means that the viewpoint is located 1 unit along the X axis, -1 unit along the Y axis, and 1 unit along the Z axis.

Command:_vpoint Rotate/ <View point>
<1.0000,-1.0000,1.0000>: 1,-1,1

2. Type in a new value of -1,-1,1; press Enter.

Figure 14-17 shows the resulting new orientation.

The object viewed from the -1,-1,1 viewpoint

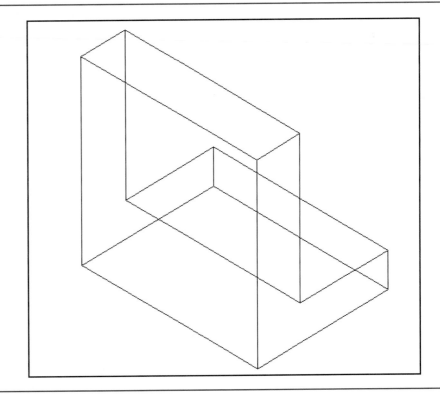

Figure 14-17

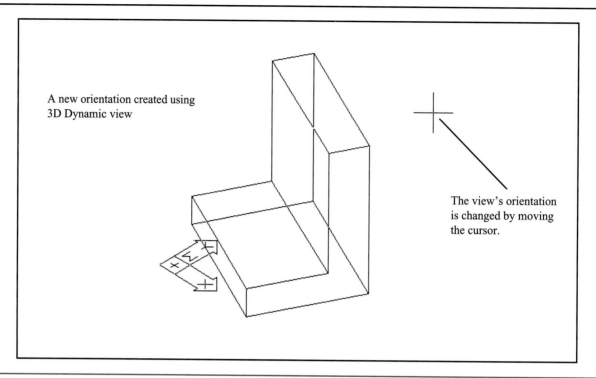

A new orientation created using
3D Dynamic view

The view's orientation
is changed by moving
the cursor.

Figure 14-18

14-5 3D DYNAMIC VIEW

1. Select the View pull-down menu, then 3D Dynamic View.

   Command: _dview
   Select objects:

2. Select the 3D object.

   Select objects:

3. Press Enter.

   CAmera/TArget/Distance/POints/PAn/Zoom/
   TWist/Clip/Hide/Off/Undo/<eXit>:

4. Type CA; press Enter.

   Toggle angle in/Enter angle from XY plane
   <35.2644>:

   The given value is for the SE Isometric viewpoint. The object may be moved by moving the cursor.

5. Once a new orientation has been selected, press the left mouse button.

   Figure 14-18 shows a new orientation.

CAmera/TArget/Distance/POints/PAn/Zoom/
TWist/Clip/Hide/Off/Undo/<eXit>:

6. Type X to exit the 3D Dynamic View command, or type CA to shift to another orientation.

   Use the UNDO command to return to the previous orientation.

14-6 VIEWPOINTS AND UCSs

It is important to understand that a change in viewpoint is not the same as a change in UCS. A viewpoint change is simply a different way of looking at an existing object. The working coordinate system remains the same. To help demonstrate this concept, consider the object shown in Figure 14-19. The axis icon has been positioned on the origin of the WCS which is the current working axis. All objects created will be created on the WCS X,Y plane. If a hole is to be added to the vertical portion of the object a UCS aligned with the vertical portion must be defined. Changing the viewpoint is not sufficient.

Figure 14-20 shows the same object from the Right View viewpoint, zoomed at a .5 factor. The broken pencil icon in the lower left corner of the screen indicates that the current working X,Y plane is almost perpendicular to the viewing direction. Any objects drawn will be viewed from a perpendicular direction. A circle will appear as a straight line and a cylinder as a rectangle. Figure 14-20 shows the resulting cylinder.

Figure 14-21 shows the object viewed from the original S.E. Isometric viewpont. Note the position of the cylinder relative to the object and to the WCS.

To create a UCS aligned with a given viewpoint preset

When working in three dimensions, it is often convenient to look at the object from several viewpoints. This can easily be achieved by using one of the 3D preset viewpoints or by using the 3D DYNAMIC VIEW command. It is also convenient to be able to reference quickly a UCS aligned with a viewpoint so that additions and modifications can be made and incorporated. Figure 14-22 shows a 3D object from the Right preset viewpoint. This is a standard 3D Viewpoint preset and can be automatically aligned with the Right UCS preset.

1. Select the Preset tool from the UCS Toolbar.

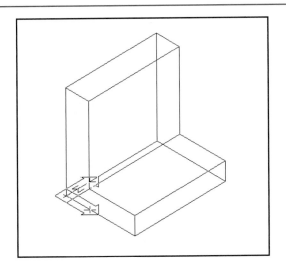

An object aligned with the WCS

Figure 14-19

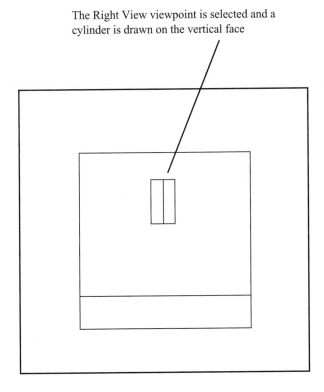

The Right View viewpoint is selected and a cylinder is drawn on the vertical face

Figure 14-20

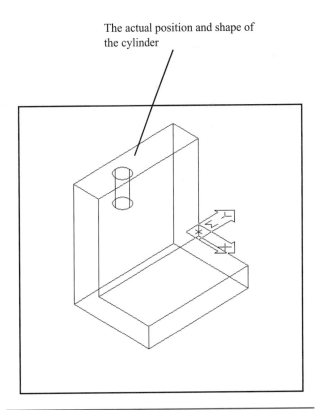

The actual position and shape of the cylinder

Figure 14-21

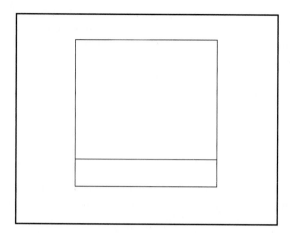

The object viewed from the Right viewpoint

Figure 14-22

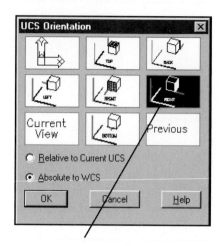

Select the preset Right UCS.

Figure 14-23

The UCS Orientation dialog box will appear. See Figure 14-23.

2. Select the Right preset UCS.

A UCS icon now appears at the origin, indicating that work can be done in that plane. See Figure 14-24.

The Right UCS has the same origin as the WCS. See Figure 14-25. The origin can be moved using the Origin tool on the UCS toolbar. Figure 14-26 shows the Right UCS origin moved to the front of the vertical section of the object.

The preset Right UCS will align with the Right viewpoint.

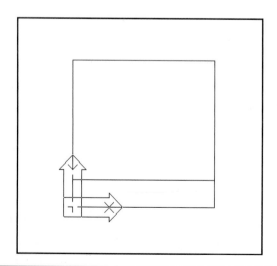

Figure 14-24

The preset Right UCS will have the same origin as the WCS.

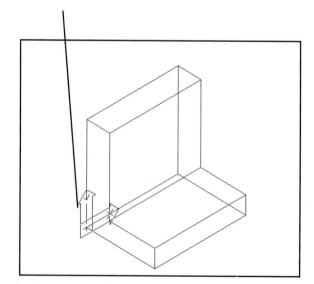

Figure 14-25

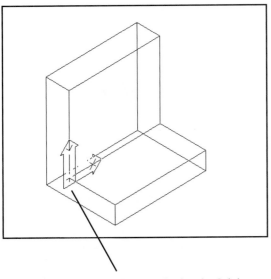

The UCS origin can be moved using the Origin tool on the UCS toolbar.

Figure 14-26

14-7 NAMED VIEWS

If an object is better displayed or viewed from a point other than one of the standard preset viewpoints, a new viewpoint can be defined using the 3D VIEWPOINT and ROTATE commands, then saved as a Named View for easy reference.

To create a named view

1. Select the View pull-down menu, then 3D Viewpoint, then Rotate.

   Command:_vpoint Rotate/ <View point> <1.0000,-1.0000,1.0000>:_r Enter angle in XY plane from X axis <315>:

2. Type 10; press Enter.

   Enter angle from XY plane <35>:

3. Press Enter.

The object will appear in the new orientation; it will be displayed over the entire screen. Use the ZOOM command to reduce its size. In this example, a zoom factor of .5 was used. See Figure 14-27.

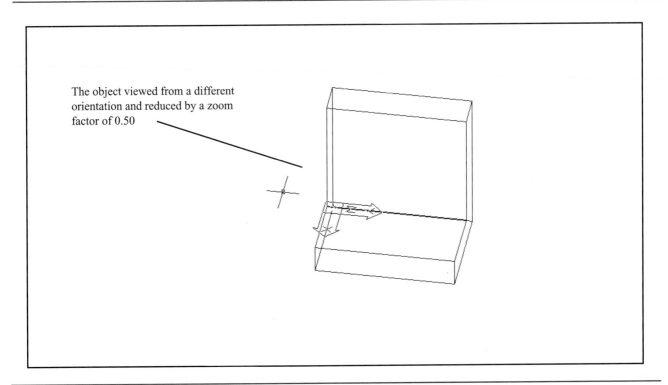

The object viewed from a different orientation and reduced by a zoom factor of 0.50

Figure 14-27

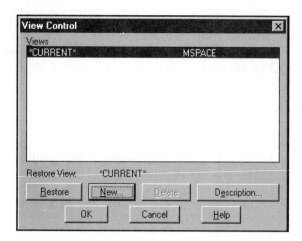

Figure 14-28

Type view name here

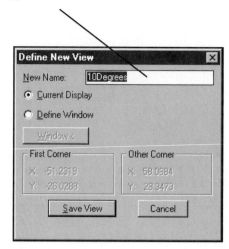

Figure 14-29

4. Select the the Named Views tool from the Viewpoint toolbar.

The View Control dialog box will appear. See Figure 14-28.

5. Select New.

The Define New View dialog box will appear. See Figure 14-29. Be sure the Current Display option is on.

To return to a named view, hightlight the view name, click Restore, then OK.

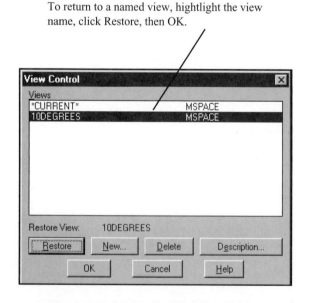

Figure 14-30

6. Type in the name of the new viewpoint.

In this example the name "10Degrees" was selected.

7. Select Save View, then OK.

The View Control dialog box will reappear. See Figure 14-30.

8. Select OK to return to the drawing.

To return to a named view

Assume that the object is in its original SE Isometric orientation.

1. Select the View pull-down menu, then Named View.

The View Control dialog box will appear. See Figure 14-30.

2. Highlight the view names 10Degrees.
3. Select the Restore command.

The named view will appear next to the Restore View heading.

4. Select OK.

The object will be displayed from the named viewpoint.

Named views can also be created using the TRIPOD, VECTOR, or 3D DYNAMIC VIEW commands.

14-8 EXERCISE PROBLEMS

Redraw the objects shown in EX14-1 through EX14-4 as solid models, then create four view ports for each. Display the objects using the SW, SE, NE, and NW Isometric viewpoints.

EX14-1 MILLIMETERS

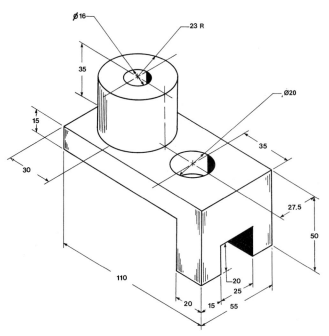

EX14-3 MILLIMETERS

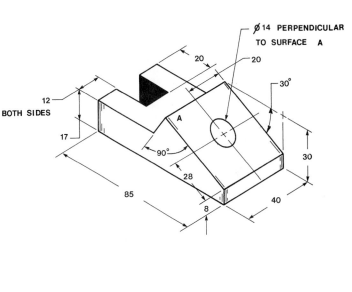

EX14-2 MILLIMETERS

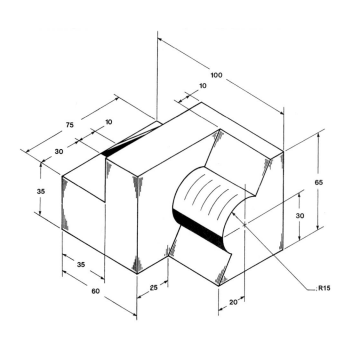

EX14-4 MILLIMETERS

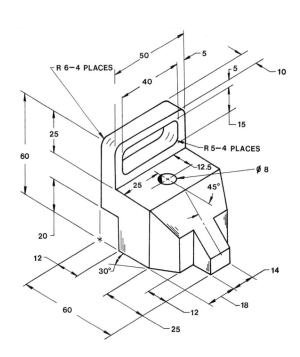

Redraw the objects shown in EX14-5 through EX14-8 as solid models, then create four view ports for each. Display the objects using the SE Isometric, Front, Top, and Right viewpoints.

EX14-5 MILLIMETERS

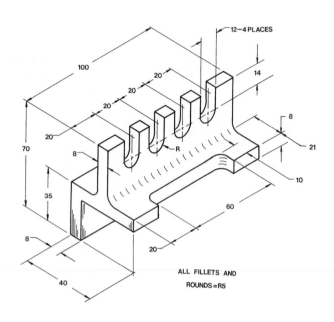

ALL FILLETS AND
ROUNDS = R5

EX14-7 INCHES (SCALE:2=1)

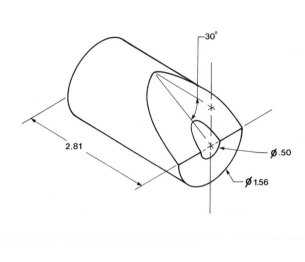

EX14-6 MILLIMETERS

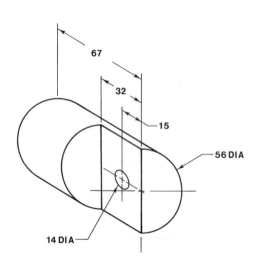

EX14-8 MILLIMETERS

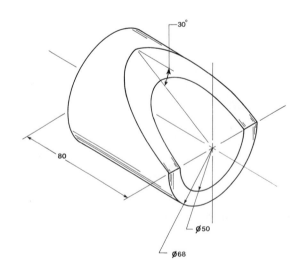

Redraw the objects shown in EX14-4 through EX 14-12 as solid models, then create four view ports for each. Display the objects using the SE Isometric, Front, Top, and Right viewports.

EX14-9 MILLIMETERS

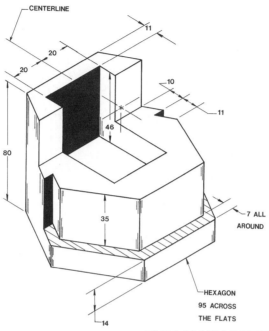

EX14-10 MILLIMETERS

EX14-11 MILLIMETERS

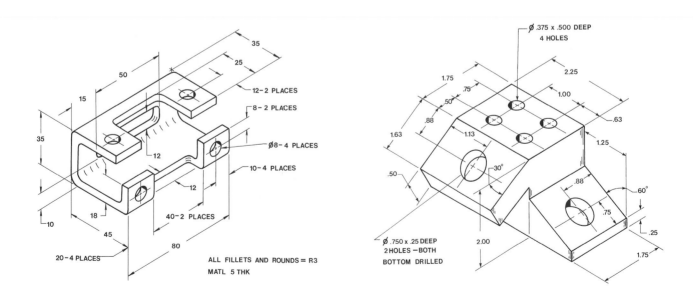

EX14-12 MILLIMETERS

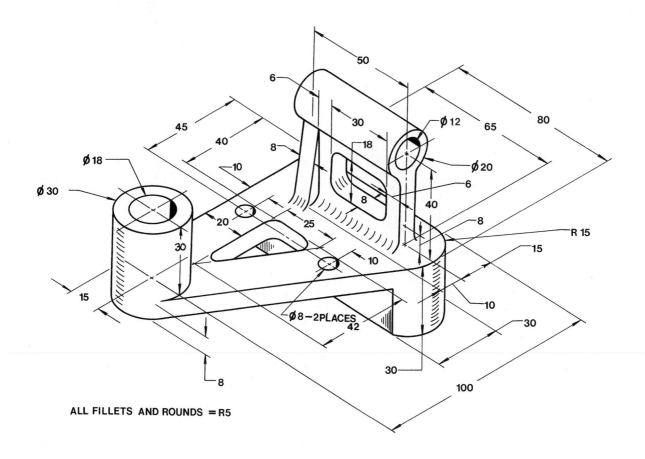

ALL FILLETS AND ROUNDS = R5

Dimensioning in 3D

15-1 INTRODUCTION

This chapter explains and demonstrates how to apply dimensions to three-dimensional solid models. In addition, the chapter will show how to apply both linear and geometric tolerances. The use of the Dimensioning toolbar was covered in Chapter 8; it is suggested that the chapter be reviewed before proceeding with this chapter.

15-2 LINEAR DIMENSIONS — SIZES

Figure 15-1 shows a dimensioned solid model. This is the same object that was used in Chapter 14 and instructions for its creation are included in section 14-2.

To size the dimensions

The Dimension Styles tool on the Dimensioning toolbar is used to size the height of the dimension values and arrowheads. Dimensional values should be large enough to be seen clearly, yet not obscure the drawing. In the example shown, an overall scale factor of 4 was selected as follows.

1. Select the Dimension Styles tool from the Dimensioning toolbar.

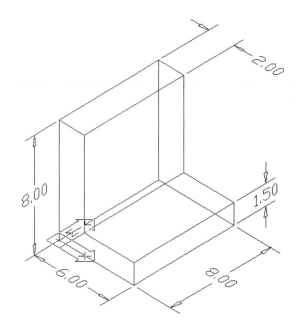

Figure 15-1

429

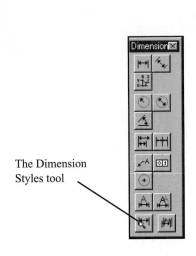

The Dimension
Styles tool

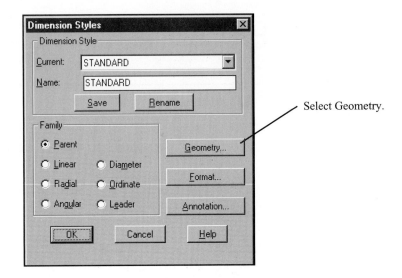

Select Geometry.

Figure 15-2

The Dimension Styles dialog box will appear. See Figure 15-2.

2. Select the Geometry box.

The Geometry dialog box will appear as shown in Figure 15-3.

3. Change the Overall Scale value to 4.0000, then select OK.

The Dimension Styles dialog box will reappear.

4. Select the Format box.

The Format dialog box will appear. See Figure 15-4.

5. Assure that the Vertical Justification is set for Centered, then select OK.

The Dimension Styles dialog box will reappear.

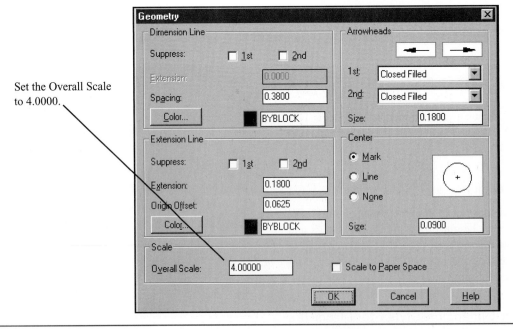

Set the Overall Scale
to 4.0000.

Figure 15-3

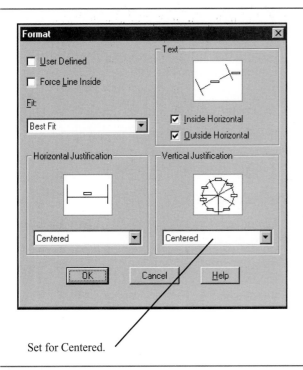

Set for Centered.

Figure 15-4

To change the dimension precision

AutoCAD's default precision for dimensional units is four decimal places (8.0000). In this example two decimal places will be used.

1. Select the Annotation box on the Dimension Styles dialog box.

The Annotation dialog box will appear. See Figure 15-5.

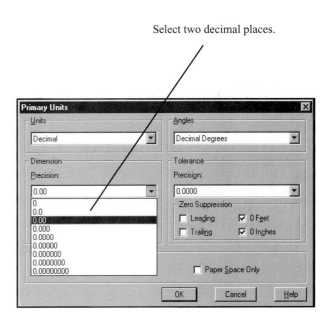

Select two decimal places.

Figure 15-6

2. Select the Units box

The Primary Units dialog box will appear. See Figure 15-6.

3. Select the arrow to the right of the Precision box, and then select two decimal places.
4. Select OK.

The original object will return to the screen. All dimensions created will have two decimal places.

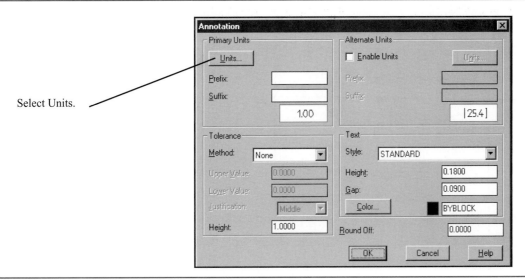

Select Units.

Figure 15-5

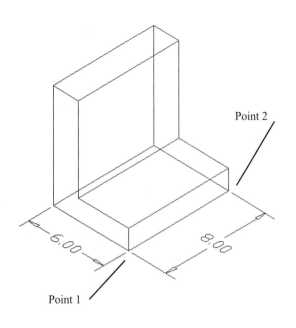

Figure 15-7

15-3 LINEAR DIMENSIONS — BASE PLANE

Linear dimensions can be applied in the base plane (the X,Y plane of the WCS) using the Linear Dimension tool from the Dimensioning toolbar.

To create a linear dimension

1. Select the Linear Dimension tool from the Dimensioning toolbar.

   Command: _dimlinear
   First extension line origin or Enter to select:

2. Select point 1 as defined in Figure 15-7.

   Second extension line origin:

3. Select point 2.

   Dimension line location (Text/ Angle/ Horizontal/ Vertical/ Rotated):

4. Position the dimension line by moving the mouse, then press the left mouse button to fix the position.

   It is suggested that you move the mouse around the screen and watch the changes that occur in the dimension.

   A second dimension was added on the base plane using the same procedure as the first dimension. See Figure 15-7.

To create a dimension in a different plane

A vertical dimension is needed to define the height of the object but cannot be created using the current WCS setup. Dimensions can be created only in the current X,Y plane, so a new UCS is needed.

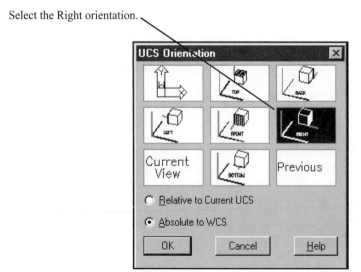

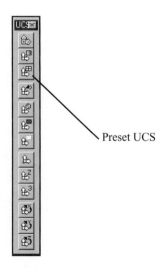

Figure 15-8

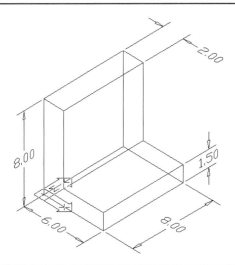

Figure 15-9

1. Select the Preset tool from the UCS toolbar.

The UCS orientation dialog box will appear. See Figure 15-8.

2. Select the Right preset.

If you had selected the Left preset orientation, the dimensions would appear backwards.

3. Select the Linear Dimension tool from the Dimensioning toolbar and create the appropriate vertical dimension.

See Figure 15-9.

To change dimensioning planes

The 1.50 thickness dimension for the thickness of the lower flange is added using the Right orientation, but requires a change in origin. The dimension could be added using the current origin but would require that the extension line origins be located using OSNAP.

1. Select the Origin tool from the UCS toolbar.

Origin point <0,0,0>:

2. Use the OSNAP, Endpoint command to select the corner indicated in Figure 15-10.
3. Select the Linear Dimension tool from the Dimensioning toolbar and create the thickness dimension.

To save a UCS

It is good practice to save any new UCS that is not a

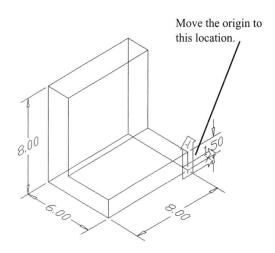

Figure 15-10

preset UCS. This makes it easy to return to that UCS for additions or modification to the dimensions.

1. Select the Named UCS tool from the UCS toolbar.

The UCS Control dialog box will appear. See figure 15-11. The new origin location will be listed as *NO NAME*.

2. Highlight the NO NAME line, then type in a name for the UCS in the Rename to box.

In this example, FRONTTHK (for front thickness) was selected.

Figure 15-11

New UCS will be listed here.

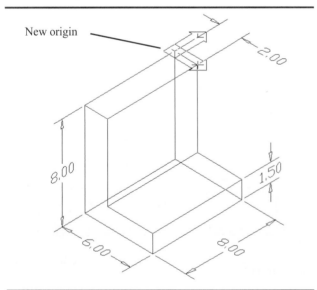

Figure 15-12

To verify that a UCS has been saved

1. Select the Named tool from the UCS toolbar.

The UCS Control dialog box will appear. See Figure 15-12. The new UCS name should be listed.

To retrieve saved UCS

1. Select the Named UCS tool from the UCS toolbar.
2. Highlight the desired UCS, then select the current box, then OK.

The origin will be shifted to the named UCS.

New origin

Figure 15-13

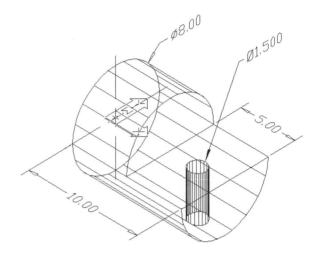

Figure 15-14

To create the top thickness dimension

See Figure 15-13.

1. Select the World tool from the UCS toolbar.
2. Select the Origin tool from the UCS toolbar.
3. Use the OSNAP, Endpoint command; select a new origin as shown in Figure 15-13.
4. Create the required dimension.
5. Save the newly created UCS.

15-4 DIMENSIONING CYLINDERS

Figure 15-14 shows a 3D cylinder. The horizontal center plane of the cylinder is located on the X,Y plane of the WCS. The cylinder's diameter is 8.00 and its length is 10.00. The small vertical hole is Ø1.500 starts at the 7.5,0,0 point and goes through the object. The Isolines value was set at 18. Add dimensions as follows.

To create the 5.00 linear dimensions

1. Select the Dimension Styles tool from the Dimensioning toolbar and set the Overall Scale to 4.0000 and the Unit precision for two decimal places.

2. Select the Linear tool from the Dimensioning toolbar and add the 5.00 dimension using the OSNAP, Endpoint command to ensure accuracy.

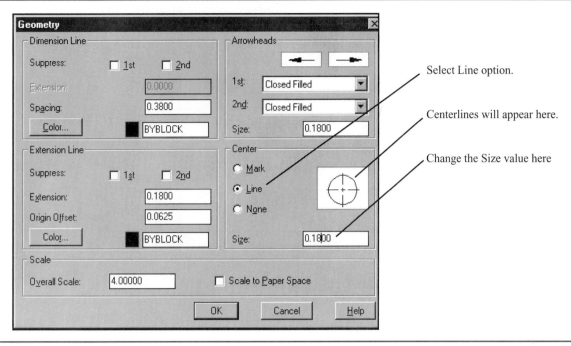

Select Line option.

Centerlines will appear here.

Change the Size value here

Figure 15-15

The OSNAP command cannot be applied to the isolines that define the surface of the cylinder as they are not real lines and therefore not recognized by the OSNAP commands.

To create a diameter dimension

The overall cylinder diameter is to be aligned with the back surface of the cylinder. This location was selected because it is aligned with the right UCS preset plane.

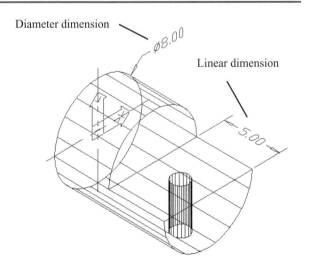

Diameter dimension

Linear dimension

Figure 15-16

1. Select the Preset tool from the UCS toolbar.

The UCS Orientation dialog box will appear. See Figure 15-8.

2. Select the Right preset UCS.

The Diameter Dimension command must be set to generate centerlines on the elliptical surface. The default setting is a center mark.

3. Select the Dimension Styles tool from the Dimensioning toolbar, then Geometry.

The Geometry dialog box will appear. See Figure 15-15.

4. Turn on the Line option within the Center box, and assure that the Size value is 0.1800.
5. Select OK, then OK again to return to the drawing.
6. Select the Diameter Dimension tool from the Dimensioning toolbar.

Select arc or circle:

7. Select the back surface ellipse.

Dimension line location (Text/Angle)

8. Locate the dimension line and press the left mouse button.

See Figure 15-16.

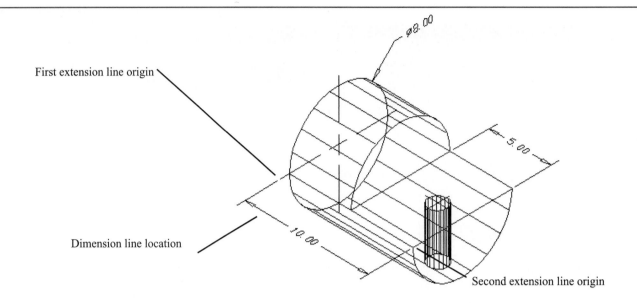

Figure 15-17

To create the 10.00 overall length dimension

1. Return to the WCS using the World tool from the UCS toolbar.
2. Select the Linear tool from the Dimensioning toolbar.

   First extension line or RETURN to select:

3. Use OSNAP and Endpoint to select the end of the horizontal center line created when adding the diameter dimension.

   Second extension line origin:

4. Use OSNAP and Endpoint to select the second origin as indicated in Figure 15-17.

   Dimension line location (Text/ Angle/ Horizontal/ Vertical/ Rotated):

5. Select the location for the dimension line and press the left mouse button.

If OSNAP cannot be used

In this example the overall length is known to be 10.00, but if the centerlines had not been added there would not be lines for the OSNAP command to grab. The isolines, used to define the surface of the cylinder, are not real lines and cannot be used with OSNAP. The location

for the first extension line origin would have to be approximated, that is, the drawing would be zoomed and the location determined by eye. Because OSNAP could not be used, the resulting dimension value will not be 10.00. The Text option within the Linear Dimension command can be used to type and enter the correct value.

1. Select the Linear Dimension tool from the Dimensioning toolbar and locate the first and second extension line origins.

   Dimension line location (MText/ Text/ Angle/ Horizontal/ Vertical/ Rotated):

2. Type M; press Enter.

   The Multiline Text Editor dialog box will appear. See Figure 15-18. A flashing cursor will appear next to a ◇ symbol. This symbol represents the current dimension value.

3. Backspace to remove the <> symbol and type in 10.00, then select OK.

   Dimension line location (Text/ Angle/ Horizontal/ Vertical/ Rotate):

4. Position the dimension line, and press the left mouse button.

   The Text option could also have been used.

1. Type T instead of M to access the Text option.

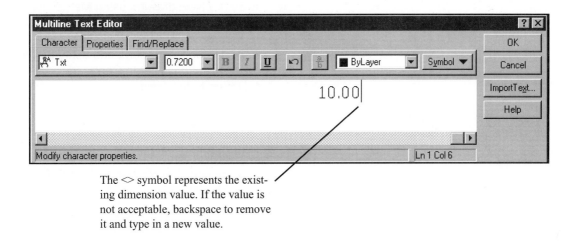

The ◇ symbol represents the existing dimension value. If the value is not acceptable, backspace to remove it and type in a new value.

Figure 15-18

A command line will appear.

Dimension text <9.87>:

2. Type 10.00; press Enter.

To create the Ø1.500 diameter dimension

The diameter dimension for the 1.50 diameter hole is added using a leader line. For the sake of consistency and to help give the drawing a uniform appearance, the text for the dimension will be made parallel to the existing Ø8.00 dimension.

1. Select the Preset UCS tool from the UCS toolbar, then Right from the UCS Orientation dialog box, then OK.

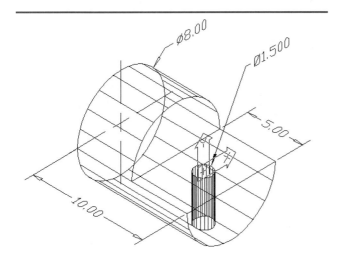

Figure 15-19

2. Select the Origin tool from the UCS toolbar and move the origin to the center of the Ø1.500 hole.
3. Select the Leader tool from the Dimensioning toolbar.

   From point:

4. Select a location on the 1.500 circle as shown in Figure 15-19.

   To point:

5. Select a point up and away from the object as shown.

The second leader line point was selected to make the leader line approximately the same angle as the Ø8.00 leader line. This gives the drawing a more organized appearance and makes it easier for the reader to follow.

To point (Format/ Annotation/ (Undo) <Annotation>:

6. Press Enter.

   Annotation (or RETURN for options):

7. Press Enter.

   Tolerance/Copy/Block/None/<MText>:

8. Press Enter.

The Edit MText dialog box will appear.

9. Type Ø1.500, then select OK.

The symbol Ø is created by using the Symbolsoptions located on the Multiline Text Editor dialog box.

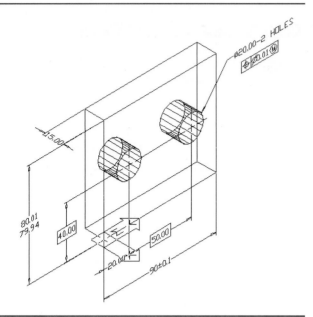

Figure 15-20

15-5 TOLERANCES

Both linear and geometric tolerances can be added to 3D drawings. Figure 15-20 shows a dimensioned and toleranced object. The object is a flat plate 90 x 80 x 15 millimeters. The drawing limits are set for 297,210 with a grid spacing of 20 and a snap spacing of 10. The front plane of the object is aligned with the Right UCS X,Y plane. The isolines value is set at 18. The dimensions and tolerances were added as follows.

1. Start a New drawing using Metric units.
2. Select the Dimension Styles tool from the Dimensioning toolbar.
3. Select the Format box and verify that the Vertical Justification is set to Centered.
4. Return to the drawing.

Adding center marks to the holes

1. Select the Center Mark tool from the Dimensioning toolbar and add center marks to each of the two holes.

See Figures 15-21 and 15-22. The Center Mark command is controlled using the Center box on the Geometry dialog box accessed using the Dimension Styles tool. The marks were added first so OSNAP could be used to locate extension line origins.

2. Select the Linear tool from the Dimensioning toolbar; add the 20.00 dimension shown in Figure 15-23.
3. Use the Linear tool to create the 50.00 dimension.

The 50.00 dimension is a base dimension, so it is enclosed in a rectangle. The rectangle is created using the LINE command from the Draw toolbar.

4. Create a rectangle around the 50.00 dimension.
5. Explode the dimension then turn on the ORTHO command and use the LINE command to extend the dimension line so it touches the rectangle.

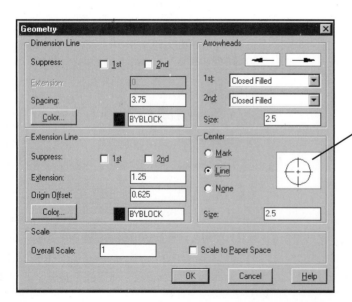

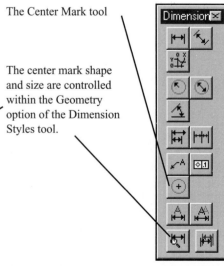

The Center Mark tool

The center mark shape and size are controlled within the Geometry option of the Dimension Styles tool.

Figure 15-21

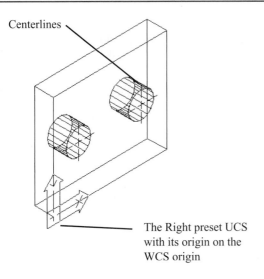

Centerlines

The Right preset UCS
with its origin on the
WCS origin

Figure 15-22

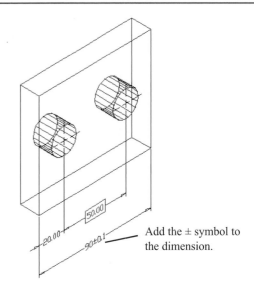

Add the ± symbol to
the dimension.

Figure 15-24

To create a plus and minus (±) dimension

The 90 ± 0.1 dimension is created as follows. See Figure 15-24.

1. Select the Linear Dimension tool from the Dimensioning toolbar and define the two extension lines' origins.

   Dimension line location (MText/ Text/ Angle/ Horizontal/ Vertical/ Rotate):

2. Type m in response to the dimension line location prompt.

The Multiline Text Editor dialog box will appear. Backspace to remove the ◇ symbol and type in 90 ± 0.1. The symbol ± was created using the Symbol option located on the Multiline Text Editor dialog box. See Figure 15-25.

Dimension line location (Text/ Angle/ Horizontal/ Vertical/ Rotate):

3. Locate the dimension and press the left mouse button.

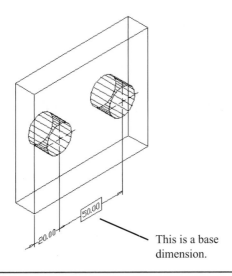

This is a base
dimension.

Figure 15-23

The Multiline Text Editor dialog box

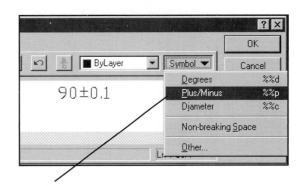

Create the ± symbol by selecting here.

Figure 15-25

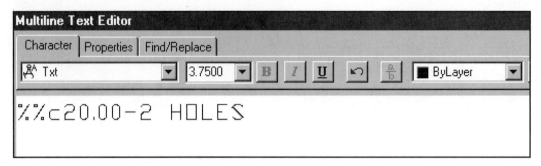

This symbol will change to Ø when it is
entered on the screen.

Figure 15-26

To create a diameter dimension

1. Select the Leader tool from the Dimensioning
   toolbar.

   From point:

2. Select a point on the edge of the right hole.

   This point will locate the tip of the leader line's
   arrow. The leader should be positioned so that it points at
   the center mark of the hole.

   To point:

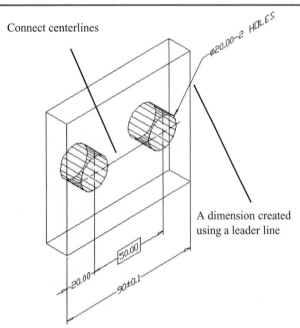

Connect centerlines

A dimension created
using a leader line

Figure 15-27

3. Draw a short line segment parallel to the X axis;
   press the right mouse button twice.

   Tolerance/Copy/Block/None/<Mtext>:

4. Press Enter.

   The Multiline Text Editor dialog box will appear.
   See Figure 15-26.

5. Type %%c - 2 HOLES, then select OK.

   The %%c was created by selecting Diameter from
   the Symbol options.

   Dimension line location (Text/Angle):

6. Locate the dimension and press the left mouse
   button.

   The value Ø20.00 will automatically be created by
   the Diameter Dimension command, but the note 2 HOLES
   must be added. Notes on drawings are usually written
   using only upper-case letters. Also remember that you are
   dimensioning holes, not circles.

7. Draw a horizontal line connecting the two centerlines to indicate that they use the same vertical
   dimension.

To create a geometric tolerance

1. Select the Tolerance tool from the Dimensioning
   toolbar.

   The Symbol dialog box will appear. See Figure
   15-28.

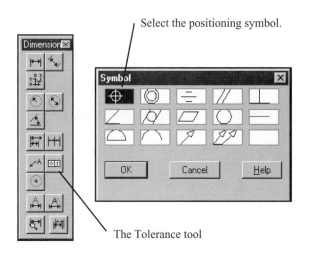

Select the positioning symbol.

The Tolerance tool

Figure 15-28

Select the Maximum Material Condition symbol.

Figure 15-30

2. Select the symbol for positional tolerance, then OK.

The Geometric Tolerance dialog box will appear. See Figure 15-29.

3. Click the Dia box within the Tolerance 1 box, then type in 0.01 in the Value box.
4. Click the MC box to the right of the Value box.

The Material Condition dialog box will appear. See Figure 15-30.

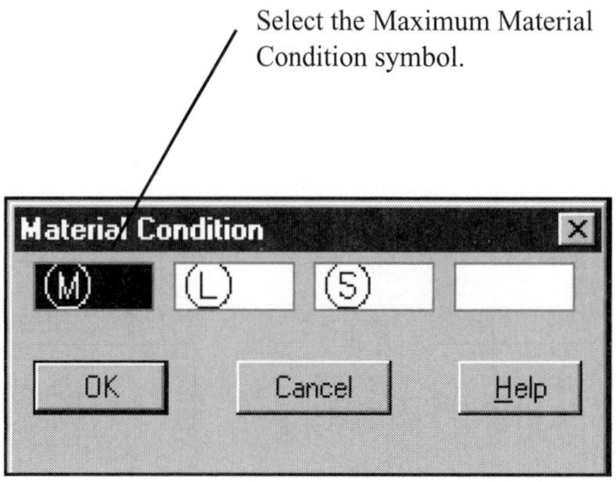

5. Select the Maximum Material Condition symbol, then OK, OK to return to the drawing.

Enter tolerance location:

6. Locate a point under the feature tolerance and press the left mouse button.

See Figure 15-31. If the geometric tolerance symbol is not located properly, reposition the symbol until it is directly under the feature dimension Ø20.00.

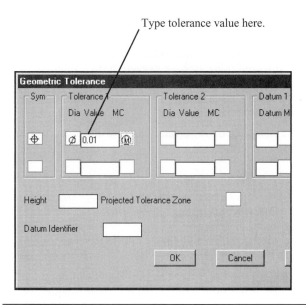

Type tolerance value here.

Figure 15-29

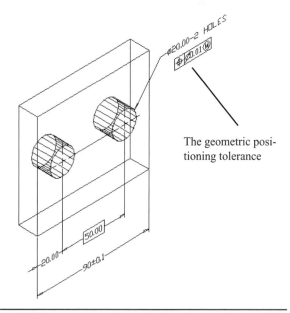

The geometric positioning tolerance

Figure 15-31

Add the vertical base dimension using the Linear tool.

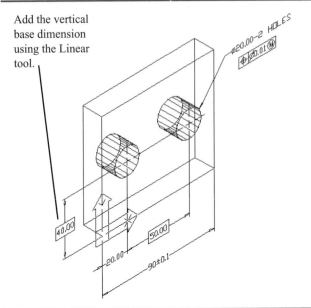

FIgure 15-32

The vertical base dimension of 40.00 was created in the same way as the horizontal 50.00 base dimension. The rectangle was created using the Line tool from the Draw toolbar. See Figure 15-32.

If the text appears in a vertical orientation, that is, aligned with the vertical dimension line, use the Text box on the Format dialog box to turn on both Inside and Outside Horizontal settings. See Figure 15-33. The Format dialog box is accessed using the Dimension Styles tool.

Turn these settings on.

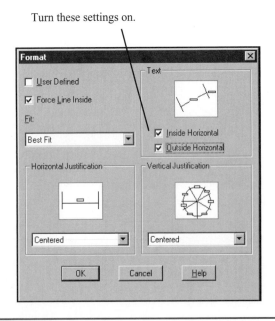

Figure 15-33

Type in limit tolerance.

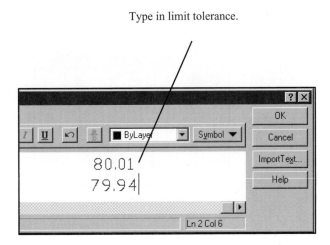

Figure 15-34

To create limit tolerances

The 80.01/79.94 limit tolerance used to define the overall height of the object is created as follows.

1. Select the Linear Dimension tool from the Dimensioning toolbar and define the two extension lines' origins.

Dimension line location (MText/ Text/ Angle/ Horizontal/ Vertical/ Rotated):

2. Type m; press Enter.

The Multiline Text Editor dialog box will appear. See Figure 15-34.

3. Backspace to remove the <> symbol and type in the limit tolerance as shown, then select OK.

Dimension line location (Text/ Angle/ Horizontal/ Vertical/ Rotated):

4. Locate the dimension line and press the left mouse button.

See Figure 15-35.

To create the thickness dimension

The thickness dimension must be created using a plane parallel to the WCS as follows.

1. Select the World tool from the UCS toolbar.
2. Select the Origin tool from the UCS toolbar.

Origin point<0,0,0>:

Add limit tolerance

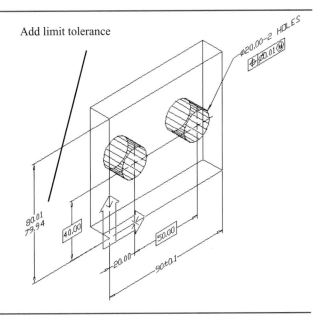

Figure 15-35

3. Use the OSNAP and Endpoint commands to select the corner of the object as shown in Figure 15-36.
4. Select the Linear Dimension tool from the Dimensioning toolbar and create the 15.00 thickness dimension.

See Figure 15-36. The arrowheads for the 15.00 dimension are aligned with the arrowheads of the height dimension to give the drawing a more organized appearance.

Locate a new origin and add thickness dimension.

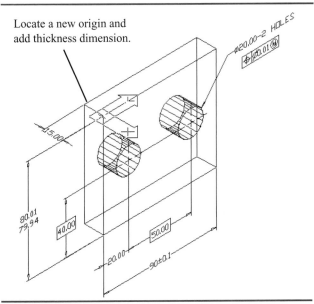

Figure 15-36

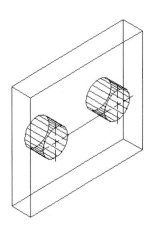

Figure 15-37

15-6 COMBINING 2D AND 3D DRAWINGS

Figure 15-37 shows the object presented in Section 15-5 without dimensions. It is often easier to understand a drawing that contains both 3D and 2D presentations. In this example, two ports will be created. One will give a 2D dimensional view, and the other a 3D drawing.

1. Select the View pull-down menu, then Model Space (Floating).

   ON/OFF/Hideplot/Fit/2/3/4/Restore/<First Point>:

2. Type 2; press Enter.

   Horizontal/<Vertical>:

3. Type Enter.

   Fit/<First point>:

4. Select a point near the upper left corner of the drawing screen.

   Second point:

5. Select a point near the lower right corner of the drawing screen.

See Figure 15-38. The port with the crosshairs is the active port. To change active ports, move the cursor into the other port and press the left mouse button.

First point selected for Model Space (Floating)

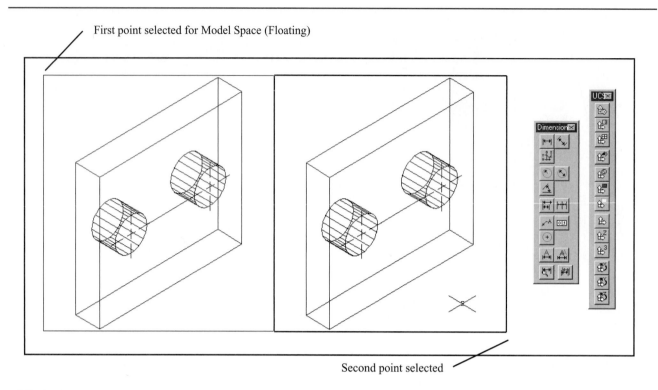

Second point selected

Figure 15-38

To create a 2D view of the object

1. Make the left port the current active port.
2. Select the Right tool from the View toolbar.

3. Type zoom, press Enter, and assign a value of 2.

Use the ZOOM command to increase the size of the view if necessary. See Figure 15-39.

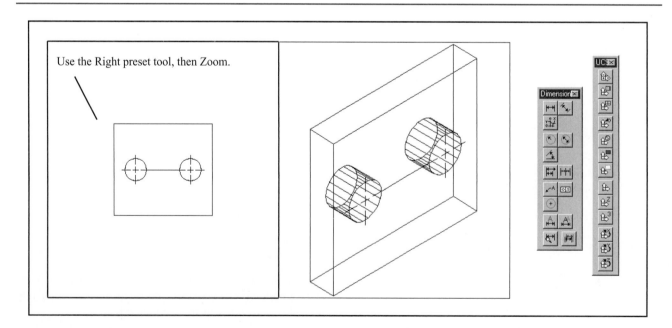

Use the Right preset tool, then Zoom.

Figure 15-39

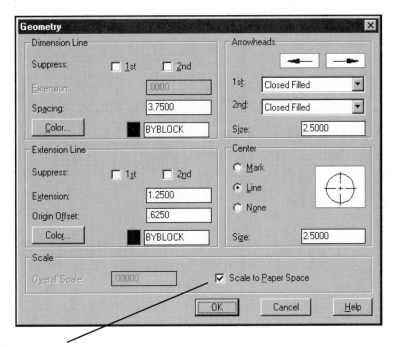

Select Scale to Paper Space option.

Figure 15-40

To add dimensions to the 2D port

Dimensions are added using Paper space. The drawing is currently in Model Space so this must be changed.

1. Double-click the Model icon at the bottom of the screen.

The word Model will be replaced with the word Paper. The Paper space icon, a right triangle, will appear in the lower left corner of the screen. The cursor is now free to move over the entire drawing screen, not just within a port as with Model space.

2. Type Dim in response to a command prompt.

Dim:

3. Type DIMLFAC; press Enter.

The dimlfac command is a systems variable that controls the global scale factor for linear dimensions.

Current value <1.0000> New value (Viewport):

4. Type V; press Enter.

Select viewport to scale:

5. Select the left vertical boundary line of the left port.

Select viewport to scale: DIMLFAC set to -0.9376

The DIMLFAC command uses absolute values, so the minus sign is not significant.

To calibrate the overall scale factor to the port

1. Select the Dimension Styles tool from the Dimensioning toolbar.

The Dimension Styles dialog box will appear.

2. Select the Geometry box.

The Geometry dialog box will appear. See Figure 15-39.

3. Select the Scale to Paper Space option within the Scale box.

4. Return to the drawing screen.

The left port is now ready for dimensions. Dimensions may be created using Dimensioning toolbar tools as before.

5. Create the appropriate dimensions using the dimensioning conventions presented in Chapter 8.

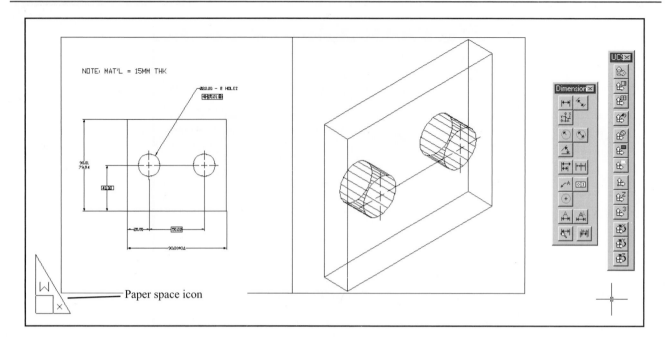

Figure 15-41

Figure 15-41 shows the dimensioned port.

To enlarge the object viewed in the right port

1. Double-click the word Paper at the bottom of the screen to return to Model space.
2. Select the right port.

3. Type zoom and define a scale factor of 2.0.
4. Select the Hide tool from the Render toolbar.

Figure 15-42 shows the resulting ports. Remember, to print the entire drawing, that is, both ports together, the drawing must be in Paper space. If the drawing is in Model space, only the active port will be printed.

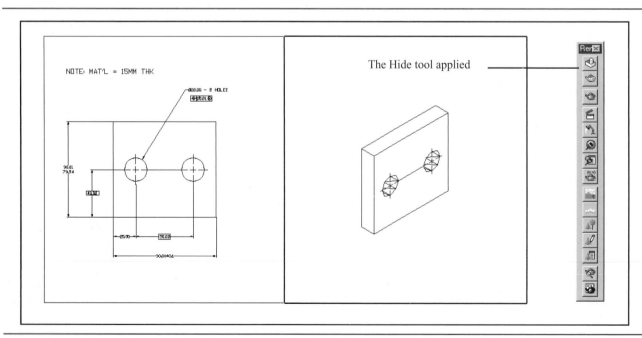

Figure 15-42

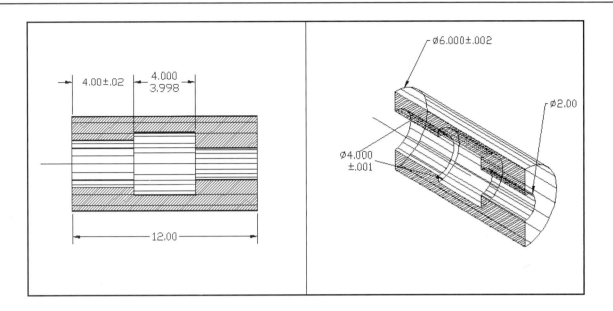

Figure 15-43

15-7 DIMENSIONED SECTIONAL VIEWS

Figure 15-43 shows a drawing which includes a sectional view and an isometric view of the same object. Both views of the object include dimensions. The drawing was created as follows.

To set up the drawing

1. Set the Grid spacing for 0.5000 and the Snap spacing for 0.2500, select the SE Isometric preset 3D viewport, and set the isolines value for 18.
2. Use the LINE command from the Draw toolbar and draw two lines, one directly aligned with the X axis and the second directly aligned with the Y axis.

See Figure 15-44. The line along the X axis will be the longitudinal centerline for the cylinder. The intersection of the two lines will be the center point of the back surface of the cylinder located on the WCS origin.

To draw the outside surface of the object

1. Select the Preset tool from the UCS toolbar, then the Right UCS.
2. Select the Cylinder tool from the Solids toolbar.

Elliptical/<center point><0,0,0>:

3. Select the 0,0,0 origin.

Diameter/<Radius>:

4. Type 3; press Enter.

Center of other end/<Height>:

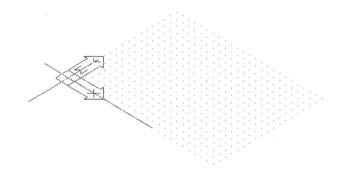

Figure 15-44

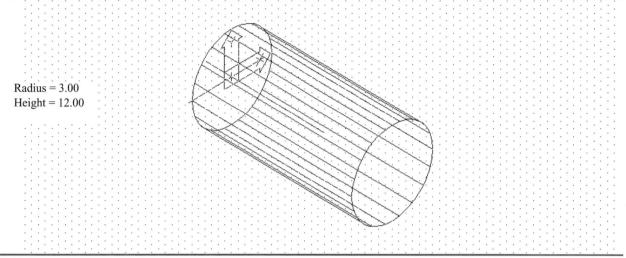

Radius = 3.00
Height = 12.00

Figure 15-45

5. Type 12; press Enter.

See Figure 15-45.

To draw the inside surfaces of the object

1. Select the Cylinder tool from the Solids toolbar.
2. Draw a solid cylinder of radius 1.50 and height 4.
3. Select the Origin tool from the UCS toolbar.

Origin point <0,0,0>;

4. Type 0,0,4; press Enter.

The origin has to be moved four inches down the longitudinal centerline, which is the Z axis in the Right UCS.

5. Select the Center Cylinder tool from the Solids toolbar; create a cylinder of radius 2 and height 4.
6. Select the Origin tool from the UCS toolbar.

Origin point <0,0,0>;

7. Type 0,0,4; press Enter.

The origin input is a relative input, that is, it is defined from the current origin. The 0,0,4 location would be stated as 0,0,8 from the WCS origin.

8. Select the Cylinder tool from the Solids toolbar; create a cylinder of radius 1 and height 4.
9. Use the Subtract tool from the Modify toolbar (the Subtract tool is located on the Modify II toolbar) and subtract the three internal cylinders

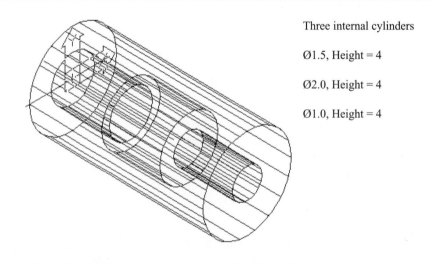

Three internal cylinders

Ø1.5, Height = 4

Ø2.0, Height = 4

Ø1.0, Height = 4

Figure 15-46

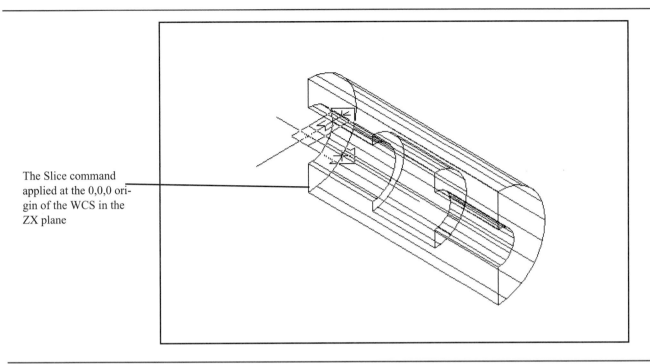

The Slice command applied at the 0,0,0 origin of the WCS in the ZX plane

Figure 15-47

from the external cylinder.

Figure 15-46 shows the resulting cylinder.

To create a sectional view

1. Return the coordinate system to the WCS.
2. Select the Slice tool from the Solids toolbar.

   Select objects:

3. Select the cylinder.

   Slicing plane by object/ Zaxis/ View/ XY/ YZ/ ZX/ <3points>:

4. Type ZX; press Enter.

The sectional view is to be taken along the longitudinal axis of the cylinder, which is the Z,X plane.

   Point on the ZX plane <0,0,0>:

The default point of 0,0,0 is acceptable, as the longitudinal axis is located on the X axis of the WCS.

5. Press Enter.

   Both sides/<Point on desired side of plane>:

6. Select a point behind the cylinder.

The point selected will determine which side of the slice will remain as part of the drawing. See Figure 15-47.

To add hatching to the sectional view

1. Select the Preset UCS tool, then Front UCS.
2. Select the Hatch tool from the Draw toolbar.

The Boundary Hatch dialog box will appear. See Figure 15-47. In this example, the ANSI31 hatch pattern was used with a scale of 1.0000.

3. Select Pick Points.

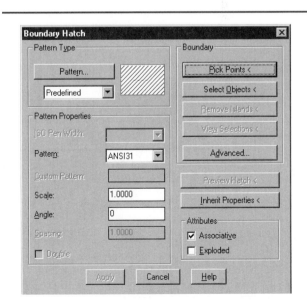

Figure 15-48

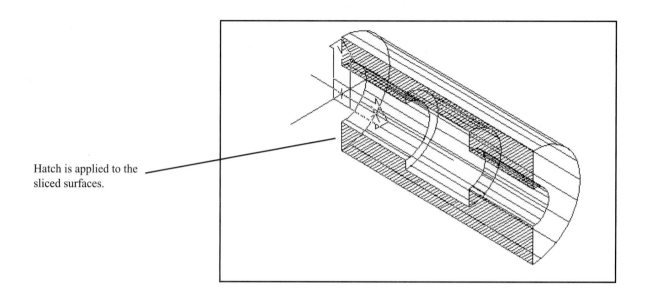

Hatch is applied to the
sliced surfaces.

Figure 15-49

Select internal point:

4. Select the sliced surfaces of the cylinder; press Enter.

The Boundary Hatch dialog box will reappear.

5. Select Apply.

The hatching will be applied over the sliced surface. See Figure 15-49. The hatch spacing in this example was made closer than normal to make it stand out in the illustrations.

To create two fitted view ports

1. Select the World icon from the UCS toolbar.
2. Select the View pull-down menu, then Model Space (Floating).

ON/OFF/Hideplot/Fit/2/3/4/Restore/<First Point>:

3. Type 2; press Enter.

Horizontal/<Vertical>:

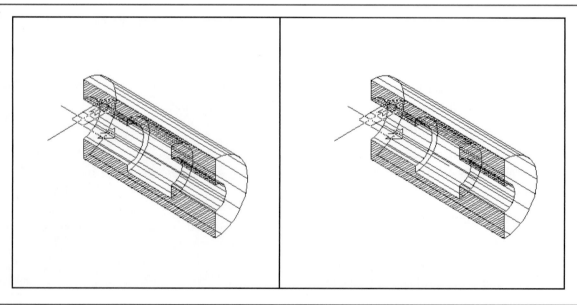

Figure 15-50

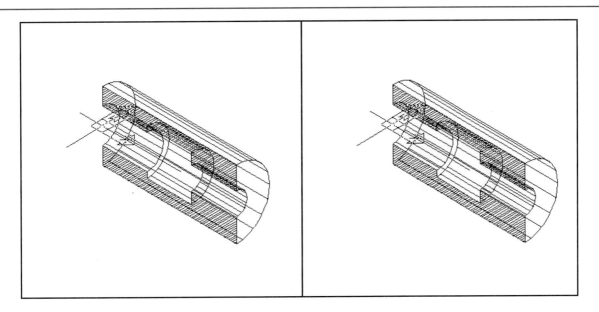

Figure 15-51

4. Press Enter.

Fit/<First point>:

5. Select a point near the upper left corner of the drawing screen.

Second point:

6. Select a point near the lower right corner of the drawing screen.
7. Type zoom and apply a zoom scale factor or 0.65 to both ports.

   See Figure 15-50.

To Create a 2D sectional view

1. Make the left port the active port.
2. Select the Front tool from the View toolbar.
3. Use the PAN command to locate the 2D sectional view in the middle of its port.
4. Use the PAN command to locate the 3D sectional view in the middle of its port if necessary.

   See Figure 15-51. The broken pencil icon in the lower left corner of the left port indicates that the viewing angle and the UCS plane are not compatible. The WCS plane is perpendicular to the viewing plane.

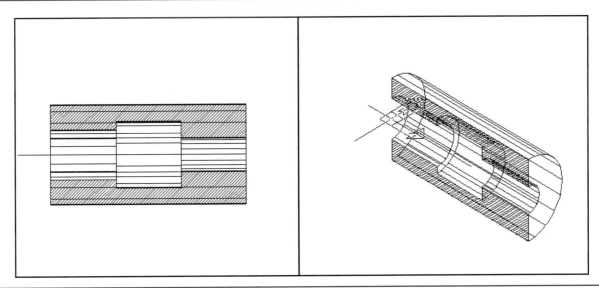

Figure 15-52

Select the Scale to Paper space option.

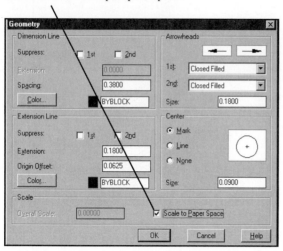

Figure 15-53

To add dimensions to the 2D sectional view

1. Double-click the Model icon at the bottom of the screen.

The word Model will be replaced with the word Paper; the Paper space icon, a right triangle, will appear in the lower left corner of the screen. The cursor is now free to move over the entire drawing screen, not just within a port as with Model space.

2. Type Dim in response to a command prompt.

Dim:

3. Type dimlfac; press Enter.

The DIMLFAC command is a systems variable that controls the global scale factor for linear dimensions.

Current value <1.0000> New value (Viewport):

4. Type V; press Enter.

Select viewport to scale:

5. Select the left vertical boundary line of the left port.

Select viewport to scale: DIMLFAC set to -2.5033

The DIMLFAC command uses absolute values, so the minus sign is not significant. The scale value will vary with the size of the viewports selected.

To calibrate the overall scale factor to the port

1. Select the Dimension Styles tool from the Dimensioning toolbar.

The Dimension Styles dialog box will appear.

2. Select the Geometry box.

The Geometry dialog box will appear.

3. Select the Scale to Paper Space option within the Scale box. See Figure 15-53

4. Use the Units command within the Annotation options and set the precision for two decimal places.

5. Return to the drawing screen.

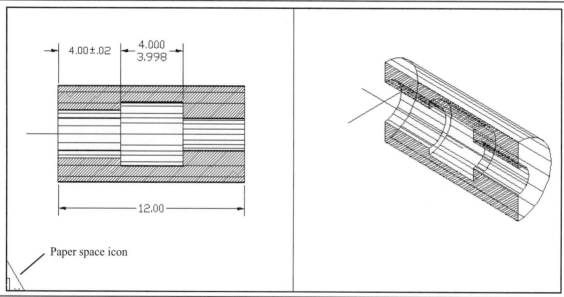

Figure 15-54

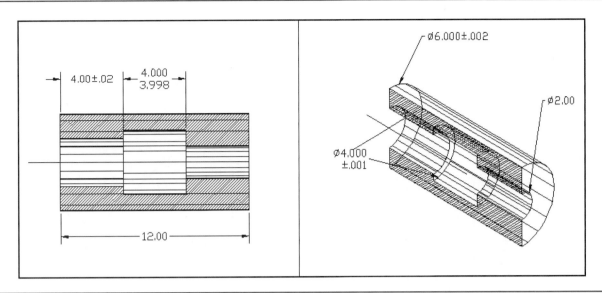

Figure 15-55

The left port is now ready for dimensions. Dimension may be created using Dimensioning toolbar tools as before.

6. Create the appropriate dimensions using the dimensioning conventions presented in Chapter 8.

Figure 15-57 shows the dimensioned port.

To dimension the 3D sectional view

1. Select the Leader tool from the Dimensioning toolbar.

   From point:

2. Select a point on the outside back surface of the 3D sectional view.

   To point:

3. Draw a short horizontal line segment; press Enter twice.

   Annotation (or RETURN for options):

4. Press Enter.

   Tolerance/Copy/Block/None/<Mtext>:

5. Press Enter.

   The Edit MText dialog box will appear.

6. Use the Symbol options to create the Ø and ± symbols.

7. Complete the dimension.

8. Change to Model space and erase the construction line aligned with the Y axis; change the X axis construction line to a center line.

   See Figure 15-55.

15-7 EXERCISE PROBLEMS

Redraw the objects shown in EX15-1 through EX 15-15 and add the appropriate dimensions. The dimensions need not be located in the same positions as shown.

EX15-1 INCHES

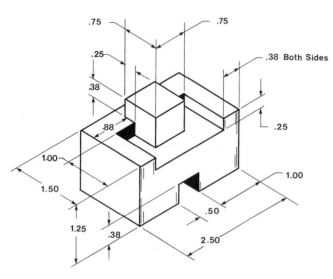

EX15-3 INCHES

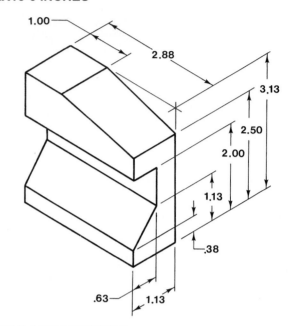

EX15-2 MILLIMETERS

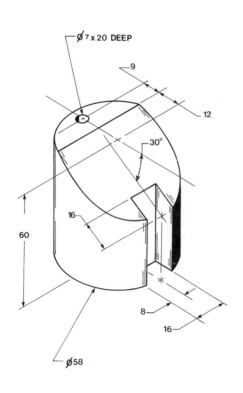

EX15-4 MILLIMETERS

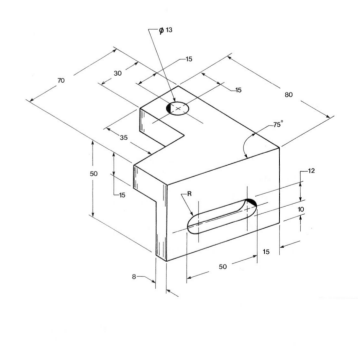

EX15-5 MILLIMETERS

EX15-7 MILLIMETERS

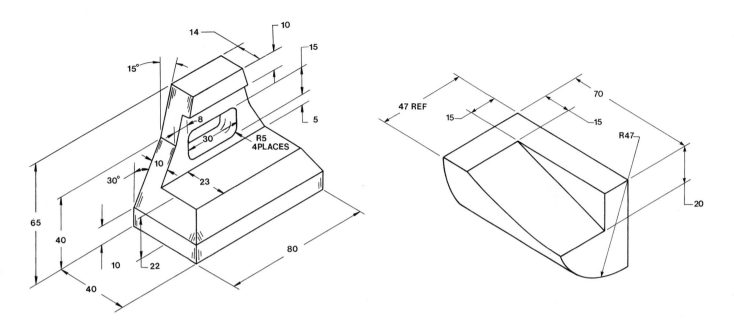

EX15-6 MILLIMETERS

EX15-8 MILLIMETERS

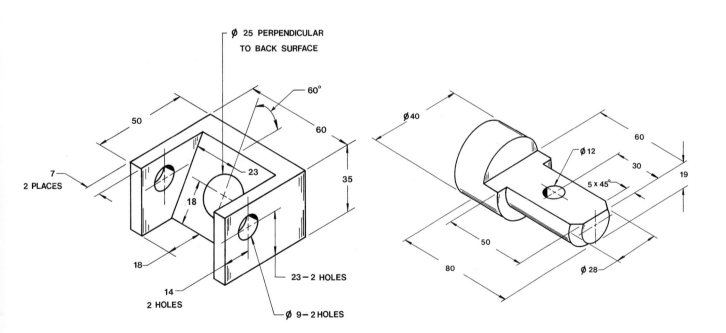

EX15-9 MILLIMETERS

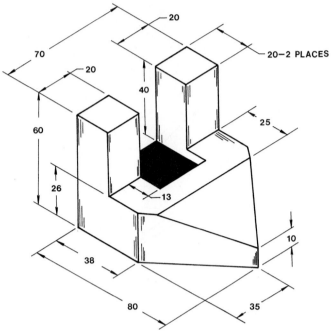

EX15-11 MILLIMETERS

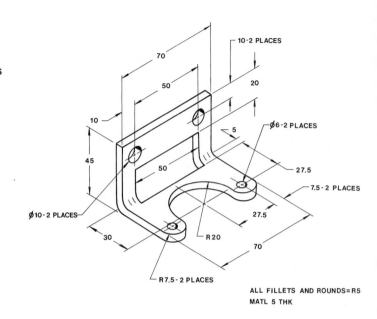

ALL FILLETS AND ROUNDS=R5
MATL 5 THK

EX15-10 MILLIMETERS

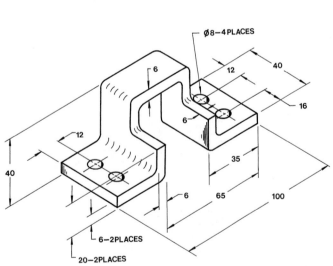

NOTE: ALL FILLET AND ROUNDS=R3

EX15-12 MILLIMETERS

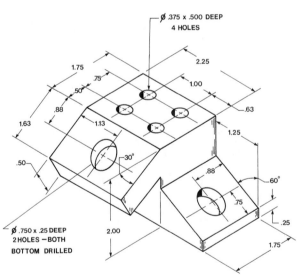

EX15-13 MILLIMETERS

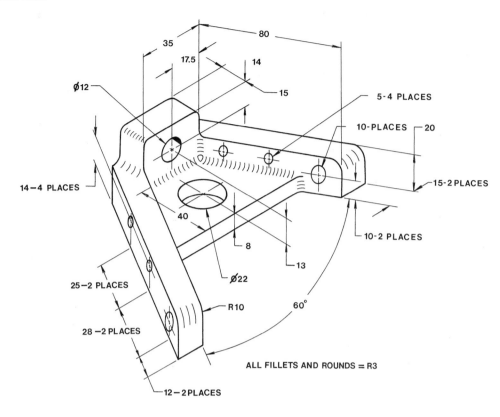

EX15-14 MILLIMETERS

EX15-15 MILLIMETERS

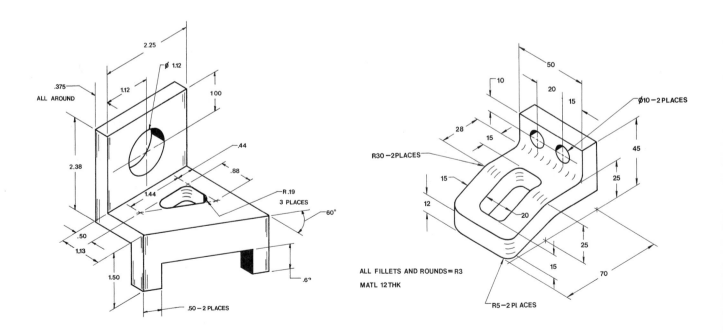

Redraw the objects shown in EX15-16 through EX15-19 as both 2D and 3D sectional views and add the appropriate dimensions.

EX15-16 MILLIMETERS

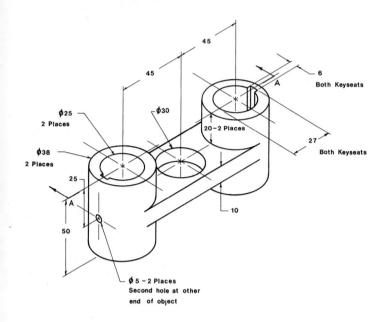

EX15-18 INCHES

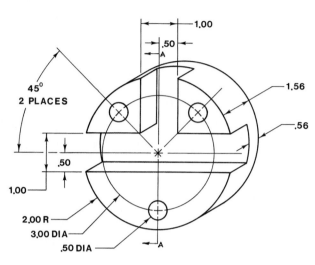

EX15-17 MILLIMETERS

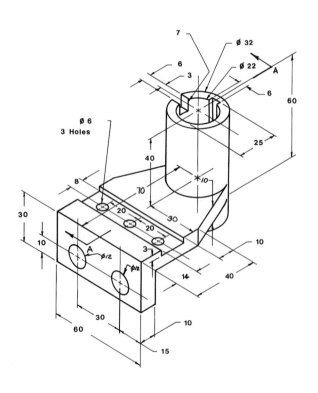

EX5-19 MILLIMETERS

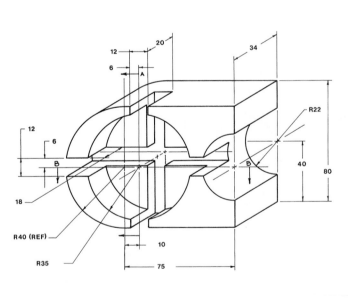

16

Design: Clips and Brackets

16-1 INTRODUCTION

The design of clips and brackets is often one of the first assignments given to new designers. Usually the problem is to create a piece that joins together two fixed locations. In this section, several examples of clip and bracket design will be presented, along with the concepts of bending allowances and several general rules associated with clips and brackets.

16-2 DESIGN EXAMPLE 16-1

Figure 16-1 shows four hole locations, two in the X,Y plane and two in the X,Z plane. The WCS was positioned so that it aligns with the two holes in the X,Y plane.

Distance between holes and the edge of an object

As a general rule, the edge of a thin object should be no closer to the edge of a hole than a distance equal to the diameter of the hole. In the example shown in Figure 16-1, the hole diameters are 1.500, so the distance from the edge of the hole to the edge of the object must be equal to or greater than 1.500. This is only a general rule and is subject to changes because of changes in the material thickness and the loading requirements.

Design a bracket that aligns with the four hole locations.

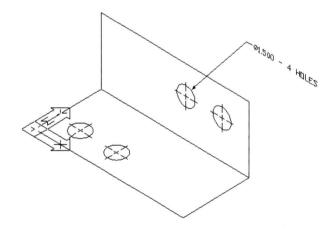

Figure 16-1

459

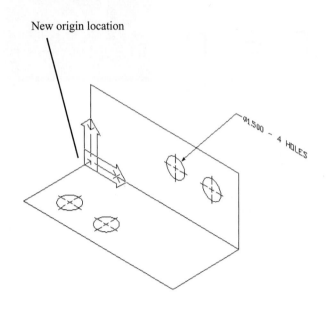

New origin location

Figure 16-2

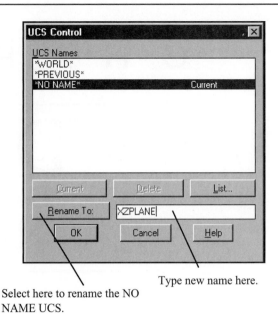

Select here to rename the NO NAME UCS.

Type new name here.

Figure 16-3

Assume the bracket is to be .125 thick and that the initial bend is 90°. The X,Y coordinates relative to the WCS are given in Figure 16-1.

To define and save a UCS

Define a new UCS for the X,Z plane and save it.

1. Select the Preset tool from the UCS toolbar.
2. Select the Front UCS.
3. Select the Origin tool from the UCS toolbar.

Origin point <0,0,0>:

4. Type 0,0,-6; press Enter.

The origin for the new UCS will be located as shown in Figure 16-2.

5. Select the Named UCS tool from the UCS toolbar.

The UCS COntrol dialog box will appear.

6. Highlight the NO NAME UCS that is the Current UCS.
7. Change the UCS name from NO NAME to XZPLANE, then select the Rename to box, then the OK box.

The UCS Control dialog box will appear with XZPLANE listed and indicated as the current UCS. See Figure 16-3.

To define the edges of the bracket

1. Select the World tool from the UCS toolbar and return to the WCS.
2. Use the Line tool from the Draw toolbar and construct perpendicular lines through the center points of the two circles in the X,Y plane as shown in Figure 16-4.

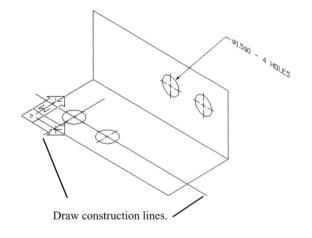

Draw construction lines.

Figure 16-4

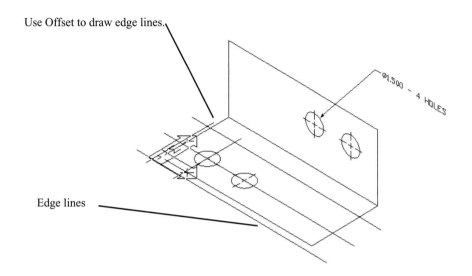

Use Offset to draw edge lines.

Ø1.500 - 4 HOLES

Edge lines

Figure 16-5

The length of the line should extend beyond the expected edges of the bracket. If the distance from the edge of the hole to the edge of the object is equal to the diameter of the hole, then the minimum distance between the hole centers and the edge of the object is (1.5)(hole diameter). In this example, the distance is (1.5)(1.500) = 2.25.

3. Select the Offset tool from the Modify toolbar.

   Offset distance or Through <Through>:

4. Type 2.25; press Enter.
5. Construct lines around the holes in the X,Y plane

as shown in Figure 16-5.
6. Select the Named UCS tool from the UCS toolbar and make the XZPLANE UCS the current UCS.
7. Add the construction lines through the holes' center points in the vertical plane.
8. Select the Offset tool from the Modify toolbar and add construction lines 2.25 from the holes' centers as shown in Figure 16-6.

The offset lines serve to define the outside edges of the bracket.

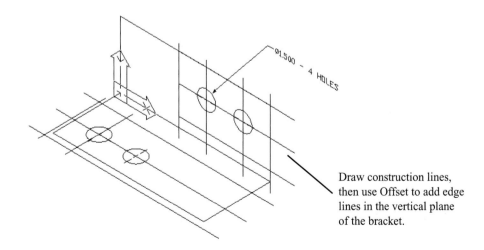

Ø1.500 - 4 HOLES

Draw construction lines, then use Offset to add edge lines in the vertical plane of the bracket.

Figure 16-6

Draw this line.

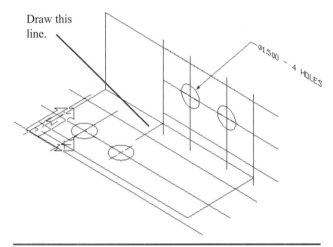

Figure 16-7

To complete the bracket shape

The edge distances around the holes have been defined, but the two individual planes must be connected.

1. Select the World tool from the UCS toolbar and return to the WCS.
2. Turn on the ORTHO command and use the LINE command to draw the lines shown in Figure 16-7.

Use the OSNAP, Intersect command to ensure accuracy.

3. Erase the lines used to outline the original X,Y and X,Z planes and any construction lines, then trim the excess lines.
4. Turn off Ortho.
5. If necessary, move the bracket so that the left corner is located on the origin of the WCS.

See Figure 16-8. The lines shown represent the minimum acceptable shape for the bracket.

To create a solid model

The solid model is created by changing the outline lines into a polyline, then extruding the polyline into a solid model.

1. Select the Edit Polyline tool from the Modify II toolbar.

Command: _pedit Select polyline:

2. Select one of the lines in the X,Y plane.

Object selected is not a polyline
Do you want to turn it into one?<Y>:

The final bracket shape.

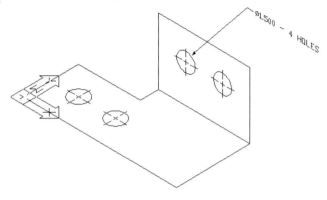

Figure 16-8

3. Type Y; press Enter.

Close/ Join/ Width/ Edit Vertex/ Fit/ Spline/ Decurve/ Ltype gen/ Undo/ eXit <x>:

4. Type J; press Enter.

Select objects:

5. Select the edge lines of the surface in the X,Y plane

See Figure 16-9.

Close/ Join/ Width/ Edit Vertex/ Fit/ Spline/ Decurve/ Ltype gen/ Undo/ eXit <x>:

6. Type X; press Enter.

The edge lines of the X,Y plane are now a single polyline.

Create a polyline.

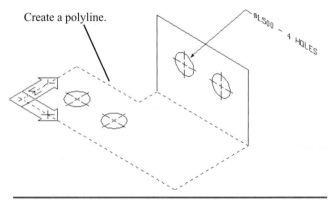

Figure 16-9

Create a solid model.

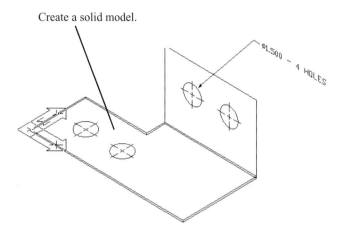

Figure 16-10

To extrude the X,Y surface

1. Select the Extrude tool from the Solids toolbar.

   Select objects:

2. Select the polyline created above.

   The entire edge of the surface in the X,Y plane should highlight.

   Path/<Height of Extrusion>:

3. Type -.125; press Enter.

   Extrusion taper angle <0>:

4. Press Enter.

   See Figure 16-10.

Draw this line, then join it to the other lines to form a polyline.

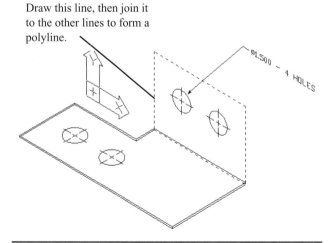

Figure 16-11

Create a solid model.

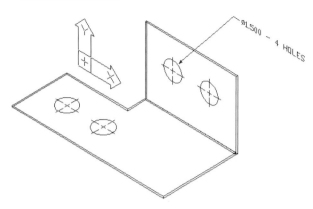

Figure 16-12

To extrude the X,Z surface

The line between the X,Y and X,Z planes was used as one of the segments in the X,Y polyline. A new line must be added before another polyline can be defined. A single line cannot be part of two different polylines.

1. Select the Line tool from the Draw toolbar and draw a new line directly over the existing bottom edge of the extruded surface.

   See Figure 16-11. Use the OSNAP, Endpoint command to ensure accuracy. The vertical lines in the plane must be extended to intersect with the new line so that a new polyline can be created.

2. Select the Extend tool from the Modify toolbar.

   Select boundary edges: (Projmode = UCS, Edgemode = No extend)
   Select objects:

3. Select the new line.

   <Select line to extend>/Project/Edge/Undo:

4. Select the two vertical lines in the X,Z plane.

   The lines will extend to form an enclosed rectangle.

5. Select the Edit Polyline tool from the Modify II toolbar and create a new polyline using the edge lines of the surface in the X,Z plane.

6. Select the Extrude tool from the Solids toolbar and extrude the X,Z surface to a height of -.125 and 0 degrees taper angle.

   See Figure 16-12.

7. Select the Union tool from the Modify II toolbar and union the two extruded surfaces.

Create the holes.

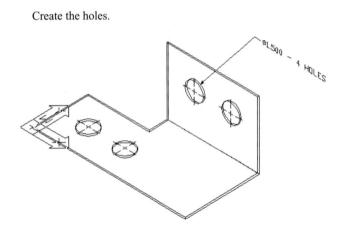

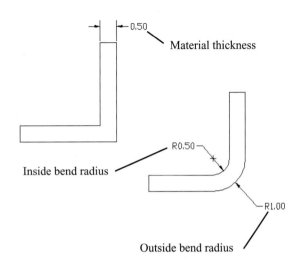

Figure 16-13

Figure 16-14

To create holes

Holes are created by subtracting solid cylinders from the bracket.

1. Select the Cylinder tool from the Solids toolbar.

    Elliptical/<center point>: <0,0,0>:

2. Select the center point of one of the circles.

    Diameter/<Radius>:

3. Type .75; press Enter.

    Center of other end/<Height>:

4. Type -.125; press Enter.
5. Repeat the procedure for the other three holes.

The holes in the X,Z plane can be properly accessed only by using the XZPLANE UCS. See Figure 16-13.

6. Select the Subtract tool from the Modify II toolbar.

    Command: _subtract Select solids and regions to subtract from...
    Select objects:

7. Select the bracket.

    Select solids and regions to subtract...
    Select objects:

8. Select the four holes, then press Enter.

There will be no visual changes in the clip, but the cylinders have been subtracted from the solid model, thus creating the holes. This may be verified by typing Hide, then pressing Enter and viewing the results. The drawing may be returned to its former settings by typing Regen, then pressing Enter.

16-3 BENDING

Objects made from thin materials are often bent into shape. A flat pattern is created and bent to form the final shape. The size of a bend may be defined using either an inside or outside bend radius. See Figure 16-14. The outside bend radius is equal to the inside bend radius plus the material thickness.

OBR = IBR + THK

The neutral axis is assumed to be in the center of the object. When bending occurs, the inside bend radius is compressed and the outside bend radius is stretched (is under tension). An actual bend radius is not perfectly symmetrical as shown on a drawing, but is slightly flattened. It is acceptable to draw symmetrical bend radii on drawings.

In the example created in Section 16-2, an inside bend radius of .25 is assigned.

Select here to create the inside bend radius.

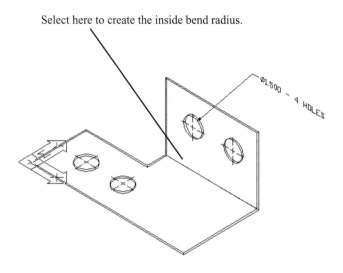

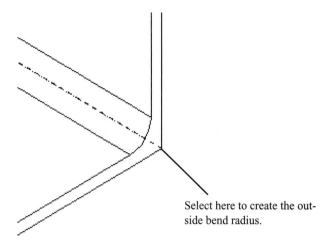

Select here to create the out-side bend radius.

Figure 16-15

Figure 16-17

To create an inside bend radius

1. Select the Fillet tool from the Modify toolbar.

   (Trim mode) Current fillet radius = 0.5000
   Polyline/Radius/Trim/<Select first object>:

2. Type R; press Enter.

   Enter fillet radius<0.5000>:

3. Type .25; press Enter.

   Command:

4. Press the right mouse button to repeat the Fillet command sequence.

(Trim mode) Current fillet radius = 0.2500
Polyline/Radius/Trim/<Select first object>:

5. Select the bending edge as shown in Figure 16-15.

   Enter radius <0.2500>:

6. Press the right mouse button to enter the default 0.2500 value.

   Chain/Radius/<Select edge>:

7. Press the right mouse button again.

The fillet will be added to the object. See Figure 16-16. The line filleted is the inside line of the corner. The outside line must also be filleted using a value equal to the inside bend radius plus the material thickness, or .250 + .125 = .375. This is the radius value for the outside bend radius.

To create an outside bend radius

1. Use the Zoom Window tool from the Standard toolbar and increase the size of the corner of the object as shown in Figure 16-17.

2. Select the Fillet tool from the Modify toolbar.

   (Trim mode) Current fillet radius = 0.2500
   Polyline/Radius/Trim/<Select first object>:

3. Type R; press Enter.

Fillet

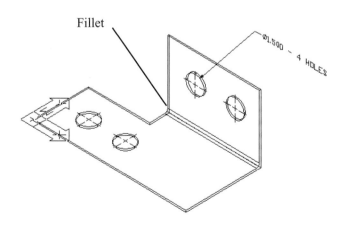

Figure 16-16

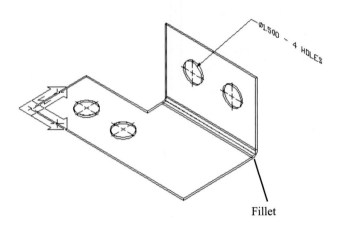

Fillet

Figure 16-18

Enter radius <0.2500>:

4. Type .375; press Enter.

   Command:

5. Press the right mouse button.

   (Trim mode) Current fillet radius = 0.375
   Polyline/Radius/Trim/<Select first object>:

6. Select the back corner of the object as shown in Figure 16-17.

   Enter radius <0.375>:

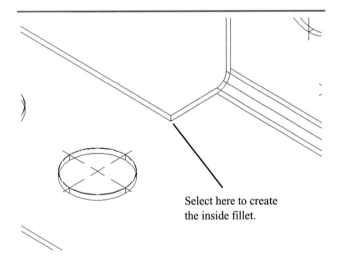

Select here to create the inside fillet.

Figure 16-19

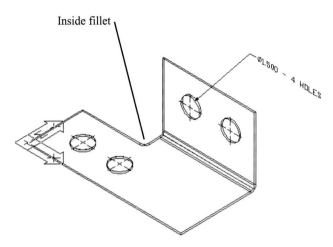

Inside fillet

Figure 16-20

7. Press the right mouse button.

   Chain/Radius/<Select edge>:

8. Press the right mouse button again.

   The outside bend radius will be formed.

9. Select the Zoom Previous command.

   Figure 16-18 shows the resulting object.

To add internal fillets

It will be slightly easier to manufacture the object if the internal 90 degree corner is rounded. For this example, a radius of .375 is selected.

1. Use the Zoom Window tool from the Standard toolbar and zoom the internal corner as shown in Figure 16-19.
2. Use the Fillet tool from the Modify toolbar and add a fillet of radius .375, then use Zoom Previous to create the figure shown in Figure 16-20.

Create the fillet using the procedure described for creating the bend radius.

To dimension a bend radius

In the example shown, the inside bend radius and material thickness are specified, so the outside bend radius need not be defined. The linear distance from the bend

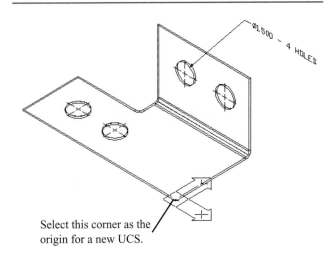

Select this corner as the origin for a new UCS.

Figure 16-21

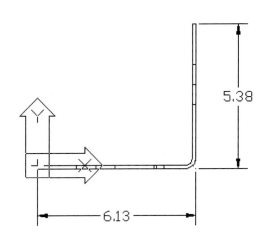

Figure 16-22

radius to the edge of the object is defined from a theoretical corner based on the inside surfaces. The dimensions are created by moving the dimensioning plane origin to the right edge of the object, then adding dimensions as explained in Chapter 15.

To create dimensions on a rounded corner

1. Select the Origin tool from the UCS toolbar.

   Origin point <0,0,0>:

2. Use the OSNAP, Endpoint command to select the lower right corner of the object as shown in Figure 16-21.

3. Select Preset from the UCS toolbar, then the Right UCS.

4. Select the View pull-down menu, then 3D Viewpoint, then Plan view, and Current UCS.

5. Use the Pan tool from the Standard toolbar to move the object to the center of the drawing screen.

6. Select the Linear tool from the Dimensioning toolbar and add the linear dimensions shown.

   See Figure 16-22.

7. Select the SE Isometric tool from the View toolbar.

8. Select the World tool from the UCS toolbar.

9. Dimension the object.

   See Figure 16-23.

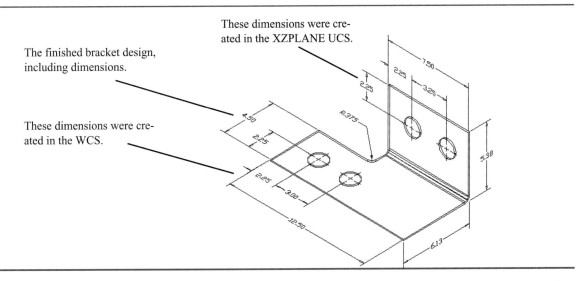

These dimensions were created in the XZPLANE UCS.

The finished bracket design, including dimensions.

These dimensions were created in the WCS.

Figure 16-23

Specify the text height here.

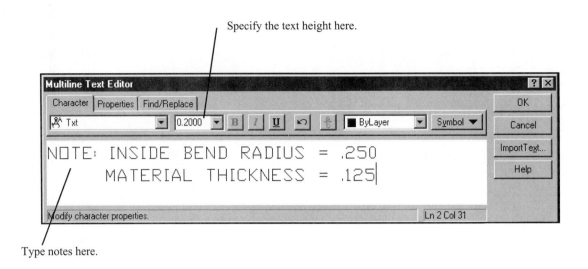

Figure 16-24

To add notes

In this example, the material thickness and inside bend radius were defined using drawing notes. They are created by changing to paper space as follows.

1. Select the View pull-down menu, then paper Space.

The bracket will dissapear from the screen and the paper space icon will appear in the lower left corner of the drawing sreen.

2. Select the Multiline Text tool from the Draw toolbar.

Specify first corner;

3. Define an area in the upper portion of the drawing screen.

The Multiline Text Editor dialog box will appear. See Figure 16-24.

4. Type the required note as shown; select OK

The text will appear on the drawing screen.

5. Select the View pull-down menu, then Model Space (Floating).

ON/OFF/Hideplot/Fit/2/3/4/Restore/<First Point>:

6. Type F; press Enter.

The dimension height must be reset for each text input. As soon as the text box is clicked, the numerical value will return to its default value, but the line of text will appear at the designated value. Figure 16-25 shows the drawing with its notes.

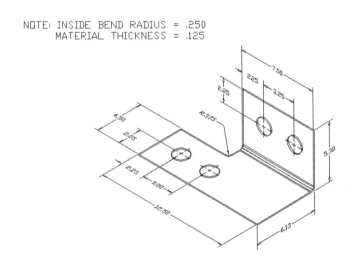

Figure 16-25

NOTE: INSIDE BEND RADIUS = .250
      MATERIAL THICKNESS = .125

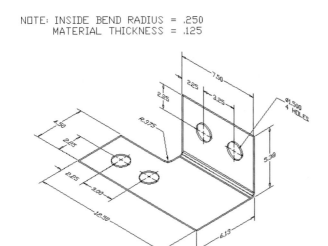

Figure 16-26

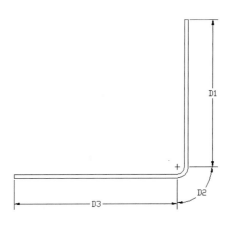

Figure 16-27

To dimension the holes

1. If necessary, Double-click the word Paper at the bottom of the screen to return to Model space.
2. Use the Named tool from the UCS toolbar to return to the XZPLANE UCS.
3. Select the Leader tool from the Dimensioning toolbar and add the hole dimension.

Figure 16-26 shows the finished dimensioned drawing.

16-4 FLAT PATTERNS

Most thin parts are fabricated by first cutting a flat pattern and then bending the pattern into its final shape. As metal bends, it stretches, meaning the length of the flat pattern must be less than the overall length of the finished object.

The correct size of a flat pattern is determined by considering the overall size requirements and the amount of material for bend allowance. Bend allowance is the amount of material needed to form bends and includes a consideration of how material will stretch during bending. Bend allowances and the lengths of straight sections are added together to determine the final length of a flat pattern.

To determine the length of a flat pattern

Figure 16-27 shows a profile of the bracket designed in Sections 16-2 and 16-3. The flat pattern length is determined by adding the three designated lengths, the two straight sections and the bent section.

Flat Pattern Length = D1 + D2 + D3

To determine the length of D1

Length D1 is found by taking the overall dimension value for the section and subtracting the inside bend radius and the material thickness. See Figure 16-28.

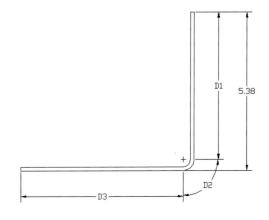

Figure 16-28

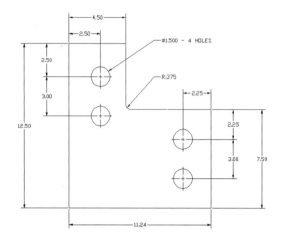

Figure 16-29

D1 = 5.38 - .125 -.250
D1 = 5.005

The length will be rounded off so that D1 = 5.00.

To determine the length of D3

D3 = 6.13 - .125 -.250
D3 = 5.755

This length will also be rounded off so that D3 = 5.75.

To determine the flat pattern length of the bent corner D2

The distance D2 is calculated using the formula taken from the ASTME Die Design Handbook.

D2 = 2p(A/360)(IR + Kt)

where

B = Bend allowance
A = Bend angle
IR = Inside bend radius
 t = Material thickness
K = Constant

The constant K is equal to 0.33 when the inside bend radius is less than 2t (two times the material thickness) and is equal to 0.50 when the inside bend radius is equal to or greater than 2t. In this example, the inside bend radius equals .25 or 2t so the constant K will have a value of 0.50. The other values are as follows.

A = 90d
IR = .250
 t = .125

Therefore

D3 = 2p(90/360)[.25 + (.50)(.125)]
    = 2p(.25)(.313)
    = .49

So

Flat pattern length = D1 + D2 + D3
        = 5.00 + .49 + 5.75
        = 11.24

Figure 16-29 shows a dimensioned flat pattern for the object.

Other Possibilities

Design problems often have more than one solution. Figure 16-30 shows another solution to the problem originally defined in Section 16-2. There are many other possible solutions.

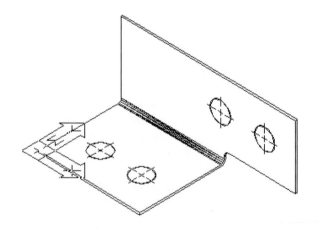

Figure 16-30

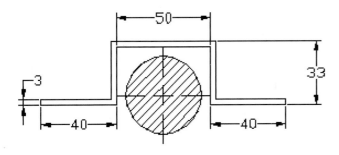

Figure 16-31

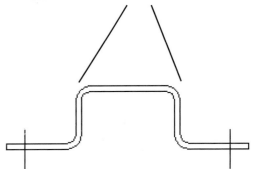

Use Fillet to add the outside bend radii.

Figure 16-33

16-5 DESIGN EXAMPLE 16-2

In this example, a clip will be created from a given profile. Figure 16-31 shows the approximate profile of a clip design to clear a circular object. The material is 3 millimeters thick and all inside bend radii are to be 5 millimeters. There are to be four 8 millimeter diameter holes, two on each side located at least 16 millimeters apart with edges no closer than 10 millimeters from the edge of the clip.

To create the bend radii

1. Select the Fillet tool from the Modify toolbar.

   Polyline/Radius/Trim/<Select first object>:

2. Type R; press Enter.

   Enter fillet radius <0.5000>:

3. Type 5; press Enter.
4. Add the appropriate inside bend radii.

   See Figure 16-32.

5. Select the Fillet tool from the Modify toolbar, set the radius for 8 (inside bend radius plus the material thickness), and add the outside bend radii.

   See Figure 16-33.

To determine the depth of the clip

The parameters of the problem require that the four 8 millimeter diameter holes be at least 16 millimeters apart. The distance from the hole's edge to its center point is 4 millimeters so the minimum distance between center points is $16 + 4 + 4 = 24$ millimeters. No maximum distance was specified in this example; the distance between center points was defined as 25 millimeters. See Figure 16-34.

The distance between the edge of the hole and the edge of the object must be at least equivalent to the diameter of the hole (8 millimeters). This means that the distance

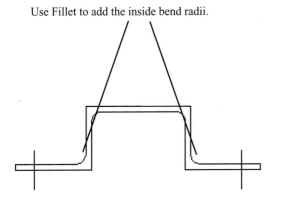

Use Fillet to add the inside bend radii.

Figure 16-32

The overall length of the clip = 55.

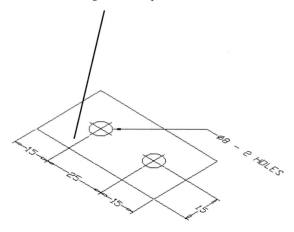

Figure 16-34

from the holes' centerpoints to the edge of the object must be at least 8 + 4 = 12 millimeters. In this example, a distance of 15 was assigned.

The total length of the object is then 15 + 25 + 15 = 55 millimeters.

To create a polyline

See Figure 16-35. If drawn, erase the lines that represent the hole locations.

1. Select the Edit Polyline tool from the Modify II toolbar.

   Command: _pedit Select polyline:

2. Select any line in the profile.

   Object selected is not a polyline
   Do you want to turn it into one? <Y>:

3. Press Enter.

   Close/ Join/ Width/ Edit vertex/ Fit/ Spline/ Decurve/ Ltupe gen/ Undo/ eXit/ <X>:

4. Type J; press Enter.

   Select objects:

5. Window the entire object.

   Close/ Join/ Width/ Edit vertex/ Fit/ Spline/ Decurve/ Ltupe gen/ Undo/ eXit/ <X>:

6. Press Enter.

Use the Edit Polyline tool to create a polyline for extrusion.

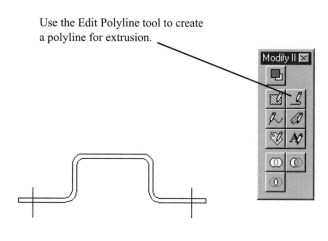

Figure 16-35

To create a solid model

1. Select the Extrude tool from the Solids toolbar.

   Select objects:

2. Select the polyline.

   Path/<Height of Extrusion>:

3. Type -55; press Enter.

   Extrusion taper angle <0>:

4. Press Enter.
5. Select the SE Isometric tool from the View toolbar.

   See Figure 16-36.

The polyline extruded to form a solid model

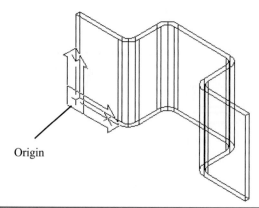

Origin

Figure 16-36

The Plan view of the Front preset UCS

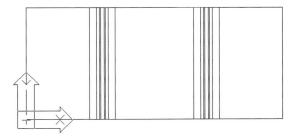

Figure 16-37

The holes are created using the Center Cylinder tool from the Solids toolbar.

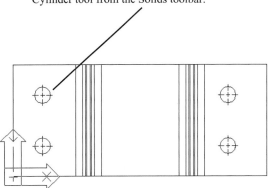

Figure 16-38

To add the holes

1. Select the Preset UCS tool from the UCS toolbar, then Front UCS.
2. Select the Origin tool from the UCS toolbar.

   Origin point <0,0,0>:

3. Select the corner of the bracket as shown in Figure 16-36.

   Use OSNAP, Endpoint to ensure accuracy.

4. Select the View pull-down menu, then 3D Viewpoint, then Plan View, and Current UCS.
5. Type Zoom, then define a scale factor of 1.00; use the Zoom Window tool from the Standard toolbar to size the object on the screen.

   See Figure 16-37.

6. Select the Center Cylinder tool from the UCS toolbar.

   Elliptical/<center point>: <0,0,0>:

7. Locate the center point of the hole at X = 15, Y = 15, and Z = 0.

   See Figure 16-38. The Snap spacing is set for 5 millimeters so the center point for this and the other three holes is located on snap points.

   Diameter/<Radius>:

8. Type 4; press Enter.

   Center of other end/<Height>:

9. Type -3; press Enter.

   To verify the minus sign, apply the right-hand rule to the current X,Y,Z axis system. See Figure 16-38.

10. Use the Copy tool from the Modify toolbar and create the other holes using the given dimensions.

To present the clip in 3D

1. Select the SE Isometric tool from the View toolbar.

   See Figure 16-39.

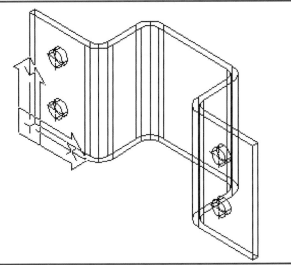

Figure 16-39

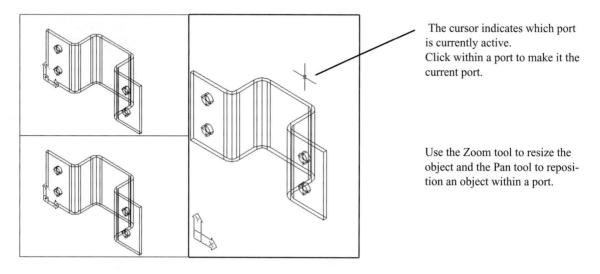

The cursor indicates which port is currently active.
Click within a port to make it the current port.

Use the Zoom tool to resize the object and the Pan tool to reposition an object within a port.

Figure 16-40

2. Select the View pull-down menu, then Model Space (Floating).

ON/OFF/Hideplot/Fit/2/3/4/Restore/<First Point>:

3. Type 3; press Enter.

Horizontal/Vertical/Above/Below/Left/<Right>:

4. Press Enter, accepting the default Right option.

Fit/<First Point>:

5. Select a point in the upper left corner of the drawing screen.

Second point:

6. Select a point in the lower right corner of the drawing screen.

To position the 3D view

See Figure 16-40. The right port on the screen is the active port from which to work.

1. Type Zoom; press Enter.

All/Center/Dynamic/Extents/Left/Previous/Vmax/Window/<Scale(X/XP)>:

2. Type 2; press Enter.

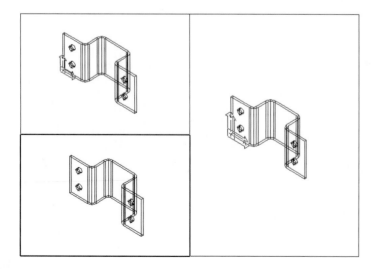

The Zoom and Pan tools were applied to all three ports.

Figure 16-41

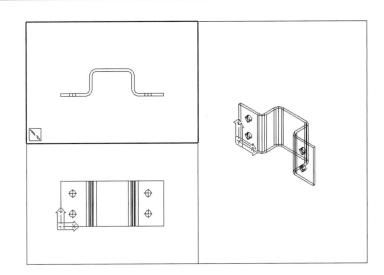

Two orthographic
views of the object

Figure 16-42

3. Use the Pan tool on the Standard toolbar to center the object within the port.

See Figure 16-41.

To create orthographic views

1. Activate the lower left port by moving the cursor arrow into the port and pressing the left mouse button.

The cross hairs will appear in the port signifying it is the current port.

2. Select the View pull-down menu, then 3D Viewpoint Presets, then Plan view, and Current.

3. Type Zoom and enter a zoom factor of 2.
4. Make the top left view the current port.
5. Select the Top View tool from the View toolbar.
6. Type Zoom and enter a zoom factor of 2.

See Figure 16-42.

To add dimensions

1. Change to paper space by double-clicking the word Model at the bottom of the screen.
2. Type Dim; press Enter.

Dim:

3. Type dimlfac; press Enter.

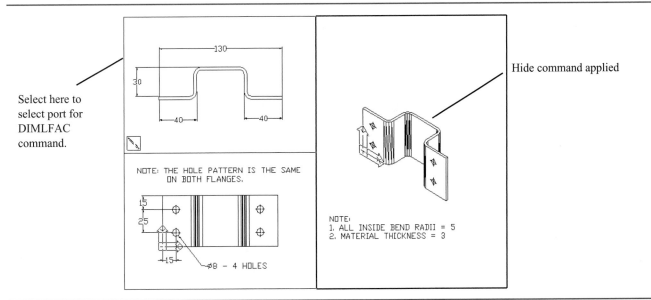

Select here to
select port for
DIMLFAC
command.

Hide command applied

Figure 16-43

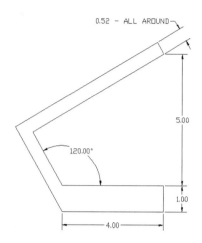

Figure 16-44

Current value <1.0000> New Value (Viewport):

4. Type V; press Enter.

Select viewport to set scale:

5. Select the left horizontal line of the viewports.
6. Use the Dimensioning toolbar to add dimensions to the drawing.
7. Apply the Hide command to the 3D port.

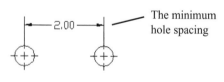

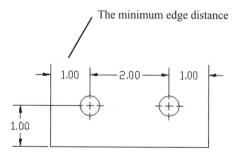

Figure 16-45

In Figure 16-43, the HIDE command was applied to the 3D port. HIDE can be applied only in Model space. After HIDE is applied, the drawing changes back to Paper space.

16-6 DESIGN EXAMPLE 16-3

Figure 16-44 shows a preliminary concept sketch for a heavy-duty bracket. The design requires that two Ø500±.001 holes be added to the slanted top surface, that these holes be no closer together than 2.00 inches, and that they be no nearer to an edge than .75.

Other design requirements specify that inside bend radii equal = 1.00, and that a dovetailed slot .25 high and 1.00 wide at its opening be added to the base.

To determine the length

The length of the bracket is determined from the requirements for the two holes in the slanted top surface.

1. Draw the two Ø.500 holes 2.00 apart as specified in the design requirements.

The edges of the object can be no closer than .75. This means that the distance from the center point of the holes to the nearest edge can be no less than 1.00 (.75 + .25).

2. Draw lines 1.00 from the center points of the two holes.

See Figure 16-45.

To add the bend radii

1. Select the Fillet tool from the Modify toolbar.

   Polyline/Radius/Trim/<Select first object>:

2. Type R; press Enter.

   Enter fillet radius <0.5000>:

3. Type 1; press Enter.

   Command:

4. Press Enter.

   Polyline/Radius/Trim/<Select first object>:

5. Select the lines on which to create the two inside bend radii.
6. Select the Fillet tool and add two 1.52 outside bend radii.

The object with bend radii applied

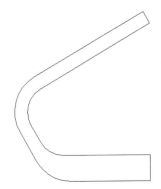

Figure 16-46

See Figure 16-46. The 1.52 value was derived from the inside radius (1.00) plus the thickness (.52). The 1.52 was also applied to the transition between the 1.00 lower surface and the .52 back surface.

To create a polyline

1. Select the Edit Polyline tool from the Modify II toolbar.

   Command: _pedit Select polyline

2. Select any line in the bracket.

   Object is not a polyline
   Do you want to turn it into one? <Y>:

3. Press Enter.

   Close/ Join/ Width/ Edit vertex/ Fit/ Spline/ Decurve/ Ltype gen/ Undo/eXit <X>:

4. Type J; press Enter.

   Select objects:

5. Window the object, then press Enter.

   Close/ Join/ Width/ Edit vertex/ Fit/ Spline/ Decurve/ Ltype gen/ Undo/ eXit <X>:

6. Press Enter.

   The object is now a polyline.

The object as a solid model in the WCS

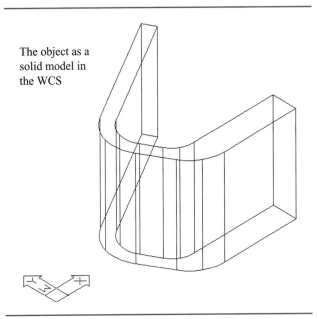

Figure 16-47

To create a solid model

1. Select the Extrude tool from the Solids toolbar.

   Select objects:

2. Select the object.

   Path/<Height of Extrusion>:

3. Type -4; press Enter.

   The minus sign is needed because of the current object orientation. Apply the right-hand rule to verify the Z direction.

   Extrusion taper angle <0>:

4. Press Enter.
5. Select the SW Isometric tool from the View toolbar.

   See Figure 16-47.

To create the needed UCSs

Two UCSs must be created to complete the design, one for the dovetail and one for the two holes in thes slanted top surface. First create the UCSs, then save them for future access.

1. Select 3 Point from the UCS toolbar.

   Origin/Zaxis/3point/OBject/View/X/Y/Z/Prev/Restore/Save/Del/?/<World>/: _3
   Origin Point <0,0,0>:

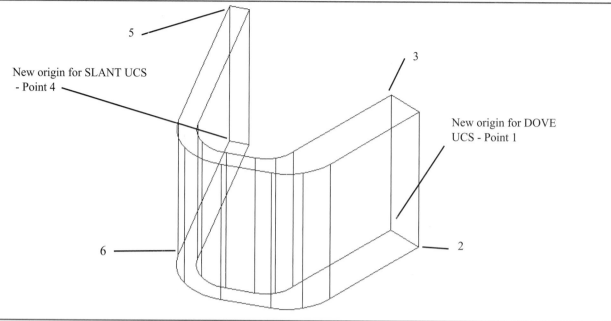

Figure 16-48

2. Use the OSNAP, Endpoint command to select the new origin point labeled 1 in Figure 16-48.

Point on the positive portion of the X-axis <11.5000,5.0000,-4.000>:

3. Select the point labeled 2 in Figure 16-48.

Point on positive-Y portion of the UCS X,Y plane <10.5000,6.0000,-4.0000>:

4. Select the point labeled 3 in Figure 16-48.
5. Select the Named UCS tool from the UCS toolbar and save the current UCS as DOVE.
6. Select the World tool from the UCS toolbar.
7. Create another UCS using the 3 Point tool.

Use OSNAP, Endpoint to ensure accuracy. Locate the origin at the point labeled 4 in Figure 16-48, the second point at the point labeled 5, and the third point at the point labeled 6.

8. Save the UCS as SLANT.
9. Return to the World UCS.

To add the dovetail

1. Select the Named tool from the UCS toolbar.

The UCS Control dialog box will appear. See Figure 16-49.

Figure 16-49

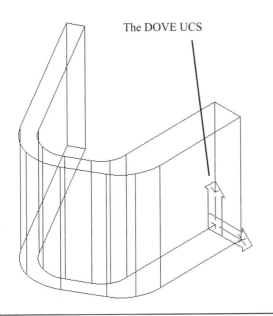

The DOVE UCS

Figure 16-50

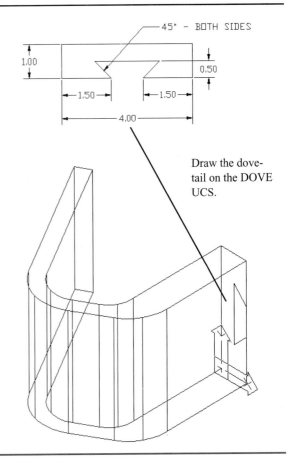

45° - BOTH SIDES

1.00

0.50

1.50

1.50

4.00

Draw the dovetail on the DOVE UCS.

Figure 16-51

2. Select the View pull-down menu, then Display, UCS Icon; select both the On and Origin options.

The UCS tool will appear at the origin of the DOVE UCS. See Figure 16-50.

3. Select the Polyline tool from the Draw toolbar and add the dovetail shape to the object as shown in Figure 16-51.

Figure 16-51 also shows the dimensions for the dovetail shape. The design requirements specified a 1.00 opening for the dovetail and a .25 height. The sides of a dovetail are always 45 degrees. The dovetail was positioned in the center of the object.

4. Select the Extrude tool from the Solids toolbar.

Select objects:

5. Select the dovetail.

Path/<Height of Extrusion>:

6. Type 6; press Enter.

The dovetail is now a solid model. The value 6 was selected because the exact length of the object is known to be somewhat greater than 4, so a value of -6 will extend well beyond the end of the object.

Extrusion taper angle <0>:

7. Press Enter.

See Figure 16-52.

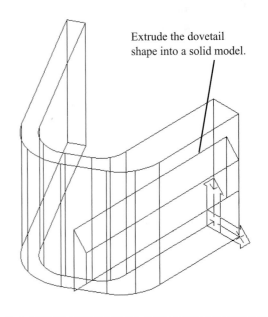

Extrude the dovetail shape into a solid model.

Figure 16-52

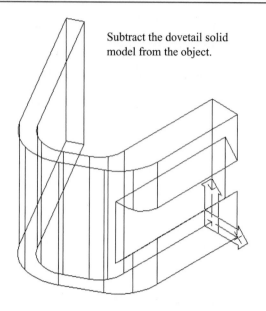

Subtract the dovetail solid model from the object.

Figure 16-53

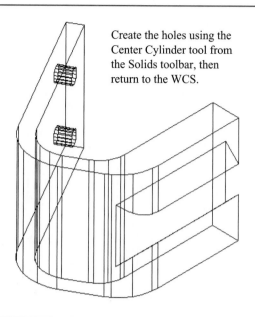

Create the holes using the Center Cylinder tool from the Solids toolbar, then return to the WCS.

Figure 16-55

8. Select the Subtract tool from the Modify II toolbar and subtract the dovetail.

See Figure 16-53.

9. Return to the WCS.

To add the holes in the slanted surface

1. Select the Named tool from the UCS toolbar and make the SLANT UCS the current UCS.

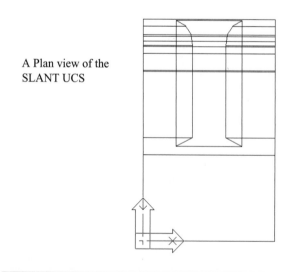

A Plan view of the SLANT UCS

Figure 16-54

2. Select the View pull-down menu, then 3D Viewpoint, then Plan View, and Current UCS.

Use the Pan tool from the Standard toolbar, if needed, to position the object near the center of the screen. See Figure 16-45.

3. Use the Center Cylinder tool from the Solids toolbar and draw the two holes.

See Figure 16-55. The dimensions were derived from the design requirements and the length calculations. The height of the two holes is .50.

4. Use the Subtract tool from the Modify II toolbar and subtract the cylinders from the object.
5. Select the SW Isometric tool from the View toolbar.
6. Select the WCS tool and return to the WCS.

See Figure 16-55.

To reposition the object

1. Type Rotate3D; press Enter.

   Select objects:

2. Select the object.

   Axis by Object/ Last/ View/ Xaxis/ Yaxis/ Zaxis/ <2points>:

3. Type x; press Enter.

The object rotated
about the WCS X
axis, then rotated
90° in the XY
plane.

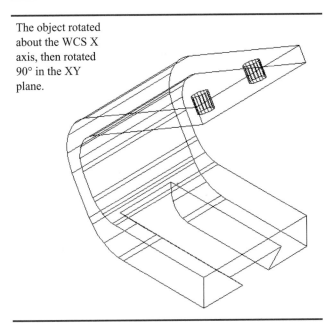

Figure 16-56

Point on X <0,0,0>:
4. Press Enter

<Rotation angle>/Reference:

4. Type 90; press Enter.
5. Repeat the proceedure rotating the object -90° about the Z axis,
6. Select the Pan tool from the Standard toolbar and locate the object near the center of the screen.

The HIDE com-
mand applied

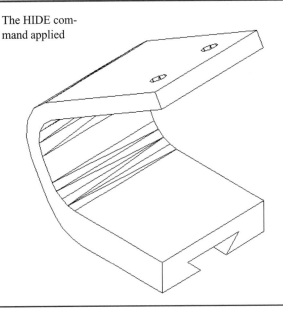

Figure 16-57

See Figure 16-56. Other orientations may be created by first changing the UCS and then rotating the object about the UCS's Z axis.

7. Select the Hide tool from the Render toolbar.

See Figure 16-57.

8. Type Regen to return to the original drawing.

If desired, multiple ports may be created and dimensions added as previously explained.

16-7 EXERCISE PROBLEMS

Use two viewing ports and create two views of each of the objects shown in EX16-1 through EX16-4. One port should show a solid model of the object and the other port a dimensioned flat pattern for the object.

EX16-1 MILLIMETERS

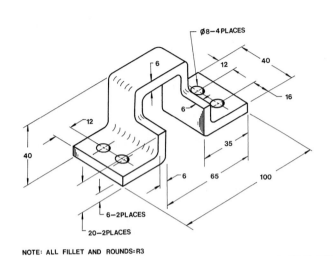

NOTE: ALL FILLET AND ROUNDS=R3

EX16-3 MILLIMETERS

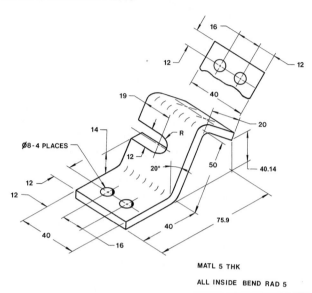

MATL 5 THK

ALL INSIDE BEND RAD 5

EX16-2 MILLIMETERS

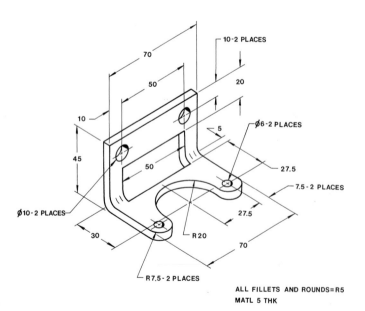

ALL FILLETS AND ROUNDS=R5
MATL 5 THK

EX16-4 MILLIMETERS

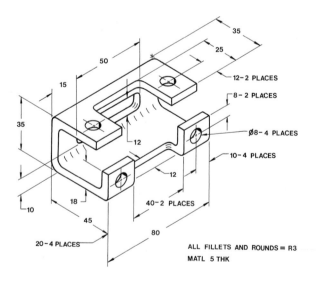

ALL FILLETS AND ROUNDS = R3
MATL 5 THK

Design clips based on the shape definitions presented in EX16-5 through EX16-10 and subject to the following parameters.

1. All flanges and sections marked with short perpendicular lines contain two centered holes.
2. The holes' center points can be no closer than two times the holes' diameters.
3. The distance from the center point of a hole and an edge can be no less than 1.5 times the diameter of the hole.

Preset each object using four viewing ports as follows.

Port 1 - A solid model of the object presented using the SE Isometric preset viewpoint
Port 2 - A dimensioned front orthographic (profile) view of the object
Port 3 - A dimensioned top orthographic view
Port 4 - A dimensioned flat pattern of the object

EX16-5 INCHES

EX16-6 MILLIMETERS

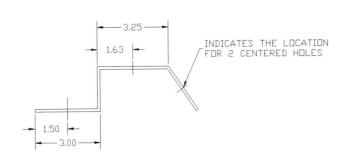

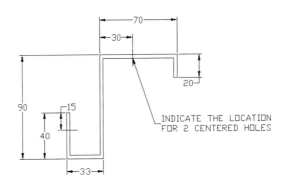

MATL = .13 THICK

ALL HOLES = Ø.50

ALL INSIDE BEND RADII = .25

MATL = 3 THICK

ALL HOLES = Ø10

ALL INSIDE BEND RADII = 10

EX16-7 MILLIMETERS

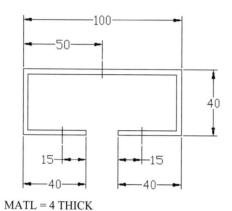

MATL = 4 THICK

ALL HOLES = Ø12

ALL INSIDE BEND RADII = 10

EX16-9 MILLIMETERS

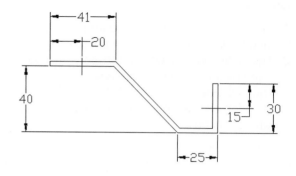

MATL = 3 THICK

ALL HOLES = Ø14

ALL INSIDE BEND RADII = 6

EX16-8 MILLIMETERS

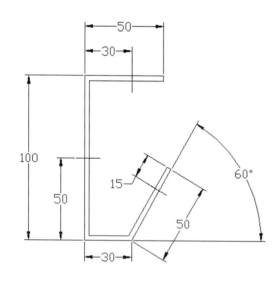

MATL = 3 THICK

ALL HOLES = Ø15

ALL INSIDE BEND RADII = 8

EX16-10 INCHES

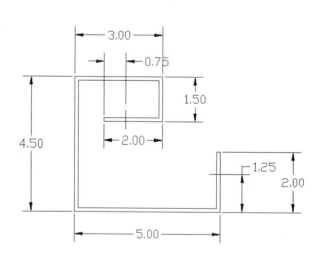

MATL = .125

ALL HOLES = Ø.375

ALL INSIDE BEND RADII = .185

Design brackets to align with the hole patterns defined in EX16-11 through EX16-14.

EX16-11 INCHES

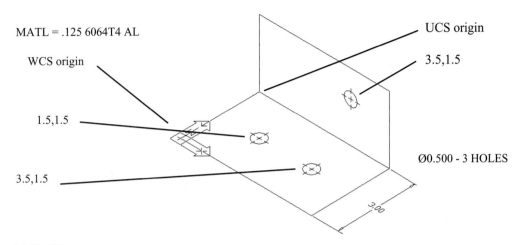

MATL = .125 6064T4 AL

WCS origin

1.5,1.5

3.5,1.5

UCS origin

3.5,1.5

Ø0.500 - 3 HOLES

3.00

EX16-12 INCHES

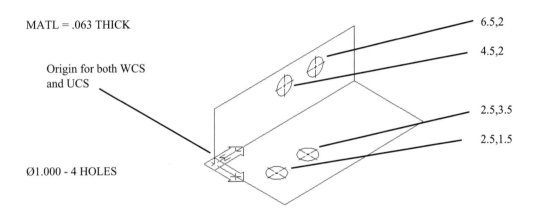

MATL = .063 THICK

Origin for both WCS and UCS

Ø1.000 - 4 HOLES

6.5,2

4.5,2

2.5,3.5

2.5,1.5

EX16-13 MILLIMETERS

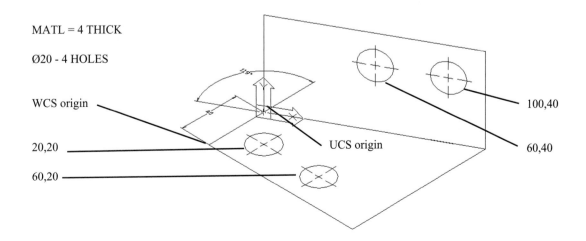

MATL = 4 THICK

Ø20 - 4 HOLES

WCS origin

20,20

60,20

UCS origin

100,40

60,40

EX16-14 MILLIMETERS

MATL = 4 THICK

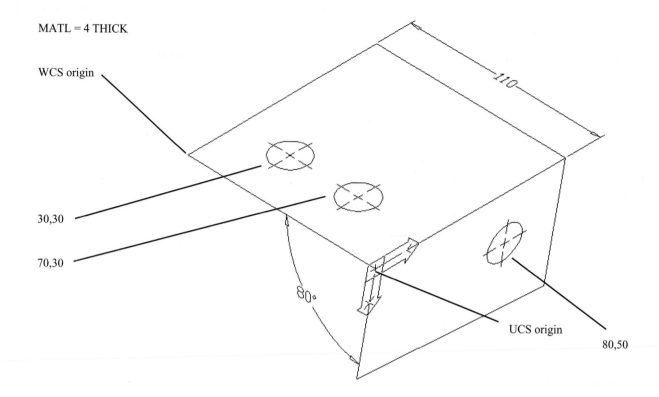

WCS origin

30,30

70,30

80°

110

UCS origin

80,50

CHAPTER 17

Design: Using Standard Fasteners

17-1 INTRODUCTION

This chapter explains how to draw representations of fasteners in 3D, how to size fasteners and their appropriate clearance holes, and how to locate fastener holes using geometric tolerances.

The chapter will also explain how to create and save Blocks and WBlocks of various thread and head shapes to create a fastener library for use on future drawings.

17-2 THREAD TERMINOLOGY

The peak of a thread is called the crest and the valley portion is called the root. See Figure 17-1. The major diameter of a thread is the distance across the thread from crest to crest. The minor diameter is the distance across the thread from root to root.

The pitch of a thread is the linear distance along the thread from crest to crest. Thread pitch is usually referenced in terms of a unit length, such as 20 threads per inch or 1.5 threads per millimeter. Pitch sizes are classified as coarse, fine, or extra fine, among other sizes, and are manufactured to standard interchangeable sizes and tolerances.

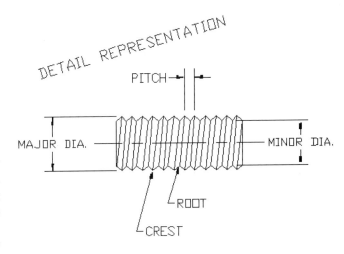

Figure 17-1

A metric thread of major diameter 10 millimeters and a length of 30 millimeters. The thread is assumed to be coarse because no pitch specification is given.

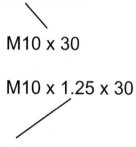

A metric thread with a thread specification. The value 1.25 means 1.25 threads per millimeter.

Figure 17-2

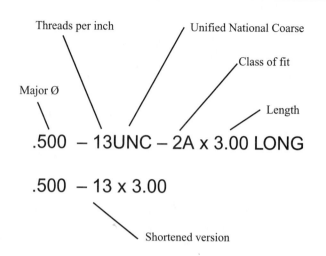

Figure 17-4

17-3 THREAD CALLOUTS (METRIC UNITS)

Threads are specified on a drawing using callouts. See Figure 17-2. The M at the front of a drawing callout specifies that the callout is for a metric thread. Holes that are not threaded use the Ø symbol.

The number adjacent to the M is the major diameter of the thread. An M10 thread has a major diameter of 10 millimeters. The pitch of a metric thread is assumed to be coarse unless otherwise stated. The callout M10 x 30 assumes a coarse thread, or 1.5 threads per millimeter. The number 30 is the thread length in millimeters. The "x" is read as "by," so the thread is called a "ten by thirty."

The callout M10 x 1.25 x 30 specifies a pitch of 1.25 threads per millimeter. This is not a standard coarse thread size, so the pitch must be specified.

When selecting thread sizes for a design, it is best to use standard or preferred sizes. Figure 17-3 is a listing of some preferred thread sizes.

17-4 THREAD CALLOUTS (ENGLISH UNITS)

English unit threads always include a thread form specification. Thread form specifications are designated by capital letters, as shown in the example in Figure 17-4, and are defined as fellows.

UNC - Unified National Coarse
UNF - Unified National Fine
UNEF - Unified National Extra Fine
UN - Unified National, or constant pitch threads

An English unit callout starts by defining the major diameter of the tread followed by the pitch specification. The callout .500 – 13 UNC means a thread whose major diameter is .500 inches with 13 threads per inch. The thread is manufactured to the Unified National Coarse (UNC) standards.

There are three classes of fit for a thread: 1, 2, and 3. The different classes specify a set of manufacturing tolerances. A class 1 thread is the loosest; class 3 is the most exact. A class 2 fit is the most common. If no class of fit is specified for a thread, it is assumed to be a class 2 fit.

Major Dia	Coarse		Fine	
	Pitch	Tap Drill Dia	Pitch	Tap Drill Dia
1.6	0.35	1.25		
2	0.4	1.6		
2.5	0.45	2.05		
3	0.5	2.5		
4	.7	3.3		
5	0.8	4.2		
6	1	5.0		
8	1.25	6.7	1	7.0
10	1.5	8.5	1.25	8.7
12	1.75	10.2	1.25	10.8
16	2	14	1.5	14.5
20	2.5	17.5	1.5	18.5
24	3	21	2	22
30	3.5	26.5	2	28
36	4	32	3	33
42	4.5	37.5	3	39
48	5	43	3	45

Figure 17-3

Major Dia	Decimal	UNC Thread/in	UNC Tap drill Dia.	UNF Thread/in	UNF Tap drill Dia.	UNEF Thread/in	UNEF Tap drill Dia.
#6	.138	40	#38	44	#37		
#8	.164	32	#29	36	#29		
#10	.190	24	#25	32	#21		
1/4	.250	20	7	28	3	32	.219
5/16	.312	18	F	24	1	32	.281
3/8	.375	16	.312	24	Q	32	.344
7/16	.438	14	U	20	.391	28	Y
1/2	.500	13	.422	20	.453	28	.469
9/16	.562	12	.484	18	.516	24	.516
5/8	.625	11	.531	18	.578	24	.578
3/4	.750	10	.656	16	.688	20	.703
7/8	.875	9	.766	14	.812	20	.828
1	1.000	8	.875	12	.922	20	.953
1 1/4	1.250	7	1.109	12	1.172	18	1.188
1 1/2	1.500	6	1.344	12	1.422	18	1.438

UNC = Unified National Coarse

UNF = Unified National Fine

UNEF = Unified National Extra Fine

Figure 17-5

The letter A designates an external thread, B an internal thread. The symbol x, read "by," precedes the thread's length. The word LONG is often included in the callout to prevent confusion about which value represents the length.

Drawing callouts for English unit threads are sometimes shortened, such as shown in Figure 17-4. The callout .500 – 13UNC – 2A x 3.00 LONG is shorted to .500 – 13 x 3.00. Only a coarse thread has 13 threads per inch, and it should be obvious whether a thread is internal or external, so the specifications may be dropped. Most threads are class 2, so it is tacitly accepted that all threads are class 2 unless otherwise specified. The shortened callout form is not universally accepted. When in doubt, use a complete thread callout.

Figure 17-5 shows a listing of some standard English unit threads.

17-5 THREAD REPRESENTATIONS IN 3D

Threads are usually presented on 3D drawings using representations, that is, inexact drawings that visually resemble a thread.

To represent an external thread in 3D

1. Select the Cylinder tool from the Solids toolbar.

   Elliptical/<center point><0,0,0>:

2. Select a center point.

A solid cylinder surrounded by arrayed circles is used to represent 3D threads.

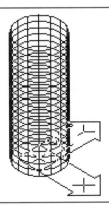

Figure 17-6

In this example, the center point was located on the 0,0,0 point of the WCS.

Diameter/<Radius>:

3. Type 1; press Enter.

Center of other end/<Height>:

4. Type 5; press Enter.

The threads will be represented by an array of circles located along the cylinder.

5. Select the Circle tool from the Draw toolbar and draw a circle of radius 1 and a center point located on the same center point as the cylinder.

6. Select the Modify pull-down menu, then 3D Operation, then 3D Array.

   Select object:

7. Select the circle drawn in step 5.

If it is difficult to select the circle, draw the circle elsewhere on the grid, array it, then move the stack of circles onto the cylnder.

Rectangular or Polar array (P/R):

8. Type R; press Enter.

Number of rows (——) <1>:

9. Press Enter.

Number of columns (|||)<1>:

10. Press Enter.

Number of levels (...)<1>:

11. Type 20; press Enter.

Distance between levels (...):

12. Type .25; press Enter.

See Figure 17-6.

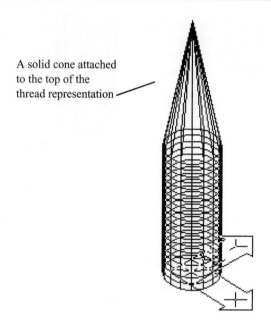

A solid cone attached to the top of the thread representation

Figure 17-7

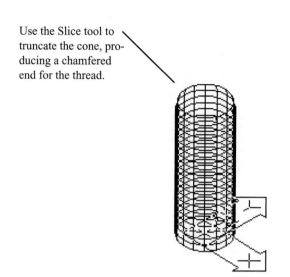

Use the Slice tool to truncate the cone, producing a chamfered end for the thread.

Figure 17-8

To add a chamfer to a thread

1. Select the Cone tool from the Solids toolbar.

   Elliptical/<center point>: <0,0,0>:

2. Type 0,0,5; press Enter.

   This is a center point at the top end of the thread.

   Diameter/<Radius>:

3. Type 1; press Enter.

   The radius 1 will match the diameter of the existing thread.

   Apex/<Height>:

4. Type 5; press Enter.

   See Figure 17-7. The cone will be sliced to form the chamfered shape.

5. Select the Slice tool from the Solids toolbar.

   Select objects:

6. Select the cone.

   Slicing plane by Object/ Zaxis/ View/ XY/ YZ/ ZX/ <3points>:

7. Press Enter to accept the 3 point default option.

   The slicing plane will be defined by specifying three

points in the plane. The plane will be parallel to the WCS X,Y plane 5.25 above it.

   1st point on plane:

8. Type 0,0,5.25; press Enter.

   2nd point on plane:

9. Type 1,0,5.25; press Enter.

   3rd point on plane

10. Type 0,1,5.25; press Enter.

    Both sides/<Point on desired side of the plane>:

11. Select a point near the top of the cone.

    See Figure 17-8.

12. Select the Union tool from the Modify II toolbar and union the truncated cone to the cylinder.

To create a WBlock for the thread

   The thread representation should be saved as a WBlock so it can be used on other drawings.

1. Select the Make Block tool from the Draw toolbar.

   The Block Definition dialog box will appear. See Figure 17-9.

Type block name here.

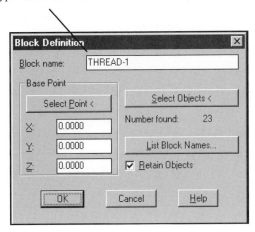

Type wblock name here

Figure 17-9

Select here to access blocks created on the drawing.

Select here to select Wblock names.

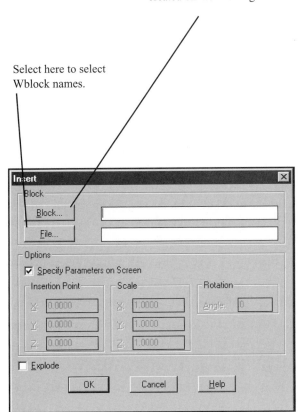

Figure 17-10

2. Type THREAD-1 in the Block name box.
3. Select the Select Objects box.

   Select object:

   Select the object by windowing the entire object. If just the object is selected, the arrayed circles that represent the threads will not appear.

   The Block Definition dialog box will reappear. The default base point setting is 0,0,0 which is acceptable.

4. Select the OK box.

   Blocks are unique to each drawing, so the object must also be saved as a WBlock so it can be inserted into other drawings.

5. Type wblock in response to a command prompt.

   The Create Drawing File dialog box will appear. See Figure 17-9.

   Block name:

6. Type THREAD-1.

   The object is now saved as a WBlock. WBlocks are saved as drawing files with .dwg roots.

To insert a block

1. Select the Insert Block tool from the Draw toolbar.

   The Insert dialog box will appear. See Figure 17-10.

2. Select the Block option.

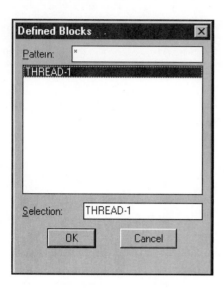

Figure 17-11

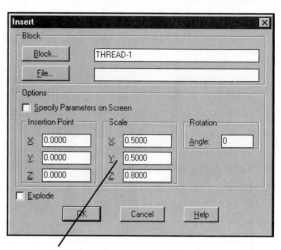

Change scale values here.

Figure 17-13

The Defined Blocks dialog box will appear. See Figure 17-11.

3. Select the THREAD-1 block, then OK.

The Insert dialog box will again appear with the selected block shown next to the Block option box. See figure 17-12.

4. Select OK to return to the drawing screen.

Insertion point:

5. Select the point where the block is to be inserted.

In this example the 0,0,0 point was selected.

X scale factor <1>/Corner/XYZ:

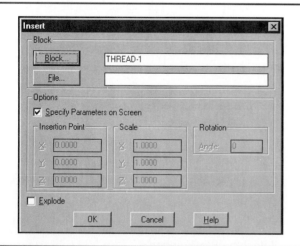

Figure 17-12

6. Press Enter to accept the default scale factor of 1.

Y scale factor (default=X):

7. Press Enter to make the Y scale factor also equal to 1.

Rotation angle <0.00>:

8. Press Enter.

To change the block insert scale factors

The scale factors can be used to change the size of the threads. For example, if the required thread were 0.50 in diameter and 4.00 long, the scale factors option could be used to create the new thread requirements from the thread block.

1. Select the Insert Block tool from the Draw toolbar.

The Insert dialog box will appear. See Figure 17-13.

2. Turn off the Specify Parameters on Screen option.
3. Change the X,Y, and Z Scale factors to 0.50, 0.50, and 0.80 respectively.
4. Define the Insertion Point as X = 5.000, Y = 0, Z = 0.

The insertion point could be defined to align with other existing objects, but in this example, a point on the X

The block with all scale
values set at 1.0000

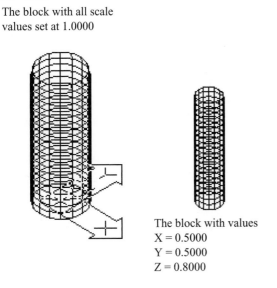

The block with values
X = 0.5000
Y = 0.5000
Z = 0.8000

Figure 17-14

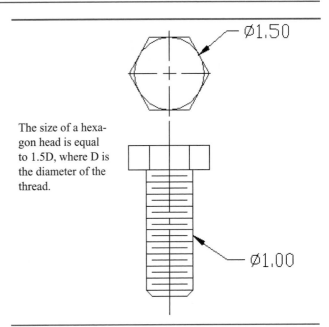

The size of a hexagon head is equal to 1.5D, where D is the diameter of the thread.

Figure 17-15

axis was selected. The Move tool was used to relocate the thread. The X and Y scale factors must be equal to maintain the circular shape of the thread. If the X and Y scale factors were not equal, an elliptical shape would result. See Figure 17-14.

17-6 HEXAGON-SHAPED HEADS

Hexagon-shaped heads are very common for fasteners. As a rule, the size of a hexagon head is determined from the major diameter of the thread. The hexagon shape is drawn as a hexagon circumscribed about a circle whose diameter is 1.5 times the major diameter of the thread. The height of the head is calculated as two thirds (.67) times the major diameter of the thread. See Figure 17-15.

To draw a hexagon-shaped head

1. Select the Polygon tool from the Draw toolbar.

   Command: _polygon Number of sides <4>:

2. Type 6; press Enter.

   Edge/<Center of polygon>:

3. Select the 0,0,0 point of the WCS.

   Inscribe in circle/Circumscribe about a circle (I/C)<C>:

4. Press Enter.

Radius of circle:

5. Type 1.5; press Enter.

The value 1.5 was used to match the threads drawn in section 17-2. That thread had a radius of 1.00, so (1.5)(1.00) = 1.50. See Figure 17-16.

A hexagon drawn with the 0,0,0 point of the WCS as its centerpoint

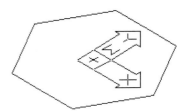

To locate the WCS icon on the origin, select the View pull down menu, then Display, UCS icon, Origin.

Figure 17-16

Use the Extrude tool to create a solid hexagon shape.

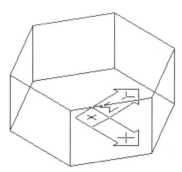

Results of using the Chamfer tool

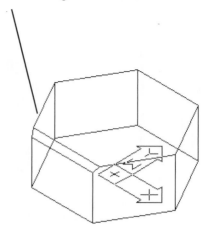

Figure 17-17

Figure 17-18

To create a solid model hexagon head

1. Select the Extrude tool from the Solids toolbar.

   Select object:

2. Select the hexagon.

   Hexagons are automatically drawn as polylines, so there is no need to use the Edit Polyline tool to join the lines into a polyline.

   Path/<Height of Extrusion>:

3. Type 1.34; press Enter.

   The value 1.34 was entered based on the general rule that the head height is .67 times the major diameter of the thread, (.67)(2.00) = 1.34. The negative sign was used to make it easier to fit the head on top of the thread block created in section 17-2. A block will also be created for the head using a 0,0,0 insertion point on the WCS, so the thread and head blocks can be joined at a common point to form a screw that can be unioned into a solid.

   Extrusion taper angle <0>:

4. Press Enter.

   See Figure 17-17.

To add chamfers to the hexagon

   Chamfers can be added to the top surface of the hexagon head to give it a realistic appearance.

1. Select the Chamfer tool from the Modify toolbar.

   (TRIM mode) Current chamfer Dist1 = 0.5000, Dist 0.5000
   Polyline/Distance/Angle/Trim/Method/<Select first line>:

2. Type D; press Enter.

   Enter first chamfer distance <0.5000>:

3. Type .125; press Enter.

   Enter second chamfer distance <0.1250>:

4. Press Enter.
5. Press Enter to start the chamfer sequence again.

   Polyline/Distance/Angle/Trim/Method/<Select first line>:

6. Select the top of one of the edge surfaces.

   Select base surface
   Next/<OK>:

7. Press Enter.

   Enter base surface distance <0.1250>:

8. Press Enter.

   Enter other surface distance <0.1250>:

9. Press Enter.

   Loop/<Select edge>:

10. Select the top edge again; press Enter.

    Figure 17-18 shows the results.

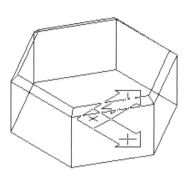

Figure 17-19

11. Repeat the procedure for the other top edges of the hexagon.

Figure 17-19 shows the results.

To create a block of the hexagon head

1. Select the Make Block tool from the Draw tool-bar.

The Block Definition dialog box will appear.

2. Type HEXHEAD in the Block name box.
3. Select the Select Object box.

Select object:

4. Select the modified hexagon.
5. Accept the 0,0,0 base point by selecting the OK box.

To create a WBlock

1. Type wblock; press Enter.

The Select Drawing File dialog box will appear.

2. Type HEXHEAD in the File Name box.

Block name:

3. Type HEXHEAD.

To save the WBlock on a disk in the A drive, change the directory to A.

17-7 SCREWS

Screws are created by combining threads with a head shape. In this section, the blocks THREAD and HEXHEAD will be combined to form a screw. The screw in turn will be saved as a new block.

To create a screw from existing blocks

It is assumed that the HEXHEAD block is already on the screen.

1. Select the Insert tool from the Draw toolbar.

The Insert dialog box will appear.

2. Select the File box.

The Select Drawing File dialog box will appear.

3. Select the Thread-1 block.

In this example, the WBlock was accessed rather than the Block. Either the Block or the WBlock could have been used. To insert an existing WBlock, select the File option on the Insert dialog box. The Select Drawing File dialog box will appear. See Figure 17-20. Select the Thread-1 drawing file located on the A: drive, then OK. The Insert dialog box will reappear as shown in Figure 17-21.

4. Select OPEN to return to the drawing screen.

The cross hairs will be aligned with the insertion point of the block.

Insertion point:

5. Select the 0,0,0 point on the WCS.

Insertion point: X scale factor <1>/Corner/XYZ:

6. Press Enter.

Y scale factor (default=X):

7. Press Enter.

Rotation angle <0.00>:

8. Press Enter.

See Figure 17-22.

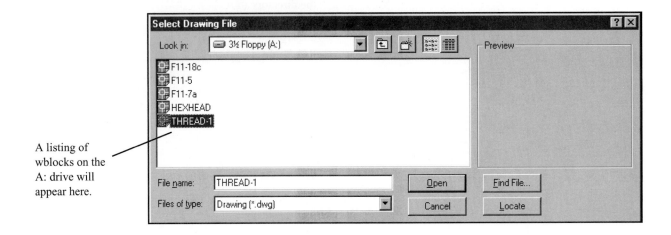

A listing of wblocks on the A: drive will appear here.

Figure 17-20

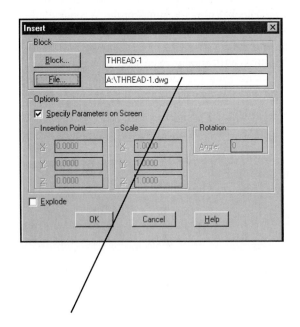

The file name of the selected wblock will appear here.

Figure 17-21

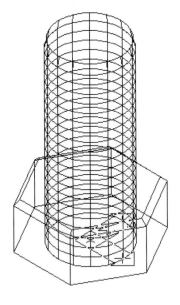

The blocks SCREW and THREAD-1 inserted on the 0,0,0 point of the WCS

Figure 17-22

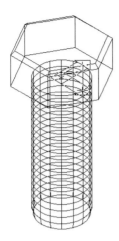

Use the 3D Rotate tool to rotate the
THREAD-1 block into position.

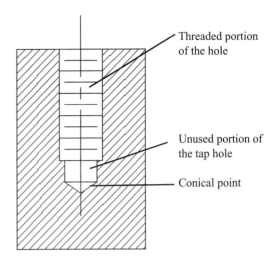

Threaded portion
of the hole

Unused portion of
the tap hole

Conical point

Figure 17-23

Figure 17-24

To rotate the thread into positions

The THREAD-1 block was created with a positive Z component value, positioning the thread through the hexhead. The 3D ROTATE command is used to rotate the thread into the correct orientation.

1. Type rotate3D; press Enter.

Select object:

2. Select the thread.

Axis by object/ Last/ View/ Xaxis/ Yaxis/ Zaxis/<2points>:

3. Type X; press Enter.

Point on X axis <0,0,0>:

4. Press Enter.

<Rotation angle>/Reference:

5. Type 180; press Enter.

The thread will be rotated under the hexhead. See Figure 17-23. Use the ZOOM and PAN commands to position the screw to the center of the screen.

To save the screw as a block

1. Select the Make Block tool from the Draw toolbar.

The Block Definition dialog box will appear.

2. Type SCREW in the Block name box.
3. Select the Select Object box.

Select object:

4. Select the screw.
5. Accept the 0,0,0 base point by selecting the OK box.

The object is now saved as a block. Repeat the procedure described above, and save the screw as a WBlock.

17-8 THREADED HOLES

Threaded holes are manufactured by first drilling a tap hole using a normal twist drill, then cutting the threads using a tapping bit. A tapping bit does not have cutting edges on its bottom surface, so threads are never cut to the bottom of the tap hole. A listing of recommended tap drill sizes for threaded holes is included in the appendix.

The general rule is to allow the tap hole to extend at least two thread lengths or the distance of 2 pitches beyond the threaded portion of the hole. See Figure 17-24.

In addition, because the tap hole is cut using a twist drill, a conical point will be created at the bottom of the hole. The conical point is represented on a drawing using slanted lines 30° to the bottom line of the hole as shown in Figure 17-24.

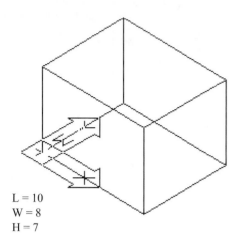

L = 10
W = 8
H = 7

Figure 17-25

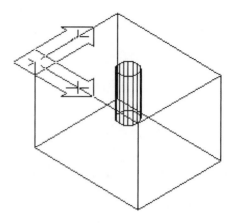

Figure 17-27

To draw a pilot hole in 3D

In this example, the hole will be created in a solid cube.

1. Select the Box tool from the Solids toolbar and create a box with its corner on the 0,0,0 point of the WCS and dimensions of length = 10, width = 8, and height = 7.

    See Figure 17-25.

2. Select the Origin tool from the UCS toolbar and create a new UCS with its origin on the top surface of the box, that is, at the 0,0,7 point relative to the WCS.

Create a new UCS.

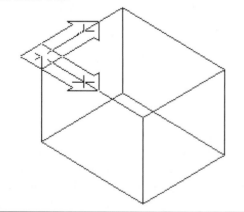

Figure 17-26

See Figure 17-26.

3. Select the Cylinder tool from the Solids toolbar.

    Elliptical/<Center point> <0,0,0>:

4. Select the 5,4,0 point on the new UCS.

    Diameter/<Radius>:

5. Type .89; press Enter.

    The value .89 was derived from the thread chart in the appendix. The chart specifies a tap drill of Ø1-25/32 (1.78) for a Ø2.00 thread. Therefore, the radius value equals 0.89.

    Center of other end/<Height>:

6. Type -4.

    The value -4 is an arbitrary hole depth. See Figure 17-27.

To add the conical point to the hole

1. Select the Cone tool from the Solids toolbar.

    Elliptical/<center point> <0,0,0>:

2. Select point 5,4,-4.

    This point is the center point of the bottom surface of the cylinder.

    Diameter/<Radius>:

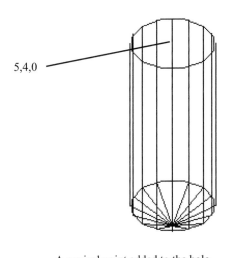

5,4,0

A conical point added to the hole

FIgure 17-28

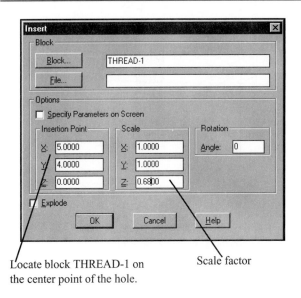

Locate block THREAD-1 on
the center point of the hole.

Scale factor

Figure 17-29

3. Type .89; press Enter.

Apex/<Height>:

4. Type -.51; press Enter.

The given radius of .89 and the 30 degree angle
requirement were used to create a 30 degree right triangle
with the longest leg equal to 0.89. The resulting shorter leg
was found to equal 0.51. Figure 17-28 shows a zoomed
view of the hole.

5. Select the Union tool from the Modify II toolbar
and union the cylinder and cone.

To create a block of the pilot hole

1. Select the Make Block tool from the Draw toolbar.

The Block Definition dialog box will appear. See
Figure 17-9.

2. Type THDHOLE in the Block name box.
3. Select the Select Object box.

Select object:

4. Select the hole.
5. Select the center point of the top surface or the
5,4,0 point on the UCS.

6. Type wblock and save the hole as a WBlock
named THDHOLE.

To determine the THREAD-1 block's scale factors

The THREAD-1 block has a length, the distance
along the Z axis, of 5.25. The original length of 5 plus the
.25 chamfer on the end.

For the thread table in the appendix, a UNC d2.00
thread has 4.5 threads per inch or a pitch of .22 (1/4.5 =
.22). Two pitches equal .44, so the threaded portion of the
hole must be at least .44 from the bottom of the 4.00 deep
tap hole.

4.00 - 0.44 = 3.56

The thread depth, the distance along the Z axis, must
equal 3.56. The current length of the THREAD-1 block is
5.25, so a scale factor must be calculated to reduce the
block's length during insertion.

5.25X = 3.56

X = .68

This scale factor will be used as the Z scale factor in
the Insert dialog box. See Figure 17-29.

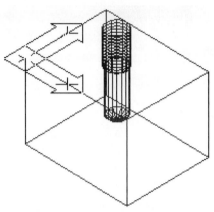

The scaled THREAD-1 block is
added at insert point 5,4,0.

Figure 17-30

To create a threaded hole

1. Select the Insert tool from the Draw toolbar.
2. Use the Block box to access the THREAD-1 block if you are still on the same drawing , or use the File box to access the THREAD-1 WBlock.

The Insert dialog box will appear.

3. Turn off the Specify Parameters on Screen option. Change the Insertion Point and Scale factors as shown in Figure 17-29.

The block is added at the 4,5,0 point on the UCS or the top centerpoint of the hole. The Z scale factor of 0.68 was derived above. See Figure 17-30.

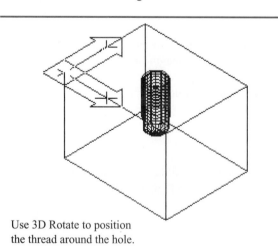

Use 3D Rotate to position
the thread around the hole.

Figure 17-31

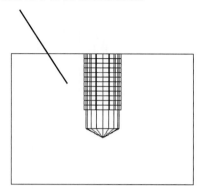

A front view of the threaded hole.

Figure 17-32

4. Type rotate3d; press Enter.

Select objects:

5. Select the THREAD-1 block.

Axis by object/ Last/ View/ Xaxis/ Yaxis/ Zaxis/ <2points>:

6. Type X; press Enter.

Point on the X axis <0,0,0>:

7. Type 5,4,0; press Enter.

<Rotation angle>/Reference:

8. Type 180; press Enter.

See Figure 17-31. Figure 17-32 shows a front view of the threaded hole

17-9 A SCREW IN A THREADED HOLE

This section will explain how to size and locate a screw to fit into a threaded hole. As can be seen in Figure 17-32, a threaded hole presents a crowded visual display, that is, the many lines may be confusing. An Isolines value of 16 was used to create both the tap hole and the threads. A smaller value could be used to create a presentation that uses fewer lines and would be somewhat less confusing, but too few lines can also be confusing.

In this example, the screw will be transferred to a different layer and drawn using a color that is different

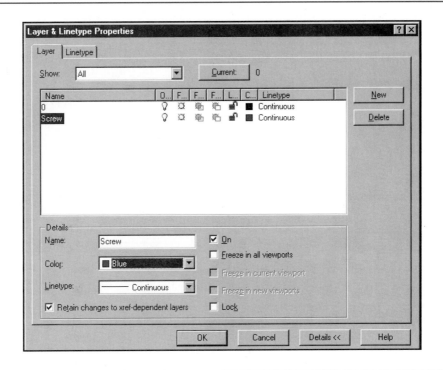

Figure 17-33

from the threaded hole. The screw can then be displayed with the threaded hole, then turned off to allow just the hole to be seen.

To create a new layer for the screw

1. Select the Layers tool from the Standard toolbar.

The Layer & Linetype Properties dialog box will appear. See Figure 17-33.

2. Select the New box and type Screw as the new layer name.
3. Select the Color box and change the screw layer's color to blue, then select OK.

To determine the scale factors for the screw

The block SCREW must be sized to fit the threaded hole. It is good design practice to have the screw length less than the length of the threaded portion of a hole. This prevents the screw from "bottoming out" and assures that all threads on the screw will be engaged. A general rule is to allow at least two unused threads beyond the end of the screw, or 2 pitch lengths. See Figure 17-34. The Ø2.00 thread has 4.5 threads per inch, or a pitch of 0.22. In the previous section the thread depth was calculated at 3.56, so

the screw length cannot exceed 3.12 (3.56 – 0.44 = 3.12).

AutoCAD controls the size of Blocks using scale factors. The original SCREW block length (distance in the Z direction) was 5.25. To reduce this length to 3.12 requires a scale factor of 0.59 determined as follows.

$$5.25Z = 3.12$$

$$Z = 0.59$$

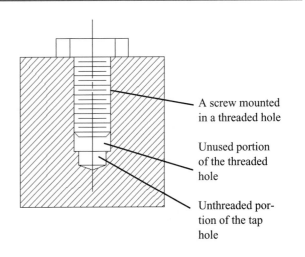

A screw mounted in a threaded hole

Unused portion of the threaded hole

Unthreaded portion of the tap hole

Figure 17-34

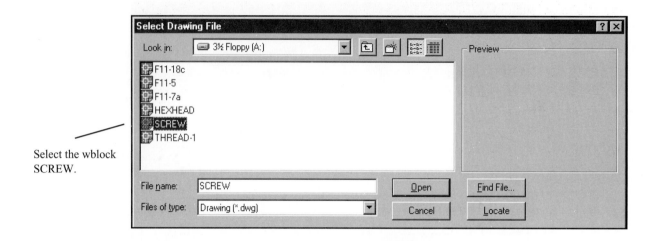

Figure 17-35

To insert the WBlock screw

1. Select the Insert Block tool from the Draw tool-bar.

The Insert dialog box will appear. The block SCREW is not part of this drawing and will not be listed as a block. It was saved as a WBlock and can be accessed using the File box.

2. Select the File box.

The Select Drawing File dialog box will appear. See Figure 17-35.

3. Select the SCREW file.

The Insert dialog box will reappear.

4. Turn off the Specify Parameters on the Screen option. Enter the Insertion Point and Scale values as shown in Figure 17-36.

The insertion point is the top center point of the threaded hole relative to the current top surface UCS and the Z scale factor as calculated above. Figure 17-37 shows the SCREW block inserted into the drawing.

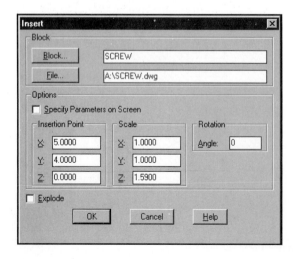

Figure 17-36

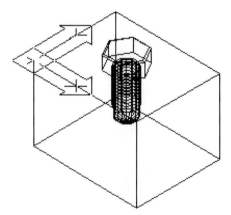

A 3D screw mounted into a threaded hole

Figure 17-37

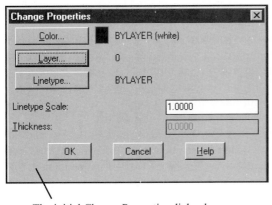

The initial Change Properties dialog box

The Change Properties dialog box after the layer change

Figure 17-38

To move the screw to a different layer

1. Select the Properties tool from the Standard toolbar.

   Select objects:

2. Select the screw.

   The Change Properties dialog box will appear. See Figure 17-38.

3. Select the Layer box.

   The Select Layer dialog box will appear. See Figure 17-39.

4. Select the SCREW layer.

   The Modify Block Insertion dialog box will reappear showing the layer as SCREW and the color as blue.

5. Select OK to return to the drawing screen.

   The screw will now be blue.

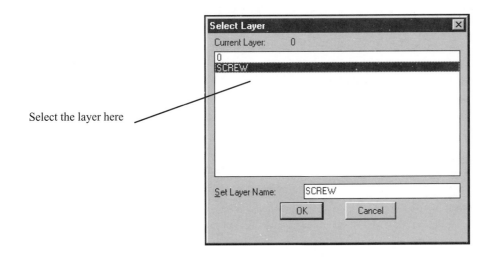

Select the layer here

Figure 17-39

17-10 STANDARD SCREW SIZES

In the previous example, a screw size was determined from the length of a threaded hole that was, in turn, determined from a tap hole depth. Normally the reverse is true; a screw length is used to determine the depth of threads needed and the depth of the tap hole. Screws are manufactured to standard lengths and it is usually more efficient to fit the threaded hole depth to match the standard sizes than to modify the screw length.

To determine a threaded hole depth: English units

Assume that a 1/4 - 20 UNC x 1.50 long hex head screw is to be mounted into a part.

1. Calculate the thread pitch.

   20 thds per inch = 1/20 = .05 inches = P

2. Determine the thread hole depth by adding 2P to the screw's length.

   1.50 + .10 = 1.60 inches

3. Determine the tap hole depth by adding 2P to the thread hole depth.

   1.60 + .1 = 1.70 inches

To determine a threaded hole depth; Metric units

Assume that an M10 x 1.5 x 40 is to be mounted into a part.

1. Calculate the thread pitch.

   1.5 threads per millimeter = 1/1.5 = .66 millimeters = P

2. Determine the thread hole depth by adding 2P to the screw's length.

   40 + 1.3 = 41.3 millimeters

3. Determine the tap hole depth by adding 2P to the thread hole depth.

   41.3 + 1.3 = 42.6 millimeters

17-11 NUTS

This section will explain how to draw a hexagon-shaped nut and how to save the nut as both a Block and a WBlock. There are many styles of nuts. A finished nut has a flat surface on one side that acts as a bearing surface when the nut is tightened against an object. As a rule, a finished nut has a thickness equal to .88D, where D is the major diameter of the nut's thread size.

A lock nut is symmetrical, with the top and bottom surfaces being identical. A lock nut generally has a thickness equal to .5D, where D is the major diameter of the nut's thread size.

To draw a solid hexagon shape

The size of the hexagon shape will be determined from the nut's thread size. As a rule, the distance across the flats of a nut is equal to 1.5D where D is the nut's thread size. In this example, the diameter of the thread is assumed to be 2.00 or a radius of 1.00 unit.

1. Select the Polygon tool from the Draw toolbar.

   Command: _polygon Number of sides <4>:

2. Type 6; press Enter.

   Edge/<Center of polygon>:

3. Select the 0,0 point on the WCS.

   Polygon is a 2D command, so coordinate values of 0,0 are used to locate the WCS origin.

   Inscribe in a circle/Circumscribe about a circle (I/C)<C>:

4. Press Enter to accept the circumscribe option or type C if circumscribe is not the default selection.

   Radius of circle:

5. Type 1.5; press Enter.

   A hexagon will appear on the screen.

6. Select the Extrude tool from the Solids toolbar.

   Select objects:

7. Select the hexagon.

   The Polygon tool draws shapes as enclosed polylines; there is no need to join the lines to form a polygon as was done in previous examples.

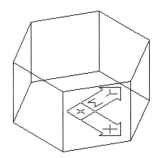

A solid hexagon-shaped nut

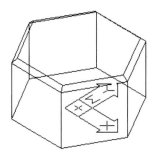

Chamfers added to the top edges

Figure 17-40

Figure 17-41

Path/<Height of Extrusion>:

8. Type 1.76; press Enter.

The value 1.76 was derived from 2.00 thread diameter, and from the general rule that the height equals .88D, where D is the nut's thread diameter.

Extrusion taper angle <0>:

9. Press Enter.

See Figure 17-40.

To chamfer the top surface

1. Select the Chamfer tool from the Modify toolbar.

(TRIM mode) Current chamfer Dist1 = 0.5000, Dist 0.5000
Polyline/Distance/Angle/Trim/Method/<Select first line>:

2. Type D; press Enter.

Enter first chamfer distance <0.5000>:

3. Type .125; press Enter.

Enter second chamfer distance <0.1250>:

4. Press Enter.
5. Press Enter to start the chamfer sequence again.

Polyline/Distance/Angle/Trim/Method/<Select first line>:

6. Select the top of one of the edge surfaces.

Select base surface
Next/<OK>:

7. Press Enter.

Enter base surface distance <0.1250>:

8. Press Enter.

Enter other surface distance <0.1250>:

9. Press Enter.

Loop/<Select edge>:

10. Select the top edge again.
11. Repeat the procedure for the other top edges of the hexagon.

Figure 17-41 shows the results.

To add the threaded hole

1. Select the Cylinder tool from the Solids toolbar.

Elliptical/<center point>: <0,0,0>:

2. Press Enter.

Diameter/<Radius>:

3. Type 1; press Enter.

Center of other end/<Height>:

4. Type 1.76; press Enter.

See Figure 17-42.

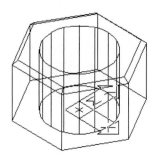

Hole added to solid nut

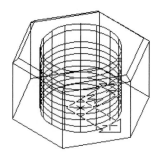

The thread representations are created by adding arrayed circles along the hole's surface.

Figure 17-42

Figure 17-43

To represent the threads

1. Select the Circle tool from the Draw toolbar.

   Command: _circle 3P/2P/TTR/<Center point>:

2. Select the 0,0,0 point.

   Diameter/Radius:

3. Type 1; press Enter.
4. Type 3darray; press Enter.

   Select objects:

5. Select the circle.

   Rectangular or Polar array (R/P):

6. Type r; press Enter.

   Number of rows (—) <1>:

7. Press Enter.

   Number of columns (|||) <1>:

8. Press Enter.

   Number of levels (...) <1>:

9. Type 10; press Enter.

   Distance between levels (...):

10. Type .176; press Enter.

    See Figure 17-43.

To save the finish nut as a block and WBlock

1. Select the Block tool from the Draw toolbar and save the finished nut as a block named FINNUT.
2. Type wblock in response to a command prompt and create a WBlock named FINNUT based on the FINNUT block.

To create a lock nut

A lock nut's height is .5D, where D is the diameter of the nut's thread. In this example, the thread diameter is Ø2.00, so .5D equals 1.00. Both sides of the nut are chamfered. The lock nut will be drawn at half height (.50), then mirrored to form the full height and shape.

1. Select the Polygon tool from the Draw toolbar and draw a hexagon of radius 1.5 centered on the 0,0 point.
2. Select the Extrude tool from the Solids toolbar and create a solid hexagon .50 high and 0d taper angle.
3. Add .125 x .125 chamfers to the top edges of the hexagon.

   See Figure 17-44.

4. Select the Cylinder tool from the Solids toolbar and create a 1.00 radius cylinder .50 high centered at the 0,0,0 point.

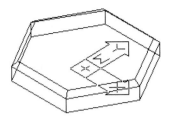

Half a lock nut

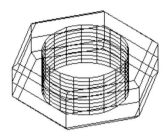

Use the 3D Mirror tool to create the complete lock nut.

Figure 17-44

Figure 17-46

5. Draw a 1.00 radius circle about the 0,0,0 point and use 3D Rectangular Array to create thread representations.
6. Define 5 levels 0.1 apart.

See Figure 17-45.

To mirror the object

1. Type mirror3d; press Enter.

Select objects:

2. Window the nut.

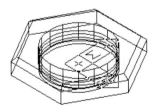

The hole representation added to the half lock nut

Figure 17-45

Plane by object/ Last/ Zaxis/ View/ XY/ YZ/ X/ <3points>:

3. Type XY; press Enter.

Point on the XY plane <0,0,0>:

4. Press Enter.

Delete old objects? <N>:

5. Press Enter.

The object will be mirrored as shown in Figure 17-46.

6. Save the lock nut as a block and a WBlock named LOCKNUT.

Make the insertion point 0,0,.5, the center point of the top surface.

17-12 STANDARD SCREWS

Figure 17-47 shows a group of standard screw shapes. The proportions given in Figure 17-47 are acceptable for general drawing purposes and represent average values. The exact dimensions for specific screws are available from manufacturers' catalogs. A partial listing of standard screw sizes is included in the appendix.

The given head shape dimensions are all in terms of D, the major diameter of the screw's thread. Information about the available standard major diameters and lengths is available in the appendix.

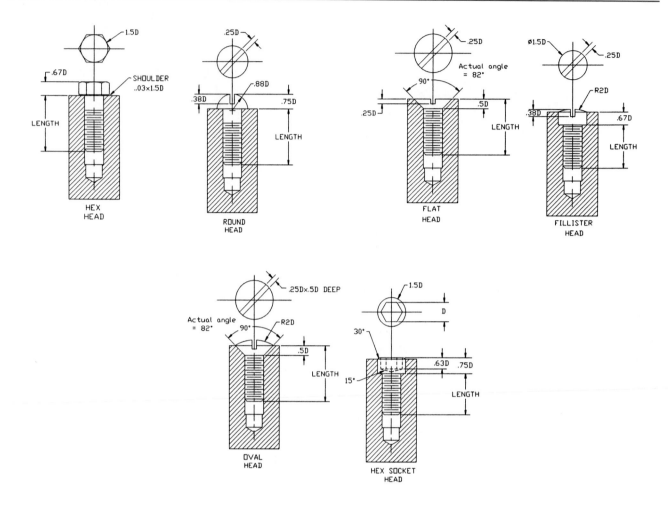

Figure 17-47

The choice of head shape is determined by the specific design requirements. For example, a flat head mounted flush with the top surface is a good choice when space is critical, when two parts butt against each other, or when aerodynamic considerations are involved. A round head can be assembled using a common blade screw driver, but it is more susceptible to damage than a hex head. The hex head, however, requires a specific wrench for assembly.

17-13 SET SCREWS

Set screws are fasteners that are used to hold parts like gears and pulleys to rotating shafts or other objects to prevent slippage between the two objects. See Figure 17-48.

Most set screws have recessed heads to help prevent interference with other parts. Many different head and point styles are available. See Figure 17-49.

Set screws are referenced on a drawing using the following format.

THREAD SPECIFICATION
HEAD SPECIFICATION POINT SPECIFICATION
SET SCREW

.250 - 20 UNC - 2Ax1.00 LONG
SLOT HEAD FLAT POINT
SET SCREW

M8 x 20 LONG
SQUARE HEAD OVAL POINT
SET SCREW

The words Long, Head, and Point are optional.

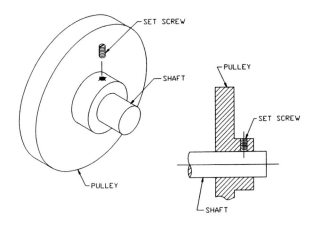

Figure 17-48

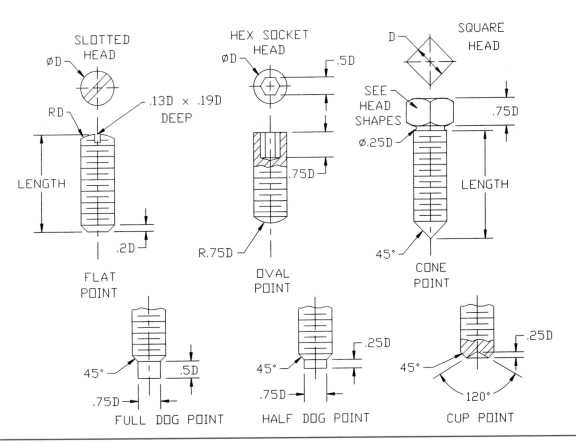

Figure 17-49

Plain flat washers

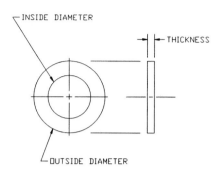

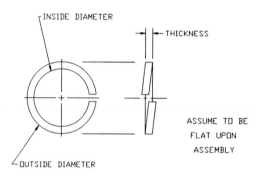

Figure 17-50

17-14 WASHERS

There are many styles of washers available for different design applications. The three most common types of washers are plain, lock, and star. Plain washers can be used to help distribute the bearing load of a fastener or used as a spacer to help align and assemble objects.

Lock and star washers help absorb vibrations and prevent fasteners from loosening prematurely. All washers are identified as follows.

Inside diameter x outside diameter x thickness

Examples of plain washers and their callouts are shown in Figure 17-50. A listing of standard washer sizes is included in the appendix.

To draw a star washer shape

Draw the following star washer.

1.50 x 1.75 x .125

1. Select the Circle tool from the Draw toolbar and draw three circles of diameter 1.500, 1.625, and 1.750.

The Ø1.500 is the inside diameter of the washer, 1.750 is the outside diameter, and 1.625 is halfway between the inside diameter and outside diameter and will be used to locate the bottom surfaces of the cutouts.

2. Select the LINE command from the Draw toolbar and draw a vertical line from the circles' center point beyond the outside diameter.

See Figure 17-51.

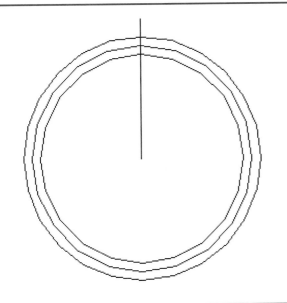

Figure 17-51

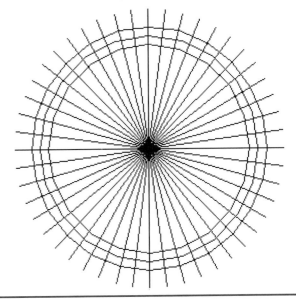

Figure 17-52

3. Select the ARRAY command from the Modify toolbar, then the polar option and array the line 48 times around the center point.

   See Figure 17-52.

4. Zoom the upper portion of the washer to a convenient size.

5. Select the Line tool from the Draw toolbar and draw the line 1-2-3-4-5-6 as shown in Figure 17-53.

   Use the OSNAP, Intersection option to ensure accuracy.

6. Return the washer to its normal size and erase the 1.625, 1.750 circles and all the arrayed lines.

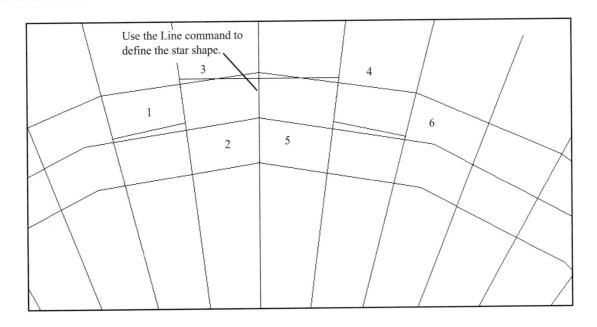

Figure 17-53

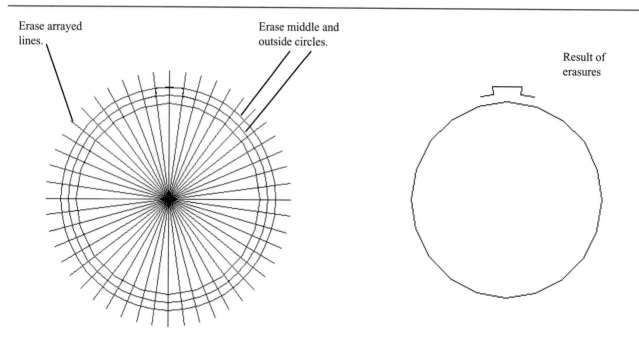

Erase arrayed lines.

Erase middle and outside circles.

Result of erasures

Figure 17-54

Figure 17-54 shows the results.

7. Use the POLAR ARRAY command and array the star shape 12 times about the circle's center point.

Figure 17-55 shows the resulting star washer.

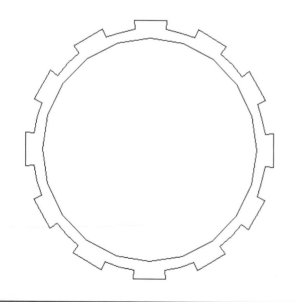

Figure 17-55

To create a solid star washer

1. Select the Edit Polyline tool from the Modify II toolbar.

   Select polyline:

2. Select any line on the outside edge of the washer.

   Object selected is not a polyline
   Do you want to turn it into one? <Y>:

3. Press Enter.

   Close/ Join/ Width/ Edit vertex/ Fit/ Spline/ Decurve/ Ltype gen/ Undo/ eXit/<X>:

4. Type J; press Enter.

   Select objects:

5. Window the entire star shape.

   Close/ Join/ Width/ Edit vertex/ Fit/ Spline/ Decurve/ Ltype gen/ Undo/ eXit/<X>:

6. Select the Extrude tool from the Solids toolbar.

   Select objects:

7. Select both the outside star pattern polyline and the inside circle; extrude them to a height of .125 and 0 degree taper.

The star washer with the HIDE command applied

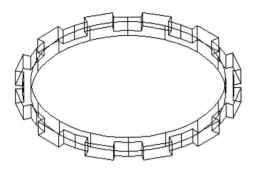

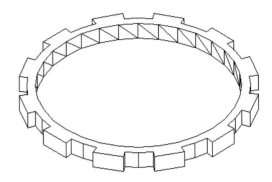

Figure 17-56

Figure 17-57

8. Select the SE Isometric view.

   See Figure 17-56.

9. Select the Subtract tool from the Modify II toolbar and subtract the inside circular shape from the star shape.

   Figure 17-57 shows the washer with the HIDE command applied.

17-15 3D ASSEMBLY DRAWINGS WITH FASTENERS

Figure 17-58 shows an assembly drawing that includes a fastener. Fasteners, washers, and nuts, saved as blocks, are often added to assembly drawings. The following example shows how to size, orient, and locate blocks of fasteners and washers.

A sample 3D assembly drawing created using solid models with the HIDE command applied

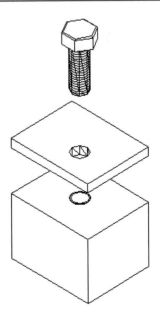

Figure 17-58

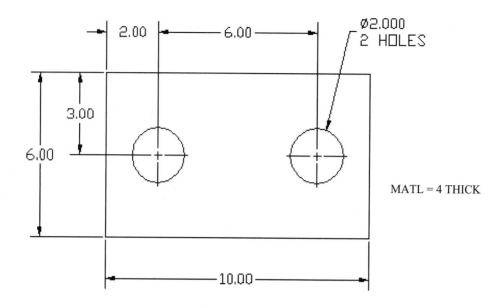

Figure 17-59

Design Problem

Two identical base plates, as defined in Figure 17-59, are to be joined by two screws with lock nuts. The assembly must also include four plain washers, 2.125 ˘ 3.00 ˘ .125.

Define the parts to be assembled

1. Select the Box and Cylinder tool from the Solids toolbar and draw one of the base plates with its corner on the 0,0,0 point of the WCS.
2. Use the Copy Object tool from the Modify toolbar and create a second base plate.

See Figure 17-60. Objects should be positioned on an assembly drawing so that each can be clearly seen. However, by locating the second base plate slightly over the first, the drawing gains visual depth. Figure 17-61 shows the drawing with the HIDE command applied. Note how one part is clearly behind the other.

Define the centerlines for the fasteners

1. Select the Presets UCS tool from the UCS toolbar.

The UCS Orientation dialog box will appear.

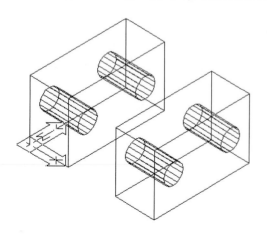

Figure 17-60

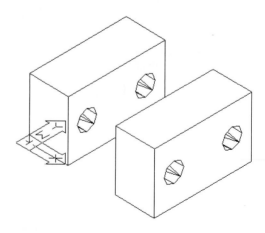

Figure 17-61

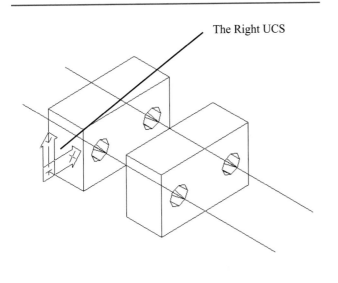

The Right UCS

Figure 17-62

2. Select the Right option.

See Figure 17-62.

3. Select the Line tool from the Draw toolbar and draw a centerline aligned with the center points

of the holes and extending well beyond the edges of the base plates.

In this example, one line was drawn between points 2,3,-6 to 2,3,24 and a second between 8,3,-6 and 8,3,24. These values were determined from the given dimensions.

To insert the washers

1. Select the Insert Block tool from the Draw toolbar and insert the block WASHER.

If a block named WASHER has not been saved, create a solid washer by drawing two solid cylinders and subtracting the smaller from the larger. Scale the block to fit the drawing requirements.

2. Define 3,2,12 as the block's insertion point.

See Figure 17-63. If the resulting location is too close or too far from the base plates, use the MOVE command to relocate the WASHER blocks as needed. The X,Y components of 3,2 and 8,2 will remain constant. Variations in the Z value will move the washer along the holes' center line.

The other three WASHER blocks were located at points 3,2,-4, 8,3,12, and 8,3,-4. Figure 17-63 shows the results.

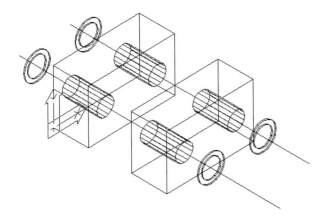
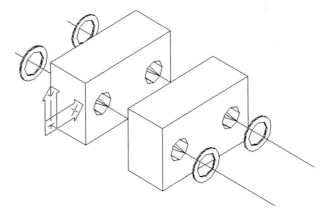

Figure 17-63

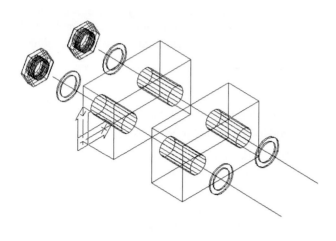

Figure 17-64

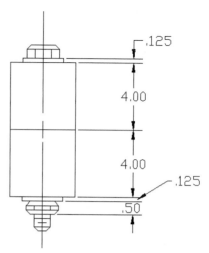

Determine the required thread length for the screws.

Figure 17-65

To add the nuts

1. Select the Insert tool from the Draw toolbar and insert two LOCKNUT blocks at insertion points 3,2,-6, and 8,3,-6.

Figure 17-64 shows the results.

To determine the required thread length

The required thread length is determined by calculating the length needed to pass through all the parts, including the washers, with at least two threads extending beyond the nut. Figure 17-65 shows a profile view of the two base plates, washers, nut, and screw. The height of the screw head is not part of the length calculation.

The height of the parts is found to equal 9.25 (.125 + 4.00 + 4.00 + .125 + 1.00 = 9.25). The pitch of the thread equals .22, so two pitches equal .44. The minimum thread length equals 9.69.

A review of manufacturer's catalogs specifies that Ø2.00-4.5 UNC threads are available in standard lengths of 9.00, 10.00, and 12.00. The 10.00 length is selected as correct for this assembly.

To determine the scale factor for the screws

The block SCREW has a length of 5.25 and must now be scaled to fit the new length requirement of 10.00.

The calculation is as follows.

$$5.25(Z) = 10.00$$
$$Z = 1.90$$

To apply the scale factor

1. Select the Insert Block tool from the Draw toolbar and select the WBlock SCREW.

The Insert dialog box will appear. See Figure 17-66.

2. Turn off the Specify Parameters on Screen option and set the insertion point for 2,3,30 and the Z scale factor for 1.90.

The screw will appear on the screen.

To create the second screw

1. Select the Copy tool from the Modify toolbar.

   Select objects:

2. Select the screw.

   <Base point or displacement>/Multiple:

3. Type 2,3,0.

   Second point of displacement:

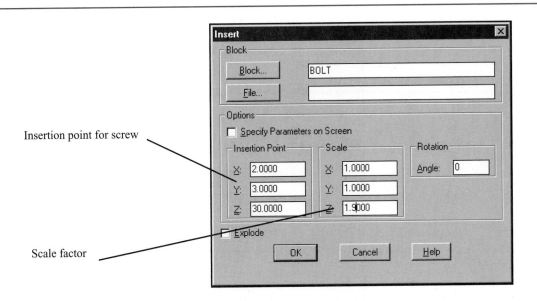

Figure 17-66

4. Type 8,3,0.

The displacement points were selected in the current X,Y plane. Figure 17-67 shows the results with the HIDE command applied.

17-16 METRIC SCALE FACTORS

The existing block SCREW is defined for a 2.00 - 4.5 UNC thread. The overall thread length, including the end chamfer, is 5.25. This block can be redefined for any size screw by using scale factors.

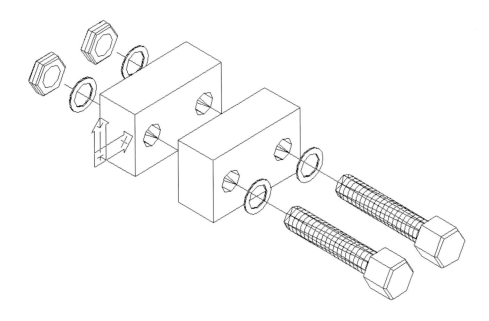

Figure 17-67

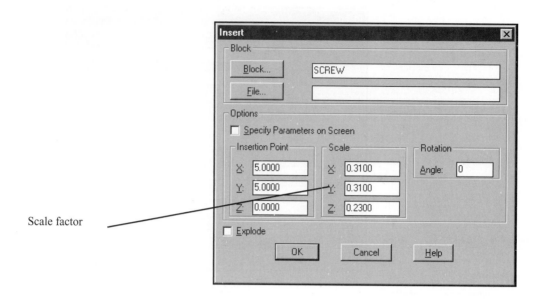

Figure 17-68

To draw an M16 x 30 hex head screw

1. Determine the metric equivalent dimensions for the original block.

The conversion factor between inches and millimeters is 1.00 inch = 25.4 millimeters. A Ø2.00 and 5.25 length in inches becomes Ø50.8 and 133 millimeters.

2. Determine the scale factors as follows.

$$50.8(X) = 16$$
$$X = .31$$

$$133(Z) = 30$$
$$X = .23$$

3. Select the Insert tool from the Draw toolbar and select the block SCREW.

The Insert dialog box will appear. See Figure 17-68.

4. Turn off the Specify Parameters on Screen option and add the appropriate scale factors.

The 5,5,0 insertion point is arbitrary. Figure 17-69 shows the two screws.

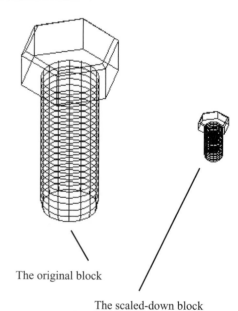

The original block

The scaled-down block

Figure 17-69

17-17 EXERCISE PROBLEMS

Draw the fasteners defined in EX17-1 through EX17-4 as solid models and save them as WBlocks.

EX17-1

A. 1.00–8UNCx2.00 LONG – HEX HEAD SCREW

B. 1.00–8UNCx2.00 LONG –ROUND HEAD SCREW

C. 1.00–8UNCx2.00 LONG – FLAT HEAD SCREW

D. 1.00–8UNCx2.00 LONG – FILLISTER HEAD SCREW

E. 1.00–8UNCx2.00 LONG – OVAL HEAD SCREW

F. 1.00–8UNCx2.00 LONG – HEX SOCKET HEAD SCREW

EX17-2

A. M10 x 20 – HEX HEAD SCREW

B. M10 x 20 – ROUND HEAD SCREW

C. M10 x 20 – FLAT HEAD SCREW

D. M10 x 20 – FILLISTER HEAD SCREW

E. M10 x 20 – OVAL HEAD SCREW

F. M10 x 20 – HEX SOCKET HEAD SCREW

EX17-3

A. .25–20UNC x 1.00
SLOTTED HEAD FLAT POINT
SET SCREW

B. .25–20UNC x 1.00
HEX SOCKET HEAD OVAL POINT
SET SCREW

C. .25–20UNC x 1.00
SQUARE HEAD CONE POINT
SET SCREW

EX17-4

A. M4 x 10
SLOTTED HEAD FLAT POINT
SET SCREW

B. M4 x 10
HEX SOCKET HEAD OVAL POINT
SET SCREW

C. M4 x 10
SQUARE HEAD CONE POINT
SET SCREW

EX17-5

The figure for EX17-5 shows a 3D assembly drawing. Create a similar 3D drawing based on the following information. Determine the length of the screws and washer sizes. The distance between the hole must be no less than 4D, and the distance from a hole's center point to any edge must be closer than 2D where D is the major diameter of the thread. Each plate is 2.5D thick.

A. .500–13UNC HEX HEAD SCREWS

B. M10 HEX HEAD SCREW

C. .250–20 UNC ROUND HEAD SCREW

D. M8 ROUND HEAD SCREW

E. #10 (.190)-24UNC HEX HEAD SCREW

F. M4 x 0.7 x 10 HEX HEAD SCREW

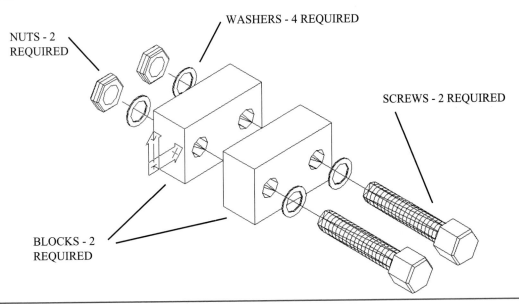

Figure EX17-5

EX17-6

Draw the given object as a solid model. Add three threaded hex head screws in the appropriate threaded holes. The screws should be long enough to extend into the Ø1.25 [30] center hole. Specify the selected screws using standard thread drawing callouts.

A. Use inches

B. Use millimeters

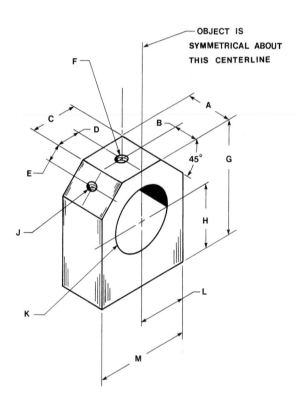

DIMENSION	INCHES	mm
A	1.00	26
B	.50	13
C	1.00	26
D	.50	13
E	.38	10
F	.190–32 UNF	M8X1
G	2.38	60
H	1.38	34
J	.164–36 UNF	M6
K	Ø1.25	Ø30
L	1.00	26
M	2.00	52

EX17-7

Draw the following object as a solid model. All parts have a depth of 3.00 inches or 75 millimeters. Select and draw standard screws and nuts to join parts 1, 2, 3, and 4 together at the indicated locations. Specify the selected screws and nuts using standard thread drawing callouts.

    A. Use inches

    B. Use millimeters

NOTE: ALL PARTS ARE 3.00 INCHES

OR 75 MILLIMETERS DEEP

(W)(X)(Y)(Z) INDICATES FASTENER LOCATIONS

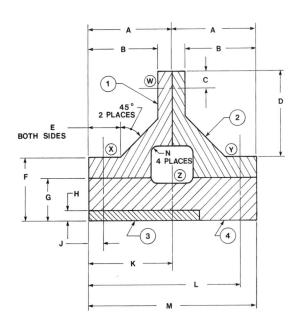

DIMENSION	INCHES	m m
A	1.00	50
B	1.63	41
C	.38	10
D	2.00	50
E	.75	19
F	1.50	38
G	1.00	25
H	.25	6
J	.38	10
K	2.00	50
L	3.63	92
M	4.00	100
N	R.13	3

EX17-8

Draw the following objects as solid models. Select and draw standard screws to join parts 23 and 24 together using the requirements listed below. Specify the selected screws using standard thread drawing callouts.

1. The hole labeled 1 is to use a screw and bolt.

2. The hole labeled 2 is to use a screw that passes through part 23 and screws into part 24. Specify the thread on part 24.

3. The hole labeled 3 is a countersunk hole in part 23. The screw passes through part 24 and is joined to a nut.

Use the following dimensional values:
A. Use inches
B. Use millimeters

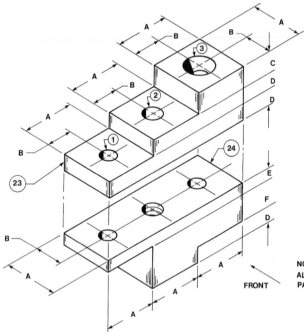

NOTE: HOLES IN PART 24 ALIGN WITH THOSE IN PART 23.

DIMENSION	INCHES	mm
A	1.25	32
B	.63	16
C	.50	13
D	.38	10
E	.25	7
F	.63	16

EX17-9

Draw the following objects as a solid model. Select and draw a standard screw and nut to join parts 12 and 13 through the hole labeled H. Specify the selected screw and nut using standard thread drawing callouts.

Select and draw set screws that fit in the holes labeled K. Add the appropriate threads to part 12. The set screws should press up against part 13.

A. Use inches

B. Use millimeters

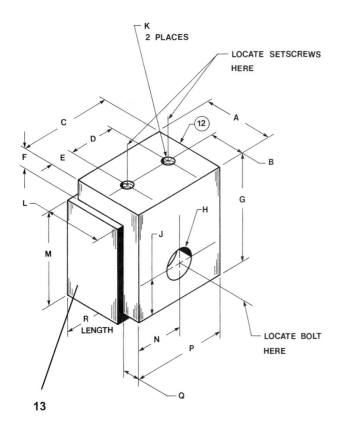

DIMENSION	INCHES	mm
A	1.50	38
B	.75	16
C	2.00	50
D	1.00	25
E	.50	12.5
F	.38	9
G	2.25	57
H	Ø.66	Ø17
J	.75	19
K	.250–20 UNC	M6
L	1.25	32
M	2.00	50
N	1.00	25
P	2.00	50
Q	.38	9
R	2.00	50

EX17-10

Draw the following objects as a solid model. Select and draw a standard screw and nut to join parts 220 and 221 through the holes labeled H and L. Specify the selected screw and nut using standard thread drawing callouts. Consider carefully any possible interference between the screws' heads as they are assembled. If necessary, specify assembly sequence.

A. Use inches
B. Use millimeters

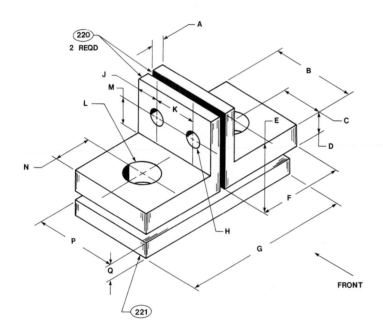

DIMENSION	INCHES	mm
A	.25	6
B	2.00	50
C	1.00	25
D	.50	13
E	1.75	45
F	2.00	50
G	4.00	100
H	Ø.438	Ø11
J	.50	12.5
K	1.00	25
L	Ø.781	Ø19
M	.63	16
N	.88	22
P	2.00	50
Q	.25	6

CHAPTER

Design: Working Drawings

18-1 INTRODUCTION

This chapter shows how to present designs using working drawings. Working drawings are a group of drawings that includes an assembly drawing, detailed drawings of each non-standard part, and a parts list that names all parts required for the design. The chapter also shows how to create a title block.

Figure 18-1 shows a 3D assembly drawing for a pivot assembly. All the sizes are nominal. The assembly is presented using an exploded format, that is, the parts are not shown in their assembled positions. This chapter will use the pivot assembly as a model for developing a set of working drawings.

18-2 SETTING UP THE DRAWINGS

Before starting a group of working drawings, it is best to determine the viewpoints, layers, and linetypes needed; the colors for the various linetypes; and the drawing's units, scale, and limits. As the drawings progress, the required UCSs will also be defined and saved.

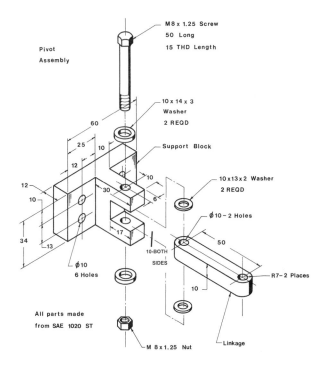

Figure 18-1

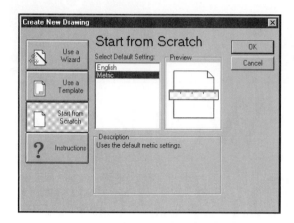 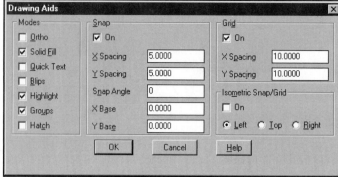

Figure 18-2

To start a new drawing and determine the drawing units and limits

The pivot assembly is dimensioned using millimeters, so decimal units and the default drawing limits of 420,297 will be used. Decimal units are the default setting, so change will not be necessary.

1. Select the New tool from the Standard toolbar.

The Create New Drawing dialog box will appear. See Figure 18-2.

2. Select the Metric option associated with the Start from Scratch option.

The drawing screen is calibrated in decimal units and has a lower left coordinate value of 0,0 and an upper limit of 410,297. This can be verified by moving the crosshairs to the upper right corner of the screen and reading the coordinate value output at the lower left corner of the screen.

To set the Grid and Snap values

1. Select the Tools pull-down menu, then the Drawing Aids command.

The Drawing Aids dialog box will appear. See Figure 18-2.

2. Turn on both Snap and Grid and set the X,Y snap spacing for 5 and the grid X,Y spacing for 10.
3. Select OK.

A 10 x 10 grid will appear on the screen.

To change to a 3D viewpoint

1. Select the SE Isometric tool from the View pull-down menu.
2. Select the Pan tool from the standard toolbar, and move the grid to the center of the screen.

To locate the coordinate system icon on the origin

1. Select the View pull-down menu, then Display, UCS Icon, Origin.

The coordinate system icon will be located on the WCS origin. See Figure 18-3.

Origin icon

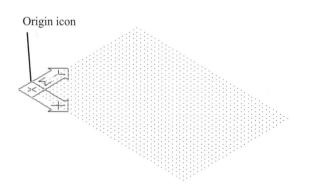

Figure 18-3

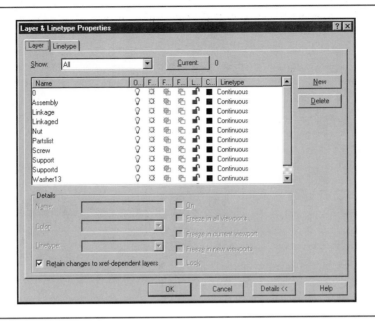

Figure 18-4

To determine and set the layers

The pivot assembly will require a separate layer for each part (6), for the assembly drawing, and for the parts list. In addition, each nonstandard part will require a layer for dimensions. There are two nonstandard parts, so 10 layers are required. Each part layer will be named using the appropriate part name as follows.

1. Select the Layers tool from the Object Properties toolbar.

The Layer & Linetype Properties dialog box will appear. See Figure 18-4.

2. Add the following layers by selecting the New box, then remove the default Layer I name and type in the new layer name.

ASSEMBLY
PARTSLIST
SUPPORT
SUPPORTD
LINKAGE
LINKAGED
SCREW
WASHER14
WASHER13
NUT

The letter D at the end of the support and linkage layers designates the layer that will be used for dimensioning.

18-3 THE SUPPORT BLOCK

The support block will be drawn as a solid model in the WCS.

To draw the support block

1. Select the Support layer and make it the current layer.
2. Select the Box toolbar from the Solids toolbar and draw a box 12 x 60 x 34 with its corner on the WCS origin.

See Figure 18-5. The values 12, 60, and 34 are the length, width, and height values for the box or the X,Y, and Z values.

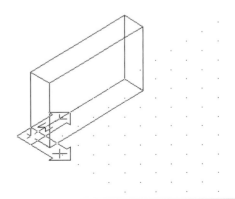

Figure 18-5

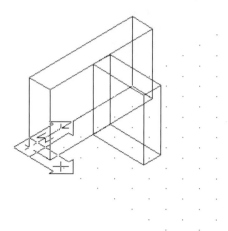

Figure 18-6

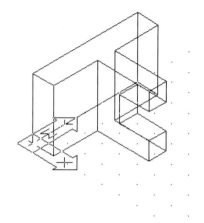

Figure 18-7

3. Select the Box tool from the Solids toolbar and draw a second box with its corner at 12,25,0 and a length equal to 30, width equal to 10, and a height of 34.

The values were determined from the given dimensions. See Figure 18-6.

4. Select the Union tool from the Modify II toolbar (the Union tool is a flyout from the Explode tool).
5. Select the Box tool from the Solids toolbar and draw a box with its corner at 25,25,10 and a length equal to 17, a width equal to 12, and a height equal to 14.

The values were determined from the given dimensions. See Figure 18-7.

6. Subtract the box from the support block.

To add the pivot screw hole

The Isolines value was set at 14 for this example.

1. Select the Cylinder tool from the Solids toolbar.

Elliptical/<center point><0,0,0>:

2. Type 32,30s,0; press Enter.

This point location was determined from the given dimensions and is relative to the WCS origin.

Diameter/<radius>:

3. Type 4; press Enter.

Center of other end/<Height>:

4. Type 34; press Enter.
5. Select the Subtract tool from the Modify II toolbar and subtract the cylinder from the support block.

See Figure 18-8. There are several methods for creating the holes in the base of the support block. The cylinders could be defined by defining the front and back center points relative to the WCS, or a Plan view of the Right option of the Preset UCS tool could be used to switch the drawing to 2D. In this example, the Right option will be used, but the origin will be moved to the front surface and saved. This new UCS will be used for dimensioning.

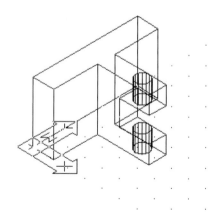

Figure 18-8

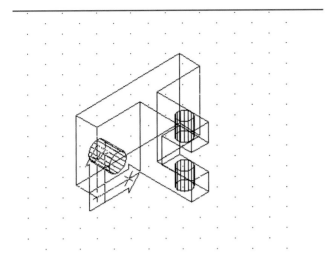

Figure 18-9

To add the four Ø10 holes

1. Select the Preset tool from the UCS toolbar, then select the Right option.

   The UCS icon will change orientations, but will have the same origin as the WCS.

2. Select the Origin tool from the UCS toolbar and use the OSNAP, Endpoint command to locate the UCS origin on the front corner of the support block.

   See Figure 18-8.

3. Select the Save tool from the UCS toolbar and save the UCS using the file name Right.

4. Select the Center Cylinder tool from the Solids toolbar and locate the Ø10 cylinder's center point at 12,11,0, and define a height of -12.

Use the right-hand rule to verify the negative height value. See Figure 18-9. The three additional holes could be created in a similar manner by selecting the Center Cylinder tool and defining their center points based on the given dimensions. The Copy tool can also be used.

5. Select the Copy tool from the Modify toolbar.

   Select object:

6. Select the Ø10 cylinder.

   <Base point or displacement>/Multiple:

7. Type M; press Enter.

   Base point:

8. Type 12,11,0; press Enter.

   Second point of displacement:

9. Type 12,24,0; press Enter.

   Second point of displacement:

10. Type 48,11,0; press Enter.

    Second point of displacement:

11. Type 48,24,0; press Enter.

    See Figure 18-10.

18-4 DIMENSIONING THE SUPPORT BRACKET

The dimensions for the support bracket will be created on the SUPPORTD layer.

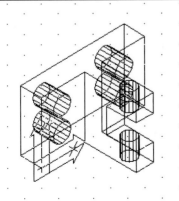

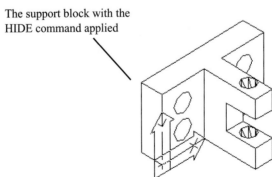

The support block with the HIDE command applied

Figure 18-10

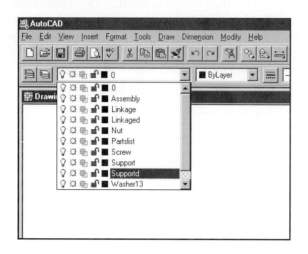

Figure 18-11

To change to the SUPPORTD layer

1. Select the Layers tool from the Objects Properties toolbar.
2. Select the SUPPORTD layer.

The layer name, SUPPORTD, will appear in the Layer Control box indicating that it is the current layer. See Figure 18-11.

To change a layer's color

1. Select the Layers tool from the Objects Properties toolbar.

The Layer Control dialog box will appear.

2. Select the SUPPORTD layer.
3. Select the Set Color box.

The Select Color dialog box will appear.

4. Select the green box, then OK.

The selection of green is arbitrary. Any color can be used for dimension lines, as long as all dimension lines are the same color. The Layer Control dialog box will appear with green listed as the SUPPORTD layer color. See Figure 18-12.

To set up the dimensions

All dimensions are whole numbers, so the units precision must be adjusted from the 0.0000 default setting. Other settings will be as follows.

1. Select the Dimensions Styles tool from the Dimensioning toolbar.
2. Select the Geometry box.

The Geometry Dialog box will appear. See Figure 18-13.

3. Turn on the Line option within the Center box, and set the size factor to 5.0000; then select OK.
4. Select the Format box.

The Format dialog box will appear. See Figure 18-14.

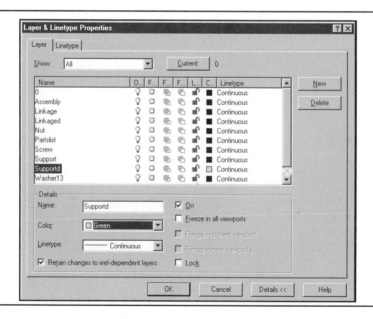

Figure 18-12

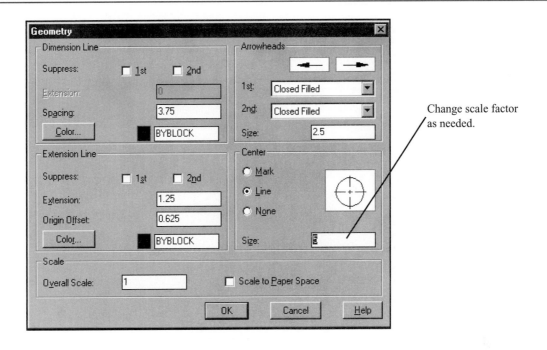

Change scale factor as needed.

Figure 18-13

Select the centered options.

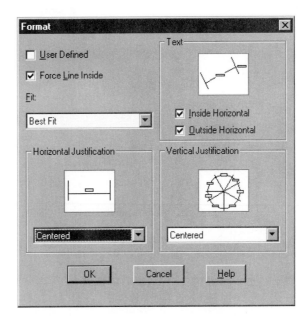

Figure 18-14

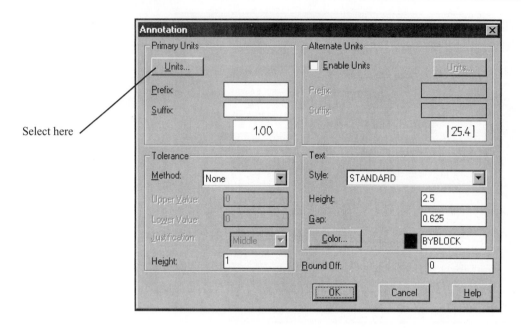

Figure 18-15

5. Set both the Horizontal and Vertical Justification boxes for Centered, select OK.
6. Select the Annotation box.

The Annotation dialog box will appear. See Figure 18-15.

7. Select the Units box.

The Primary Units dialog box will appear. See Figure 18-16.

8. Select Decimal units, then select the arrow box to the right of the Precision box, then select the 0 option.

See Figure 18-17.

9. Select OK and return to the drawing screen.

To move the UCS origin

Dimensions can be drawn only within the current

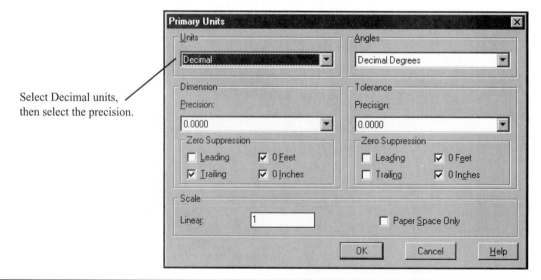

Figure 18-16

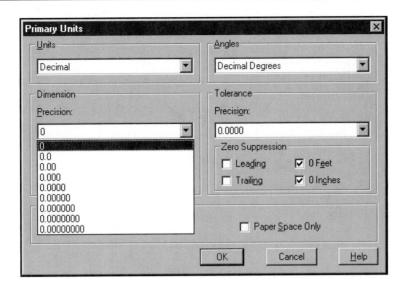

Figure 18-17

UCS and at the orientation of the current UCS. Most of the dimensions are on the front face of the support bracket so a new UCS will be created aligned with the front face. It is assumed that the drawing is currently in the WCS.

1. Select the Presets tool from the UCS toolbar.

   The UCS Orientation dialog box will appear.

2. Select the Right option.

   The Origin icon will shift to the right orientation.

3. Select the Origin tool from the UCS toolbar.

   Origin point <0,0,0>:

4. Use OSNAP, Endpoint to move the origin to the front face of the support bracket.

   See Figure 18-18.

To change to a 2D orientation

1. Select the View pull-down menu, then 3D Viewpoint, then Plan view, and Current UCS.
2. Type zoom; press Enter.

   All/Center/Dynamic/Extents/Left/Previous/Vmax/ Window/<Scale(X/XP)>:

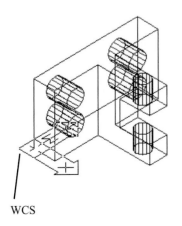
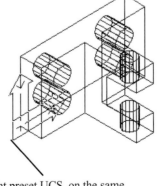
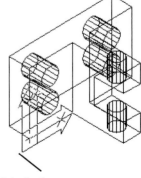
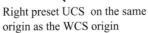

WCS

Right preset UCS on the same origin as the WCS origin

New origin for the Right UCS

Figure 18-18

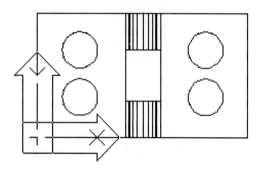

Figure 18-19

3. Type 2; press Enter.

See Figure 18-19.

To draw the circle's center lines

1. Use the Zoom tool from the Standard toolbar and enlarge the support bracket to a comfortable size.
2. Select the Center Mark tool from the Dimensioning toolbar.

Select arc or circle:

3. Select the four circles.

See Figure 18-20. The size of the centerlines is determined by the 5.0000 size factor set above in the Geometry dialog box. Change this scale factor if necessary.

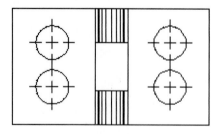

Figure 18-20

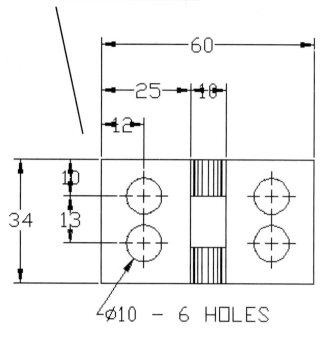

Interference between dimensions

Figure 18-21

To create dimensions on the front face

The UCS icon was turned off using the Options and UCS commands so that the surface could be more clearly seen.

1. Select the Linear Dimensions tool from the Dimensioning toolbar and add the appropriate dimensions.

Use the OSNAP, Endpoint option to ensure accuracy. If, as the dimensions are being applied, an interference between dimensions occurs, such as that shown in Figure 18-21, use the Explode tool to change the dimensions to individual entities, then use the Erase and Move tools as necessary. See Figure 18-22. Some of the dimensions are applied differently than they were in the original drawing, which is a hand-drawn ink drawing that uses press-on letters.

2. Select the Leader tool from the Dimensioning toolbar and dimension the four holes as shown.

See Figure 18-23.

3. Use the Move tool to position the leader text if necessary.

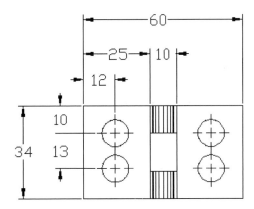

Figure 18-22

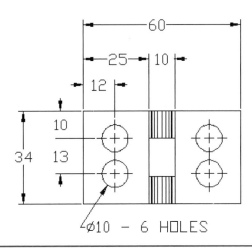

Figure 18-23

To add a drawing note

In this example, a vertical centerline was created and a drawing note added, specifying that the object is symmetrical about the centerline. This means that only half of the object needs to be dimensioned.

The centerline was created by first drawing a vertical line aligned with the left edge of the support bracket, then moving the line 31 millimeters to the right. The Edit and Properties commands were used to change the line into a centerline. The Linetype scale was set at 25.4 and the color red was selected.

1. Select the Pan tool from the Standard toolbar and move the support bracket down and to the right.
2. Type Dtext; press Enter.

   Justify/Style/<Start point>:

3. Select a point in the upper left of the drawing screen.

   <Height>:

4. Type 3; press Enter.

   Rotation angle <0.00>:

5. Press Enter.

   Text:

6. Type NOTE: THE OBJECT IS SYMMETRICAL ABOUT THE VERTICAL CENTERLINE.

   Text:

7. Press Enter.
8. Use the Move tool to position the drawing note if necessary.

   See Figure 18-24.

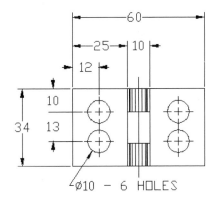

Figure 18-24

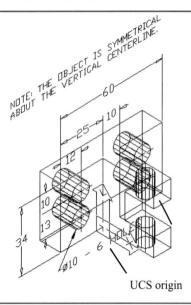

Figure 18-25

To add the other dimensions

The other dimensions are added to the support bracket by selecting the Left and Top Preset UCS, moving the origin to the plane of the dimensions, and adding the appropriate dimensions.

1. Select the World tool from the UCS toolbar.
2. Select the View pull-down menu, then 3D Viewpoint and SE Isometric UCS.
3. Select the Preset tool from the UCS toolbar, then the Front option.

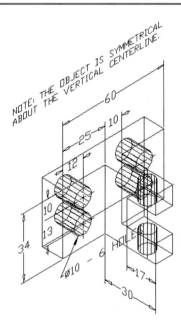

Figure 18-26

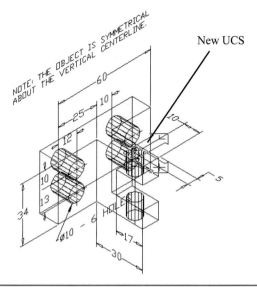

Figure 18-27

4. Select the Origin tool from the UCS toolbar, and position the origin of the Front UCS as shown.

See Figure 18-25.

5. Add the dimensions as shown.

See Figure 18-26. In this example, the dimensions are added while the drawing is in the SE Isometric viewpoint. The drawing could have been dimensioned in 2D by using the plan view orientation.

6. Select the World tool from the UCS toolbar and return to the WCS.
7. Select the Center Mark tool from the Dimensioning toolbar and add centerlines to the top surface of the pivot hole.
8. Select the Origin tool from the UCS toolbar and move the origin as shown.

See Figure 18-27. The HIDE command was applied to hide some of the excess lines and to make it easier to add the top surface dimensions.

9. Add the dimensions to the top surface.

See Figure 18-28.

To dimension the cutout

1. Select the WCS tool from the UCS toolbar.
2. Select the Preset tool from the UCS toolbar, then the Right option.
3. Select the Origin tool from the UCS toolbar and create a new UCS as shown.

See Figure 18-29.

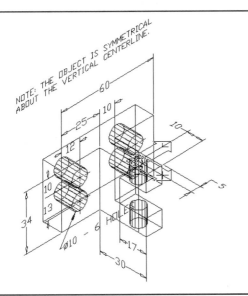

Figure 18-28

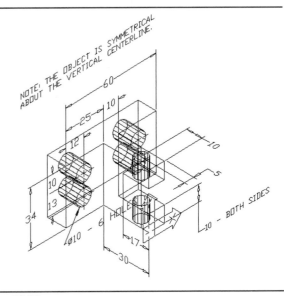

Figure 18-30

4. Add the dimensions for the cutout.

See Figure 18-30.

18-5 THE PIVOT CENTERLINE

The pivot screw, the washers, and the nut are all centered on the same centerline, so a UCS will be created with that centerline as its origin.

To create a new UCS

1. Select the Origin tool from the UCS toolbar and move the origin to the center point on the top surface of the pivot hole.

See Figure 18-31.

2. Save the new UCS as PIVOT.

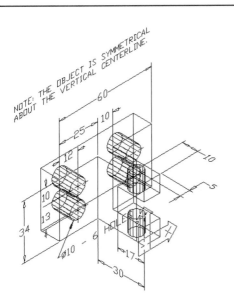

Figure 18-29

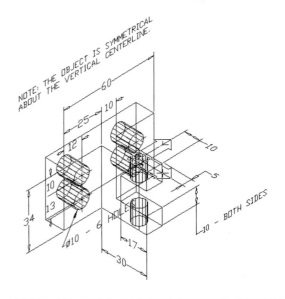

Figure 18-31

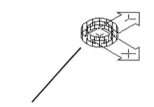

A washer drawn about the origin of the
PIVOT UCS

Figure 18-32

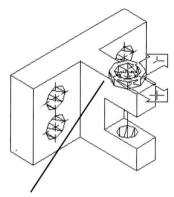

The washer and support block

Figure 18-33

18-6 WASHER — 10 x 14 x 3

This example assumes that the PIVOT UCS is the current UCS.

To draw a 10 x 14 x 3 washer

1. Turn off the SUPPORT and SUPPORTD layers, and turn on the WASHER14 layer.

   Only the PIVOT UCS should appear on the screen.

2. Select the Cylinder tool from the Solids toolbar and draw a cylinder centered at the origin of the PIVOT UCS of diameter 14 and height 3.

3. Select the Cylinder tool again from the Solids toolbar and draw a cylinder centered at the origin of the PIVOT UCS of diameter 10 and height 3.

4. Subtract the inside cylinder from the outside cylinder.

   See Figure 18-32.

To create the second 10 x 14 x 3 washer

1. Turn the SUPPORT layer on, but keep the WASHER14 layer as the current layer.

   See Figure 18-33. The Hide command was applied in Figure 18-31 for clarity.

2. Select the Copy tool from the Modify toolbar.

Select objects:

3. Select the washer.

   <Base point or displacement>/Multiple:

4. Type 0,0,0; press Enter.

   Second point of displacement:

5. Type 0,0,-37; press Enter.

The 37 value was derived from the 34 height of the support block and the 3 thickness of the washer. The bottom surface of the top washer is located on the top surface, and the top surface of the bottom washer is aligned with the bottom surface of the support block, locating the 0,0,0 displacement point to 0,0,-37.

If desired, change the color of the WASHER14 layer.

18-7 WASHER — 10 x 13 x 2

This section assumes that the PIVOT UCS is the current UCS.

1. Make the WASHER13 layer the current layer.
2. Type Hide to create a clearer view of the object.

To create a 10 x 13 x 2 washer

1. Select the Cylinder tool from the Solids toolbar.

   Elliptical/<Center point><0,0,0>:

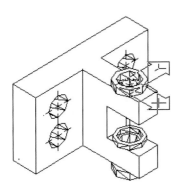

FIgure 18-34

2. Type 0,0,-24; press Enter.

The -24 value was determined from the given dimensions; the top flange has a height of 10 and the cutout 14 (10 + 14 = 24).

Diameter/<Radius>:

3. Type 6.5; press Enter.

Center of other end/<Height>:

4. Type 2; press Enter, Enter (double-click Enter).

Elliptical/<Center point><0,0,0>:

5. Type 0,0,-24; press Enter.

Diameter/<Radius>:

6. Type 5; press Enter.

Center of other end/<Height>:

7. Type 2; press Enter.
8. Select the Subtract tool from the Modify II toolbar and subtract the Ø10 cylinder from the Ø13 cylinder.
9. Type Hide; press Enter.

See Figure 18-34.

To copy the 10 x 13 x 2 washer

This section assumes that the PIVOT UCS is the current UCS.

1. Select the Copy tool from the Modify toolbar.

Select objects:

2. Select the 10 x 13 x 2 washer.

<Basepoint or displacement>/Multiple:

3. Type 0,0,-24; press Enter.

Second base point of displacement:

4. Type 0,0,-12; press Enter.
5. Type Hide; press Enter.

See Figure 18-35. The second point of displacement value of -12 moves the washer up the Z axis 12 units from its original -24 position. The total height of the cutout is 14

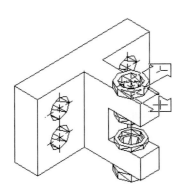

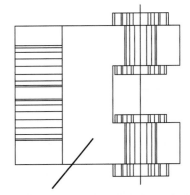

Use the Front option of the 3D Viewpoint Preset command to verify that the washers are in the correct positions.

Figure 18-35

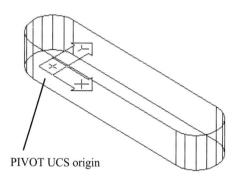

PIVOT UCS origin

Figure 18-36

and the washer is 2 thick, so a change of 12 units will align the top surface of the second washer with the bottom surface of the top flange.

The position of the washer may be verified by selecting the View pull-down menu, then 3D Viewpoint Preset, Right. See Figure 18-35.

18-8 THE LINKAGE

In this example, the linkage will be drawn on the origin of the PIVOT UCS and moved into its correct position. The dimensions will be located on a separate layer.

To create the rounded ends

1. Select the Layer Control tool from the Object Properties toolbar and turn off all the layers; then select LINKAGE as the current layer.

The drawing screen should be blank, except for the PIVOT UCS icon.

2. Select the Cylinder tool from the Solids toolbar.

   Elliptical/<Center point>:

3. Press Enter, accepting the 0,0,0 default center-point.

   Diameter/<Radius>:

4. Type 7; press Enter.

   Center of other end/<Height>:

5. Type 10; press Enter.
6. Select the Copy tool from the Modify toolbar.

   Select objects:

7. Select the cylinder.

   <Base point or displacement>/Multiple:

8. Type 0,0,0; press Enter.

   Second point of displacement:

9. Type 50,0,0; press Enter.

   The value 50 came from the given dimensions, that is, the distance between holes on the linkage.

To create the center section

1. Select the Box tool from the Solids toolbar.

   Center/<Corner of box><0,0,0>:

2. Type 0,-7,0; press Enter.

   The -7 value is equal to the radius of the end section.

   Cube/Length/<Other corner>:

3. Type 50,7,0; press Enter.

   Height:

4. Type 10; press Enter.
5. Select the Union tool from the Modify II toolbar and union the end and center sections.

   See Figure 18-36.

To add the holes

1. Select the Cylinder tool from the Solids toolbar.

   Diameter/<Radius>:

2. Type 5; press Enter.

   Center of other end/<Height>:

3. Type 10; press Enter.
4. Select the Copy tool from the Modify toolbar.

   Select objects:

5. Select the cylinder.

   <Base point or displacement>/Multiple:

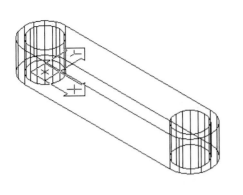

Figure 18-37

6. Type 0,0,0; press Enter.

Second point of displacement:

7. Type 50,0,0; press Enter.
8. Select the Subtract tool from the Modify II toolbar and subtract the two cylinders from the linkage.

See Figure 18-37.

To position the linkage within the assembly

1. Turn the SUPPORT layer on, but keep the LINK-AGE layer as the current layer.
2. Select the Move tool from the Modify toolbar.

   Select objects:

3. Select the linkage.

   Base point of displacement:

4. Type 0,0,0; press Enter.

   The 0,0,0 coordinate value is the origin of the PIVOT UCS.

   Second point of displacement:

5. Type 0,0,-22; press Enter.

   Figure 18-38 shows the results.

6. Turn on the WASHER14 and WASHER13 layers.

   Figure 18-39 shows the assembly with the HIDE command applied.

18-9 DIMENSIONING THE LINKAGE

To create a UCS on the upper surface of the linkage

1. Turn off all layers but the LINKAGE layer.

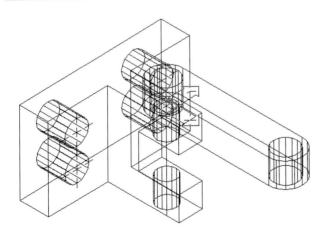

Wire frame of the assembly

Figure 18-38

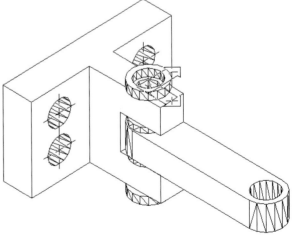

Assembly with the HIDE command applied

Figure 18-39

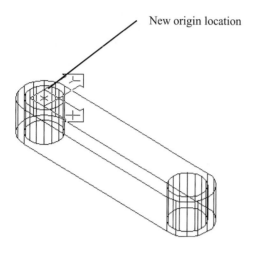

New origin location

Figure 18-40

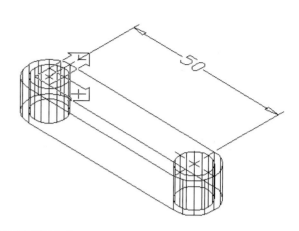

Figure 18-41

2. Select the Layer Control tool from the Objects Properties toolbar and make LINKAGED the current layer.
3. Select the Origin tool from the UCS toolbar.

   Origin point <0,0,0>:

4. Type 0,0,-12; press Enter.

   See Figure 18-40.

5. Select the Save tool from the UCS toolbar and save the new UCS as LINKAGE.

To add dimensions to the linkage

1. Select the Center Mark tool from the Dimensioning toolbar and add centerlines to the two holes.
2. Select the Linear tool from the Dimensioning toolbar and add the 50 dimension.

   See Figure 18-41. The UCS icon is turned off for visual convenience.

3. Select the Leader tool from the Dimensioning toolbar and add the Ø10 dimension.

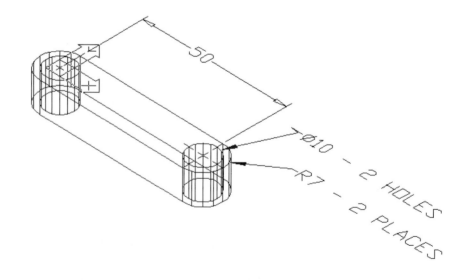

Figure 18-42

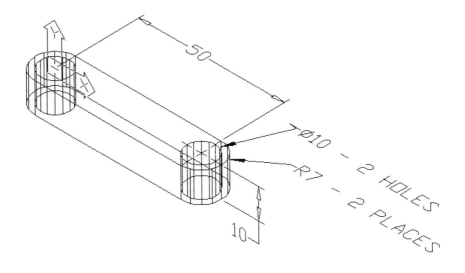

Figure 18-43

4. Select the Leader tool from the Dimensioning toolbar and add the R7 dimension.

See Figure 18-42.

5. Select the Preset tool from the UCS toolbar and select the Front UCS.

This UCS orientation will be used for the linkage height dimension.

6. Select the Linear tool from the Dimensioning toolbar and add the thickness dimension.

See Figure 18-43.

18-10 THE PIVOT SCREW

This section shows how to draw the pivot screw. The head of the pivot screw will be created from an existing WBlock (HEAD) created in Chapter 17.

To set up the drawing

1. Turn off all layers except the SUPPORT layer and make the SCREW layer the current layer.
2. Select the Named tool from the UCS toolbar and make the PIVOT UCS the current UCS.

The M8 pivot screw will be created on the origin of the PIVOT UCS and then moved into its correct position.

To create the body of the screw

1. Select the Cylinder tool from the Solids toolbar.

   Elliptical/<Center point><0,0,0>:

2. Press Enter.

   Diameter/<Radius>:

3. Type 4; press Enter.

   Center of other end/<Height>:

4. Type -50; press Enter.

The screw must pass through the support block, two washers, and a nut (the nut height equals .88D). The screw length must also include an allowance for threads to extend beyond the nut. The calculation is as follows.

```
34 = support block height
 3 = top washer
 3 = bottom washer
 7 = nut height
 2 = threads beyond the nut
49 = minimum height of screw
```

A length of 50 will be used, as 50 is a standard length.

To draw the hexagon head

In Section 17-3 a WBlock of a hexagon head was created and saved. The major diameter of the thread used

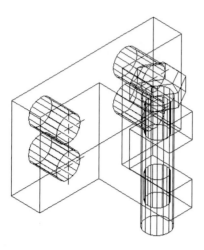

Figure 18-44

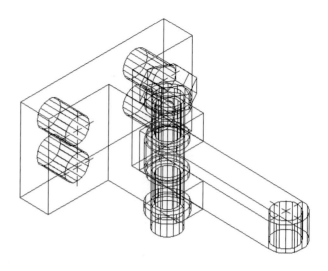

Figure 18-45

to create the head WBlock was Ø2.00 inches, or 2.00 decimal units. The diameter of the pivot screw is 8mm, or 8 decimal units. The hexagon head block can therefore be inserted into the current drawing using a scale factor of 4 (2 x 4 = 8).

1. Select the Insert Block tool from the Draw toolbar and access the HEAD WBlock.

If you do not have a saved hexagon head block, see Section 17-8 for instructions on how to create a hexagon shaped screw head.

Insertion point:

2. Type 0,0,0; press Enter.

X scale factor <1>/Corner/XYZ:

3. Type 4; press Enter.

Y scale factor (default=X):

4. Press Enter.

Rotation angle<0.00>:

5. Press Enter.

See Figure 18-44. The pivot screw, including the head, must now be moved upward along the pivot center line 3 units to allow for the top 10 x 14 x 3 washer.

6. Select the Move tool from the Modify toolbar.

Select objects:

7. Select the pivot screw and hex head.

Base point of displacement:

8. Type 0,0,0; press Enter.

Second point of displacement:

9. Type 0,0,3; press Enter.

Figure 18-45 shows the total assembly with the HIDE command applied.

18-11 M8 NUT

A WBlock for a nut was created in Section 17-6 and will be used in this section. If you do not have a nut block, see Section 17-6 for instructions to create one.

The major diameter of the thread used to create the nut WBlock was Ø2.00 inches, or 2.00 decimal units. The diameter of the pivot screw is 8mm, or 8 decimal units. The nut block can therefore be inserted into the current drawing using a scale factor of 4 (2 x 4 = 8).

To prepare the drawing

1. The PIVOT UCS should be the current UCS.
2. Turn off all the layers except the NUT layer.
3. Make the NUT layer the current layer.

The drawing screen should be blank except for the PIVOT UCS origin icon.

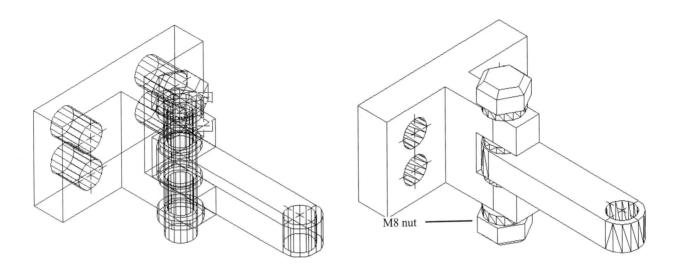

Figure 18-46

To insert the nut WBlock

1. Select the Insert Block tool from the Draw toolbar and access the NUT WBlock.

WBlocks will be listed under the File option of the Insert dialog box.

   Insertion point:

2. Type 0,0,0; press Enter.

   X scale factor <1>/Corner/XYZ:

3. Type 4; press Enter.

   Y scale factor (default=X):

4. Press Enter.

   Rotation angle <0.00>:

5. Press Enter.

To position the nut

The nut must be rotated 180°, then moved to its correct position.

1. Type rotate3d; press Enter.

   Select objects:

2. Select the nut.

   Axis by object/ Last/ View/ Xaxiz/ Yaxis/ Zaxis/ <2points>:

3. Type X; press Enter.

   Point on the X axis <0,0,0>:

4. Press Enter.

   <Rotation angle>/Reference:

5. Type 180; press Enter.
6. Select the Move tool from the Modify toolbar.

   Select objects:

7. Select the nut.

   Base point or displacement:

8. Type 0,0,0; press Enter.

   Second point of displacement:

9. Type 0,0,-37; press Enter.

Figure 18-46 shows the finished pivot assembly drawing in wire frame format and with the HIDE command applied.

18-12 THE ASSEMBLY DRAWING

Each part on an assembly drawing must be assigned an assembly number. In this example, these numbers will be assigned and applied in the ASSEMBLY layer. In general, the largest parts are assigned the smallest numbers, so the support block will be part 1.

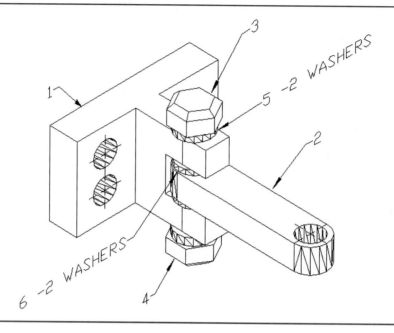

Figure 18-47

To number the assembly drawing

1. Select the Layer Control tool from the Object Properties toolbar and make the ASSEMBLY layer the current layer.
2. Type Hide; press Enter.

The HIDE command is applied to make it easier to see the various parts of the assembly. The drawing is currently in the PIVOT UCS.

3. Select the Preset tool from the UCS toolbar, then Right UCS.

4. Select the Leader tool from the Dimensioning toolbar and create the assembly parts numbers.

See Figure 18-47.

5. Select the Circle tool from the Draw toolbar and draw a circle around assembly number 1.
6. Select the Multiple option of the Copy tool and locate a circle around all the assembly numbers.

See Figure 18-48.

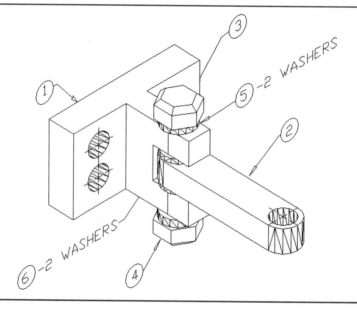

Figure 18-48

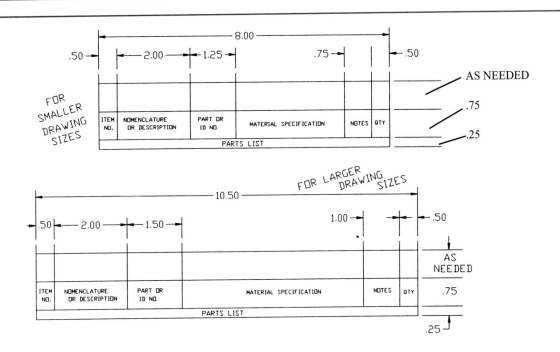

Figure 18-49

18-13 PARTS LIST

A parts list is a listing of all parts used in an assembly. The information included in a parts list varies greatly from company to company. Figure 18-49 shows formats for two different parts lists; Figure 18-50 shows a parts list for the pivot assembly using a different format. The listed items are as follows.

PARTS LIST				
ITEM NO	DESCRIPTION	PART NO	MATL	QTY
1	SUPPORT BLOCK	AM311	SAE1020	1
2	LINKAGE	AM312	SAE1020	1
3	PIVOT SCREW M8X1.25X50 HEXHEAD SCREW-15 LONG THREAD		STEEL	1
4	M8 HEX NUT		STEEL	1
5	10X14X3 PLAIN WASHER		STEEL	2
6	10X13X2 PLAIN WASHER		STEEL	2

Figure 18-50

Item Number

Item numbers are the assembly numbers assigned in the last section. Item numbers reference only individual assembly drawings.

Description

A description can be either a part's name or a drawing callout that completely defines the part. The support block and linkage are described using their names. Standard parts that are purchased from an outside vendor are identified using their drawing callouts. For example, the 10 x 14 x 3 plain washer is sufficient to describe the required washer.

Part Number

A part number is an individual number assigned to a part. Standard purchase parts, such as screws and washers, need not be assigned parts numbers. Each part that is to be manufactured must have a part number. The part number will also be used as the part's drawing number.

Part numbers are different from item numbers. An individual part may be used on more than one assembly. Each assembly part may be assigned a different item number, but the part number will always be the same.

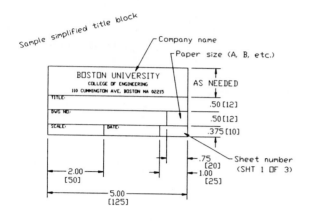

Figure 18-51

Material

The material requirements for each part must be listed. The designation SAE1020 means Society of Automotive Engineers 1020 steel designation. The number 1020 refers to the carbon content of the steel. Material specification should be in accordance with nationally accepted standards.

Quantity

List the quantity of each part required for the assembly.

To draw parts list

1. Select the Layer Control tool from the Objects Properties toolbar and turn off all layers.
2. Make the PARTSLIST layer the current layer.
3. Draw one of the formats shown in Figure 18-49 or define your own.
4. Type Dtext; type the needed information.

18-14 TITLE BLOCKS

A title block includes a minimum of the drawing's name and number, the company's name, the drawing scale, the release date of the drawing, and the sheet number of the drawing. Other information may be included. Figure 18-51 shows a sample title block. Figure 18-52 shows two general formats. Title blocks may be designed with attributes. See Chapter 9.

Drawing titles (names)

Drawing titles should be chosen so they clearly define the function of the part. They should be presented in the following word sequence.

Noun, modifier, modifying phrase

For example

SHAFT,HIGH SPEED,LEFT HAND
GASKET,LOWER

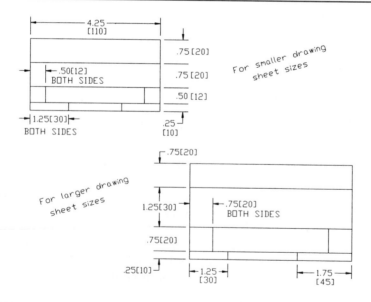

Figure 18-52

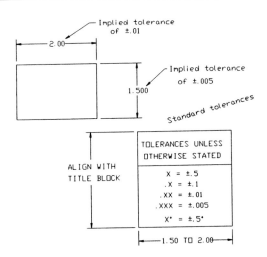

Figure 18-53

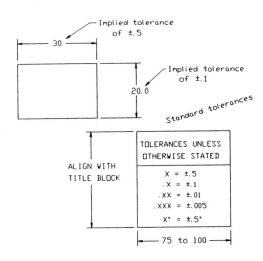

Figure 18-54

Noun names may be two words if the normal usage includes the two words.

GEAR BOX,COMPOSITE
SHOCK ABSORBER,LEFT

Drawing numbers

Drawing numbers are assigned by companies according to their usage requirements. Usually drawing numbers are recorded in a log book to prevent duplication of numbers. Drawing numbers are the same as the parts numbers listed in the parts list.

Company name

The company's name and logo are preprinted on drawing paper or are included as a WBlock so they can be inserted on each drawing.

Scale

Define the scale of the drawing.

SCALE: FULL or 1 = 1

The drawing's size is the same as the part's size.

SCALE: 2 = 1

The drawing's size is twice as large as the part's size.

SCALE: 1 = 2

The drawing's size is half as large as the part's size.

Release date

A drawing is released only after all persons required by the company's policy have reviewed the drawing. Once released, a drawing becomes a legal document. Copies of drawings that have not been released are often stamped with statements like "NOT RELEASED" or "FOR REFERENCE ONLY."

Sheet

The number of the sheet relative to the total number of sheets that make up the drawing should be stated clearly.

SHEET 2 of 3

or

SH 2 OF 3

18-15 TOLERANCE BLOCKS

Most drawings include a tolerance block next to the title block that lists the standard tolerances that apply to the dimensions on the drawing. A dimension that does not have a specific tolerance is assumed to have the appropriate standard tolerance.

Figure 18-53 shows a sample tolerance block for inch values; Figure 18-54 shows a sample tolerance block for millimeter values. In Figure 18-53 the dimension 2.00 has an implied tolerance of ±.01. In Figure 18-54 the 30 dimension has an implied tolerance of ±.5.

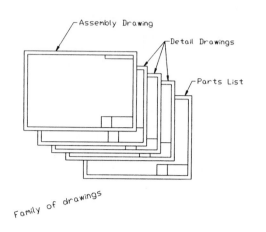

Figure 18-55

18-16 DRAWING PRESENTATION

An assembly drawing, along with its parts list and appropriate detail drawings, is referred to as a family of drawings. See figure 18-55. Detail drawings are dimensioned drawings of individual nonstandard parts. In this example there are two detailed drawings required: the support block and the linkage.

To create a drawing format

1. Draw a title block and save it as a Block and a WBlock.
2. Draw a tolerance block and save it as a Block and a WBlock.
3. Combine both the title and tolerance block to form a new block and WBlock, with the lower right corner of the block as the insertion point.

To create an assembly drawing

1. Select the Layer Control tool from the Object Properties toolbar; turn on all layers except the dimension and parts list layers.
2. Select the View pull-down menu, then Model Space (Floating).

ON/OFF/Hideport/Fit/2/3/4/Restore/ <First point>:

3. Select a point near the upper left corner of the screen.

Other corner:

4. Select a corner in the lower right corner of the screen.

5. Type Hide; press Enter.

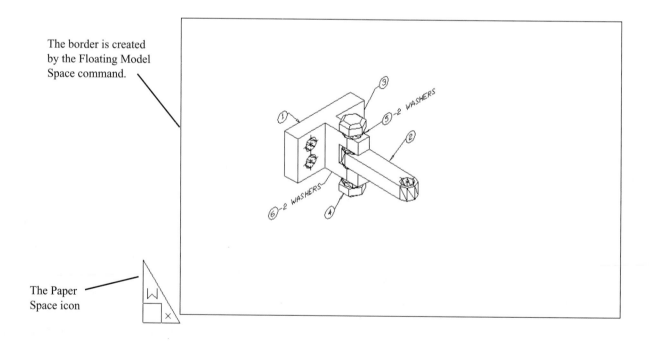

The border is created by the Floating Model Space command.

The Paper Space icon

Figure 18-56

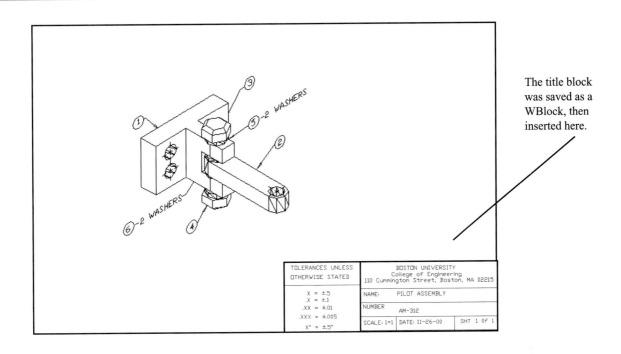

The title block was saved as a WBlock, then inserted here.

Figure 18-57

To add the title block

1. Double-click the word Model at the bottom of the screen to change the drawing to Paper space.

   See figure 18-56.

2. Select the Insert Block tool from the Draw toolbar and access the combined title and tolerance block.

   Insertion point:

3. Use the OSNAP, Endpoint command to select the lower right corner of the rectangular border created as part of Floating Model Space.
4. Define the X and Y scale factors of the block as .5 with a 0 rotation angle.

   See Figure 18-57. The drawing may be saved using the Save As option as a new drawing.

To create a detailed drawing of the support block

1. Select the UNDO command from the Standard toolbar and return to the original assembly drawing in Model space.
2. Select the Layer Control tool from the Object Properties toolbar and turn off all layers except SUPPORT and SUPPORTD; make the SUPPORT layer the current layer.

   Select the Zoom All tool, if necessary, to fit the object and its dimensions within the drawing screen, and the Zoom tool to enlarge the object.

3. Repeat the procedure outlined above and add a title and tolerance block to the drawing.

   See Figure 18-58. The drawing can be saved as a new drawing using the Save As command.

To present the linkage

1. Use the Undo tool from the Standard toolbar and return to the original assembly drawing in Model space.
2. Turn off all layers except the LINKAGE and LINKAGED layers: make the LINKAGE layer the current layer.
3. Repeat the procedure outlined above and add a title and tolerance block to the drawing.

   See Figure 18-59. The drawing can be saved as a new drawing using the Save As command.

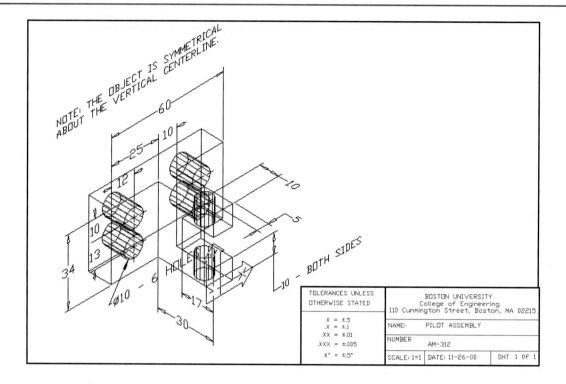

Figure 18-58

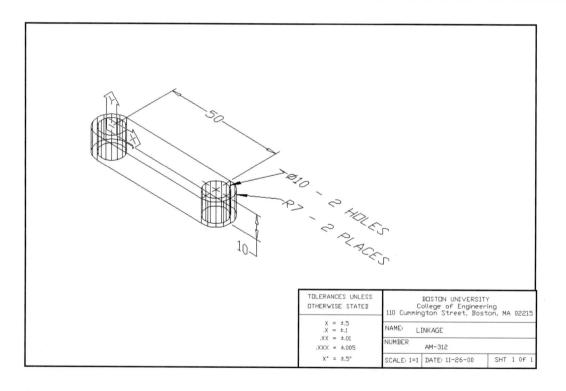

Figure 18-59

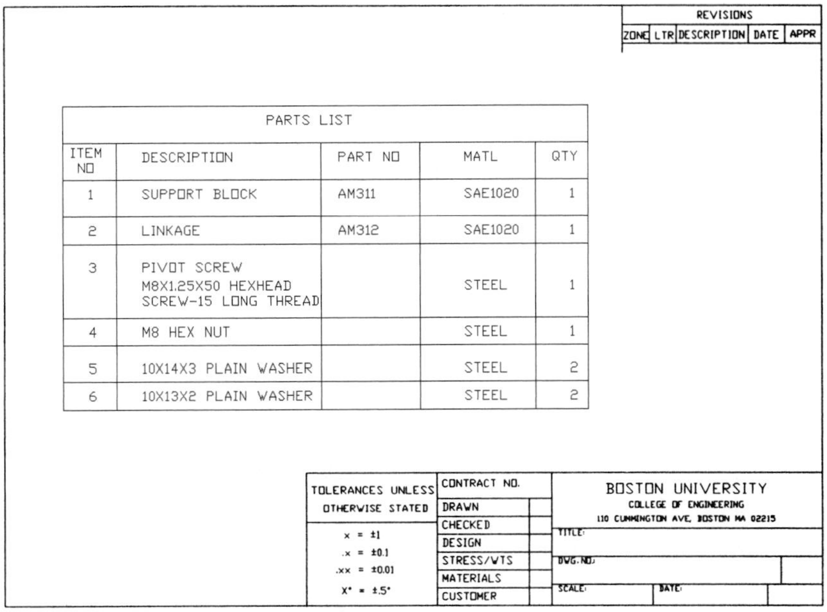

PARTS LIST				
ITEM NO	DESCRIPTION	PART NO	MATL	QTY
1	SUPPORT BLOCK	AM311	SAE1020	1
2	LINKAGE	AM312	SAE1020	1
3	PIVOT SCREW M8X1.25X50 HEXHEAD SCREW-15 LONG THREAD		STEEL	1
4	M8 HEX NUT		STEEL	1
5	10X14X3 PLAIN WASHER		STEEL	2
6	10X13X2 PLAIN WASHER		STEEL	2

REVISIONS
ZONE | LTR | DESCRIPTION | DATE | APPR

TOLERANCES UNLESS OTHERWISE STATED
x = ±1
.x = ±0.1
.xx = ±0.01
X° = ±.5°

CONTRACT NO.
DRAWN
CHECKED
DESIGN
STRESS/WTS
MATERIALS
CUSTOMER

BOSTON UNIVERSITY
COLLEGE OF ENGINEERING
110 CUMMINGTON AVE, BOSTON MA 02215
TITLE:
DWG.NO.:
SCALE: DATE:

Figure 18-60

To present the parts list

Parts lists are usually presented using a drawing format as shown in Figure 18-60. Some companies have a separate format for parts lists that are included with the required drawings.

18-17 EXERCISE PROBLEMS

Draw the assemblies shown in EX18-1 through EX18-7 as solid models. Prepare the following drawings.

A. An assembly drawing
B. A detailed drawing of each nonstandard part
C. A parts list of all parts in the assembly

EX18-1 MILLIMETERS

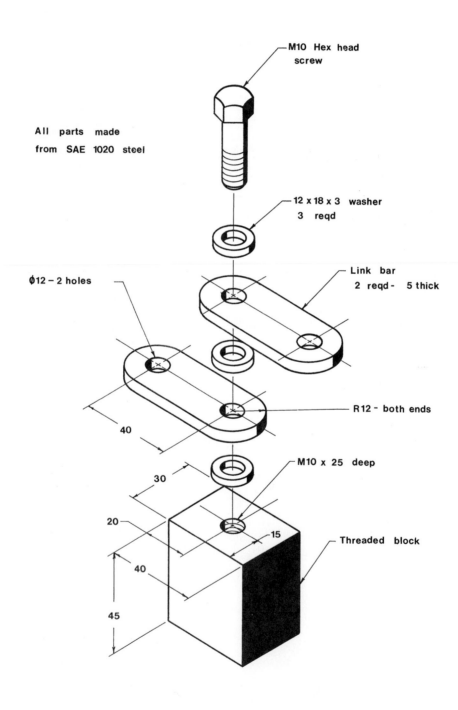

EX18-2 MILLIMETERS

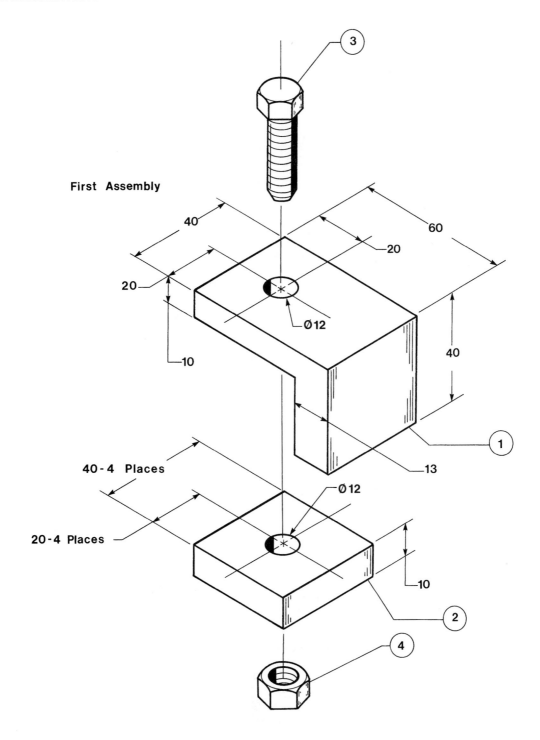

First Assembly

EX18-3 MILLIMETERS

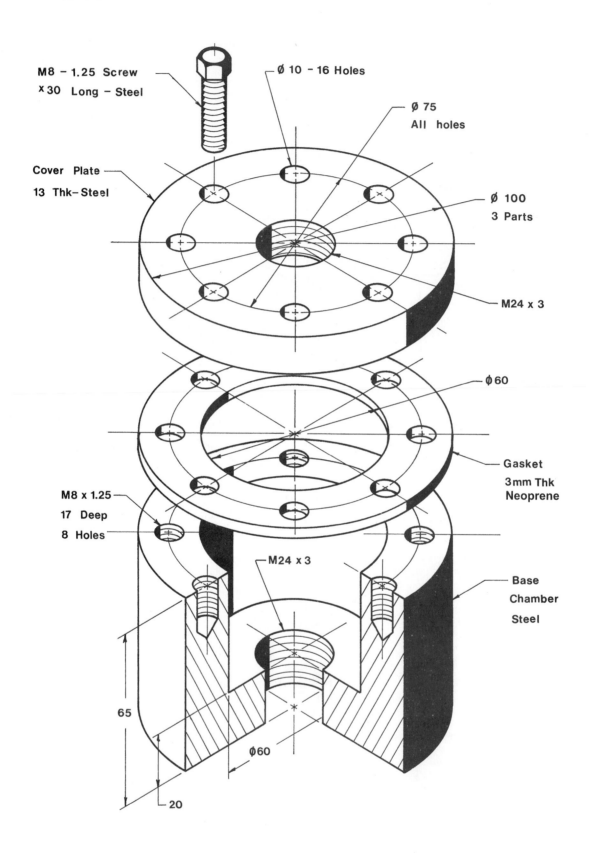

M8 – 1.25 Screw
× 30 Long – Steel

∅ 10 – 16 Holes

∅ 75
All holes

Cover Plate
13 Thk–Steel

∅ 100
3 Parts

M24 x 3

∅ 60

Gasket
3mm Thk
Neoprene

M8 x 1.25
17 Deep
8 Holes

M24 x 3

Base
Chamber
Steel

65

20

∅ 60

EX18-4 MILLIMETERS

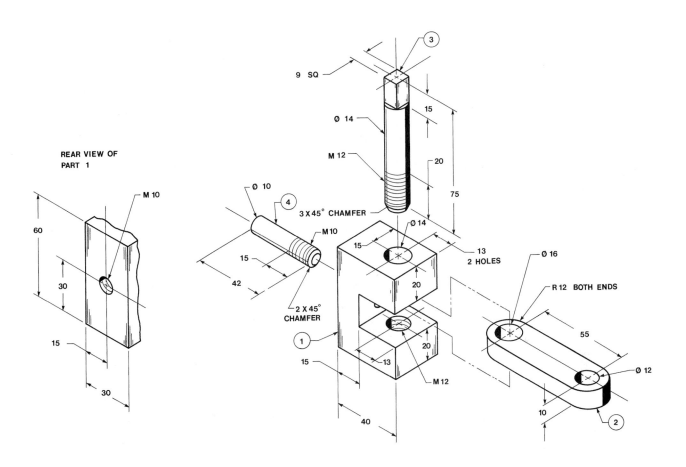

REAR VIEW OF
PART 1

9 SQ

Ø 14

M 12

15

20

75

3 X 45° CHAMFER

Ø 10

M 10

15

42

2 X 45°
CHAMFER

M 10

60

30

15

30

Ø 14
13
2 HOLES

15

20

20

13

M 12

15

40

Ø 16

R 12 BOTH ENDS

55

Ø 12

10

1

2

3

4

EX18-5 MILLIMETERS

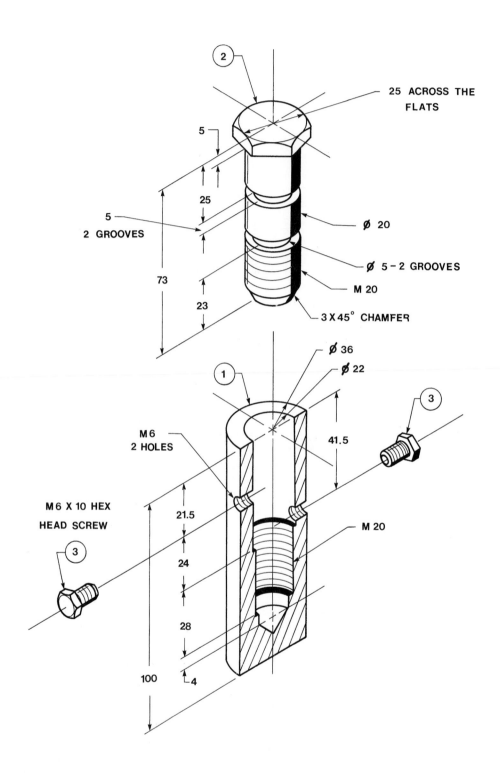

EX18-6 INCHES

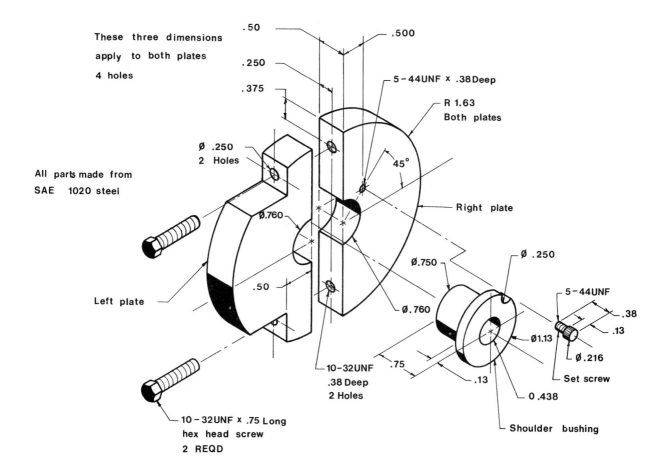

These three dimensions
apply to both plates
4 holes

.50

.500

.250

.375

5 – 44UNF x .38 Deep

R 1.63
Both plates

Ø .250
2 Holes

45°

All parts made from
SAE 1020 steel

Ø.760

Right plate

Ø.750

Ø .250

5 – 44UNF

.38

.13

.50

Left plate

Ø.760

Ø1.13

Ø .216

Set screw

10 – 32UNF
.38 Deep
2 Holes

.75

.13

0 .438

10 – 32UNF x .75 Long
hex head screw
2 REQD

Shoulder bushing

EX18-7 INCHES

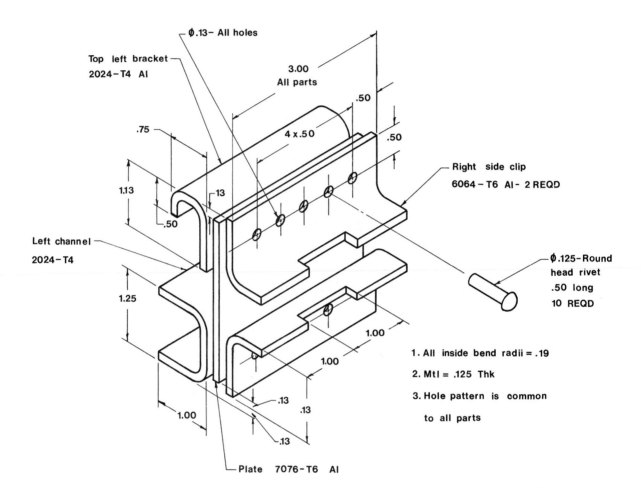

Φ.13– All holes

Top left bracket
2024–T4 Al

3.00
All parts

.50

.75

4 x .50

.50

1.13

.13

.50

Right side clip
6064–T6 Al– 2 REQD

Left channel
2024–T4

Φ.125– Round
head rivet
.50 long
10 REQD

1.25

1.00

1.00

1.00

1. All inside bend radii = .19

2. Mtl = .125 Thk

3. Hole pattern is common

   to all parts

.13

.13

.13

Plate 7076–T6 Al

Redraw the assemblies shown in EX18-8 through EX18-13 as solid models. Select the appropriate standard fasteners and supply all missing dimensions. Use either millimeters or inches. Clearly state your choice of units. Create the following drawings for each assembly.

    A. An assembly drawing

    B. Detailed drawings for all nonstandard parts

    C. A parts list of all parts in the assembly

EX18-8

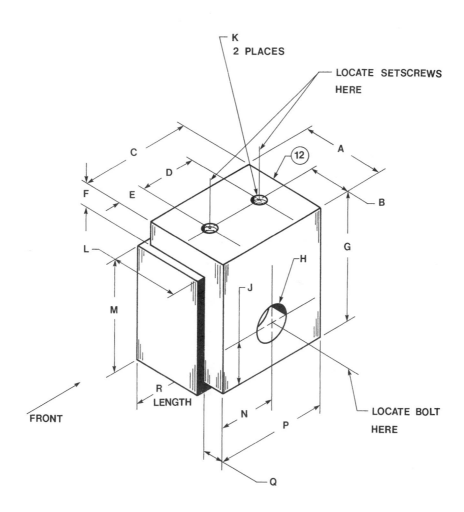

EX18-9

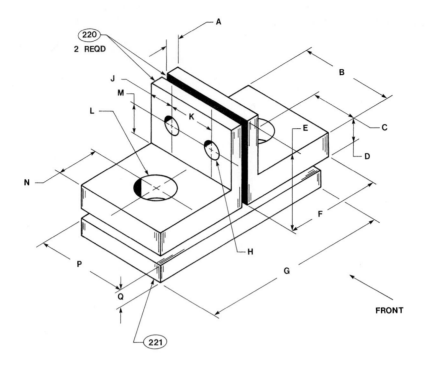

EX18-10

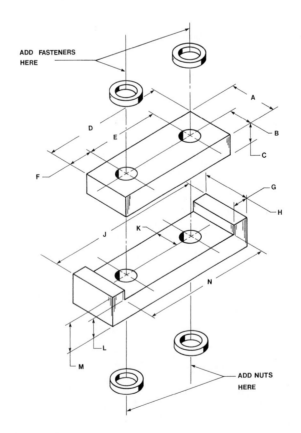

EX18-11

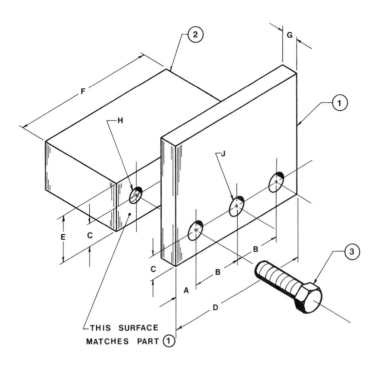

EX18-12

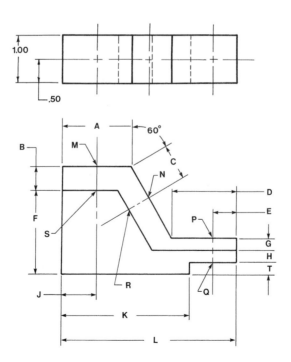

EX18-13

SIMPLIFIED SURFACE GAGE

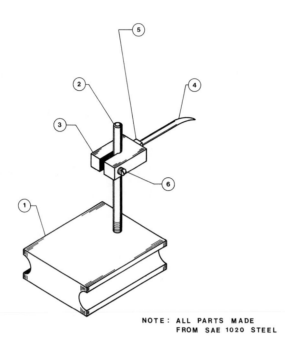

NOTE: ALL PARTS MADE
FROM SAE 1020 STEEL

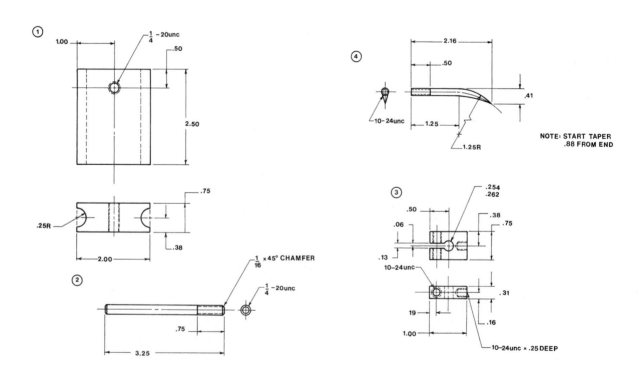

Appendix

Wire and Sheet Metal Gages

Gage		Thickness	Gage	Thickness
000	000	0.5800	18	0.0403
00	000	0.5165	19	0.0359
0	000	0.4600	20	0.0320
	000	0.4096	21	0.0285
	00	0.3648	22	0.0253
	0	0.3249	23	0.0226
	1	0.2893	24	0.0201
	2	0.2576	25	0.0179
	3	0.2294	26	0.0159
	4	0.2043	27	0.0142
	5	0.1819	28	0.0126
	6	0.1620	29	0.0113
	7	0.1443	30	0.0100
	8	0.1285	31	0.0089
	9	0.1144	32	0.0080
	10	0.1019	33	0.0071
	11	0.0907	34	0.0063
	12	0.0808	35	0.0056
	13	0.0720	36	0.0050
	14	0.0641	37	0.0045
	15	0.0571	38	0.0040
	16	0.0508	39	0.0035
	17	0.0453	40	0.0031

Figure A-1

American Standard Clearance Locational Fits

Nominal Size Range Inches Over To	Limits of Clearance	Class LC1 Standard Limits Hole H6	Shaft h5	Limits of Clearance	Class LC2 Standard Limits Hole H7	Shaft h6	Limits of Clearance	Class LC3 Standard Limits Hole H8	Shaft h7	Limits of Clearance	Class LC4 Standard Limits Hole H10	Shaft h9
0 - 0.12	0 0.45	+0.25 0	0 -0.2	0 0.65	+0.4 0	0 -0.25	0 1	+0.6 0	0 -0.4	0 2.6	+1.6 0	0 -1.0
0.12 - 0.24	0 0.5	+0.3 0	0 -0.2	0 0.8	+0.5 0	0 -0.3	0 1.2	+0.7 0	0 -0.5	0 3.0	+1.8 0	0 -1.2
0.24 - 0.40	0 0.65	+0.4 0	0 -0.25	0 1.0	+0.6 0	0 -0.4	0 1.5	+0.9 0	0 -0.6	0 3.6	+2.2 0	0 -1.4
0.40 - 0.71	0 0.7	+0.4 0	0 -0.3	0 1.1	+0.7 0	0 -0.4	0 1.7	+1.0 0	0 -0.7	0 4.4	+2.8 0	0 -1.6
0.71 - 1.19	0 0.9	+0.5 0	0 -0.4	0 1.3	+0.8 0	0 -0.5	0 2	+1.2 0	0 -0.8	0 5.5	+3.5 0	0 -2.0
1.19 - 1.97	0 1.0	+0.6 0	0 -0.4	0 1.6	+1.0 0	0 -0.6	0 2.6	+1.6 0	0 -1.0	0 6.5	+4.0 0	0 -2.5

Figure A-2A

Nominal Size Range Inches Over To	Limits of Clearance	Class LC5 Standard Limits Hole H7	Shaft g6	Limits of Clearance	Class LC6 Standard Limits Hole H9	Shaft f8	Limits of Clearance	Class LC7 Standard Limits Hole H10	Shaft e9	Limits of Clearance	Class LC8 Standard Limits Hole H10	Shaft d9
0 - 0.12	0.1 0.75	+0..4 0	-0.1 -0.35	0.3 1.9	+1.0 0	-0.3 -0.9	0.6 3.2	+1.6 0	-0.6 -1.6	1.0 3.6	+1.6 0	-1.0 -2.0
0.12 - 0.24	0.15 0.95	+0.5 0	-0.15 -0.45	0.4 2.3	+1.2 0	-0.4 -1.1	0.8 3.8	+1.8 0	-0.8 -2.0	1.2 4.2	+1.8 0	-1.2 -2.4
0.24 - 0.40	0.2 1.2	+0.6 0	-0.2 -0.6	0.5 2.8	+1.4 0	-0.5 -1.4	1.0 4.6	+2.2 0	-1.0 -2.4	1.6 5.2	+2.2 0	-1.6 -3.0
0.40 - 0.71	0.25 1.35	+0.7 0	-0.25 -0.65	0.6 3.2	+1.6 0	-0.6 -1.6	1.2 5.6	+2.8 0	-1.2 -2.8	2.0 6.4	+2.8 0	-2.0 -3.6
0.71 - 1.19	0.3 1.6	+0.8 0	-0.3 -0.8	0.8 4.0	+2.0 0	-0.8 -2.0	1.6 7.1	+3.5 0	-1.6 -3.6	2.5 8.0	+3.5 0	-2.5 -4.5
1.19 - 1.97	0.4 2.0	+1.0 0	-0.4 -1.0	1.0 5.1	+2.5 0	-1.0 -2.6	2.0 8.5	+4.0 0	-2.0 -4.5	3.0 9.5	+4.0 0	-3.0 -5.5

Figure A-2B

American Standard Running and Sliding Fits
(Hole Basis)

Nominal Size Range Inches Over — To	Limits of Clearance	Class RC1 Standard Limits Hole H5	Shaft g4	Limits of Clearance	Class RC2 Standard Limits Hole H6	Shaft g5	Limits of Clearance	Class RC3 Standard Limits Hole H7	Shaft f6	Limits of Clearance	Class RC4 Standard Limits Hole H8	Shaft f7
0 — 0.12	0.1 / 0.45	+0.2 / 0	−0.1 / −0.25	0.1 / 0.55	+0.25 / 0	−0.1 / −0.3	0.3 / 0.95	+0.4 / 0	−0.3 / −0.55	0.3 / 1.3	+0.6 / 0	−0.3 / −0.7
0.12 — 0.24	0.15 / 0.5	+0.2 / 0	−0.15 / −0.3	0.15 / 0.65	+0.3 / 0	−0.15 / −0.35	0.4 / 1.12	+0.5 / 0	−0.4 / −0.7	0.4 / 1.5	+0.7 / 0	−0.4 / −0.0
0.24 — 0.40	0.2 / 0.6	+0.25 / 0	−0.2 / −0.35	0.2 / 0.85	+0.4 / 0	−0.2 / −0.45	0.5 / 1.5	+0.6 / 0	−0.5 / −0.9	0.5 / 2.0	+0.9 / 0	−0.5 / −1.1
0.40 — 0.71	0.25 / 0.75	+0.3 / 0	−0.25 / −0.45	0.25 / 0.95	+0.4 / 0	−0.25 / −0.55	0.6 / 1.7	+0.7 / 0	−0.6 / −1.0	0.6 / 2.3	+1.0 / 0	−0.6 / −1.3
0.71 — 1.19	0.3 / 0.95	+0.4 / 0	−0.3 / −0.55	0.3 / 1.2	+0.5 / 0	−0.3 / −0.7	0.8 / 2.1	+0.8 / 0	−0.8 / −1.3	0.8 / 2.8	+1.2 / 0	−0.8 / −1.6
1.19 — 1.97	0.4 / 1.1	+0.4 / 0	−0.4 / −0.7	0.4 / 1.4	+0.6 / 0	−0.4 / −0.8	1.0 / 2.6	+1.0 / 0	−1.0 / −1.6	1.0 / 3.6	+1.6 / 0	−1.0 / −2.0

Figure A-3A

Nominal Size Range Inches Over — To	Limits of Clearance	Class RC5 Standard Limits Hole H8	Shaft e7	Limits of Clearance	Class RC6 Standard Limits Hole H9	Shaft e8	Limits of Clearance	Class RC7 Standard Limits Hole H9	Shaft d8	Limits of Clearance	Class RC8 Standard Limits Hole H10	Shaft c9
0 — 0.12	0.6 / 1.6	+0.6 / 0	−0.6 / −1.0	0.6 / 2.2	+1.0 / 0	−0.6 / −1.2	1.0 / 2.6	+1.0 / 0	−1.0 / −1.6	2.5 / 5.1	+1.6 / 0	−2.5 / −3.5
0.12 — 0.24	0.8 / 2.0	+0.7 / 0	−0.8 / −1.3	0.8 / 2.7	+1.2 / 0	−0.8 / −1.5	1.2 / 3.1	+1.2 / 0	−1.2 / −1.9	2.8 / 5.8	+1.8 / 0	−2.8 / −4.0
0.24 — 0.40	1.0 / 2.5	+0.9 / 0	−1.0 / −1.6	1.0 / 3.3	+1.4 / 0	−1.0 / −1.9	1.6 / 3.9	+1.4 / 0	−1.6 / −2.5	3.0 / 6.6	+2.2 / 0	−3.0 / −4.4
0.40 — 0.71	1.2 / 2.9	+1.0 / 0	−1.2 / −1.9	1.2 / 3.8	+1.6 / 0	−1.2 / −2.2	2.0 / 4.6	+1.6 / 0	−2.0 / −3.0	3.5 / 7.9	+2.8 / 0	−3.5 / −5.1
0.71 — 1.19	1.6 / 3.6	+1.2 / 0	−1.6 / −2.4	1.6 / 4.8	+2.0 / 0	−1.6 / −2.8	2.5 / 5.7	+2.0 / 0	−2.5 / −3.7	4.5 / 10.0	+3.5 / 0	−4.5 / −6.5
1.19 — 1.97	2.0 / 4.6	+1.6 / 0	−2.0 / −3.0	2.0 / 6.1	+2.5 / 0	−2.0 / −3.6	3.0 / 7.1	+2.5 / 0	−3.0 / −4.6	5.0 / 11.5	+4.0 / 0	−5.0 / −7.5

Figure A-3B

American Standard Transition Locational Fits

Nominal Size Range Inches		Class LT1				Class LT2				Class LT3	
	Fit	Standard Limits		Fit	Standard Limits			Fit	Standard Limits		
Over — To		Hole H7	Shaft js6		Hole H8	Shaft js7				Hole H7	Shaft k6
0 — 0.12	-0.10 +0.50	+0.4 0	+0.10 -0.10	-0.2 +0.8	+0.6 0	+0.2 -0.2				
0.12 — 0.24	-0.15 -0.65	+0.5 0	+0.15 -0.15	-0.25 +0.95	+0.7 0	+0.25 -0.25				
0.24 — 0.40	-0.2 +0.5	+0.6 0	+0.2 -0.2	-0.3 +1.2	+0.9 0	+0.3 -0.3	-0.5 +0.5		+0.6 0	+0.5 +0.1
0.40 — 0.71	-0.2 +0.9	+0.7 0	+0.2 -0.2	-0.35 +1.35	+1.0 0	+0.35 -0.35	-0.5 +0.6		+0.7 0	+0.5 +0.1
0.71 — 1.19	-0.25 +1.05	+0.8 0	+0.25 -0.25	-0.4 +1.6	+1.2 0	+0.4 -0.4	-0.6 +0.7		+0.8 0	+0.6 +0.1
1.19 — 1.97	-0.3 +1.3	+1.0 0	+0.3 -0.3	-0.5 +2.1	+1.6 0	+0.5 -0.5	+0.7 +0.1		+1.0 0	+0.7 +0.1

Figure A-4A

Nominal Size Range Inches		Class LT4				Class LT5				Class LT6	
	Fit	Standard Limits		Fit	Standard Limits			Fit	Standard Limits		
Over — To		Hole H8	Shaft k7		Hole H7	Shaft n6				Hole H7	Shaft n7
0 — 0.12				-0.5 +0.15	+0.4 0	+0.5 +0.25	-0.65 +0.15			+0.4 0	+0.65 +0.25
0.12 — 0.24				-0.6 +0.2	+0.5 0	+0.6 +0.3	-0.8 +0.2			+0.5 0	+0.8 +0.3
0.24 — 0.40	-0.7 +0.8	+0.9 0	+0.7 +0.1	-0.8 +0.2	+0.6 0	+0.8 +0.4	-1.0 +0.2			+0.6 0	+1.0 +0.4
0.40 — 0.71	-0.8 +0.9	+1.0 0	+0.8 +0.1	-0.9 +0.2	+0.7 0	+0.9 +0.5	-1.2 +0.2			+0.7 0	+1.2 +0.5
0.71 — 1.19	-0.9 +1.1	+1.2 0	+0.9 +0.1	-1.1 +0.2	+0.8 0	+1.1 +0.6	-1.4 +0.2			+0.8 0	+1.4 +0.6
1.19 — 1.97	-1.1 +1.5	+1.6 0	+1.1 +0.1	-1.3 +0.3	+1.0 0	+1.3 +0.7	-1.7 +0.3			+1.0 0	+1.7 +0.7

Figure A-4B

American Standard Interference Locational Fits

Nominal Size Range Inches Over — To	Limits of Interference	Class LN1		Limits of Interference	Class LN2		Limits of Interference	Class LN3	
		Standard Limits			Standard Limits			Standard Limits	
		Hole H6	Shaft n5		Hole H7	Shaft p6		Hole H7	Shaft r6
0 — 0.12	0 0.45	+0.25 0	+0.45 +0.25	0 0.65	+0.4 0	+0.63 +0.4	0.1 0.75	+0.4 0	+0.75 +0.5
0.12 — 0.24	0 0.5	+0.3 0	+0.5 +0.3	0 0.8	+0.5 0	+0.8 +0.5	0.1 0.9	+0.5 0	+0.9 +0.6
0.24 — 0.40	0 0.65	+0.4 0	+0.65 +0.4	0 1.0	+0.6 0	+1.0 +0.6	0.2 1.2	+0.6 0	+1.2 +0.8
0.40 — 0.71	0 0.8	+0.4 0	+0.8 +0.4	0 1.1	+0.7 0	+1.1 +0.7	0.3 1.4	+0.7 0	+1.4 +1.0
0.71 — 1.19	0 1.0	+0.5 0	+1.0 +0.5	0 1.3	+0.8 0	+1.3 +0.8	0.4 1.7	+0.8 0	+1.7 +1.2
1.19 — 1.97	0 1.1	+0.6 0	+1.1 +0.6	0 1.6	+1.0 0	+1.6 +1.0	0.4 2.0	+1.0 0	+2.0 +1.4

Figure A-5

American Standard Force and Shrink Fits

Nominal Size Range Inches Over – To	Class FN 1 Limits of Interference	Class FN 1 Standard Limits Hole	Class FN 1 Standard Limits Shaft	Class FN 2 Limits of Interference	Class FN 2 Standard Limits Hole	Class FN 2 Standard Limits Shaft	Class FN 3 Limits of Interference	Class FN 3 Standard Limits Hole	Class FN 3 Standard Limits Shaft	Class FN 4 Limits of Interference	Class FN 4 Standard Limits Hole	Class FN 4 Standard Limits Shaft
0 – 0.12	0.05 / 0.5	+0.25 / 0	+0.5 / +0.3	0.2 / 0.85	+0.4 / 0	+0.85 / +0.6				0.3 / 0.95	+0.4 / 0	+0.95 / +0.7
0.12 – 0.24	0.1 / 0.6	+0.3 / 0	+0.6 / +0.4	0.2 / 1.0	+0.5 / 0	+1.0 / +0.7				0.4 / 1.2	+0.5 / 0	+1.2 / +0.9
0.24 – 0.40	0.1 / 0.75	+0.4 / 0	+0.75 / +0.5	0.4 / 1.4	+0.6 / 0	+1.4 / +1.0				0.6 / 1.6	+0.6 / 0	+1.6 / +1.2
0.40 – 0.56	0.1 / 0.8	+0.4 / 0	+0.8 / +0.5	0.5 / 1.6	+0.7 / 0	+1.6 / +1.2				0.7 / 1.8	+0.7 / 0	+1.8 / +1.4
0.56 – 0.71	0.2 / 0.9	+0.4 / 0	+0.9 / +0.6	0.5 / 1.6	+0.7 / 0	+1.6 / +1.2				0.7 / 1.8	+0.7 / 0	+1.8 / +1.4
0.71 – 0.95	0.2 / 1.1	+0.5 / 0	+1.1 / +0.7	0.6 / 1.9	+0.8 / 0	+1.9 / +1.4				0.8 / 2.1	+0.8 / 0	+2.1 / +1.6
0.95 – 1.19	0.3 / 1.2	+0.5 / 0	+1.2 / +0.8	0.6 / 1.9	+0.8 / 0	+1.9 / +1.4	0.8 / 2.1	+0.8 / 0	+2.1 / +1.6	1.0 / 2.3	+0.8 / 0	+2.1 / +1.8
1.19 – 1.58	0.3 / 1.3	+0.6 / 0	+1.3 / +0.9	0.8 / 2.4	+1.0 / 0	+2.4 / +1.8	1.0 / 2.6	+1.0 / 0	+2.6 / +2.0	1.5 / 3.1	+1.0 / 0	+3.1 / +2.5
1.58 – 1.97	0.4 / 1.4	+0.6 / 0	+1.4 / +1.0	0.8 / 2.4	+1.0 / 0	+2.4 / +1.8	1.2 / 2.8	+1.0 / 0	+2.8 / +2.2	1.8 / 3.4	+1.0 / 0	+3.4 / +2.8

Figure A-6

Preferred Clearance Fits — Cylindrical Fits (Hole Basis; ANSI B4.2)

Basic Size		Loose Running			Free Running			Close Running			Sliding			Locational Clear.		
		Hole H11	Shaft c11	Fit	Hole H9	Shaft d9	Fit	Hole H8	Shaft f7	Fit	Hole H7	Shaft g6	Fit	Hole H7	Shaft h6	Fit
4	Max	4.075	3.930	0.220	4.030	3.970	0.090	4.018	3.990	0.040	4.012	3.996	0.024	4.012	4.000	0.020
	Min	4.000	3.855	0.070	4.000	3.940	0.030	4.000	3.978	0.010	4.000	3.988	0.004	4.000	3.992	0.000
5	Max	5.075	4.930	0.220	5.030	4.970	0.090	5.018	4.990	0.040	5.012	4.996	0.024	5.012	5.000	0.020
	Min	5.000	4.855	0.070	5.000	4.940	0.030	5.000	4.978	0.010	5.000	4.988	0.004	5.000	4.992	0.000
6	Max	6.075	5.930	0.220	6.030	5.970	0.090	6.018	5.990	0.040	6.012	5.996	0.024	6.012	6.000	0.020
	Min	6.000	5.885	0.070	6.000	5.940	0.030	6.000	5.978	0.010	6.000	5.988	0.004	6.000	5.992	0.000
8	Max	8.090	7.920	0.260	8.036	7.960	0.112	8.022	7.987	0.050	8.015	7.995	0.029	8.015	8.000	0.024
	Min	8.000	7.830	0.080	8.000	7.924	0.040	8.000	7.972	0.013	8.000	7.986	0.005	8.000	7.991	0.000
10	Max	10.090	9.920	0.260	10.036	9.960	0.112	10.022	9.987	0.050	10.015	9.995	0.029	10.015	10.000	0.024
	Min	10.000	9.830	0.080	10.000	9.924	0.040	10.000	9.972	0.013	10.000	9.986	0.005	10.000	9.991	0.000
12	Max	12.112	11.905	0.315	12.043	11.950	0.136	12.027	11.984	0.061	12.018	11.994	0.035	12.018	12.000	0.029
	Min	12.000	11.795	0.095	12.000	11.907	0.050	12.000	11.966	0.016	12.000	11.983	0.006	12.000	11.989	0.000
16	Max	16.110	15.905	0.315	16.043	15.950	0.136	16.027	15.984	0.061	16.018	15.994	0.035	16.018	16.000	0.029
	Min	16.000	15.795	0.095	16.000	15.907	0.050	16.000	15.966	0.016	16.000	15.983	0.006	16.000	15.989	0.000
20	Max	20.130	19.890	0.370	20.052	19.935	0.169	20.033	19.980	0.074	20.021	19.993	0.041	20.021	20.000	0.034
	Min	20.000	19.760	0.110	20.000	19.883	0.065	20.000	19.959	0.020	20.000	19.980	0.007	20.000	19.987	0.000
25	Max	25.130	24.890	0.370	25.052	24.935	0.169	25.033	24.980	0.074	25.021	24.993	0.041	25.021	25.000	0.034
	Min	25.000	24.760	0.110	25.000	24.883	0.065	25.000	24.959	0.020	25.000	24.980	0.007	25.000	24.987	0.000
30	Max	30.130	29.890	0.370	30.052	29.935	0.169	30.033	29.980	0.074	30.021	29.993	0.041	30.021	30.000	0.034
	Min	30.000	29.760	0.110	30.000	29.883	0.065	30.000	29.959	0.020	30.000	29.980	0.007	30.000	29.987	0.000

Figure A-7

Preferred Transition and Interference Fits — Cylindrical Fits (Hole Basis; ANSI B4.2)

Basic Size		Locational Trans.			Locational Trans.			Locational Inter.			Medium Drive			Force		
		Hole H7	Shaft k6	Fit	Hole H7	Shaft n6	Fit	Hole H7	Shaft p6	Fit	Hole H7	Shaft s6	Fit	Hole H7	Shaft u6	Fit
4	Max	4.012	4.009	0.011	4.012	4.016	0.004	4.012	4.020	0.000	4.012	4.027	-0.007	4.012	4.031	-0.011
	Min	4.000	4.001	-0.009	4.000	4.008	-0.016	4.000	4.012	-0.020	4.000	4.019	-0.027	4.000	4.023	-0.031
5	Max	5.012	5.009	0.011	5.012	5.016	0.004	5.012	5.020	0.000	5.012	5.027	-0.007	5.012	5.031	-0.011
	Min	5.000	5.001	-0.009	5.000	5.008	-0.016	5.000	5.012	-0.020	5.000	5.019	-0.027	5.000	5.023	-0.031
6	Max	6.012	6.009	0.011	6.012	6.016	0.004	6.012	6.020	0.000	6.012	6.027	-0.007	6.012	6.031	-0.011
	Min	6.000	6.001	-0.009	6.000	6.008	-0.016	6.000	6.012	-0.020	6.000	6.019	-0.027	6.000	6.023	-0.031
8	Max	8.015	8.010	0.014	8.015	8.019	0.005	8.015	8.024	0.000	8.015	8.032	-0.008	8.015	8.037	-0.013
	Min	8.000	8.001	-0.010	8.000	8.010	-0.019	8.000	8.015	-0.024	8.000	8.023	-0.032	8.000	8.028	-0.037
10	Max	10.015	10.010	0.014	10.015	10.019	0.005	10.015	10.024	0.000	10.015	10.032	-0.008	10.015	10.037	-0.013
	Min	10.000	10.001	-0.010	10.000	10.010	-0.019	10.000	10.015	-0.024	10.000	10.023	-0.032	10.000	10.028	-0.037
12	Max	12.018	12.012	0.017	12.018	12.023	0.006	12.018	12.029	0.000	12.018	12.039	-0.010	12.018	12.044	-0.015
	Min	12.000	12.001	-0.012	12.000	12.012	-0.023	12.000	12.018	-0.029	12.000	12.028	-0.039	12.000	12.033	-0.044
16	Max	16.018	16.012	0.017	16.018	16.023	0.006	16.018	16.029	0.000	16.018	16.039	-0.010	16.018	16.044	-0.015
	Min	16.000	16.001	-0.012	16.000	16.012	-0.023	16.000	16.018	-0.029	16.000	16.028	-0.039	16.000	16.033	-0.044
20	Max	20.021	20.015	0.019	20.021	20.028	0.006	20.021	20.035	-0.001	20.021	20.048	-0.014	20.021	20.054	-0.020
	Min	20.000	20.002	-0.015	20.000	20.015	-0.028	20.000	20.022	-0.035	20.000	20.035	-0.048	20.000	20.041	-0.054
25	Max	25.021	25.015	0.019	25.021	25.028	0.006	25.021	25.035	-0.001	25.021	25.048	-0.014	25.021	25.061	-0.027
	Min	25.000	25.002	-0.015	25.000	25.015	-0.028	25.000	25.022	-0.035	25.000	25.035	-0.048	25.000	25.048	-0.061
30	Max	30.021	30.015	0.019	30.021	30.028	0.006	30.021	30.035	-0.001	30.021	30.048	-0.014	30.021	30.061	-0.027
	Min	30.000	30.002	-0.015	30.000	30.015	-0.028	30.000	30.022	-0.035	30.000	30.035	-0.048	30.000	30.048	-0.061

Figure A-8

Preferred Clearance Fits — Cylindrical Fits (Shaft Basis; ANSI B4.2)

Basic Size		Loose Running			Free Running			Close Running			Sliding			Locational Clear.		
		Hole C11	Shaft h11	Fit	Hole D9	Shaft h9	Fit	Hole F8	Shaft h7	Fit	Hole G7	Shaft h6	Fit	Hole H7	Shaft h6	Fit
4	Max	4.145	4.000	0.220	4.060	4.000	0.090	4.028	4.000	0.040	4.016	4.000	0.024	4.012	4.000	0.020
	Min	4.070	3.925	0.070	4.030	3.970	0.030	4.010	3.988	0.010	4.004	3.992	0.004	4.000	3.992	0.000
5	Max	5.145	5.000	0.220	5.060	5.000	0.090	5.028	5.000	0.040	5.016	5.000	0.024	5.012	5.000	0.020
	Min	5.070	4.925	0.070	5.030	4.970	0.030	5.010	4.988	0.010	5.004	4.992	0.004	5.000	4.992	0.000
6	Max	6.145	6.000	0.220	6.060	6.000	0.090	6.028	6.000	0.040	6.016	6.000	0.024	6.012	6.000	0.020
	Min	6.070	5.925	0.070	6.030	5.970	0.030	6.010	5.988	0.010	6.004	5.992	0.004	6.000	5.992	0.000
8	Max	8.170	8.000	0.260	8.076	8.000	0.112	8.035	8.000	0.050	8.020	8.000	0.029	8.015	8.000	0.024
	Min	8.080	7.910	0.080	8.040	7.964	0.040	8.013	7.985	0.013	8.005	7.991	0.005	8.000	7.991	0.000
10	Max	10.170	10.000	0.260	10.076	10.000	0.112	10.035	10.000	0.050	10.020	10.000	0.029	10.015	10.000	0.024
	Min	10.080	9.910	0.080	10.040	9.964	0.040	10.013	9.985	0.013	10.005	9.991	0.005	10.000	9.991	0.000
12	Max	12.205	12.000	0.315	12.093	12.000	0.136	12.043	12.000	0.061	12.024	12.000	0.035	12.018	12.000	0.029
	Min	12.095	11.890	0.095	12.050	11.957	0.050	12.016	11.982	0.016	12.006	11.989	0.006	12.000	11.989	0.000
16	Max	16.205	16.000	0.315	16.093	16.000	0.136	16.043	16.000	0.061	16.024	16.000	0.035	16.018	16.000	0.029
	Min	16.095	15.890	0.095	16.050	15.957	0.050	16.016	15.982	0.016	06.006	15.989	0.006	16.000	15.989	0.000
20	Max	20.240	20.000	0.370	20.117	20.000	0.169	20.053	20.000	0.074	20.028	20.000	0.041	20.021	20.000	0.034
	Min	20.110	19.870	0.110	20.065	19.948	0.065	20.020	19.979	0.020	20.007	19.987	0.007	20.000	19.987	0.000
25	Max	25.240	25.000	0.370	25.117	25.000	0.169	25.053	25.000	0.074	25.028	25.000	0.041	25.021	25.000	0.034
	Min	25.110	24.870	0.110	25.065	24.948	0.065	25.020	24.979	0.020	25.007	24.987	0.007	25.000	24.987	0.000
30	Max	30.240	30.000	0.370	30.117	30.000	0.169	30.053	30.000	0.074	30.028	30.000	0.041	30.021	30.000	0.034
	Min	30.110	29.870	0.110	30.065	29.948	0.065	30.020	29.979	0.020	30.007	29.987	0.007	30.000	29.987	0.000

Figure A-9

Preferred Transition and Interference Fits — Cylindrical Fits
(Shaft Basis; ANSI B4.2)

Basic Size		Locational Trans. Hole K7	Shaft h6	Fit	Locational Trans. Hole N7	Shaft h6	Fit	Locational Inter. Hole P7	Shaft h6	Fit	Medium Drive Hole S7	Shaft h6	Fit	Force Hole U7	Shaft h6	Fit
4	Max	4.003	4.000	0.011	3.996	4.000	0.004	3.992	4.000	0.000	3.985	4.000	-0.007	3.981	4.000	-0.011
	Min	3.991	3.992	-0.009	3.984	3.992	-0.016	3.980	3.992	-0.020	3.973	3.992	-0.027	3.969	3.992	-0.031
5	Max	5.003	5.000	0.011	4.996	5.000	0.004	4.992	5.000	0.000	4.985	5.000	-0.007	4.981	5.000	-0.011
	Min	4.991	4.992	-0.009	4.984	4.992	-0.016	4.980	4.992	-0.020	4.973	4.992	-0.027	4.969	4.992	-0.031
6	Max	6.003	6.000	0.011	5.996	6.000	0.004	5.992	6.000	0.000	5.985	6.000	-0.007	5.981	6.000	-0.011
	Min	5.991	5.992	-0.009	5.984	5.992	-0.016	5.980	5.992	-0.020	5.973	5.992	-0.027	5.969	5.992	-0.031
8	Max	8.005	8.000	0.014	7.996	8.000	0.005	7.991	8.000	0.000	7.983	8.000	-0.008	7.978	8.000	-0.013
	Min	7.990	7.991	-0.010	7.981	7.991	-0.019	7.976	7.991	-0.024	7.968	7.991	-0.032	7.963	7.991	-0.037
10	Max	10.005	10.000	0.014	9.996	10.000	0.005	9.991	10.000	0.000	9.983	10.000	-0.008	9.978	10.000	-0.013
	Min	9.990	9.991	-0.010	9.981	9.991	-0.019	9.976	9.991	-0.024	9.968	9.991	-0.032	9.963	9.991	-0.037
12	Max	12.006	12.000	0.017	11.995	12.000	0.006	11.989	12.000	0.000	11.979	12.000	-0.010	11.974	12.000	-0.015
	Min	11.988	11.989	-0.012	11.977	11.989	-0.023	11.971	11.989	-0.029	11.961	11.989	-0.039	11.956	11.989	-0.044
16	Max	16.006	16.000	0.017	15.995	16.000	0.006	15.989	16.000	0.000	15.979	16.000	-0.010	15.974	16.000	-0.015
	Min	15.988	15.989	-0.012	15.977	15.989	-0.023	15.971	15.989	-0.029	15.961	15.989	-0.039	15.956	15.989	-0.044
20	Max	20.006	20.000	0.019	19.993	20.000	0.006	19.986	20.000	-0.001	19.973	20.000	-0.014	19.967	20.000	-0.020
	Min	19.985	19.987	-0.015	19.972	19.987	-0.028	19.965	19.987	-0.035	19.952	19.987	-0.048	19.946	19.987	-0.054
25	Max	25.006	25.000	0.019	24.993	25.000	0.006	24.986	25.000	-0.001	24.973	25.000	-0.014	24.960	25.000	-0.027
	Min	24.985	24.987	-0.015	24.972	24.987	-0.028	24.965	24.987	-0.035	24.952	24.987	-0.048	24.939	24.987	-0.061
30	Max	30.006	30.000	0.019	29.993	30.000	0.006	29.986	30.000	-0.001	29.973	30.000	-0.014	29.960	30.000	-0.027
	Min	29.985	29.987	-0.015	29.972	29.987	-0.028	29.965	29.987	-0.035	29.952	29.987	-0.048	29.939	29.987	-0.061

Figure A-10

American National Standard Type A Plain Washers
(ANSI B18.22.1-1965, R1975)

Nominal Washer Size		Series	Inside Diameter			Outside Diameter			Thickness		
			Basic	Tolerance		Basic	Tolerance		Basic	Max.	Min.
				Plus	Minus		Plus	Minus			
#6	.138		.156	.008	.005	.375	.015	.005	.049	.065	.036
#8	.164		.188	.008	.005	.438	.015	.005	.049	.065	.036
#10	.190		.219	.008	.005	.500	.015	.005	.049	.065	.036
1/4	.250	N	.281	.015	.005	.625	.015	.005	.065	.080	.051
1/4	.250	W	.312	.015	.005	.734	.015	.007	.065	.080	.051
5/16	.312	N	.344	.015	.005	.688	.015	.007	.065	.080	.051
5/16	.312	W	.375	.015	.005	.875	.030	.007	.083	.104	.064
3/8	.375	N	.406	.015	.005	.812	.015	.007	.065	.080	.051
3/8	.375	W	.438	.015	.005	1.000	.030	.007	.083	.104	.064
7/16	.438	N	.469	.015	.005	.922	.015	.007	.065	.080	.051
7/16	.438	W	.500	.015	.005	1.250	.030	.007	.083	.104	.064
1/2	.500	N	.531	.015	.005	1.062	.030	.007	.095	.121	.074
1/2	.500	W	.562	.015	.005	1.375	.030	.007	.109	.132	.086
9/16	.562	N	.594	.015	.005	1.156	.030	.007	.095	.121	.074
9/16	.562	W	.625	.015	.005	1.469	.030	.007	.109	.132	.086
5/8	.625	N	.656	.030	.007	1.312	.030	.007	.095	.121	.074
5/8	.625	W	.688	.030	.007	1.750	.030	.007	.134	.160	.108
3/4	.750	N	.812	.030	.007	1.469	.030	.007	.134	.160	.108
3/4	.750	W	.812	.030	.007	2.000	.030	.007	.148	.177	.122
7/8	.875	N	.938	.030	.007	1.750	.030	.007	.134	.160	.108
7/8	.875	W	.938	.030	.007	2.250	.030	.007	.165	.192	.136
1	1.000	N	1.062	.030	.007	2.000	.030	.007	.134	.160	.108
1	1.000	W	1.062	.030	.007	2.500	.030	.007	.165	.192	.136
1 1/8	1.125	N	1.250	.030	.007	2.250	.030	.007	.134	.160	.108
1 1/8	1.125	W	1.250	.030	.007	2.750	.030	.007	.165	.192	.136
1 1/4	1.250	N	1.375	.030	.007	2.500	.030	.007	.165	.192	.136
1 1/4	1.250	W	1.375	.030	.007	3.000	.030	.007	.165	.192	.136
1 3/8	1.375	N	1.500	.030	.007	2.750	.030	.007	.165	.192	.136
1 3/8	1.375	W	1.500	.045	.010	3.250	.045	.010	.180	.213	.153
1 1/2	1.500	N	1.625	.030	.007	3.000	.030	.007	.165	.192	.136
1 1/2	1.500	W	1.625	.045	.010	3.500	.045	.010	.180	.213	.153
1 5/8	1.625		1.750	.045	.010	3.750	.045	.010	.180	.213	.153
1 3/4	1.750		1.875	.045	.010	4.000	.045	.010	.180	.213	.153
1 7/8	1.875		2.000	.045	.010	4.250	.045	.010	.180	.213	.153
2	2.000		2.125	.045	.010	4.500	.045	.010	.180	.213	.153
2 1/4	2.250		2.375	.045	.010	4.750	.045	.010	.220	.248	.193
2 1/2	2.500		2.625	.045	.010	5.000	.045	.010	.238	.280	.210
2 3/4	2.750		2.875	.065	.010	5.250	.065	.010	.259	.310	.228
3	3.000		3.125	.065	.010	5.500	.065	.010	.284	.327	.249

Figure A-11

American National Standard Helical Spring Lock Washers (ANSI B18.21.1-1972)

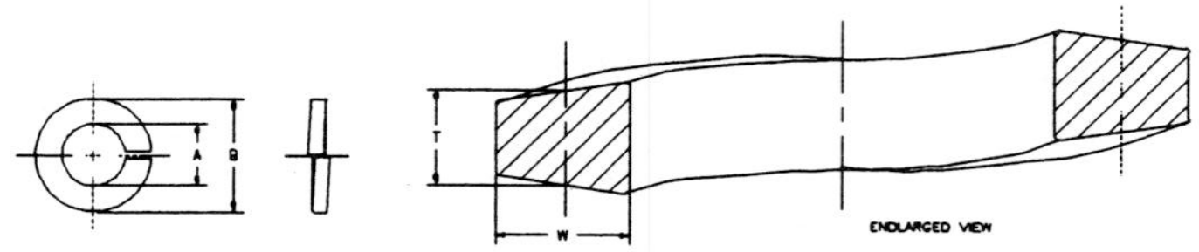

ENLARGED VIEW

Nominal Washer Size		Inside Diameter, A		Regular			Heavy			Extra Duty		
		Max	Min	O.D., B Max	Section Width	Section Thickness	O.D., B Max	Section Width	Section Thickness	O.D., B Max	Section Width	Section Thickness
#2	.086	.094	.088	.172	.035	.020				.208	.053	.027
#3	.099	.107	.101	.195	.040	.025	.182	.040	.025	.239	.062	.034
#4	.112	.120	.114	.209	.040	.025	.209	.047	.031	.253	.062	.034
#5	.125	.133	.127	.236	.047	.031	.253	.047	.031	.300	.079	.045
#6	.138	.148	.141	.250	.047	.031	.266	.055	.040	.314	.079	.045
#8	.164	.174	.167	.293	.055	.040	.307	.062	.047	.375	.096	.057
#10	.190	.200	.193	.334	.062	.047	.350	.070	.056	.434	.112	.068
#12	.216	.227	.220	.377	.070	.056	.391	.077	.063	.497	.130	.080
1/4	.250	.262	.254	.489	.109	.062	.491	.110	.077	.535	.132	.084
5/16	.312	.326	.317	.586	.125	.078	.596	.130	.097	.622	.143	.108
3/8	.375	.390	.380	.683	.141	.094	.691	.145	.115	.741	.170	.123
7/16	.438	.455	.443	.779	.156	.109	.787	.160	.133	.839	.186	.143
1/2	.500	.518	.506	.873	.171	.125	.883	.176	.151	.939	.204	.162
9/16	.562	.582	.570	.971	.188	.141	.981	.193	.170	1.041	.223	.182
5/8	.625	.650	.635	1.079	.203	.156	1.093	.210	.189	1.157	.242	.202
11/16	.688	.713	.698	1.176	.219	.172	1.192	.227	.207	1.258	.260	.221
3/4	.750	.775	.760	1.271	.234	.188	1.291	.244	.226	1.361	.279	.241
13/16	.812	.843	.824	1.367	.250	.203	1.391	.262	.246	1.463	.298	.261
7/8	.875	.905	.887	1.464	.266	.219	1.494	.281	.266	1.576	.322	.285
15/16	.938	.970	.950	1.560	.281	.234	1.594	.298	.284	1.688	.345	.308
1	1.000	1.042	1.017	1.661	.297	.250	1.705	.319	.306	1.799	.366	.330
1 1/16	1.062	1.107	1.080	1.756	.312	.266	1.808	.338	.326	1.910	.389	.352
1 1/8	1.125	1.172	1.144	1.853	.328	.281	1.909	.356	.345	2.019	.411	.375
1 3/16	1.188	1.237	1.208	1.950	.344	.297	2.008	.373	.364	2.124	.431	.396
1 1/4	1.250	1.302	1.271	2.045	.359	.312	2.113	.393	.384	2.231	.452	.417
1 5/16	1.312	1.366	1.334	2.141	.375	.328	2.211	.410	.403	2.335	.472	.438
1 3/8	1.375	1.432	1.398	2.239	.391	.344	2.311	.427	.422	2.439	.491	.458
1 7/16	1.438	1.497	1.462	2.334	.406	.359	2.406	.442	.440	2.540	.509	.478
1 1/2	1.500	1.561	1.525	2.430	.422	.375	2.502	.458	.458	2.638	.526	.496

Figure A-12

American National Standard Internal-External Tooth Lock Washers
(ANSI B18.21.1-1972)

Size	A — Inside Diameter Max.	Min.	B — Outside Diameter Max.	Min.	C — Thickness Max.	Min.
#4	.123	.115	.475	.460	.021	.021
	.123	.115	.510	.495	.021	.017
			.610	.580		
#6	.150	.141	.510	.495	.028	.023
			.610	.580		
			.690	.670		
#8	.176	.168	.610	.580	.034	.028
			.690	.670		
			.760	.740		
#10	.204	.195	.610	.580	.034	.028
	.204	.195	.690	.670	.040	.032
			.760	.740		
			.900	.880		
#12	.231	.221	.690	.670	.040	.032
	.231	.221	.760	.725	.045	.037
			.900	.880		
			.985	.965		
1/4	.267	.256	.760	.725	.040	.032
	.267	.256	.900	.880	.045	.037
			.985	.965		
			1.070	1.045		
5/16	.332	.320	.900	.865	.040	.032
	.332	.320	.985	.965	.045	.037
	.332	.320	1.070	1.045	.050	.042
			1.155	1.130		
3/8	.398	.384	.985	.965	.045	.037
	.398	.384	1.070	1.045	.050	.042
			1.155	1.130		
			1.260	1.220		
7/16	.464	.448	1.070	1.045	.050	.042
	.464	.448	1.155	1.130	.055	.047
			1.260	1.220		
			1.315	1.290		
1/2	.530	.512	1.260	1.220	.055	.047
	.530	.512	1.315	1.290	.060	.052
	.530	.512	1.410	1.380	.067	.059
			1.620	1.590		
9/16	.596	.576	1.315	1.290	.055	.047
	.596	.576	1.430	1.380	.060	.052
	.596	.576	1.620	1.590	.067	.059
			1.830	1.797		
5/8	.663	.640	1.410	1.380	.060	.052
	.663	.640	1.620	1.590	.067	.059
			1.830	1.797		
			1.975	1.935		

Figure A-13

British Standard Bright Metal Washers – Metric Series (BS 4320:1968)

NORMAL DIAMETER SIZES										
Nominal Size of Bolt or Screw	Inside Diameter			Outside Diameter			Thickness			
							Form A (Normal Range)			Form B (Light Range)
	Nom.	Max.	Min.	Nom.	Max.	Min.	Nom.	Max.	Min.	Nom.	Max.	Min.
M 1.0	1.1	1.25	1.1	2.5	2.5	2.3	.3	.4	.2			
M 1.2	1.3	1.45	1.3	3.0	3.0	2.8	.3	.4	.2			
M 1.4	1.5	1.65	1.5	3.0	3.0	2.8	.3	.4	.2			
M 1.6	1.7	1.85	1.7	4.0	4.0	3.7	.3	.4	.2			
M 2.0	2.2	2.35	2.2	5.0	5.0	4.7	.3	.4	.2			
M 2.2	2.4	2.55	2.4	5.0	5.0	4.7	.5	.6	.4			
M 2.5	2.7	2.85	2.7	6.5	6.5	6.2	.5	.6	.4			
M 3	3.2	3.4	3.2	7	7	6.7	.5	.6	.4			
M 3.5	3.7	3.9	3.7	7	7	6.7	.5	.6	.4			
M 4	4.3	4.5	4.3	9	9	8.7	.8	.9	.7			
M 4.5	4.8	5.0	4.8	9	9	8.7	.8	.9	.7			
M 5	5.3	5.5	5.3	10	10	9.7	1.0	1.1	.9			
M 6	6.4	6.7	6.4	12.5	12.5	12.1	1.6	1.8	1.4	.8	.9	.7
M 7	7.4	7.7	7.4	14	14	13.6	1.6	1.8	1.4	.8	.9	.7
M 8	8.4	8.7	8.4	17	17	16.6	1.6	1.8	1.4	1.0	1.1	.9
M 10	10.5	10.9	10.5	21	21	20.5	2.0	2.2	1.8	1.25	1.45	1.05
M 12	13.0	13.4	13.0	24	24	23.5	2.5	2.7	2.3	1.6	1.80	1.40
M 14	15.0	15.4	15.0	28	28	27.5	2.5	2.7	2.3	1.6	1.8	1.4
M 16	17.0	17.4	17.0	30	30	29.5	3.0	3.3	2.7	2.0	2.2	1.8
M 18	19.0	19.5	19.0	34	34	33.2	3.0	3.3	2.7	2.0	2.2	1.8
M 20	21	21.5	21	37	37	36.2	3.0	3.3	2.7	2.0	2.2	1.8
M 22	23	23.5	23	39	39	38.2	3.0	3.3	2.7	2.0	2.2	1.8
M 24	25	25.5	25	44	44	43.2	4.0	4.3	3.7	2.5	2.7	2.3
M 27	28	28.5	28	50	50	49.2	4.0	4.3	3.7	2.5	2.7	2.3
M 30	31	31.6	31	56	56	55.0	4.0	4.3	3.7	2.5	2.7	2.3
M 33	34	34.6	34	60	60	59.0	5.0	5.6	4.4	3.0	3.3	2.7
M 36	37	37..6	37	66	66	65.0	5.0	5.6	4.4	3.0	3.3	2.7
M 39	40	40.6	40	72	72	71.0	6.0	6.6	5.4	3.0	3.3	2.7

Figure A-14

American National Standard and Unified Square Bolts (ANSI B18.2.1-1972)

SQUARE BOLTS

Nominal Size or Basic Product Diameter		Body Diam., E	Width Across Flats, F			Width Across Corners, G		Height, H			Radius of Fillet, R
		Max.	Basic	Max.	Min.	Max.	Min.	Basic	Max.	Min.	Max.
1/4	.2500	.260	3/8	.375	.362	.530	.498	11/64	.188	.156	.03
5/16	.3125	.324	1/2	.500	.484	.707	.665	13/64	.220	.186	.03
3/8	.3750	.388	9/16	.562	.544	.795	.747	1/4	.268	.232	.03
7/16	.4375	.452	5/8	.625	.603	.884	.828	19/64	.316	.278	.03
1/2	.5000	.515	3/4	.750	.725	1.061	.995	21/64	.348	.308	.03
5/8	.6250	.642	15/16	.938	.906	1.326	1.244	37/64	.444	.400	.06
3/4	.7500	.768	1 1/8	1.125	1.088	1.591	1.494	1/2	.524	.476	.06
7/8	.8750	.895	1 5/16	1.312	1.269	1.856	1.742	19/32	.620	.568	.06
1	1.0000	1.022	1 1/2	1.500	1.450	2.121	1.991	21/32	.684	.628	.09
1 1/8	1.1250	1.149	1 11/16	1.688	1.631	2.386	2.239	3/4	.780	.720	.09
1 1/4	1.2500	1.277	1 7/8	1.875	1.812	2.652	2.489	27/32	.876	.812	.09
1 3/8	1.3750	1.404	2 1/16	2.062	1.994	2.917	2.738	29/32	.940	.872	.09
1 1/2	1.5000	1.531	2 1/4	2.250	2.175	3.182	2.986	1	1.036	.964	.09

Figure A-15

American National Standard and Unified Standard Hex Head Screws
(ANSI B18.2.1-1972)

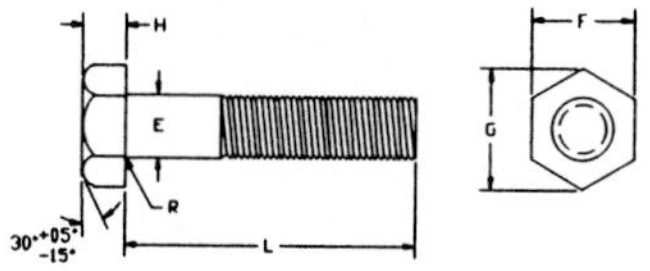

Nominal Size or Basic Diam.		Body Diam., E	Width Across Flats, F			Width Across Corners, G		Height, H			Radius of Fillet, R	
		Max.	Basic	Max.	Min.	Max.	Min.	Basic	Max.	Min.	Max.	Min.
HEX BOLTS												
1/4	.2500	.260	7/16	.438	.425	.505	.484	11/64	.188	.150	.03	.01
5/16	.3125	.324	1/2	.500	.484	.577	.552	7/32	.235	.195	.03	.01
3/8	.3750	.388	9/16	.562	.544	.650	.620	1/4	.268	.226	.03	.01
7/16	.4375	.452	5/8	.625	.603	.722	.687	19/64	.316	.272	.03	.01
1/2	.5000	.515	3/4	.750	.725	.866	.826	11/32	.364	.302	.03	.01
5/8	.6250	.642	15/16	.938	.906	1.083	1.033	27/64	.444	.378	.06	.02
3/4	.7500	.768	1 1/8	1.125	1.088	1.299	1.240	1/2	.524	.455	.06	.02
7/8	.8750	.895	1 5/16	1.312	1.269	1.516	1.447	37/64	.604	.531	.06	.02
1	1.0000	1.022	1 1/2	1.500	1.450	1.732	1.653	43/64	.700	.591	.09	.03
1 1/8	1.1250	1.149	1 11/16	1.688	1.631	1.949	1.859	3/4	.780	.658	.09	.03
1 1/4	1.2500	1.277	1 7/8	1.875	1.812	2.165	2.066	27/32	.876	.749	.09	.03
1 3/8	1.3750	1.404	2 1/16	2.062	1.994	2.382	2.273	29/32	.940	.810	.09	.03
1 1/2	1.5000	1.531	2 1/4	2.250	2.175	2.598	2.480	1	1.036	.902	.09	.03
1 3/4	1.7500	1.785	2 5/8	2.625	2.538	3.031	2.893	1 5/32	1.196	1.054	.12	.04
2	2.0000	2.039	3	3.000	2.900	3.464	3.306	1 11/32	1.388	1.175	.12	.04
2 1/4	2.2500	2.305	3 3/8	3.375	3.262	3.897	3.719	1 1/2	1.548	1.327	.19	.06
2 1/2	2.5000	2.559	3 3/4	3.750	3.625	4.330	4.133	1 21/32	1.708	1.479	.19	.06
2 3/4	2.7500	2.827	4 1/8	4.125	3.988	4.763	4.546	1 13/16	1.869	1.632	.19	.06
3	3.0000	3.081	4 1/2	4.500	4.350	5.196	4.959	2	2.060	1.815	.19	.06
3 1/4	3.2500	3.335	4 7/8	4.875	4.712	5.629	5.372	2 3/16	2.251	1.936	.19	.06
3 1/2	3.5000	3.589	5 1/4	5.250	5.075	6.062	5.786	2 5/16	2.380	2.057	.19	.06
3 3/4	3.7500	3.858	5 5/8	5.625	5.437	6.495	6.198	2 1/2	2.572	2.241	.19	.06
4	4.0000	4.111	6	6.000	5.800	6.982	6.612	2 11/16	2.764	2.424	.19	.06

Figure A-16

Coarse-Thread Series, UNC, UNRC, and NC — Basic Dimensions

Sizes	Basic Major Diam., D	Thds. per Inch, n	Basic Pitch Diam., E	Minor Diameter		Lead Angle at Basic P.D.		Area of Minor Diam. at D-2h	Tensile Stress Area
				Ext. Thds., Ks	Int. Thds., Kn	Deg.	Min.		
	Inches		Inches	Inches	Inches			Sq. In.	Sq. In.
1 (.073)	.0730	64	.0629	.0538	.0561	4	31	.00218	.00263
2 (.086)	.0860	56	.0744	.0641	.0667	4	22	.00310	.00370
3 (.099)	.0990	48	.0855	.0734	.0764	4	26	.00406	.00487
4 (.112)	.1120	40	.0958	.0813	.0849	4	45	.00496	.00604
5 (.125)	.1250	40	.1088	.0943	.0979	4	11	.00672	.00796
6 (.138)	.1380	32	.1177	.0997	.1042	4	50	.00745	.00909
8 (.164)	1.640	32	.1437	.1257	.1302	3	58	.01196	.0140
10 (.190)	.1900	24	.1629	.1389	.1449	4	39	.01450	.0175
12 (.216)	.2160	24	.1889	.1649	.1709	4	1	.0206	.0242
1/4	.2500	20	.2175	.1887	.1959	4	11	.0269	.0318
5/16	.3125	18	.2764	.2443	.2524	3	40	.0454	.0524
3/8	.3750	16	.3344	.2983	.3073	3	24	.0678	.0775
7/16	.4375	14	.3911	.3499	.3602	3	20	.0933	.1063
1/2	.5000	13	.4500	.4056	.4167	3	7	.1257	.1419
9/16	.5625	12	.5084	.4603	.4723	2	59	.162	.182
5/8	.6250	11	.5660	.5135	.5266	2	56	.202	.226
3/4	.7500	10	.6850	.6273	.6417	2	40	.302	.334
7/8	.8750	9	.8028	.7387	.7547	2	31	.419	.462
1	1.0000	8	.9188	.8466	.8647	2	29	.551	.606
1 1/8	1.1250	7	1.032	.9497	.9704	2	31	.693	.763
1 1/4	1.2500	7	1.572	1.0747	1.0954	2	15	.890	.969
1 3/8	1.3750	6	1.2667	1.1705	1.1946	2	24	1.054	1.155
1 1/2	1.5000	6	1.3917	1.2955	1.3196	2	11	1.294	1.405

Figure A-17

Fine-Thread Series, UNC, UNRC, and NC — Basic Dimensions

Sizes	Basic Major Diam., D	Thds. per Inch, n	Basic Pitch Diam., E	Minor Diameter		Lead Angle at Basic P.D.		Area of Minor Diam. at D-2h	Tensile Stress Area
				Ext. Thds., Ks	Int. Thds., Kn	Deg.	Min.		
	Inches		Inches	Inches	Inches			Sq. In.	Sq. In.
1 (.073)	.0730	72	.0640	.0560	.0580	3	57	.00237	.00278
2 (.086)	.860	64	.0759	.0668	.0691	3	45	.00339	.00394
3 (.099)	.990	56	.0874	.0771	.0797	3	43	.00451	.00523
4 (.112)	.1120	48	.0985	.0864	.0894	3	51	.00566	.00661
5 (.125)	.1250	44	.1102	.0971	.1004	3	45	.00716	.00830
6 (.138)	.1380	40	.1218	.1073	.1109	3	44	.00874	.01015
8 (.164)	.1640	36	.1460	.1299	.1339	3	28	.01285	.01474
10 (.190)	.1900	32	.1697	.1517	.1562	3	21	.0175	.0200
12 (.216)	.2160	28	.1928	.1722	.1773	3	22	.0226	.0258
1/4	.2500	28	.2268	.2062	.2113	2	52	.0326	.0364
5/16	.3125	24	.2854	.2614	.2674	2	40	.0524	.0580
3/8	.3750	24	.3479	.3239	.3299	2	11	.0809	.0878
7/16	.4375	20	.4050	.3762	.3834	2	15	.1090	.1187
1/2	.5000	20	.4675	.4387	.4459	1	57	.1486	.1599
9/16	.5625	18	.5264	.4943	.5024	1	55	.189	.203
5/8	.6250	18	.5889	.5568	.5649	1	43	.240	.256
3/4	.7500	16	.7094	.6733	.6823	1	36	.351	.373
7/8	.8750	14	.8286	.7874	.7977	1	34	.480	.509
1	1.0000	12	.9459	.8978	.9098	1	36	.625	.663
1 1/8	1.1250	12	1.0709	1.0228	1.0348	1	25	.812	.856
1 1/4	1.2500	12	1.1959	1.1478	1.1598	1	16	1.024	1.073
1 3/8	1.3750	12	1.3209	1.2728	1.2848	1	9	1.260	1.315
1 1/2	1.5000	12	1.4459	1.3978	1.4098	1	3	1.521	1.581

Figure A-18

American National Standard General-Purpose Acme Screw Thread Form—
Basic Dimensions (ANSI B1.5-1977)

Thds. per Inch	Pitch	Height of Thread (Basic)	Total Height of Thread	Thread Thickness (Basic)	Width of Flat	
					Crest of Internal Thread (Basic)	Root of Internal Thread
16	.06250	.03125	.0362	.03125	.0232	.0206
14	.07143	.03571	.0407	.03571	.0265	.0239
12	.08333	.04167	.0467	.04167	.0309	.0283
10	.10000	.05000	.0600	.05000	.0371	.0319
8	.12500	.06250	.0725	.06250	.0463	.0411
6	.16667	.08333	.0933	.08333	.0618	.0566
5	.20000	.10000	.1100	.10000	.0741	.0689
4	.25000	.12500	.1350	.12500	.0927	.0875
3	.33333	.16667	.1767	.16667	.1236	.1184
2 1/2	.40000	.20000	.2100	.20000	.1483	.1431
2	.50000	.25000	.2600	.25000	.1853	.1802
1 1/2	.66667	.33333	.3433	.33333	.2471	.2419
1 1/3	.75000	.37500	.3850	.37500	.2780	.2728
1	1.0000	.50000	.5100	.50000	.3707	.3655

Figure A-19

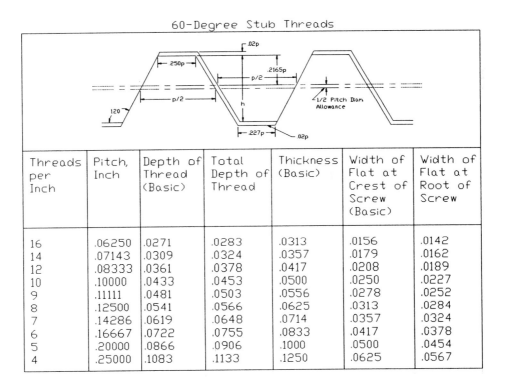

60-Degree Stub Threads

Threads per Inch	Pitch, Inch	Depth of Thread (Basic)	Total Depth of Thread	Thickness (Basic)	Width of Flat at Crest of Screw (Basic)	Width of Flat at Root of Screw
16	.06250	.0271	.0283	.0313	.0156	.0142
14	.07143	.0309	.0324	.0357	.0179	.0162
12	.08333	.0361	.0378	.0417	.0208	.0189
10	.10000	.0433	.0453	.0500	.0250	.0227
9	.11111	.0481	.0503	.0556	.0278	.0252
8	.12500	.0541	.0566	.0625	.0313	.0284
7	.14286	.0619	.0648	.0714	.0357	.0324
6	.16667	.0722	.0755	.0833	.0417	.0378
5	.20000	.0866	.0906	.1000	.0500	.0454
4	.25000	.1083	.1133	.1250	.0625	.0567

Figure A-20

American National Standard Slotted 100° Flat Countersunk
Head Machine Screws (ANSI B18.6.3-1972, R1977)

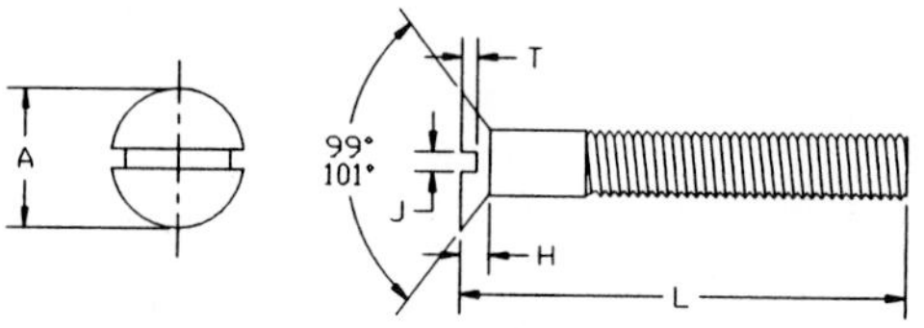

Nominal Size or Basic Screw Diam.		Head Diam., A		Head Height, H	Slot Width, J		Slot Depth, T	
		Max., Edge Sharp	Min., Edge Rounded or Flat	Ref.	Max.	Min.	Max.	Min.
0000	.0210	.043	.037	.009	.008	.005	.008	.004
000	.0340	.064	.058	.014	.012	.008	.011	.007
00	.0470	.093	.085	.020	.017	.010	.013	.008
0	.0600	.119	.096	.026	.023	.016	.013	.008
1	.0730	.146	.120	.031	.026	.019	.016	.010
2	.0860	.172	.143	.037	.031	.023	.019	.012
3	.0990	.199	.167	.043	.035	.027	.022	.014
4	.1120	.225	.191	.049	.039	.031	.024	.017
6	.1380	.279	.238	.060	.048	.039	.030	.022
8	.1640	.332	.285	.072	.054	.045	.036	.027
10	.1900	.385	.333	.083	.060	.050	.042	.031
1/4	.2500	.507	.442	.110	.075	.064	.055	.042
5/16	.3125	.635	.556	.138	.084	.072	.069	.053
3/8	.3750	.762	.670	.165	.094	.081	.083	.065

Figure A-21

American National Standard Slotted Truss Head Machine Screws
(ANSI B18.6.3-1972, R1977)

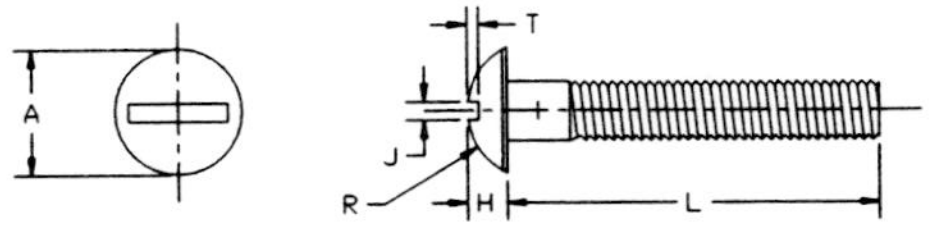

Nominal Size or Basic Screw Diam.	Head Diam., A		Head Height, H		Head Radius, R	Slot Width, J		Slot Depth, T	
	Max.	Min.	Max.	Min.	Max.	Max.	Min.	Max.	Min.
0000 .0210	.049	.043	.014	.010	.032	.009	.005	.009	.005
000 .0340	.077	.071	.022	.018	.051	.013	.009	.013	.009
00 .0470	.106	.098	.030	.024	.070	.017	.010	.018	.012
0 .0600	.131	.119	.037	.029	.087	.023	.016	.022	.014
1 .0730	.164	.149	.045	.037	.107	.026	.019	.027	.018
2 .0860	.194	.180	.053	.044	.129	.031	.023	.031	.022
3 .0990	.226	.211	.061	.051	.151	.035	.027	.036	.026
4 .1120	.257	.241	.069	.059	.169	.039	.031	.040	.030
5 .1250	.289	.272	.078	.066	.191	.043	.035	.045	.034
6 .1380	.321	.303	.086	.074	.211	.048	.039	.050	.037
8 .1640	.384	.364	.102	.088	.254	.054	.045	.058	.045
10 .1900	.448	.425	.118	.103	.283	.060	.050	.068	.053
12 .2160	.511	.487	.134	.118	.336	.067	.056	.077	.061
1/4 .2500	.573	.546	.150	.133	.375	.075	.064	.087	.070
5/16 .3125	.698	.666	.183	.162	.457	.084	.072	.106	.085
3/8 .3750	.823	.787	.215	.191	.538	.094	.081	.124	.100
7/16 .4375	.948	.907	.248	.221	.619	.094	.081	.142	.116
1/2 .5000	1.073	1.028	.280	.250	.701	.106	.091	.161	.131
9/16 .5625	1.198	1.149	.312	.279	.783	.118	.102	.179	.146
5/8 .6250	1.323	1.269	.345	.309	.863	.133	.116	.196	.162
3/4 .7500	1.573	1.511	.410	.368	1.024	.149	.131	.234	.182

Figure A-22

American National Standard Plain and Slotted Hexagon Head Machine Screws (ANSI B18.6.3-1972, R1977)

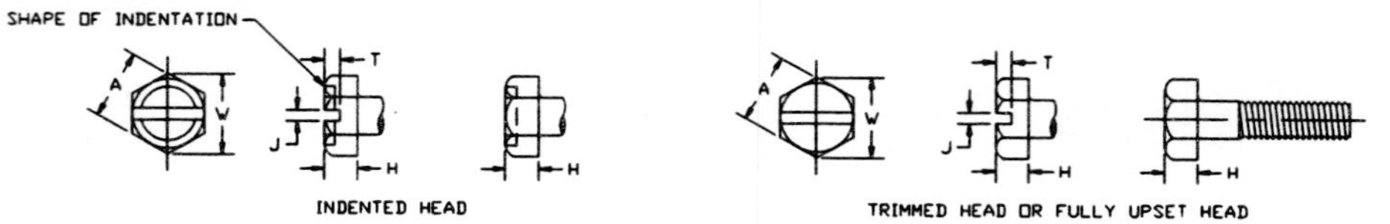
SHAPE OF INDENTATION
T
A
W
J
H
INDENTED HEAD
TRIMMED HEAD OR FULLY UPSET HEAD

Nominal Size	or Basic Screw Diam.	Regular Head Width Across Flats A		Regular Head Across Corn. W	Large Head Width Across Flats A		Large Head Across Corn. W	Head Height H		Slot Width J		Slot Depth T	
		Max.	Min.	Min.	Max.	Min.	Min.	Max.	Min.	Max.	Min.	Max.	Min.
1	.0730	.125	.120	.134				.044	.036				
2	.0860	.125	.120	.134				.050	.040				
3	.0990	.188	.181	.202				.055	.044				
4	.1120	.188	.181	.202	.219	.213	.238	.060	.049	.039	.031	.036	.025
5	.1250	.188	.181	.202	.250	.244	.272	.070	.058	.043	.035	.042	.030
6	.1380	.250	.244	.272				.093	.080	.048	.039	.046	.033
8	.1640	.250	.244	.272	.312	.305	.340	.110	.096	.054	.045	.066	.052
10	.1900	.312	.305	.340				.120	.105	.060	.050	.072	.057
12	.2160	.312	.305	.340	.375	.367	.409	.155	.139	.067	.056	.093	.077
1/4	.2500	.375	.367	.409	.438	.428	.477	.190	.172	.075	.064	.101	.083
5/16	.3125	.500	.489	.545				.230	.208	.084	.072	.122	.100
3/8	.3750	.562	.551	.614				.295	.270	.094	.081	.156	.131

Figure A-23

Slotted Round Head Machine Screws
(ANSI B18.6.3-1972, R1977 Appendix)

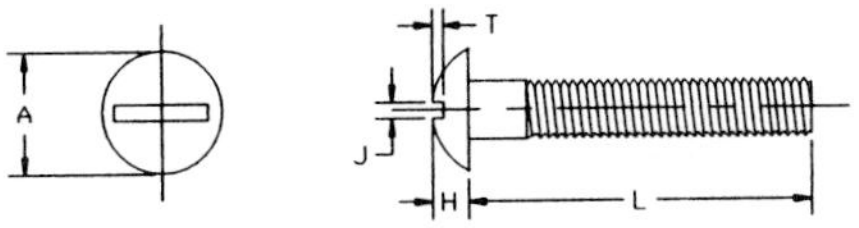

Nominal Size or Basic Screw Diam.		Head Diameter, A		Head Height, H		Slot Width, J		Slot Depth, T	
		Max.	Min.	Max.	Min.	Max.	Min.	Max.	Min.
0000	.0210	.041	.035	.022	.016	.008	.004	.017	.013
000	.0340	.062	.056	.031	.025	.012	.008	.018	.012
00	.0470	.089	.080	.045	.036	.017	.010	.026	.018
0	.0600	.113	.099	.053	.043	.023	.016	.039	.029
1	.0730	.138	.122	.061	.051	.026	.019	.044	.033
2	.0860	.162	.146	.069	.059	.031	.023	.048	.037
3	.0990	.187	.169	.078	.067	.035	.027	.053	.040
4	.1120	.211	.193	.086	.075	.039	.031	.058	.044
5	.1250	.236	.217	.095	.083	.043	.035	.063	.047
6	.1380	.260	.240	.103	.091	.048	.039	.068	.051
8	.1640	.309	.287	.120	.107	.054	.045	.077	.058
10	.1900	.359	.334	.137	.123	.060	.050	.087	.065
12	.2160	.408	.382	.153	.139	.067	.056	.096	.073
1/4	.2500	.472	.443	.175	.160	.075	.064	.109	.082
5/16	.3125	.590	.557	.216	.198	.084	.072	.132	.099
3/8	.3750	.708	.670	.256	.237	.094	.081	.155	.117
7/16	.4375	.750	.707	.328	.307	.094	.081	.196	.148
1/2	.5000	.813	.766	.355	.332	.106	.091	.211	.159
9/16	.5625	.938	.887	.410	.385	.118	.102	.242	.183
5/8	.6250	1.000	.944	.438	.411	.133	.116	.258	.195
3/4	.7500	1.250	1.185	.547	.516	.149	.131	.320	.242

Figure A-24

AMERICAN NATIONAL STANDARD SQUARE HEAD SET SCREWS (ANSI B18.6.2)

Nominal Size of Basic Screw Diameter		Width Across Flats		Width Across Corners		Head Height		Neck Relief Diameter		Max Neck Relief Fillet Radius	Min Neck Relief Width	Min Head Radius
		Max.	Min.	Max.	Min.	Max.	Min.	Max.	Min.			
10	0.1900	0.188	0.180	0.265	0.247	0.148	0.134	0.145	0.140	0.027	0.083	0.48
1/4	0.2500	0.250	0.241	0.354	0.331	0.196	0.178	0.185	0.170	0.032	0.100	0.62
5/16	0.3125	0.312	0.302	0.442	0.415	0.245	0.224	0.240	0.225	0.036	0.111	0.78
3/8	0.3750	0.375	0.362	0.530	0.497	0.293	0.270	0.294	0.279	0.041	0.125	0.94
7/16	0.4375	0.438	0.423	0.619	0.581	0.341	0.315	0.345	0.330	0.046	0.143	1.09
1/2	0.5000	0.500	0.484	0.707	0.665	0.389	0.361	0.400	0.385	0.050	0.154	1.25
9/16	0.5625	0.562	0.545	0.795	0.748	0.437	0.407	0.454	0.439	0.054	0.167	1.41
5/8	0.6250	0.625	0.606	0.884	0.833	0.485	0.452	0.507	0.492	0.059	0.182	1.56
3/4	0.7500	0.750	0.729	1.060	1.001	0.582	0.544	0.620	0.605	0.065	0.200	1.88
7/8	0.8750	0.875	0.852	1.237	1.170	0.678	0.635	0.731	0.716	0.072	0.222	2.19
1	1.0000	1.000	0.974	1.414	1.337	0.774	0.726	0.838	0.823	0.081	0.250	2.50
1 1/8	1.1250	1.125	1.096	1.591	1.505	0.870	0.817	0.939	0.914	0.092	0.283	2.81
1 1/4	1.2500	1.250	1.219	1.768	1.674	0.966	0.908	1.064	1.039	0.092	0.283	3.12
1 3/8	1.3750	1.375	1.342	1.945	1.843	1.063	1.000	1.159	1.134	0.109	0.333	3.44
1 1/2	1.5000	1.500	1.464	2.121	2.010	1.159	1.091	1.284	1.259	0.109	0.333	3.75

Figure A-25

AMERICAN NATIONAL STANDARD SQUARE HEAD SET SCREWS (ANSI B18.6.2)

Nominal Size or Basic Screw Diameter		Cup and Flat Point Diameters		Dog and Half-Dog Point Diameters		Point Length				Oval Point Radius +0.031 −0.000	Cone Point Angle 90° ± 2° for these Nominal Lengths or Longer, 118° ± 2° for Shorter Screws
						Dog		Half-Dog			
		Max.	Min.	Max.	Min.	Max.	Min.	Max.	Min.		
10	0.1900	0.102	0.088	0.127	0.120	0.095	0.085	0.050	0.040	0.142	1/4
1/4	0.2500	0.132	0.118	0.156	0.149	0.130	0.120	0.068	0.058	0.188	5/16
5/16	0.3125	0.172	0.156	0.203	0.195	0.161	0.151	0.083	0.073	0.234	3/8
3/8	0.3750	0.212	0.194	0.250	0.241	0.193	0.183	0.099	0.089	0.281	7/16
7/16	0.4375	0.252	0.232	0.297	0.287	0.224	0.214	0.114	0.104	0.328	1/2
1/2	0.5000	0.291	0.270	0.344	0.334	0.255	0.245	0.130	0.120	0.375	9/16
9/16	0.5625	0.332	0.309	0.391	0.379	0.287	0.275	0.146	0.134	0.422	5/8
5/8	0.6250	0.371	0.347	0.469	0.456	0.321	0.305	0.164	0.148	0.469	3/4
3/4	0.7500	0.450	0.425	0.562	0.549	0.383	0.367	0.196	0.180	0.562	7/8
7/8	0.8750	0.530	0.502	0.656	0.642	0.446	0.430	0.227	0.221	0.656	1
1	1.0000	0.609	0.579	0.750	0.734	0.510	0.490	0.260	0.240	0.750	1 1/8
1 1/8	1.1250	0.689	0.655	0.844	0.826	0.572	0.552	0.291	0.271	0.844	1 1/4
1 1/4	1.2500	0.767	0.733	0.938	0.920	0.635	0.615	0.323	0.303	0.938	1 1/2
1 3/8	1.3750	0.848	0.808	1.031	1.011	0.698	0.678	0354	0.334	1.031	1 5/8
1 1/2	1.5000	0.926	0.886	1.125	1.105	0.760	0.740	0.385	0.365	1.125	1 3/4

Figure A-26

American National Standard Slotted Headless
Set Screws (ANSI B18.6.2)

Nominal Size or Basic Screw Diameter		Crown Radius Basic	Slot Width		Slot Depth		Cup and Flat Point Diameters		Dog Point Diameters		Point Length				Oval Point Radius Basic	Cone Point Angle 90°±2° For These Nominal Lengths or Longer 118°±2° For Shorter
											Dog		Half-Dog			
			MAX	MIN	MAX	MIN	MAX	MIN	MAX	MIN	MAX	MIN	MAX	MIN		
0	0.0600	0.060	0.014	0.0010	0.020	0.016	0.033	0.027	0.040	0.037	0.032	0.028	0.017	0.013	0.045	5/64
1	0.0730	0.073	0.016	0.012	0.020	0.016	0.040	0.033	0.049	0.045	0.040	0.036	0.021	0.017	0.055	3/32
2	0.0860	0.086	0.018	0.014	0.025	0.019	0.047	0.039	0.057	0.053	0.046	0.042	0.024	0.020	0.064	7/64
3	0.0990	0.099	0.020	0.016	0.028	0.022	0.054	0.045	0.066	0.062	0.052	0.048	0.027	0.023	0.074	1/8
4	0.1120	0.112	0.024	0.018	0.031	0.025	0.061	0.051	0.075	0.070	0.058	0.054	0.030	0.026	0.084	5/32
5	0.1250	0.125	0.026	0.020	0.036	0.026	0.067	0.057	0.083	0.078	0.063	0.057	0.033	0.027	0.094	3/16
6	0.1380	0.138	0.028	0.022	0.040	0.030	0.074	0.064	0.092	0.087	0.073	0.067	0.038	0.032	0.104	3/16
8	0.1640	0.164	0.032	0.026	0.046	0.036	0.087	0.076	0.109	0.103	0.083	0.077	0.043	0.037	0.123	1/4
10	0.1900	0.190	0.035	0.029	0.053	0.043	0.102	0.088	0.127	0.120	0.095	0.085	0.050	0.040	0.142	1/4
12	0.2160	0.216	0.042	0.035	0.061	0.051	0.115	0.101	0.144	0.137	0.115	0.105	0.060	0.050	0.162	5/16
1/4	0.2500	0.250	0.049	0.041	0.068	0.058	0.132	0.118	0.156	0.149	0.130	0.120	0.068	0.058	0.188	5/16
5/16	0.3125	0.312	0.055	0.047	0.083	0.073	0.172	0.156	0.203	0.195	0.161	0.151	0.083	0.073	0.234	3/8
3/8	0.3750	0.375	0.068	0.060	0.099	0.089	0.212	0.194	0.250	0.241	0.193	0.183	0.099	0.089	0.281	7/16
7/16	0.4375	0.438	0.076	0.068	0.114	0.104	0.252	0.232	0.297	0.287	0.224	0.214	0.114	0.104	0.328	1/2
1/2	0.5000	0.500	0.086	0.078	0.130	0.120	0.291	0.270	0.344	0.334	0.255	0.245	0.130	0.120	0.375	9/16
9/16	0.5625	0.562	0.096	0.088	0.146	0.136	0.332	0.309	0.391	0.379	0.287	0.275	0.146	0.134	0.422	5/8
5/8	0.6250	0.625	0.107	0.097	0.161	0.151	0.371	0.347	0.469	0.456	0.321	0.305	0.164	0.148	0.469	3/4
3/4	0.7500	0.750	0.134	0.124	0.193	0.183	0.450	0.425	0.562	0.549	0.383	0.367	0.196	0.180	0.562	7/8

Figure A-27

Lengths for Threaded Fasteners

DIAMETER \ LENGTHS	.250	.313	.375	.438	.500	.563	.625	.750	.875	1.000	1.250	1.500	1.750	2.000	2.500	3.000	3.500	4.000
5(.125)	●	●	●	●	●	●	●	●	●	●		●						
6(.138)	●	●	●	●	●	●	●	●	●	●	●	●		●				
8(.164)	●	●	●	●	●	●	●	●	●	●	●	●		●				
10(.190)	●	●	●	●	●	●	●	●	●	●	●	●	●	●				
12(.216)	●	●	●	●	●	●	●	●	●	●		●		●				
.250	●	●	●	●	●	●	●	●	●	●	●	●	●	●	●			
.313	●	●	●	●	●	●	●	●	●	●	●	●	●	●	●			
.375	●	●	●	●	●	●	●	●	●	●	●	●		●	●	●		
.438	●	●	●	●	●	●	●	●	●	●	●	●		●	●	●		
.500	●	●	●	●	●	●	●	●	●	●	●	●		●	●	●	●	●
.563			●	●	●	●	●	●	●	●		●		●	●	●	●	●
.625			●	●	●	●	●	●	●	●		●		●		●		●
.750					●	●	●	●	●	●		●		●		●	●	●
.875							●	●	●	●		●		●		●	●	●
1.000												●		●	●	●		●

Figure A-28

Lengths for Metric Threaded Fasteners

DIAMETER \ LENGTHS	4	5	8	10	12	16	20	24	30	36	40	45	50	60	70
1.6	●	●	●												
2	●	●	●												
2.5	●	●	●	●	●										
3		●	●	●	●										
4			●	●	●	●	●								
5			●	●	●	●	●	●							
6				●	●	●	●	●							
8					●	●	●	●	●	●	●				
10						●	●	●	●	●	●	●	●	●	
12						●	●	●	●	●	●	●	●	●	●
16							●	●	●	●	●	●	●	●	●
20								●	●	●	●	●	●	●	●
24									●	●	●	●	●	●	●
30										●	●	●	●	●	●

Figure A-29

American National Standard Square and Hexagon Machine Screw Nuts
(ANSI B18.6.3-1972, R1977)

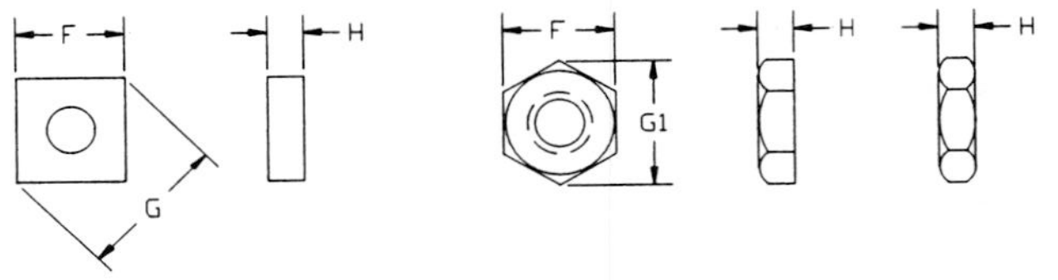

Nom. Size	Basic Diam.	Basic F	Max. F	Min. F	Max. G	Min. G	Max. G1	Min. G1	Max. H	Min. H
0	.0600	5/32	.156	.150	.221	.206	.180	.171	.050	.043
1	.0730	5/32	.156	.150	.221	.206	.180	.171	.050	.043
2	.0860	3/16	.188	.180	.265	.247	.217	.205	.066	.057
3	.0990	3/16	.188	.180	.265	.247	.217	.205	.066	.057
4	.1120	1/4	.250	.241	.354	.331	.289	.275	.098	.087
5	.1250	5/16	.312	.302	.442	.415	.361	.344	.114	.102
6	.1380	5/16	.312	.302	.442	.415	.361	.344	.114	.102
8	.1640	11/32	.344	.332	.486	.456	.397	.378	.130	.117
10	.1900	3/8	.375	.362	.530	.497	.433	.413	.130	.117
12	.2160	7/16	.438	.423	.619	.581	.505	.482	.161	.148
1/4	.2500	7/16	.438	.423	.619	.581	.505	.482	.193	.178
5/16	.3125	9/16	.562	.545	.795	.748	.650	.621	.225	.208
3/8	.3750	5/8	.625	.607	.884	.833	.722	.692	.257	.239

Figure A-30

Standard Twist Drill Sizes (Inches)

SIZE	DIAMETER	SIZE	DIAMETER	SIZE	DIAMETER	SIZE	DIAMETER
40	.098	19	.166	C	.242	U	.368
39	.0995	18	.1695	D	.246	3/8	.375
38	.1015	11/64	.1719	1/4(E)	.250	V	.377
37	.104	17	.173	F	.257	W	.386
36	.1065	16	.177	G	.261	25/64	.3906
7/16	.1094	15	.180	17/64	.2656	X	.397
35	.110	14	.182	H	.266	Y	.404
34	.111	13	.185	I	.272	13/32	.4062
33	.113	3/16	.1875	J	.277	Z	.413
32	.116	12	.189	K	.281	27/64	.4219
31	.120	11	.191	9/32	.2812	7/16	.4375
1/8	.125	10	.1935	L	.290	29/64	.4531
30	.1285	9	.196	M	.295	15/32	.4688
29	.136	8	.199	19/64	.2969	31/64	.4844
28	.1405	7	.201	N	.302	1/2	.5000
9/64	.1406	13/64	.2031	5/16	.3125	9/16	.5625
27	.144	6	.204	O	.316	5/8	.625
26	.147	5	.2055	P	.323	11/16	.6875
25	.1495	4	.209	21/64	.3281	3/4	.750
24	.152	3	.213	Q	.332	13/16	.8125
23	.154	7/32	.2188	R	.339	7/8	.875
5/32	.1562	2	.221	11/32	.3438	15/16	.9375
22	.157	1	.228	S	.348		
21	.159	A	.234	T	.358		
20	.161	B	.238	23/64	.3594		

NOTES FOR TWIST DRILL SIZES - INCHES
1. This is only a partial list of standard drill sizes.
2. Whenever possible, specify holes sizes that correspond to standard drill sizes.
3. Drill sizes are available in 1/64 increments between .5000 and 1.2500.
4. Drill sizes are available in 1/32 increments between 1.2500 and 1.500.

Figure A-31

Standard Twist Drill Sizes (Millimeters)

0.40	2.05	5.10	8.60	15.25	30.00
0.42	2.10	5.20	8.70	15.50	30.50
0.45	2.15	5.30	8.80	15.75	31.00
0.48	2.20	5.40	8.90	16.00	31.50
0.50	2.25	5.50	9.00	16.25	32.00
0.55	2.30	5.60	9.10	16.50	32.50
0.60	2.35	5.70	9.20	16.75	33.00
0.65	2.40	5.80	9.30	17.00	33.50
0.70	2.45	5.90	9.40	17.25	34.00
0.75	2.50	6.00	9.50	17.50	34.50
0.80	2.60	6.10	9.60	17.75	35.00
0.85	2.70	6.20	9.70	18.00	35.50
0.90	2.80	6.30	9.80	18.50	36.00
0.95	2.90	6.40	9.90	19.00	36.50
1.00	3.00	6.50	10.00	19.50	37.00
1.05	3.10	6.60	10.20	20.00	37.50
1.10	3.20	6.70	10.50	20.50	38.00
1.15	3.30	6.80	10.80	21.00	40.00
1.20	3.40	6.90	11.00	21.50	42.00
1.25	3.50	7.00	11.20	22.00	44.00
1.30	3.60	7.10	11.50	22.50	46.00
1.35	3.70	7.20	11.80	23.00	48.00
1.40	3.80	7.30	12.00	23.50	50.00
1.45	3.90	7.40	12.20	24.00	
1.50	4.00	7.50	12.50	24.50	
1.55	4.10	7.60	12.80	25.00	
1.60	4.20	7.70	13.00	25.50	
1.65	4.30	7.80	13.20	26.00	
1.70	4.40	7.90	13.50	26.50	
1.75	4.50	8.00	13.80	27.00	
1.80	4.60	8.10	14.00	27.50	
1.85	4.70	8.20	14.25	28.00	
1.90	4.80	8.30	14.50	28.50	
1.95	4.90	8.40	14.75	29.00	
2.00	5.00	8.50	15.00	29.50	

Figure A-32

Index